VOLUME NINETY FOUR

CURRENT TOPICS IN DEVELOPMENTAL BIOLOGY

Cancer and Development

VOLUME NINETY FOUR

Current Topics in DEVELOPMENTAL BIOLOGY

Cancer and Development

Edited by

MICHAEL A. DYER

Department of Developmental Neurobiology
St. Jude Children's Research Hospital
Memphis, Tennessee, USA, and
Howard Hughes Medical Institute
Early Career Scientist, USA

ELSEVIER

AMSTERDAM • BOSTON • HEIDELBERG • LONDON
NEW YORK • OXFORD • PARIS • SAN DIEGO
SAN FRANCISCO • SINGAPORE • SYDNEY • TOKYO
Academic Press is an imprint of Elsevier

Academic Press is an imprint of Elsevier
525 B Street, Suite 1900, San Diego, CA 92101-4495, USA
30 Corporate Drive, Suite 400, Burlington, MA 01803, USA
32, Jamestown Road, London NW1 7BY, UK
Linacre House, Jordan Hill, Oxford OX2 8DP, UK

First edition 2011

ISBN: 978-0-12-380916-2
ISSN: 0070-2153

For information on all Academic Press publications
visit our website at elsevierdirect.com

Printed and bound by CPI Group (UK) Ltd, Croydon, CR0 4YY

Transferred to Digital Print 2012

Contents

Contributors

Sheila R. Alcantara Llaguno
The University of Texas Southwestern Medical Center at Dallas, Dallas, Texas, USA

Rachel Brennan
Department of Developmental Neurobiology, and Department of Hematology/ Oncology, St. Jude Children's Research Hospital, Memphis, Tennessee, USA

Yuntao Chen
The University of Texas Southwestern Medical Center at Dallas, Dallas, Texas, USA

Meenalakshmi Chinnam
Department of Pharmacology and Therapeutics, Roswell Park Cancer Institute, Buffalo, New York, USA

J. Racquel Collins-Underwood
Department of Pathology, St. Jude Children's Research Hospital, Memphis, Tennessee, USA

Nikkilina R. Crouse
Department of Neurology, Washington University School of Medicine, St. Louis, Missouri, USA

Sonika Dahiya
Department of Pathology, Washington University School of Medicine, St. Louis, Missouri, USA

Michael A. Dyer
Department of Developmental Neurobiology, St. Jude Children's Research Hospital, Memphis, Tennessee, and Howard Hughes Medical Institute Early Career Scientist, Chevy Chase, Maryland, USA

Sara Federico
Department of Developmental Neurobiology, and Department of Hematology/ Oncology, St. Jude Children's Research Hospital, Memphis, Tennessee, USA

David W. Goodrich
Department of Pharmacology and Therapeutics, Roswell Park Cancer Institute, Buffalo, New York, USA

David H. Gutmann
Department of Neurology, Washington University School of Medicine, St. Louis, Missouri, USA

Mary E. Hatten
Laboratory of Developmental Neurobiology, The Rockefeller University, New York, USA

Manrong Jiang
Department of Genetics and Tumor Cell Biology, St. Jude Children's Research Hospital, Memphis, Tennessee, USA

Jill M. Lahti
Department of Genetics and Tumor Cell Biology, St. Jude Children's Research Hospital, Memphis, Tennessee, USA

Jean-Christophe Marine
VIB–K.U.Leuven, Department of Molecular and Developmental Genetics, Laboratory for Molecular Cancer Biology, Leuven, Belgium

Renée M. McKay
The University of Texas Southwestern Medical Center at Dallas, Dallas, Texas, USA

Charles G. Mullighan
Department of Pathology, St. Jude Children's Research Hospital, Memphis, Tennessee, USA

Peter J. Murray
Departments of Infectious Diseases and Immunology, St. Jude Children's Research Hospital, Memphis, TN, USA

Luis F. Parada
The University of Texas Southwestern Medical Center at Dallas, Dallas, Texas, USA

Joseph E. Qualls
Departments of Infectious Diseases and Immunology, St. Jude Children's Research Hospital, Memphis, TN, USA

Martine F. Roussel
Department of Tumor Cell Biology and Genetics, St. Jude Children's Research Hospital, Memphis, Tennessee, USA

Raya Saab
Children's Cancer Center of Lebanon, Department of Pediatrics, American University of Beirut, Beirut, Lebanon

Stephen X. Skapek
Department of Pediatrics, Section of Hematology/Oncology, University of Chicago and Comer Children's Hospital, Chicago, Illinois, USA

Sheri L. Spunt
Department of Oncology, St. Jude Children's Research Hospital, Memphis, Tennessee, USA

Jennifer Stanke
Department of Genetics and Tumor Cell Biology, and Department of Developmental Neurobiology, St. Jude Children's Research Hospital, Memphis, Tennessee, USA

PREFACE

The integration of disciplines has resulted in many landmark discoveries in biomedical research. One of the best examples of this is the bridging of developmental biology and cancer biology over the past few decades, which has led to great advances in our understanding of oncogenesis and the development of treatments that are now beginning to improve cancer survival rates.

When I first became interested in the coordination of proliferation, cell fate specification, and differentiation in the developing retina 15 years ago, I was surprised that researchers in developmental neurobiology were so separated from those interested in cell cycle regulation. At that time, the key biological questions at the forefront of the cell cycle field were related to proliferation, terminal cell cycle exit, cell cycle length, and proliferative competence. Those questions were, for the most part, afterthoughts in the minds of developmental neurobiologists. Similarly, questions related to cell fate specification, differentiation, lineage relationships, changing competence, pluripotency, and multipotency were not studied in detail in model systems used to study cell cycle regulation or cancer.

Fortunately, today, developmental biologists are using sophisticated approaches to better understand the dynamics of the cell cycle during normal development and how perturbations in those processes can contribute to disease. Many investigators are applying cutting-edge technologies to study the relations between cell cycle regulation and development in a variety of model organisms. Similarly, cancer biologists now recognize the importance of understanding the principles of developmental biology that serve as the foundation for their discipline. As a result of this integration, many fundamental advances have been made in both fields.

The definition and distinction between immature progenitor cells and mature differentiated cells is one of the most remarkable discoveries made of late. Cancer biology has taught developmental neurobiologists to question our assumptions about what makes a cell differentiated and whether that state is reversible or mutable. In my view, cell fate specification and differentiation are not as black and white as we thought they were 15–20 years ago. The line between differentiated cells and progenitor cells has become blurred, and I believe that this provides us with a better understanding of cancer and more insight to develop novel approaches to treating this disease. Indeed, many labs are now exploring what it means to be differentiated at the molecular and cellular level and whether perturbations in this plasticity have any implications in normal development or disease.

In many ways, tumors resemble normal organs in that they are rapidly changing, developing systems. Tumors are heterogeneous, they change over time, and they respond to powerful selective pressure. Intrinsic and extrinsic factors drive tumorigenesis, just as they drive normal development, and much of what we have learned about the normal processes that underlie developing systems is now being applied to pathogenic models.

In this volume, we have drawn on the expertise of many leaders in the field of cancer biology and developmental biology. As expected, much of the work reported here focuses on pediatric cancers, which by definition are developmental tumors, and provides the greatest insight into the connection between cell cycle regulation during development and the types of perturbations that occur in pediatric cancer. Although the chapters focus on diverse organs and tumor types ranging from muscle tumors to brain tumors, common themes become readily apparent in their juxtaposition, such as the way in which tumors utilize tissue-specific signaling pathways that help to coordinate proliferation during organogenesis.

Here we are, a decade into the new millennium, and I believe we are just now beginning to take full advantage of evolving technologies, incredible expertise, and vast knowledge gained in the fields of cancer biology, cell cycle regulation, and development. I look forward to the next decade and the landmark discoveries that will most likely arise from studies using this multidisciplinary approach. On a personal note, my thanks to all of the authors for taking time out of their busy schedules to contribute to this volume.

MICHAEL A. DYER

CHAPTER ONE

Childhood Cancer and Developmental Biology: A Crucial Partnership

Sara Federico,*,† Rachel Brennan,*,† *and* Michael A. Dyer*,‡

Contents

Abstract

For many years, there were relatively few research efforts that bridged the fields of developmental biology and cancer genetics. However, in the past decade, we have witnessed a dramatic shift and now these two fields are intertwined. Part of the impetus for this transition came from the discovery that regulatory pathways that were previously thought to be uniquely important for developmental processes were also perturbed in cancer. In addition, the conceptual framework for understanding how cells self-renew or undergo unidirectional changes in competence during development has proven to be very useful in cancer biology as researchers explore tumor initiation and progression. Finally, a deeper understanding of the process of terminal differentiation and how that relates to cellular plasticity may have important implications for both cancer biology and developmental biology. Here we highlight some of the important connections between developmental neurobiology and cancer biology in retinoblastoma. By bridging these fields, important advances have been

* Department of Developmental Neurobiology, St. Jude Children's Research Hospital, Memphis, Tennessee, USA
† Department of Hematology/Oncology, St. Jude Children's Research Hospital, Memphis, Tennessee, USA
‡ Howard Hughes Medical Institute Early Career Scientist, Chevy Chase, Maryland, USA

Current Topics in Developmental Biology, Volume 94
ISSN 0070-2153, DOI: 10.1016/B978-0-12-380916-2.00001-2

made in modeling retinoblastoma in mice, elucidating the cell-of-origin for retinoblastoma and identifying novel therapeutic approaches.

1. Introduction

Over the past several decades, we have learned a great deal about the molecular, cellular, and genetic properties of the most common human malignancies such as breast, colon, lung, and prostate cancer. However, less is known about pediatric cancers because they are rare and often form in a complex microenvironment during development. Indeed, it is not uncommon for pediatric cancers to initiate in developing organs characterized by rapid proliferative expansion, growth factor signaling, developmental angiogenesis, programmed cell death, tissue reorganization and cell migration. In addition, the microenvironment is changing as development progresses and this may directly or indirectly impact tumor initiation and growth. Not only is it more difficult to establish the etiology of pediatric cancer because it occurs in the context of development, but also makes it more difficult to treat. For example, molecular targeted therapies that perturb developmental pathways that are deregulated in the tumor may have devastating effects on normal tissues in a developmental stage and tissue specific manner. Therefore, it is essential to use developmental biology as the foundation for translational research in pediatric cancer. In this chapter, we will use retinoblastoma and retinal development as an example of the importance of developmental biology in translational research and we will draw parallels to other pediatric cancers.

2. Developmental Biology and Cancer Genetics

At a time when molecular cloning was in its infancy and high throughput genomic sequence analysis was decades away, the research of Eric Wieschaus and Christiane Nusslein-Volhard revolutionized the fields of developmental biology and cancer genetics and ultimately earned a Nobel prize (Ingham and Placzek, 2006). Their initial goal was to study the mechanisms of cell fate specification and differentiation during development by taking advantage of the tractable genetics of *Drosophila*. Wieschaus and Nusslein-Volhard used phenotype-based mutant screening approaches combined with molecular genetics to link the genetic loci and genes to developmental processes on a large scale. Their results led to the discovery of several of the major developmental pathways that are now widely studied in a variety of invertebrate, vertebrate, and mammalian species. Importantly, we now understand that many of these pathways are perturbed in human diseases, including cancer. One of the best examples of a developmental pathway that also contributes to tumorigenesis is the hedgehog (Hh) pathway. In *Drosophila,* Hh signaling controls the segmentation pattern and other cell–cell signaling events in development (Ingham and Placzek, 2006; Lum and Beachy, 2004). In mammals, there are three orthologues of *Drosophilia* hedgehog: sonic hedgehog, desert hedgehog, and Indian hedgehog (Echelard *et al.*, 1993). There are unique and overlapping functions of these hedgehog family members in mammals. Indeed, hedgehog signaling has been implicated in cell proliferation, cell survival, angiogenesis, and the epithelial-mesenchyme transition that is a hallmark of tumor metastasis (Scales and de Sauvage, 2009). Based on these important functions in normal development, it is not surprising that the hedgehog pathway is deregulated in some forms of pediatric cancer.

Some of the first evidence that aberrant hedgehog signaling contributed to tumorigenesis in humans came from the discovery that patients with Gorlin syndrome had lesions in a gene (*Patched 1*) important for hedgehog signaling (Bale *et al.*, 1998; Gailani *et al.*, 1992). Gorlin syndrome is a rare disease characterized by larger body size, developmental and skeletal anomalies, soft tissue fibromas, radiation sensitivity, and predisposition to cancers such as basal cell carcinoma, medulloblastoma, and rhabdomyosarcoma (Bale *et al.*, 1995, 1998; Cajaiba *et al.*, 2006; Hahn *et al.*, 1996; Johnson *et al.*, 1996). The pleiotropic phenotype of these patients further highlights the important role that hedgehog signaling plays in tissue development and homeostasis. Translating this important discovery into the clinical realm, targeted therapies that can silence aberrant and/or uncontrolled signaling through the hedgehog pathway could be effective for the treatment of pediatric cancers such as medulloblastoma. However, implementing this unique opportunity comes with major challenges. If the hedgehog pathway is silenced using a molecular targeted therapy in children with medulloblastoma, how will this affect all the cells and organs that rely on hedgehog signaling for normal development and homeostasis? Indeed, a recent study of hedgehog inhibitors in juvenile mice showed defects in the development of several organ systems providing critical evidence for the importance of understanding developmental biology in targeting pediatric cancer (Gorlick *et al.*, 2009). Therefore, while the knowledge of this important pathway has moved from bench to bedside in the form of a phase I/II study with hedgehog pathway inhibitors, the trial is limited to older patients who may suffer fewer side effects of targeting the hedgehog pathway (ClinicalTrials.gov Identifier: NCT00939484).

3. Clinical Features of Retinoblastoma

Retinoblastoma is the most frequent ocular neoplasm that occurs during childhood and the third most common form of cancer in infants after leukemia and neuroblastoma. In the USA, approximately 300 children are diagnosed with the disease each year (Gurney *et al.*, 1995). Internationally, physicians diagnose between 5000 and 10,000 children with retinoblastoma each year, and the majority of these patients present with advanced stage disease. Retinoblastoma is a disease of infants and young children. This reflects the developmental origin of the tumor. It is often mistakenly assumed that the tumor arises near the time that the disease first presents clinically in the first few years of life. However, it is likely that the *RB1* gene inactivation occurs during DNA replication in proliferating retinal progenitor cells and retinal progenitor cell proliferation occurs only in the fetal retina (DiCiommo *et al.*, 2000; Dyer, 2004). Consistent with this model, there are reports of premature babies with well-established retinoblastoma (Abramson *et al.*, 2002). If this is true then the latency from tumor initiation to diagnosis may reflect the time it takes for tumor progression and the accumulation of secondary and tertiary genetic lesions in retinoblastoma.

There are two distinct forms of retinoblastoma that differ in the pattern of *RB1* gene inactivation. A hereditary or germline form of the disease tends to present at earlier age with bilateral disease and multifocal tumors (DiCiommo *et al.*, 2000). A nonhereditary form of the disease is typically associated with later onset, is unilateral in presentation, and tends to be unifocal. Both forms of disease initiate with *RB1* gene inactivation. One of the unique features of retinoblastoma is that diagnosis of retinoblastoma is often made without pathologic confirmation. Instead, multiple modalities are used to make the diagnosis. A physician performs an exam under anesthesia during which he/she uses a retinal camera to identify the tumor. An ultrasound of the orbits and MRI of the orbits and brain are also obtained. Information gathered from these studies leads to the diagnosis, after which children are treated according to an individualized plan based on the stage of their disease.

The primary goals of treatment for retinoblastoma are to save life and preserve vision. The principles of treatment depend on the nature of the disease. Multiple factors contribute to the development of a treatment strategy, including tumor locality, presence of an *RB* mutation, potential for vision, and stage of disease. Treatment modalities include surgery, focal therapy, radiotherapy, and chemotherapy.

There are many different approaches in the management of retinoblastoma based on the individual, laterality and stage of disease. Patients with advanced unilateral intraocular disease typically undergo enucleation. If the patient has a high-risk tumor then he/she may receive adjuvant chemotherapy. Patients with early low-risk unilateral or multifocal retinoblastoma are often offered conservative management with a combination of chemotherapy and focal therapy. These patients are typically treated with 6–8 courses of chemotherapy given at 3-week intervals. Their disease is evaluated prior to each course of chemotherapy with an exam performed under anesthesia, measurements of intraocular pressure and an ultrasound of the orbit. While sedated for the studies, these patients will often receive focal therapy and local chemotherapy (subconjunctival administration) if that is part of the treatment protocol. The majority of patients with advanced bilateral disease are treated conservatively with a combination of chemotherapy, focal therapy, and enucleation if the tumor progresses.

In the USA, the current survival rate for retinoblastoma is greater than 90% in patients who present with low-risk disease (Broaddus *et al.*, 2009). However, in developing countries the survival rate plunges to 50% (Rodriguez-Galindo *et al.*, 2008). In spite of aggressive therapy, the morbidity associated with the disease remains significant. Blindness and enucleation occur in 50% of all patients with advanced bilateral retinoblastoma (Shields and Shields, 2004). Thus further advances must be made to improve the survival rate and quality of life of patients. Preclinical models of retinoblastoma offer the best hope of moving new therapies into clinical trials as for other pediatric cancers.

4. Mouse Models of Pediatric Cancer

One of the first attempts to model human cancer in the mouse using the same genetic lesion seen in human patients was focused on the *Rb1* gene in retinoblastoma. Children who inherit a defective copy of the *RB1* gene have an increased susceptibility to develop retinoblastoma through inactivation of the second allele. Tyler Jacks and colleagues reasoned that a mouse lacking 1 copy of the *Rb1* gene would similarly have a predisposition to retinoblastoma (Clarke *et al.*, 1992; Jacks *et al.*, 1992; Lee *et al.*, 1992). Interestingly, these mice have perfectly normal retinae and never develop retinoblastoma. Subsequent molecular, cellular, and genetic studies demonstrated that the Rb protein plays similar roles in humans and mice and the species specific difference in tumor susceptibility was not due to evolutionary differences in the *Rb1* gene or protein itself (Donovan *et al.*, 2006). A series of developmental studies focused on retinal progenitor cells in mice and humans revealed that there is a difference in the intrinsic genetic compensation among the Rb family members in human and mouse retinae (Donovan *et al.*, 2006). In the human retina, the RB1 protein is the major family member expressed in proliferating retinal progenitor cells while in mice there is a complex interplay between Rb1 and p107. When the *Rb1* gene is inactivated in the mouse retina, there is intrinsic compensation by p107 and this prevents retinoblastoma in mice (Donovan *et al.*, 2006). By inactivating the *Rb* and *p107* genes in the developing mouse retina, researchers were able to model retinoblastoma in mice for the first time (Chen *et al.*, 2004; MacPherson *et al.*, 2004; Zhang *et al.*, 2004). This provided unprecedented opportunities to begin to understand the cellular etiology of retinoblastoma using a tractable experimental system and to begin to develop new therapies for this devastating childhood cancer through preclinical testing programs.

Despite the early setback in the retinoblastoma field with respect to preclinical models, mouse models of cancer have become increasingly important in the field of cancer genetics and translational research. Fundamental advances in our understanding of cancer etiology, tumor progression, and therapeutic efficacy have come from murine models of human cancers ranging from prostate and breast cancer to brain tumors and leukemia. Such studies are particularly important for pediatric cancer because the patient population is relatively small and preclinical models are essential for validating the efficacy of new combinations of chemotherapy before initiating clinical trials in children. In addition, animal models of cancer can provide important new insights into the molecular, cellular, and genetic mechanisms underlying tumor initiation and progression.

One of the challenges with modeling pediatric cancer in mice is that these tumors initiate in the context of rapidly changing developing tissues where the cells that give rise to the tumors can display an extraordinary degree of plasticity and heterogeneity. This is further complicated by the fact that many tumor suppressor genes and proto-oncogenes play essential roles in regulating cell fate specification and differentiation during development. Specifically, the genetic lesions that contribute to tumor initiation and progression may also alter the intrinsic cell fate specification and differentiation programs in the tumor cells themselves making it very difficult to infer the cell-of-origin for that tumor.

As new chemotherapeutic agents are developed to target particular molecular pathways in cancer cells, the identification of the cell-of-origin becomes increasingly important. For example, if the cell-of-origin for a pediatric cancer is a progenitor cell, a very different strategy may be employed to target particular pathways in that cell than if it were a highly specialized differentiated cell from the same tissue. This is particularly true for tumors of the central nervous system such as retinoblastoma. Neurons are incredibly diverse and employ a wide variety of cell-type specific death pathways which may be activated under conditions of neuronal stress that are not relevant in other cell types.

5. Translational Research: Is It Clinically Relevant?

The slow process of translating molecular and genetic discoveries into clinical medicine can be attributed in part to the lack of preclinical models that faithfully recapitulate human disease. In addition, few investigators use those models in ways that recapitulate the schedule, dose, or diagnostic tests that are used clinically. Therefore, it is difficult if not impossible to predict if a new therapeutic approach will be better than the conventional treatment. Indeed, preclinical testing of oncology drugs is widely viewed unfavorably by the pharmaceutical industry because of these limitations. The extremely limited use of preclinical testing programs that have any predictive power for patient outcome creates a reliance on pharmaceutical industry animal studies which focus on toxicity and marginal readouts of biological activity. As we begin to acquire more and more molecular and genetic data on human cancers, mouse models of cancer, human cancer xenografts, and cancer cell lines, we will be uniquely positioned to take full advantage of the strengths of each of those reagents while avoiding pitfalls associated with their limitations. As mentioned above, this is important from the perspective of the cell-of-origin for cancer, the differentiated features of the tumors and the microenvironment where they develop must be considered as it relates to normal development. Some of the most important challenges with modeling human cancer in the preclinical setting relate to understanding the developmental milieu where the tumors initiate and progress.

6. Immortalized Cell Lines

The vast majority of laboratory research on cancer cell biology, molecular, and biochemical studies has come from the analysis of immortalized cancer cell lines. Many fundamental and important discoveries have been made from studying cancer cell lines. However, there are obvious limitations, especially related to the developmental microenvironment that is so important for pediatric cancer. Clearly, a clonal population of cells that has been immortalized and selected for growth in culture does not recapitulate the complex three-dimensional tumor structure or the positive and negative feedback characteristic of the interplay between the normal host cells and tumor cells. Nonetheless, it is remarkable how much has been learned from immortalized cancer cell lines and the degree to which some of these lines recapitulate the primary tumors from which they came. The strength of cell culture systems is their ease of manipulation, their often-rich literature and extensive characterization and their flexibility for use in drug screening and large-scale genomics efforts. Their limitations include genetic perturbations in culture, selection for rare and/or unusual variants that are not representative of the human tumors and the lack of environmental factors and host interactions. For most investigators, cell lines are a useful experimental system that can help to generate hypotheses that are further tested and validated in more biologically relevant *in vivo* models. Unfortunately, we are relatively limited in the number of well-characterized and validated cell lines for pediatric cancers. For example, there are only two widely used human retinoblastoma cell lines (Y79 and Weri1) available from the ATCC that were first generated in the 1970s (McFall *et al.*, 1977). Pediatric cancer in general and retinoblastoma in particular would benefit from investment in the development of new human cell lines.

7. Xenograft Models

For over 30 years scientists have been using immunocompromised mice to xenograft primary human tumors. This is a very attractive approach since it allows us to begin to overcome some of the limitations of studying immortalized cell lines in culture. Some of the host–tumor interactions are recapitulated (with the exception of the immune system) and the tumors will often be exposed to many of the complex physiological changes and variations that occur in humans such as changes in hormone levels and other signaling molecules like insulin. Unfortunately, there are still some major limitations to the way in which these models are currently used that are a particular challenge for investigation of pediatric cancers (Carol *et al.*, 2009, 2010; Gorlick *et al.*, 2009; Houghton *et al.*, 2000, 2010; Keshelava *et al.*, 2009; Kurmasheva *et al.*, 2009; Maris *et al.*, 2010; Morton *et al.*, 2009; Smith *et al.*, 2010). Rarely do researchers attempt to recapitulate the developmental environment (stage and organ site) of endogenous tumors. This is due in part to a lack of a detailed understanding of the actual cell-of-origin or the developmental stage when pediatric tumors first initiate. Most xenograft studies are performed in the flank of adult immunocompromised mice and the tumor response is measured using calipers. The lack of a detailed analysis regarding the developmental environment and the failure to recapitulate that in the xenograft models poses a major challenge for interpreting any therapeutic data for such studies such as the data generated from the Pediatric Preclinical Cancer Testing Program. Retinoblastoma is an excellent example of this concept. It is well known that retinoblastomas do not grow in the flank of immunocompromised mice, however they engraft with virtually 100% efficiency in the vitreous of the eye of immunocompromised mice (Flores-Otero and Dyer, unpublished). Moreover, cells can be injected into the eyes of newborn pups and thereby recapitulate the developmental milieu of human retinoblastomas (Laurie *et al.*, 2006).

8. Genetically Engineered Mouse Models

As mentioned above, there is a long history of efforts to model retinoblastoma in mice. Developmental biology played an essential role in helping us understand why *Rb+/−* mice do not develop retinoblastoma and these studies have been extended to begin to model secondary genetic mutations in this cancer. In 2004, three research groups generated the first knockout mouse models of retinoblastoma by conditionally inactivating Rb^{Lox} in the developing retinae of *p107*-deficient mice (Chen *et al.*, 2004; MacPherson *et al.*, 2004; Zhang *et al.*, 2004). Each of these strains mirrored the histopathologic features of human retinoblastoma and created unique opportunities to integrate studies on mouse models into human cancer research.

The newly developed knockout mouse models provided a key tool to study genetic pathways important for retinoblastoma progression. It was discovered that $Chx10\text{-}Cre;Rb^{Lox/-};\ p107^{-/-};\ p53^{Lox/-}$ mice develop bilateral aggressive retinoblastoma with ~95% penetrance (Chen *et al.*, 2004; MacPherson *et al.*, 2004; Zhang *et al.*, 2004). This finding is in contrast to $Chx10\text{-}Cre;\ Rb^{Lox/-};\ p107^{-/-}$ mice that develop unilateral, minimally invasive retinoblastoma with ~50% penetrance (Zhang *et al.*, 2004). These results suggested that the p53 pathway suppresses retinoblastoma in mice. However, the *p53* gene is intact in human retinoblastoma (Kato *et al.*, 1996; Nork *et al.*, 1997). Subsequent molecular genetic analyses revealed an amplified *MDMX* gene in 65–70% of human cases, and amplified *MDM2* in 10% of cases (Laurie *et al.*, 2006). This results in a functionally silent p53 pathway. Mouse models of retinoblastoma, retinoblastoma cell lines, human fetal retinae, and primary retinoblastoma tumors from enucleated eyes were used to confirm that amplification of *MDMX* indeed suppresses *p53*-mediated cell death in *RB1*-deficient retinoblasts, thereby promoting clonal expansion of the tumors (Laurie *et al.*, 2006) and the ectopic expression of MDMX has now been successfully modeled in retinoblastoma in combination with *Rb*; *p107* inactivation.

Genetically engineered mouse models provide several advantages over cultured immortalized cell lines or xenograft studies. The genetic lesions can be engineered to initiate in the presumed cell-of-origin *in vivo* and the tumors progress through similar stages as those seen in patients. That is, tumors initiate from focal hyperplasia, progress to malignant tumor foci and eventually expand and metastasize. If the timing of the genetically engineered oncogenic lesions is properly modeled after human cancer development, then such mice may prove to be very accurate models of human disease. However, even with the best genetic manipulations in mouse models there are some limitations.

First, most conditional inactivation approaches are relatively crude with respect to the cells where the genetic lesion occurs. For example, *Nestin-cre* has been used to inactivate the *Rb* gene in the developing mouse retina but it also inactivates the *Rb* gene throughout much of the CNS (MacPherson *et al.*, 2004). Second, mouse tumors do not metastasize as often as human tumors. It is not clear if this is due to a fundamental difference between the two species or if we have simply failed to recapitulate the appropriate genetic lesions that drive metastasis in mice. Third, there are many cases where the initiating genetic lesion is followed by a series of secondary genetic lesions that contribute to tumorigenesis. It is possible that those secondary lesions are not the same in mouse models as in human tumors and this could have a profound impact on how the tumors respond to therapy in preclinical trials. Finally, there may be physiological differences that make it difficult or impossible to perform clinically relevant preclinical trials in mice. For example, some chemotherapeutic drugs (i.e., cisplatin) are not tolerated at the same dose and schedule as in humans making it difficult to directly compare across species.

If we consider the strengths and limitations of each of the experimental models it may be most prudent to integrate all of these systems while keeping in mind the critical limitations of each approach. That is, the strengths of one system may help to complement the limitation of another system and by integrating across experimental paradigms we can hope to gain the most insight into the efficacy of new therapeutic agents. However, even with the most accurate preclinical testing programs using pharmacokinetic guided dosing and delivery of drugs, the experimental paradigms are not valuable unless there is predictive power in the clinic. For example, if studies in cell culture provide a perfect correlate to identify agents that will be effective in the clinic, then the more extensive studies in mice would not be needed. Clearly, cell culture models are not sufficient as there are many examples of drugs that are very effective in culture but fail to show efficacy in animals or in humans. One example for retinoblastoma is vincristine. This is the most toxic agent used to date to treat retinoblastoma cells in culture but it has little if any effect in animal models of retinoblastoma and its contribution to clinical management of the disease is limited at best (Laurie *et al.*, 2005). We do not yet know if orthotopic xenografts or genetic mouse models will have greater or equal predictive power for retinoblastoma. The best way to establish predictive power is to compare a new combination of therapy to a standard of care for that disease and then perform clinical trials to determine if the new drug combination is also better in the patients. A positive correlation between the animal models and the clinical outcome would provide the essential predictive power needed to justify the use of a particular preclinical model. It is reasonable to assume that the preclinical models which recapitulate the developmental milieu, the genetic lesions, and the focal nature of the tumors will be the best models for testing new combinations in preparation for clinical trials.

9. Comprehensive Preclinical Trials

Even with the best models, most preclinical testing programs use metrics for tumor response that are unrelated to the criteria used in clinical trials. For example, in flank xenografts calipers are used to measure tumor response (Carol *et al.*, 2009, 2010; Gorlick *et al.*, 2009; Houghton *et al.*, 2000, 2010; Keshelava *et al.*, 2009; Kurmasheva *et al.*, 2009; Maris *et al.*, 2010; Morton *et al.*, 2009; Smith *et al.*, 2010). We propose that there are several advantages to using the same diagnostic test and functional assessments that are used in the clinic. In this way, the data from the preclinical testing will more closely parallel the data from clinical trials.

For retinoblastoma, the tumors are diagnosed with a digital retinal camera. Once enrolled, a mouse undergoes baseline studies, including optometry to measure visual acuity and tonometry to measure intraocular pressure. Additionally, each mouse has a baseline complete blood count with differential (CBCD) obtained by facial vein blood draw. Chemotherapy is started after all baseline tests have been performed. Similar to pediatric dosing, the study includes six courses of chemotherapy given on a 21-day cycle. The AUC guided doses are based on human and mouse pharmacokinetic studies. Each mouse has surveillance monitoring every 3 weeks prior to the next course of chemotherapy to track tumor growth, intraocular pressure, visual acuity, and blood counts. If a mouse experiences tumor progression, an enucleation is performed. Any mouse that requires enucleation has an ultrasound of the orbits and an MRI of the orbits and brain performed prior to the surgery. Bilateral or unilateral enucleation can be performed as needed. Once removed, the eye is fixed in 4% paraformaldehyde by immersion, embedded in paraffin, and sectioned (5 μm) through the optic nerve for hematoxylin and eosin staining and histopathologic analysis. Following recovery, the mouse continues chemotherapy, diagnostic imaging, and functional assessments until it reaches 1 year of age and completes the study.

Ultimately, multiple chemotherapeutic regimens can be tested in such standardized preclinical trials and compared to the standard of care. The goal is to determine which therapeutic regimen in the mouse is more effective and carries fewer side effects than the prevailing standard of care and then translate this to the clinical setting. Additionally, targeted and/or novel agents directed at a specific molecular pathway could be investigated. As discussed above, it is very important to test the toxicity in juvenile animals as the ongoing development of some organ systems may lead to different toxicities than seen in adult populations.

10. Summary

Developmental biology and cancer biology are inseparable when we consider the development and progression of pediatric cancer. A thorough understanding of the cell-of-origin and the unique developmental microenvironment where tumors form is essential for understanding tumor emergence and for identifying key developmental pathways that may provide valuable targets for therapy. A deep understanding of developmental biology is also important for modeling preclinical studies. The best animal models will recapitulate the developmental environment where the tumors form. Moreover, effective use of targeted therapies will require a deep understanding of developmental processes in a variety of tissues. Therefore, pediatric cancer in general and retinoblastoma in particular presents some distinct challenges for translational research. However, unique opportunities exist. We understand a great deal about the genetics of retinoblastoma and the long tradition of clinical trials has resulted in over 90% of pediatric cancer patients enrolling on clinical studies. This allows researchers and clinicians to directly compare therapy. If we develop preclinical testing programs with proved predictive power, we will have an unprecedented opportunity over the next decade to effectively impact both retinoblastoma and pediatric cancer treatment.

REFERENCES

Abramson, D. H., *et al.* (2002). Rapid growth of retinoblastoma in a premature twin. *Arch. Ophthalmol.* **120**(9), 1232–1233.

Bale, A. E., Gailani, M. R., and Leffell, D. J. (1995). The Gorlin syndrome gene: a tumor suppressor active in basal cell carcinogenesis and embryonic development. *Proc. Assoc. Am. Physicians.* **107**(2), 253–257.

Bale, S. J., Falk, R. T., and Rogers, G. R. (1998). Patching together the genetics of Gorlin syndrome. *J. Cutan. Med. Surg.* **3**(1), 31–34.

Broaddus, E., Topham, A., and Singh, A. D. (2009). Survival with retinoblastoma in the USA: 1975-2004. *Br. J. Ophthalmol.* **93**(1), 24–27.

Cajaiba, M. M., Bale, A. E., Alvarez-Franco, M., McNamara, J., and Reyes-Mugica, M. (2006). Rhabdomyosarcoma, Wilms tumor, and deletion of the patched gene in Gorlin syndrome. *Nat. Clin. Pract. Oncol.* **3**(10), 575–580.

Carol, H., *et al.* (2009). Initial testing (stage 1) of the kinesin spindle protein inhibitor ispinesib by the pediatric preclinical testing program. *Pediatr. Blood Cancer* **53**(7), 1255–1263.

Carol, H., *et al.* (2010). Initial testing of topotecan by the pediatric preclinical testing program. *Pediatr. Blood Cancer* **54**(5), 707–715.

Chen, D., *et al.* (2004). Cell-specific effects of RB or RB/p107 loss on retinal development implicate an intrinsically death-resistant cell-of-origin in retinoblastoma. *Cancer Cell* **5**(6), 539–551.

Clarke, A. R., *et al.* (1992). Requirement for a functional Rb-1 gene in murine development. *Nature* **359**(6393), 328–330.

DiCiommo, D., Gallie, B. L., and Bremner, R. (2000). Retinoblastoma: the disease, gene and protein provide critical leads to understand cancer. *Semin. Cancer Biol.* **10**(4), 255–269.

Donovan, S. L., Schweers, B., Martins, R., Johnson, D., and Dyer, M. A. (2006). Compensation by tumor suppressor genes during retinal development in mice and humans. *BMC Biol.* **4,** 14.

Dyer, M. A. (2004). Mouse models of childhood cancer of the nervous system. *J. Clin. Pathol.* **57**(6), 561–576.

Echelard, Y., *et al.* (1993). Sonic hedgehog, a member of a family of putative signaling molecules, is implicated in the regulation of CNS polarity. *Cell* **75**(7), 1417–1430.

Gailani, M. R., *et al.* (1992). Developmental defects in Gorlin syndrome related to a putative tumor suppressor gene on chromosome 9. *Cell* **69**(1), 111–117.

Gorlick, R., *et al.* (2009). Initial testing (stage 1) of lapatinib by the pediatric preclinical testing program. *Pediatr. Blood Cancer* **53**(4), 594–598.
Gurney, J. G., Severson, R. K., Davis, S., and Robison, L. L. (1995). Incidence of cancer in children in the United States. Sex-, race-, and 1-year age-specific rates by histologic type. *Cancer* **75**(8), 2186–2195.
Hahn, H., *et al.* (1996). A mammalian patched homolog is expressed in target tissues of sonic hedgehog and maps to a region associated with developmental abnormalities. *J. Biol. Chem.* **271**(21), 12125–12128.
Houghton, P. J., *et al.* (2000). Initial testing of a monoclonal antibody (IMC-A12) against IGF-1R by the Pediatric Preclinical Testing Program. *Pediatr. Blood Cancer* **54**(7), 921–926.
Houghton, P. J., *et al.* (2010). Stage 2 combination testing of rapamycin with cytotoxic agents by the Pediatric Preclinical Testing Program. *Mol. Cancer Ther.* **9**(1), 101–112.
Ingham, P. W., and Placzek, M. (2006). Orchestrating ontogenesis: variations on a theme by sonic hedgehog. *Nat. Rev. Genet.* **7**(11), 841–850.
Jacks, T., *et al.* (1992). Effects of an Rb mutation in the mouse. *Nature* **359**(6393), 295–300.
Johnson, R. L., *et al.* (1996). Human homolog of patched, a candidate gene for the basal cell nevus syndrome. *Science* **272**(5268), 1668–1671.
Kato, M. V., *et al.* (1996). Loss of heterozygosity on chromosome 17 and mutation of the p53 gene in retinoblastoma. *Cancer Lett.* **106**(1), 75–82.
Keshelava, N., *et al.* (2009). Initial testing (stage 1) of vorinostat (SAHA) by the pediatric preclinical testing program. *Pediatr. Blood Cancer* **53**(3), 505–508.
Kurmasheva, R. T., *et al.* (2009). The insulin-like growth factor-1 receptor-targeting antibody, CP-751, 871, suppresses tumor-derived VEGF and synergizes with rapamycin in models of childhood sarcoma. *Cancer Res.* **69**(19), 7662–7671.
Laurie, N. A., *et al.* (2005). Topotecan combination chemotherapy in two new rodent models of retinoblastoma. *Clin. Cancer Res.* **11**(20), 7569–7578.
Laurie, N. A., *et al.* (2006). Inactivation of the p53 pathway in retinoblastoma. *Nature* **444**(7115), 61–66.
Lee, E. Y., *et al.* (1992). Mice deficient for Rb are nonviable and show defects in neurogenesis and haematopoiesis. *Nature* **359**(6393), 288–294.
Lum, L., and Beachy, P. A. (2004). The Hedgehog response network: sensors, switches, and routers. *Science* **304**(5678), 1755–1759.
MacPherson, D., *et al.* (2004). Cell type-specific effects of Rb deletion in the murine retina. *Genes Dev.* **18**(14), 1681–1694.
Maris, J. M., *et al.* (2010). Initial testing of the aurora kinase a inhibitor MLN8237 by the Pediatric Preclinical Testing Program (PPTP). *Pediatr. Blood Cancer* **55,** 26–34.
McFall, R. C., Sery, T. W., and Makadon, M. (1977). Characterization of a new continuous cell line derived from a human retinoblastoma. *Cancer Res.* **37**(4), 1003–1010.
Morton, C. L., *et al.* (2009). Initial testing of aplidin by the pediatric pre-clinical testing program. *Pediatr. Blood Cancer* **53**(3), 509–512.
Nork, T. M., Poulsen, G. L., Millecchia, L. L., Jantz, R. G., and Nickells, R. W. (1997). p53 regulates apoptosis in human retinoblastoma. *Arch. Ophthalmol.* **115**(2), 213–219.
Rodriguez-Galindo, C., *et al.* (2008). Retinoblastoma: one world, one vision. *Pediatrics* **122**(3), e763–e770.
Scales, S. J., and de Sauvage, F. J. (2009). Mechanisms of Hedgehog pathway activation in cancer and implications for therapy. *Trends Pharmacol. Sci.* **30**(6), 303–312.
Shields, C. L., and Shields, J. A. (2004). Diagnosis and management of retinoblastoma. *Cancer Control.* **11**(5), 317–327.
Smith, M. A., *et al.* (2010). Initial testing (stage 1) of mapatumumab (HGS-ETR1) by the pediatric preclinical testing program. *Pediatr. Blood Cancer* **54**(2), 307–310.
Zhang, J., Schweers, B., and Dyer, M. A. (2004). The first knockout mouse model of retinoblastoma. *Cell Cycle.* **3**(7), 952–959.

CHAPTER TWO

Stem Cells in Brain Tumor Development

Sheila R. Alcantara Llaguno, Yuntao Chen, Renée M. McKay, *and* Luis F. Parada

Contents

Abstract

Gliomas are highly infiltrative and aggressive brain tumors that are resistant to conventional therapies. Recent studies have implicated neural stem cells (NSCs) in brain tumor initiation and development. Subpopulations of stem-like cancer cells have also been isolated from brain tumors, and are purported to be important mediators of malignant behavior and therapeutic resistance. Similar signaling pathways may be operative in both neural and cancer stem cells, suggesting that neural developmental systems may provide important clues on brain tumorigenesis. Transcriptional regulators such as microRNAs

The University of Texas Southwestern Medical Center at Dallas, Dallas, Texas, USA

Current Topics in Developmental Biology, Volume 94
ISSN 0070-2153, DOI: 10.1016/B978-0-12-380916-2.00002-4

may also contribute to NSC and brain tumor development. Understanding the biology of neural and cancer stem cells and their regulatory mechanisms may directly impact current efforts for more directed therapeutics against these highly aggressive tumors.

1. Introduction

Gliomas are the most common primary brain cancers, accounting for about 80% of all central nervous system (CNS) malignancies (CBTRUS, 2010). The most aggressive form, glioblastoma multiforme (GBM), is also the most prevalent and considered one of the most lethal cancers. Despite intensive treatment strategies, the median survival is only about a year from the time of diagnosis, and most patients, if not all, face certain death (Stupp *et al.*, 2009).

Gliomas represent a heterogeneous group of intracranial neoplasms that bear histologic resemblance to glia, the support cells of the CNS. Astrocytomas, which histologically resemble astrocytes, comprise the majority of these tumors. Astrocytomas represent a wide range of malignancies, and are traditionally classified based on histopathologic features into World Health Organization (WHO) grades I–IV (Kleihues *et al.*, 2002). Grade I or pilocytic astrocytomas, which are common in children, are considered benign. Grade II tumors are low-grade malignant astrocytomas that appear differentiated and are slow growing but exhibit diffuse infiltration, rendering them surgically incurable. The higher grade gliomas include Grade III and Grade IV tumors. Grade III (anaplastic astrocytomas) are characterized by dense cellularity, mitotic activity, and nuclear atypia while Grade IV (GBM) are the most aggressive tumors with advanced features of malignancy, including microvascular proliferation and necrosis (Louis, 2006).

GBMs are further designated as either primary or secondary based on clinical presentation. Primary GBMs arise *de novo* with no evidence of prior clinical history and are predominantly seen in older patients whereas secondary GBMs tend to appear in younger patients and initiate as lower grade lesions that progress over variable time to Grade III or IV. Despite these differences, primary and secondary GBMs are histologically indistinguishable and carry equally poor prognosis when adjusted for patient age (Maher *et al.*, 2001; Tso *et al.*, 2006). On the other hand, gliomas are highly infiltrative and can spread to most brain regions but rarely metastasize outside of the CNS. Hence, tumor grade serves as the primary determinant of clinical outcome (Furnari *et al.*, 2007; Maher *et al.*, 2001).

The overall survival for patients with malignant astrocytomas has not improved substantially over the last 50 years. The characteristic diffuse infiltration, widespread invasion, and malignant progression render these tumors resistant to conventional forms of chemotherapy, radiation, and surgery. Patients with grade II tumors have a 5- to 10-year survival period, while Grade III tumors progress more rapidly, with a median survival of 2–5 years. GBM carries the worst prognosis, with a median survival of about 12–15 months from time of diagnosis (Stupp *et al.*, 2005, 2009).

2. Genetic Pathways in Malignant Glioma

A variety of gene mutations potentially related to cancer development and progression have been described in gliomas. Genetic alterations in human GBMs typically target pathways involved in growth factor receptor signaling and cell cycle and apoptosis regulation (Alcantara Llaguno *et al.*, 2009b; Zhu and Parada, 2002), as illustrated in Fig. 2.1. Frequent mutations in genes involved in these processes underscore the importance of mitogenic signaling through receptor tyrosine kinases (RTKs) coupled with inactivation of critical negative regulators of cell proliferation and senescence in the acquisition of the malignant phenotype (Furnari *et al.*, 2007; Louis, 2006; Zhu and Parada, 2002). This was validated by a recent large scale tumor genomic sequencing effort by The Cancer Genome Atlas Research Network (TCGA) which revealed that the most common cancer-related somatic gene mutations in human GBMs occur in *P53*, *PTEN*, *NF1*, *EGFR*, *ERBB2*, and *RB1* (TCGA, 2008). Genomic profiling also revealed the prevalence of mutations in metabolic enzyme genes such as the isocitrate dehydrogenase gene *IDH1* in a subset of human gliomas (Parsons *et al.*, 2008; Yan *et al.*, 2009), underscoring the significance of metabolic alterations in gliomagenesis. *IDH1* mutations are particularly interesting as studies suggest that it loses its normal enzymatic activity in tumors while gaining a new one, leading to production of an oncometabolite (Dang *et al.*, 2009; Yan *et al.*, 2009; Zhao *et al.*, 2009). Other molecular studies characterizing the glioblastoma genome have provided further insight into its genetic changes, core pathways and molecular subtypes (Carro *et al.*, 2010; Parsons *et al.*, 2008; Phillips *et al.*, 2006).

2.1. Cell cycle and apoptosis regulation

The retinoblastoma (RB)-mediated binding of the E2F family of transcription factors is an important block to unrestrained proliferation. RB-mediated cell cycle inhibition is overcome by genetic alterations in the *RB* gene itself, which is mutated in ~24% of high-grade astrocytomas. Functional inactivation of *RB* is also accomplished by amplification of the cyclin-dependent kinases *CDK4* and *CDK6*, and inactivation of its negative regulator $p16^{Ink4a}$ (Furnari *et al.*, 2007). On the other hand, the p53 tumor

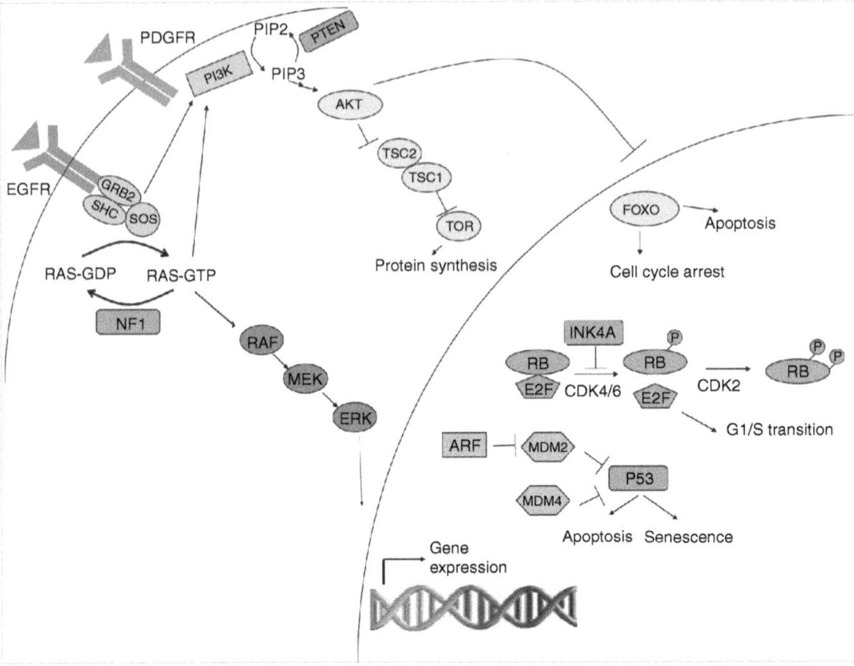

Figure 2.1 Genetic pathways in malignant glioma. Simplified schematic of genetic alterations frequently found in malignant gliomas, including growth factor receptor signaling and cell cycle and apoptosis regulation. Overactivation of RTKs such as epidermal growth factor receptor (EGFR) and platelet-derived growth factor receptor (PDGFR) leads to stimulation of its downstream effectors, mainly Ras and phosphatidylinositide-3-kinase (PI3K) pathways, which are frequently activated in malignant gliomas. Ras signals through the Raf–Mek–Erk effector arm, among others, and can also directly activate the PI3K pathway. PI3K-mediated phosphorylation of PIP2 (phosphatidylinositol-4,5-biphosphate) to PIP3 (phosphatidylinositol-3,4,5-triphosphate) activates Akt, leading to activation of TOR (target of rapamycin) and inhibition of FOXO transcription factors, promoting cellular growth and survival. Nf1 and Pten are tumor suppressors that negatively regulate Ras and PI3K signaling, respectively. In the nucleus, G1/S progression is regulated by the retinoblastoma gene Rb, which binds the E2F family of transcription factors and is inhibited by cyclin-dependent kinases Cdk4/6 and Cdk2. On the other hand, the stability of the transcription factor p53, a master regulator of apoptosis and senescence pathways, is controlled by its ubiquitin ligase MDM2 and related gene MDM4. Ink4A and Arf positively regulate the activity of Rb and p53, which are commonly inactivated in astrocytomas. (See Color Insert.)

suppressor, a critical checkpoint regulator of cell cycle progression upon DNA damage, is frequently mutated in its DNA-binding region in both low- and high-grade astrocytomas (Zhu and Parada, 2002). Contrary to what was previously thought, *TP53* (*P53*) mutations are common not only in secondary GBMs but also in primary GBMs (TCGA, 2008). *P53* loss of function also occurs by amplification of the p53 ubiquitin ligase *MDM2* and its related gene *MDM4*, or loss of function of p14^{Arf}, which is encoded by the *CDKN2A* locus and antagonizes MDM2 (Furnari *et al.*, 2007; TCGA, 2008).

2.2. Growth factor receptor signaling

Persistent activation of growth factor receptors, particularly RTKs, is a common feature of malignant gliomas. Epidermal growth factor receptor (*EGFR*)-activating mutations include gene amplification, point mutations, and deletions, including the most common, variant III deletion of the extracellular domain (EGFR-vIII mutant; TCGA, 2008). Platelet-derived growth factor (PDGF) receptor and its ligands PDGF-A and PDGF-B are also overexpressed in low- and high-grade astrocytomas. These are also frequently coexpressed, suggesting a probable autocrine or paracrine loop of growth stimulation (Zhu and Parada, 2002). Mutations in other RTK genes including *ERBB2*, a member of the EGF receptor family, and *MET*, which encodes the hepatocyte growth factor receptor, can also be found in GBMs (TCGA, 2008).

RTKs signal through adaptor proteins to downstream effectors such as Ras and phosphatidylinositide-3-kinase (PI3K). Active Ras signals through a variety of downstream effectors, including the mitogen-activated protein kinase (MAPK) pathway regulating cell growth and proliferation, and the PI3K pathway. Depending on the state of signaling receptor activity, Ras cycles between the active guanosine triphosphate (GTP)-bound and the inactive guanosine diphosphate (GDP)-bound states. GTPase-activating proteins (GAPs) stimulate Ras deactivation while guanosine exchange factors (GEFs) mediate its activation (Schubbert *et al.*, 2007). The neurofibromatosis tumor suppressor gene, *NF1*, encodes the protein neurofibromin, which functions as a RasGAP, thereby controlling Ras activity (Le and Parada, 2007). Almost a quarter of all mutations screened by TCGA are *NF1* inactivating mutations and deletions, identifying *NF1* as a bona fide glioblastoma tumor suppressor (TCGA, 2008).

The PI3K pathway is an essential survival pathway for a variety of cancer cells. PI3K-mediated phosphorylation of phosphatidylinositol-4,5-biphosphate (PIP2) to phosphatidylinositol-3,4,5-triphosphate (PIP3) establishes a cascade of events leading to activation of Akt and its downstream progrowth and survival signals. This includes release of inhibition of TSC1/TSC2 (tuberous sclerosis 1 and 2) complex on the GTPase Rheb, and thus on TOR (target of rapamycin) and negative regulation of the forkhead (FOXO) transcription factors, which mediate apoptosis and cell cycle arrest. The tumor suppressor Pten (phosphatase and tensin homologue on chromosome 10) negatively regulates the PI3K pathway by dephosphorylating PIP3 back to PIP2 (Cully *et al.*, 2006). *PTEN* mutations that target the phosphatase domain are frequent, as are mutations involving the catalytic (p110α) and regulatory (p85α) domains of PI3K (TCGA, 2008).

2.3. Tumor suppressor pathways in glioma

The Nf1, p53, and Pten tumor suppressors are present in germline cancer syndromes as well as being frequently involved in human gliomas. Patients with germline mutations in *NF1*, called neurofibromatosis Type I, have increased susceptibility to gliomas, as much as fivefold higher compared to the general population (Gutmann *et al.*, 2002). On the other hand, patients with astrocytomas have a 20-fold higher incidence of *NF1* mutations (Gutmann *et al.*, 2003). Molecular analysis of *NF1*-associated astrocytomas shows genetic changes such as *P53* mutations and *CDKN2A/P16* deletions, which are likewise found in sporadic malignant astrocytomas (Gutmann *et al.*, 2003). Individuals with germline mutations in *P53* (Li Fraumeni syndrome) and *PTEN* (Cowden disease) also have increased incidence of astrocytoma compared to the general population (Gutmann *et al.*, 2002; Ichimura *et al.*, 2004; Rasmussen *et al.*, 2001). These findings underscore a central role for these tumor suppressors in the development of malignant astrocytomas.

3. Mouse Models of Malignant Glioma

Signature mutations found in human gliomas form the basis of genetic lesions introduced into the mouse to generate genetically engineered animal models. Activated oncogenes and tumor suppressor deletions, growth factors, and even viral antigens, have been introduced alone or in combination in order to generate gliomas in mice. These include various active forms of *Ras*, *Akt*, *EGFR*, *ErbB*, *PDGF*, v-*src*, and polyoma T-antigen, or mutations in tumor suppressors such as *Nf1*, *Pten*, *p53*, *Ink4A*, and *Arf* (Bachoo *et al.*, 2002; Ding *et al.*, 2001; Holland *et al.*, 2000; Huse and Holland, 2009; Uhrbom *et al.*, 2002; Weissenberger *et al.*, 1997). Expression of these genetic changes in the germline or in embryonic or early postnatal cells gives rise to tumors with varying degrees of penetrance.

Mouse models based on tumor suppressor loss of function alone include conditional knockout mice harboring mutations in *Nf1*, *p53*, and/or *Pten*. The first genetic tumor suppressor mouse model was based on heterozygous mice carrying *cis* germline mutations in *Nf1* and *p53* (Reilly *et al.*, 2000). Further refinements led to the use of cre–lox technology in order to generate conditional mutants harboring somatic mutations in the tumor suppressors (Feil, 2007). Using a mouse transgenic expressing cre recombinase under the control of the human glial fibrillary acidic protein (*hGFAP*) promoter (Zhuo *et al.*, 2001), conditional knockout mice harboring combinations of somatic *Nf1* with germline *p53* mutations in *cis* were established. These mice were found to develop varying levels of astrocytic malignancy with full penetrance (Zhu *et al.*, 2005). The addition of a conditional *Pten* allele to the *Nf1-p53* models accelerated the formation of high-grade astrocytomas (Kwon *et al.*, 2008). Using the same cre transgene, somatic mutations in *p53* and *Pten* in combination were also shown to develop astrocytic malignancies (Zheng *et al.*, 2008). Conditional loss of *Nf1* and *Pten*, however, does not lead to astrocytoma formation (Kwon *et al.*, 2008, unpublished observation). These genetic mouse experiments reveal the common requirement for *p53* loss of function with either *Nf1* or *Pten* mutations in order to drive astrocytoma formation. These three mutations are among the most commonly mutated cancer-associated genes in human GBM (TCGA, 2008).

4. Neural Stem Cells in the Mammalian Brain

The development and morphogenesis of the CNS is characterized by substantial periods of active cell division, migration, differentiation, and remodeling. Neural stem cells (NSCs) play crucial roles during early brain formation and maturation, which continue throughout the lifetime of many eukaryotic organisms (Zhao *et al.*, 2008). The capacity of these undifferentiated cells for unlimited self-renewal makes them vulnerable targets for mutagenesis and cancer development.

In mammals and birds, neural stem cells are immature cells in the CNS that are capable of unlimited self-renewal potential and multilineage differentiation into neurons, astrocytes and oligodendrocytes (Gage, 2000). NSCs can divide symmetrically to increase the stem cell pool or asymmetrically to give rise to more restricted progenitors that can undergo several divisions before differentiating into mature cells. The balanced coordination of stem cell function is essential for its important roles in brain formation and adult tissue homeostasis (Alvarez-Buylla and Lim, 2004).

During early embryonic development, NSCs in the form of neuroepithelial cells derived from ectoderm line the ventricular surface. These cells can generate more neuroepithelial cells or give rise to early neurons. Neuroepithelial cells later elongate and give rise to astrocyte-like cells called radial glia. Radial glia have been shown to serve as neural precursor cells throughout the entire brain and also to produce intermediate progenitor cells that give rise to glia. The production of new neural cells continues until early postnatal stages (Gotz and Huttner, 2005; Kriegstein and Alvarez-Buylla, 2009). However, continuous production in the adult was not widely accepted until the 1990s despite earlier reports of such cells in the brains of mammals (Altman, 1969; Altman and Das, 1965). It is now appreciated that NSCs in the adult mammalian brain exist in several restricted regions. Like radial glia, these relatively quiescent stem cells, also known as type B cells in the subventricular zone, display ultrastructural characteristics and marker expression of astrocytes. Type B cells give rise to type C lineage-restricted or intermediate progenitor cells that undergo limited cell divisions before differentiating into mature cell types. These adult self-renewing cells also appear to retain a similar heterogeneity as well as regional specification and organization as radial glia cells (Alvarez-Buylla and Lim, 2004; Doetsch *et al.*, 2002; Merkle *et al.*, 2004, 2007).

NSCs in the adult are found primarily in two neurogenic regions: the subventricular zone (SVZ) of the lateral ventricles and the subgranular zone (SGZ) of the dentate gyrus. The SVZ represents an extensive germinal layer adjacent to the ependymal cell layer that concentrates neural precursor cells on the walls of the lateral ventricles. NSCs in the SVZ give rise to new neurons that migrate through a defined pathway called the rostral migratory stream, and into the olfactory bulb. These cells continuously replace local interneurons and thus contribute to long-term neurogenesis in the olfactory bulb (Doetsch *et al.*, 1999). In the SGZ of the dentate gyrus, on the other hand, adult SGZ stem cells give rise to new neurons that locally migrate within the granule cell layer. These new granular neurons integrate into functional circuits within the adult hippocampus (Suh *et al.*, 2007). Quiescent stem cells in both neurogenic niches exhibit radial glia morphology, and express markers such as Gfap, Nestin, Blbp, and Sox2. The more restricted progenitors do not express Gfap, but have been shown to express Nestin, Mash1/Ascl1, and Dlx2 (Doetsch *et al.*, 2002; Kim *et al.*, 2007; Suh *et al.*, 2009).

In humans, neurogenic radial glia cells have been shown to exist in the outer subventricular zone during fetal development (Hansen *et al.*, 2010), while adult NSCs have also been demonstrated in the subventricular zone of the lateral ventricles (Sanai *et al.*, 2004) and the dentate gyrus (Eriksson *et al.*, 1998). Numerous studies suggest that NSCs may play important roles in olfaction and learning and memory processes in rodents (Deng *et al.*, 2010). However, their role in humans is not well understood.

The presence of self-renewing cells throughout the lifespan of an organism ensures continuous production of new cells for growth, remodeling and repair. While the physiologic role of NSCs in the adult human brain is not yet known, these cells have been implicated in different aspects of gliomagenesis. Studies in animal models have suggested that tumors arise from the subventricular zone, a major source of NSCs (Zhu *et al.*, 2005). The presence of undifferentiated and mature cell types in the tumor bulk likewise suggests the existence of stem-like cancer cells in gliomas (Llaguno *et al.*, 2008). NSCs have been shown to be closely related to the vasculature (Tavazoie *et al.*, 2008) and may contribute to angiogenesis. These stem cells have also been investigated as vehicles for the delivery of anticancer agents, taking advantage of their migratory capacity, especially to tumor regions (Aboody *et al.*, 2000; Benedetti *et al.*, 2000). These findings suggest an important role for these cells in the continuum of brain tumor development. We will focus on the role of NSCs as glioma-initiating cells and the existence of cancer stem cells in brain tumors.

5. NSCs as Cells of Origin of Malignant Astrocytoma

The identity of the cell of origin, the original cell in which the initial transforming events occur, is a classic question in cancer biology. Because the postnatal brain has long been considered a postmitotic organ, brain tumors were thought to arise from the cell types that they best resembled. Hence, it was suggested that gliomas were derived from mature glia that undergo a process of dedifferentiation (Sanai *et al.*, 2005). With the rapid growth of stem cell biology in the past several decades, it has become increasingly evident that stem cells possess a variety of traits that are recapitulated in cancer cells, thus making them attractive targets for tumorigenesis. For example, NSCs are dividing cells that exist throughout the lifetime of an organism, allowing them to acquire mutations required for malignant transformation. Like cancer cells, stem cells also possess the cellular machinery to proliferate and migrate in response to environmental cues, to evade apoptosis, and to lengthen telomeres (Dalerba *et al.*, 2007).

The use of animal models that faithfully recapitulate glioma formation has greatly facilitated the study of the origins of these tumors. Large-scale sequencing efforts contribute to the growing list of genetic mutations found in human tumors that can be tested in animals. These advances have allowed researchers to directly target initiating mutations in different cell populations in the brain.

Improved techniques for temporal and spatial control of gene ablation are now available. The Cre recombinase protein can be fused with a modified estrogen receptor and expressed under the control of a cell type-specific promoter. Ligand administration of the estrogen analog, tamoxifen, induces nuclear translocation of the fusion protein receptor thus conveying the cre recombinase to the DNA where it can mediate targeted recombination (Feil *et al.*, 1996). This method allows for targeting of specific mutations in discrete cell populations at designated time points. Since most cell type-specific promoters are active early on during embryonic or early postnatal ages, inducible cre/lox systems allow induction at adult ages. This approach is particularly useful in studies addressing the cell of origin of cancer.

Clues to the origin of gliomas can be derived from analysis of glioma mouse models prior to full-blown tumor formation. The earliest lesion described in tumor suppressor-based mouse models is increased cellularity in the SVZ (Zhu *et al.*, 2005). Increased proliferation and expansion of nestin- and Gfap-expressing cells were also found in the SVZ. Pulse chase with bromodeoxyuridine (BrdU) demonstrated aberrant migration of SVZ cells and their progeny (Kwon *et al.*, 2008). These data consistently point to the neural stem cell niche as the first site for appearance of gliomas either from NSCs or from more actively dividing progenitor cells.

Definitive evidence, however, requires direct targeting of initiating mutations in specific cell types. Targeting of glioma-associated mutations using the promoters for Gfap and nestin, which are expressed in NSCs, during embryonic or early postnatal ages results in glioma formation (Holland *et al.*, 2000; Kwon *et al.*, 2008; Zheng *et al.*, 2008; Zhu *et al.*, 2005). It should be noted, however, that Gfap is also expressed in mature, differentiated astrocytes. Nevertheless, these results are not surprising given that active cell division and rapid cell turnover in most regions of the brain occur during these early developmental stages. Proliferation at adult ages, however, is confined to distinct brain regions. The capacity of *adult* neural cells to initiate glioma formation was demonstrated using genetic and stereotactic viral delivery methods that allow specific targeting of NSCs and their progeny. Induction of tumor suppressor mutations such as *Nf1*, *p53* and/or *Pten* in neural stem/progenitor niches using a tamoxifen-inducible nestin-driven cre transgene leads to malignant astrocytoma formation (Alcantara Llaguno *et al.*, 2009a). Furthermore, stereotactic delivery of cre-expressing adenovirus into the SVZ of adult mutant mice carrying conditional tumor suppressor alleles likewise results in tumor development (Alcantara Llaguno *et al.*, 2009a; Jacques *et al.*, 2010). Targeting glioma oncogene-expressing viruses to the adult SVZ or dentate gyrus produces tumors as well (Marumoto *et al.*, 2009). On the other hand, directing mutations in differentiated brain regions such as the cortex does not lead to tumor development (Alcantara Llaguno *et al.*, 2009a; Jacques *et al.*, 2010; Marumoto *et al.*, 2009). These data provide evidence that neural stem and progenitor cells can give rise to malignant astrocytomas using tumor suppressor mutations that are most prevalent in human tumors (Fig. 2.2). By analogy, these data suggest that human astrocytomas are more likely to originate from self-renewing neural stem and progenitor cells than mature, differentiated cell types. Lineage-tracing experiments on mutant NSCs in these studies also showed the capacity of these cells for multilineage differentiation. Migration of labeled mutant cells in these mouse models likewise recapitulates the infiltrative and invasive nature of these tumors (Alcantara Llaguno *et al.*, 2009a).

The identification of the cell of origin of gliomas makes possible investigations into the mechanisms involved in cellular transformation and tumor initiation. Development of malignant gliomas in both humans and mice are likely the consequence of processive genetic and epigenetic events including secondary mutations, activation of growth-promoting, apoptosis-evading pathways, and acquisition of more malignant phenotypes (see Fig. 2.2). Evolution of these requisite processes may precede the appearance of the aggressive nature and therapeutic resistance of these tumors.

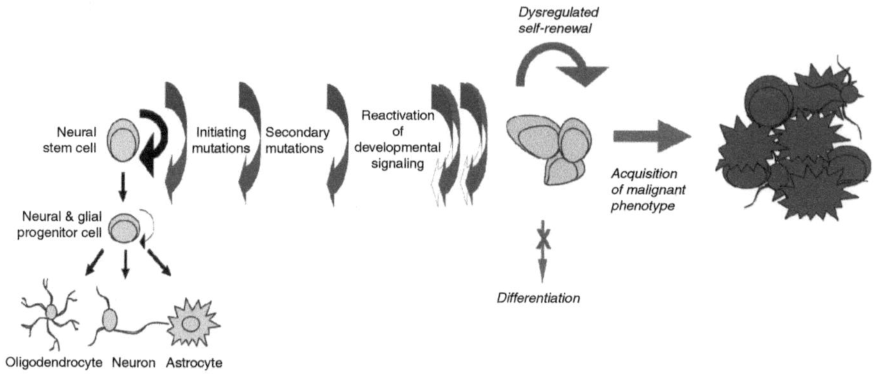

Figure 2.2 Neural stem cell origin and pathways to gliomagenesis. A model for the neural stem cell origin of malignant gliomas and possible mechanisms for multi-step tumorigenesis. Gliomas are initiated by mutations in self-renewing neural stem cells or progenitor cells. These mutant cells acquire additional mutations that lead to increasing levels of dysregulation. This allows them to hijack cellular mechanisms for continued growth and propagation, including developmental signaling pathways. The more undifferentiated tumor cells at the top of the hierarchy undergo unregulated self-renewal and acquire other phenotypic properties that allow them to advance to higher levels of malignancy.

6. Cancer Stem Cells in Malignant Glioma Progression

Malignant gliomas are highly heterogeneous tumors. While predominantly glial in character, there are a variety of cells types, including cells expressing undifferentiated markers within the tumor mass, indicating the presence of cancer cells with an immature phenotype that are capable of giving rise to different neural lineages. This suggests that a subpopulation of tumor cells may exhibit more stem cell-like properties compared to the rest of the tumor bulk. Indeed somatic tumor suppressor-based mouse models of glioma in which genetic cell lineage tracing can be performed demonstrate the spontaneous differentiation of active tumor cells into nondividing cells that acquire the differentiated morphology and markers of neurons and glia (Alcantara Llaguno *et al.*, 2009a).

The cancer stem cell hypothesis posits that cancer cells are organized in a hierarchical manner, with stem-like cancer cells at the summit, giving rise to the more differentiated, less tumorigenic cells (Reya *et al.*, 2001). Cancer stem cells are operationally defined based on the prospective isolation from tumors of a subpopulation of cells that can give rise to tumors when serially transplanted into immunodeficient mice. Because these cells represent the most malignant cell types, cancer stem cells are thought to play important roles in resistance against radiation and chemotherapy, metastasis, and tumor recurrence (Dalerba *et al.*, 2007; Wang and Dick, 2005).

Cancer stem cells have been isolated from most solid tumors and leukemias. In humans, several reports have indicated the presence of stem cell-like cancer cells in GBMs. It was originally reported that glioblastoma cells expressing the cell surface marker CD133 identified the cancer stem cells (Singh *et al.*, 2004). It was demonstrated that as few as 100 CD133+ tumor cells were sufficient to give rise to glioblastoma in immunodeficient mice, whereas the CD133− population was not as tumorigenic. Since then, numerous other markers, such as CD15/LeX/SSEA-1 (Son *et al.*, 2009), or other methods such as determining the side population (Bleau *et al.*, 2009), have been used to identify these cells. Glioma stem cells, similar to NSCs, can be cultured as suspended spheres in serum-free environment with growth factors, and differentiate during growth factor withdrawal. At the present, however, there seems to be no consensus on a universal marker for cancer stem cells in gliomas. Nevertheless, these studies are consistent with the hypothesis that there exists a subpopulation of self-renewing cells that are more tumorigenic compared with the rest of the population (see Fig. 2.3).

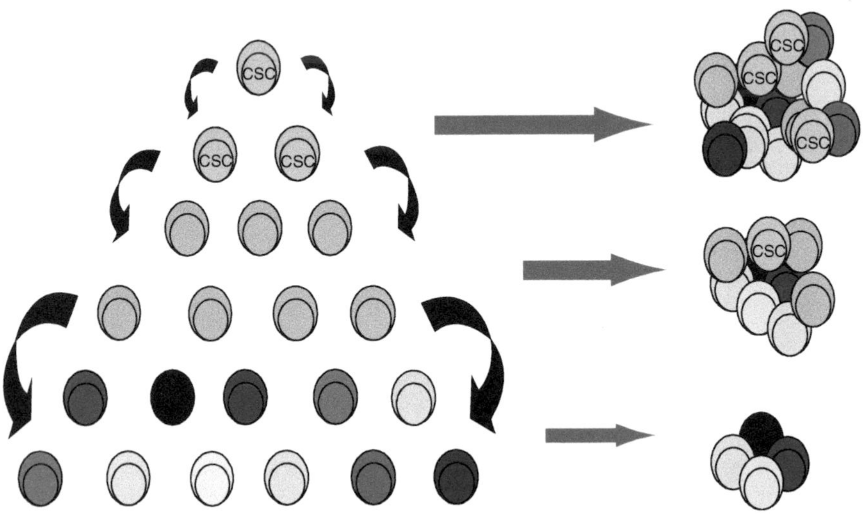

Figure 2.3 A modified cancer stem cell hypothesis in malignant gliomas. Cancer stem cells sit atop the hierarchy of glioma cells. These cells (designated as CSCs) represent a subpopulation of the most tumorigenic and undifferentiated cancer cells. Cancer stem cells can self-renew and give rise to less undifferentiated but actively proliferating cancer cells (rows of green cells). These progenitor-like cancer cells in turn give rise to more differentiated cancer cells. In this schema, cancer stem cells possess the highest tumorigenic potential and greatest malignant properties. Progenitor-like cancer cells have less tumorigenic potential than cancer stem cells and the differentiated tumor cell types are at the bottom of the totem pole. (For interpretation of the references to color in this figure legend, the reader is referred to the Web version of this chapter.)

Cancer stem cells from human GBMs have been shown to be resistant to ionizing radiation, which is currently used as adjunctive treatment in human GBM patients (Bao *et al.*, 2006). Radioresistance of glioma stem cells was shown to be in part due to preferential activation of the DNA damage response resulting in increased DNA repair capacity. On the other hand, resistance against conventional chemotherapeutic agents may be attributed to expression of drug resistance proteins such as ATP-binding cassette (ABC) transporters (Bleau *et al.*, 2009; Martin *et al.*, 2009). Hypoxia-inducible factor (HIF) and HIF-related genes have also been reported to be expressed by glioma stem cells but not nonstem tumor cells (Li *et al.*, 2009b). Since hypoxia is a well-known regulator of the angiogenic switch, this may represent yet another mechanism by which cancer stem cells drive tumorigenicity in gliomas.

Do glioma stem cells arise from NSCs? That question remains to be experimentally addressed. Recent evidence points to a neural stem/progenitor origin of gliomas, which suggests that cancer stem cells may ultimately arise from NSCs. Regardless of their lineage, however, both NSCs and cancer stem cells share a variety of fundamental properties such as the capacity for self-renewal and differentiation, and even more specialized traits such as expression of HIF genes, ABC transporters, and DNA damage repair components (Bleau *et al.*, 2009; Li *et al.*, 2009a; Sii-Felice *et al.*, 2008).

Recent findings suggest that NSCs and cancer stem cells are regulated by similar signaling pathways, particularly developmental processes that are involved in brain formation and neurogenesis. These classical developmental pathways are commonly employed by most organ systems during growth and adult homeostasis of tissue stem cells. Another form of regulation that may be shared by both neural and cancer stem cells is transcriptional regulation by microRNAs. These small RNAs are interesting molecules and potential candidates for therapeutics, as microRNAs exert transcriptional control over a whole host of genes, including self-renewal and differentiation genes, proto-oncogenes, tumor suppressors, and other genes that are important in neurogenesis and cancer. These two regulatory systems may thus play important roles in brain tumor development.

7. Classical Developmental Pathways in Neurogenesis and Gliomagenesis

Developmental signaling pathways play a critical role in embryonic patterning, cell proliferation, and cell differentiation. Several classical development pathways have also been shown to play a role in regulating adult stem cell self-renewal, proliferation, and survival, as shown in Fig. 2.4. For example, developmental signaling pathways such as Notch, bone morphogenetic protein (BMP), and Sonic hedgehog have been shown to play

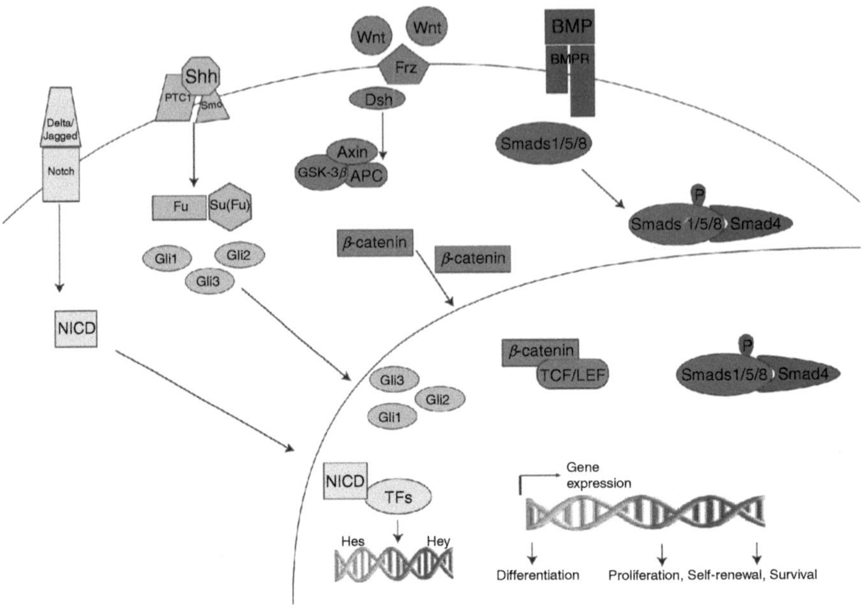

Figure 2.4 Classical developmental signaling pathways in neurogenesis and gliomagenesis. Simplified schematic diagram of developmental pathways that play a functional role in neural development as well as in adult neurogenesis and glioma formation. Expression of a number of Notch, Sonic Hedgehog (Shh), Wnt, and TGF-β/Smad signaling pathway components has been shown to be dysregulated in glioma, and overexpression or knockdown of some of these components has been shown to alter the properties of glioma stem cells and/or glioma cell lines. These pathways therefore offer potential molecular targets for therapeutic intervention. (See Color Insert.)

important roles in maintaining the neurogenic niche, the source of adult NSCs (Alvarez-Buylla and Lim, 2004). A number of recent studies have further demonstrated that dysregulation of these pathways may also be operative in stem-like cancer cells (Ligon *et al.*, 2007; Piccirillo *et al.*, 2006), with reactivation of developmental pathways a possible mechanism contributing to the aggressive and treatment-resistant phenotype of cancers such as GBM (Ben-Porath *et al.*, 2008). In this section, we will briefly describe some of these pathways and the evidence supporting a role for their involvement in brain tumorigenesis.

7.1. Notch signaling

The Notch signaling pathway is involved in a number of cell fate decisions during development and plays an important role in neuronal development and function. Notch signaling has also been shown to promote neural stem cell expansion while inhibiting differentiation (Androutsellis-Theotokis *et al.*, 2006). The Notch receptor family and its ligands, the Delta-like and Jagged protein families, are transmembrane proteins with large extracellular domains consisting of multiple EGF-like repeats. Ligand binding induces enzymatic cleavage, after which the Notch Intracellular Domain (NICD) translocates to the nucleus where it interacts with DNA-binding proteins and coactivators to ultimately activate downstream target genes such as the Hairy/Enhancer of Split (Hes) genes and the Hey genes. Aberrant Notch signaling has been demonstrated in a number of human neoplasms including brain tumors. It has been demonstrated that components of the Notch pathway such as Notch-1, Delta-like-1, and Jagged, are expressed in GBM cell lines and primary tumor tissue, and a number of findings indicate that their activation may also contribute to tumorigenesis. Human GBM cell lines that exhibited Notch overexpression and/or activation had reduced clonogenicity and cell growth properties following genetic or pharmacological inhibition of the Notch pathway (Kanamori *et al.*, 2007; Purow *et al.*, 2005; Stockhausen *et al.*, 2010). More recently, inhibition of Notch signaling using gamma-secretase inhibitors was shown to render glioma stem cells more sensitive to radiation and reduce GBM neurosphere growth and clonogenicity (Fan *et al.*, 2010; Wang *et al.*, 2010). Additionally, knockdown of either Notch1 or Notch2 also sensitized glioma stem cells to radiation and impaired tumor formation in a xenograft model (Wang *et al.*, 2010). These findings implicate a functional role for Notch signaling in gliomagenesis and offer a potential therapeutic target.

7.2. Sonic hedgehog signaling

Another important developmental signaling pathway, the Sonic hedgehog (Shh) pathway, has also been shown to play a critical role in adult neurogenesis (Han *et al.*, 2008; Lai *et al.*, 2003; Palma *et al.*, 2005). In Shh signaling, the secreted Shh protein binds to the Patched (PTC1) receptor resulting in derepression of the Smoothened (Smo) protein. This in turn results in release of sequestered Gli proteins, transcription factors that then translocate to the nucleus and activate hedgehog target genes. Like the Notch signaling pathway, components of the Shh pathway, such as Gli1, Gli2, and Ptch1, have been shown to be expressed in both GBM cell lines and primary tumor tissue (Dahmane *et al.*, 2001), and inhibition of hedgehog signaling using cyclopamine, a steroid alkyloid that binds directly to the Smo protein, or RNAi knockdown of Smo has been shown to inhibit tumor growth and survival (Clement *et al.*, 2007; Dahmane *et al.*, 2001). Additionally, cyclopamine treatment of glioma cells *in vitro* followed by xenotransplantation, or cyclopamine treatment of already established orthotopic xenografts resulted in decreased tumor growth and increased survival (Clement *et al.*, 2007; Sarangi *et al.*, 2009). These findings demonstrate a likely role for the Shh pathway in gliomagenesis.

7.3. Wnt signaling

The Wnt signaling pathway is yet another pathway involved in a number of developmental processes that is known to be dysregulated in a variety of cancers. Like the Notch and Shh pathways, Wnt signaling has also been shown to play a role in adult neurogenesis; exogenous administration of Wnt ligands positively regulates the proliferation and survival of NSCs (Kalani *et al.*, 2008; Lie *et al.*, 2005). Canonical Wnt signaling involves the binding of Wnt ligands to Frizzled (Frz) receptors, that then signal through Dishevelled (Dsh) to ultimately stabilize the β-catenin protein. The accumulated cytosolic β-catenin then translocates to the nucleus where it binds to the TCF/Lef transcription factors and activates Wnt target genes. While overactivation of the Wnt pathway is known to be a hallmark of several types of human cancers, the status of Wnt signaling in glioma was unknown. Recently, however, Sareddy *et al.* examined the expression of Wnt pathway components in human astrocytic tumors and found several, including Dvl-3 and β-catenin, to be overexpressed and constitutively active as evidenced by the upregulation of Wnt downstream target genes such as c-myc and c-jun (Sareddy *et al.*, 2009b). Similar results were observed in a rodent model of glioma (Sareddy *et al.*, 2009a). Additionally, analysis of the tumors in this rodent model showed a higher level of expression and activation of Wnt pathway components in higher grade compared to lower grade tumors, raising the possibility that aberrant Wnt signaling could be a causative mechanism underlying tumor progression (Sareddy *et al.*, 2009a). In support of this, reports by two different groups have demonstrated that siRNA-mediated knockdown of Wnt-2 or β-catenin in a human GBM cell line inhibited cell proliferation and invasiveness, and caused cell cycle arrest and apoptosis (Liu *et al.*, 2010; Pu *et al.*, 2009).

7.4. Transforming growth factor beta (TGF-β) signaling

Finally, the TGF-β pathway has multiple roles in development and physiology, including roles in cell differentiation, hematopoiesis, angiogenesis, and immune response. Like the previously mentioned pathways, it also been shown to play a role in neurogenesis and has been implicated in a large number of cancers including glioma. The TGF-β ligands, which are a subset of a larger superfamily that includes BMPs and activins, bind to Type I and Type II transmembrane serine/threonine kinase receptors and propagate their signal through the Smads. Components of the pathway, including TGF-β ligands, receptors, and Smads, are expressed in human astrocytoma tissue and, like components of the Wnt pathway, expression levels correlate with the degree of malignancy (Kjellman *et al.*, 2000). Treatment of glioma stem cells with TGF-β increases self-renewal and inhibits differentiation through the Smad-dependent induction of leukemia inhibitory factor (LIF), which activates the

JAK–STAT pathway. *In vivo*, increased TGF-β signaling results in increased tumorigenicity (Penuelas *et al.*, 2009). Recently, a report from Ikushima *et al.* demonstrated that the TGF-β pathway via signaling through Sox2, plays a critical role in retaining the stem-like properties of glioma-initiating cells (Ikushima *et al.*, 2009). It has also been shown that inhibiting TGF-β signaling either chemically or by siRNA knockdown of Sox2 significantly reduced the tumorigenic properties of primary glioma cells *in vitro* (Ikushima *et al.*, 2009) and reduced the angiogenic and invasive properties of tumors in a glioma model, while also reversing TGF-β-mediated immune suppression (Tran *et al.*, 2007). On the other hand, exogenous administration of BMP ligands such as BMP4 has been shown to induce differentiation and inhibit proliferation *in vitro* as well as decrease the *in vivo* tumorigenicity of cancer stem cells from human GBMs (Piccirillo *et al.*, 2006). Thus, the TGF-β signaling pathway offers an attractive target for therapeutic intervention and a number of clinical trials are currently underway.

The finding that neoplastic stem cells "hijack" or reactivate these developmental pathways is not surprising—these pathways regulate a variety of molecular processes known to underlie cancer progression and recurrence including cell cycle control, DNA repair mechanisms, activation of anti-apoptotic (prosurvival) factors such as Bcl-2, MYC, NF-κB, and survivin, and epigenetic alteration (Beachy *et al.*, 2004; Dominguez, 2006; Rich, 2007; Xu *et al.*, 2010). Dysregulation of these processes results in uncontrolled cell growth, increased angiogenesis, immune suppression, increased tumor aggressiveness, and chemotherapy resistance. Therapeutic targeting of key components of these developmental pathways could lead to the identification of successful therapies for preventing the development, progression, invasion, and drug resistance of these brain neoplasms.

8. Role of MicroRNAs in NSCs and Brain Tumors

MicroRNAs (miRNAs) are a class of endogenous noncoding RNAs that are ~22 nucleotides in size. In 1993, Ambros and colleagues reported that *lin-4*, a gene essential for postembryonic development in *Caenorhabditis elegans*, encodes a small RNA that negatively regulates *lin-14* expression, thus identifying the first microRNA (Lee *et al.*, 1993). Since then, thousands of miRNAs have been identified across different species with the majority of mammalian miRNAs being evolutionarily conserved. Moreover, in recent years, miRNAs have been recognized as fundamental regulators of various physiological and pathological processes including human cancer.

In general, the biogenesis of microRNAs involves two major steps. In the first step, the long primary transcripts, termed pri-miRNAs, are processed by the Drosha/DGCR8 complex within the nucleus into ~70 nucleotide precursor miRNAs (pre-miRNAs; Lee *et al.*, 2003). Following translocation into the cytoplasm, pre-miRNAs are further digested by the Dicer complex, where Dicer functions as an RNase III enzyme, into stem-loop structured mature miRNAs (Hutvagner *et al.*, 2001; Ketting *et al.*, 2001). Mature miRNA duplexes are then unwound and the single-stranded miRNAs loaded into the RNA-induced silencing complex (RISC). Through partial sequence complementary base pairing, miRNA–RISC complexes bind to the 3′-untranslated region (UTR) of target mRNA transcripts and negatively regulate gene expression by translational repression and/or mRNA degradation (Bartel, 2004).

The potential of miRNAs to simultaneously regulate multiple target genes makes them attractive candidates for pivotal controllers during the development of NSCs. In the SVZ, the major neurogenic niche of adult mammalian brain, the expression level of certain miRNA species has been shown to be developmentally regulated (Cheng *et al.*, 2009). Moreover, a number of studies indicate that hallmark properties of neural progenitor cells, such as proliferation, migration, and lineage progression, are controlled by miRNAs (Cheng *et al.*, 2009; Delaloy *et al.*, 2010; Krichevsky *et al.*, 2006).

The involvement of miRNAs in cancer is supported by multiple lines of evidence. First, human miRNA genes are not randomly distributed in the human genome and frequently map within fragile sites or cancer-associated genomic regions (Calin *et al.*, 2004). Second, high-throughput studies revealed that global downregulation of miRNA expression is commonly observed in multiple human cancers and miRNA expression profiles are very informative in tumor stratification (Lu *et al.*, 2005). Third, the unique expression pattern of a small group of miRNAs can serve as indicators of cancer patient prognosis (Yu *et al.*, 2008). Finally, many miRNA candidates have been demonstrated to function as either oncogenes or tumor suppressor genes during tumor development (Garzon *et al.*, 2009). Taken together, these findings suggest that miRNAs are a class of novel regulators that play a role in human tumorigenesis.

In human glioblastoma, several profiling studies comparing global miRNA expression in primary tumor tissues versus normal brain tissues have been carried out (Chan *et al.*, 2005; Ciafre *et al.*, 2005; Godlewski *et al.*, 2008; Silber *et al.*, 2008). Although each report identified a handful of miRNAs that were either up- or downregulated in tumor tissues (Table 2.1), only one miRNA, miR-21, was found to be consistently upregulated in all experiments. This result is not surprising given the notoriously heterogeneous nature of glioblastoma. Although these early profiling studies helped to reveal some miRNA candidates involved in brain tumor pathogenesis, further characterization of miRNA expression with regard to tumor classification and tumor stem cells is needed to better define the role of miRNAs in tumor diagnosis, prognosis, and therapy.

Of all the miRNAs found to be dysregulated in glioma so far, several candidates have been functionally validated as regulators of tumor cell

Table 2.1 Differentially expressed miRNAs in recent glioma profiling studies

	Ciafre *et al.* (2005)	Chan *et al.* (2005)	Godlewski *et al.* (2008)	Silber *et al.* (2008)		
				Common	GBM only	Anaplastic astrocytoma only
Upregulated miRNAs	miR-10b	miR-21	miR-383	miR-10b	miR-21	miR-296
	miR-130a	miR-138	miR-519d		miR-155	miR-320
	miR-221	miR-347	miR-21		miR-210	
	miR-125b-1	miR-291-5p	miR-516-3-5p			
	miR-125b-2	miR-135	miR-26a			
	miR9-2		miR-10b			
	miR-21		miR-486			
	miR-25		miR-451			
	miR-123					
Downregulated miRNAs	miR-128a	miR-198	miR-124a	miR-7	miR-101	miR-125b
	miR-181c	miR-188	miR-137	miR-31	miR-128a	miR-126
	miR-181a	miR-202	miR-323	miR-107	miR-132	miR-127
	miR-181b		miR-139	miR-124	miR-133a	miR-134
			miR-218	miR-124b	miR-133b	
			miR-128-2	miR-129	miR-149	
			miR-483	miR-137	miR-153	
			miR-128-1	miR-138	miR-154*	
			miR-299	miR-139	miR-185	
			miR-511-1	miR-187	miR-29b	
			miR-190	miR-203	miR-323	
				miR-218	miR-328	
					miR-330	

* is the official name of this particular microRNA species. Please see http://www.mirbase.org/help/nomenclature.shtml for further explanation.

proliferation, differentiation, apoptosis, and invasion. These candidates can be further divided into two subgroups: oncogenic miRNAs and tumor suppressor miRNAs. Oncogenic miRNAs function in promoting the tumorigenic potential of tumor cells and are often expressed at higher levels compared to that of negative controls. As mentioned above, miR-21 has been demonstrated to be the most consistently upregulated miRNA among all the profiling studies in glioma. Moreover, it is highly expressed in more than ten cancer types with different tissue origins (Krichevsky and Gabriely, 2009), suggesting that upregulation of miR-21 may be a common feature in human cancers. In human glioma cell lines, treatment with miR-21-specific inhibitors leads to repressed cell proliferation and increased caspase-3-dependent apoptosis (Chan *et al.*, 2005), reduced cell motility and invasiveness (Gabriely *et al.*, 2008), and diminished xenograft tumor growth in nude mice (Corsten *et al.*, 2007). These findings suggest that miR-21 may function as an oncogene in promoting glioblastoma tumorigenesis. However, as glioma cell lines poorly recapitulate tumor cell growth *in vitro* and *in vivo*, further studies are needed to characterize the precise role of miR-21 in more physiologically relevant settings.

Another example of an oncogenic miRNA in glioblastoma is miR-26a, which is also overexpressed in a subset of glioblastoma tumor tissues (Huse *et al.*, 2009). Targeted delivery of both PDGF-β and miR-26a into nestin promoter-driven tv-a viral receptor-expressing cells in the brain results in reduced tumor-free survival as compared to that of mice injected with PDGF-β and a control miR (Huse *et al.*, 2009). However, the distribution profiles of low-grade and high-grade tumors are not significantly altered between test and control groups. These results suggest miR-26a is involved in promoting tumor initiation but not tumor progression.

A greater number of tumor suppressor miRNAs have been functionally validated than oncogenic miRNAs, partly due to the observation that the majority of miRNAs are downregulated in glioblastoma (Godlewski *et al.*, 2008; Silber *et al.*, 2008). Previous studies identified miR-124 as playing an important role in inducing neuronal differentiation of normal NSCs (Makeyev *et al.*, 2007; Visvanathan *et al.*, 2007), leading to the possibility that its downregulation in glioblastoma may contribute to tumor stem cell maintenance. Consistently, miR-124 together with another downregulated miRNA, miR-137, were shown to promote differentiation of tumor-derived neurosphere-forming cells *in vitro* (Silber *et al.*, 2008). Reduced miR-128 levels have also been implicated in high-grade gliomas and miR-128 overexpression in glioma cell lines leads to reduced cell proliferation (Godlewski *et al.*, 2008). Similar effects were observed for miR-7 (Kefas *et al.*, 2008), miR-34a (Li *et al.*, 2009a), and miR-326 (Kefas *et al.*, 2009), suggesting that loss of function of these miRNAs may contribute to uncontrolled tumor cell growth.

Translational repression and/or mRNA degradation have been shown to be the core mechanisms through which miRNAs regulate target gene expression. Given that only partial sequence complementarity is required for the miRNA:mRNA recognition, each miRNA is predicted to target hundreds of transcripts in animal cells. Therefore, the identification of physiologically relevant downstream mediators of miRNAs remains a challenge. Every miRNA that is functionally involved in glioblastoma tumorigenesis has been proposed to regulate one or more downstream effectors. For example, classic oncogenes and tumor suppressor genes in glioblastoma, such as EGFR and Pten, have been implicated as the regulatory targets of miR-7 (Kefas *et al.*, 2008) and miR-26a (Huse *et al.*, 2009), respectively. Bmi-1, which functions in promoting stem cell self-renewal, has been suggested as a downstream target of miR-128 (Godlewski *et al.*, 2008). These results indicate that miRNAs influence tumor cell fate by regulating different pathways in glioma development. In addition, miRNAs can serve as master regulators of a network of gene targets, which in turn, contribute to multiple aspects of tumor growth. MiR-21, for example, has been shown to target multiple components of the p53, TGF-β, and mitochondrial apoptosis pathways (Papagiannakopoulos *et al.*, 2008). Metalloprotease inhibitors TIMP3 and RECK have also been shown to be regulated by miR-21 (Gabriely *et al.*, 2008). These findings not only help to establish the role of miR-21 as a key oncogenic miRNA during glioma tumorigenesis, but also suggest that miR-21 may serve as a promising therapeutic target for glioma as its inhibition could result in the modulation of multiple signaling pathways involved in tumor progression.

The discovery of miRNAs has opened a new avenue toward understanding underlying mechanisms of neural stem cell self-renewal and differentiation, glioma pathogenesis, identification of biomarkers for tumor classification and prognosis, and development of novel therapeutic targets for cancer treatment. We are still at the early stages of understanding the function and regulation of miRNAs in neural stem cell and glioma development. The development of comprehensive miRNA profiling analyses, genetically engineered animal models, and improved design of miRNA target prediction programs will definitely contribute to our understanding of the precise roles of miRNAs in glioma tumorigenesis.

9. Summary and Perspective: CNS and Brain Tumor Development

NSCs and cancer stem cells play important roles in the spectrum of brain tumor development. In animal models, NSCs have been demonstrated to give rise to gliomas whereas cancer stem cells have been shown to drive the malignant behavior and therapeutic resistance of both human and rodent brain tumors. These self-renewing cells are regulated by a number of mechanisms including developmental signaling pathways, such as Notch, Shh, Wnt and TGF-β, that control self-renewal and lineage commitment of NSCs and may be hijacked during tumor progression by cancer stem cells. Transcriptional control by microRNAs is also another form of regulation that is essential for neural stem cell function that becomes dysregulated during tumor development. A greater understanding of the roles of NSCs and cancer stem cells in glioma development and how these cells are regulated may lead to better designs of brain cancer therapies.

Solving the intricate puzzle of human cancer requires a multi-faceted approach to investigating tumorigenesis. Neural developmental studies have tried to map the origins and evolution of the brain and provided significant insight into the mechanisms of brain tumors. Advances in the fields of developmental biology and cancer will likely provide more sophisticated tools in dissecting the complex processes involved in neural and brain tumor development. The use of genetic mouse models and human clinical data will also accelerate the transition of animal studies into practical applications. A more integrated understanding of the molecular and physiologic processes underlying brain and brain tumor development will hopefully translate into more effective therapeutic strategies against these incurable cancers.

ACKNOWLEDGMENTS

This work was supported in part by a Children's Tumor Foundation Young Investigator Award to S.R.A.L. and NIH (5P50NS052606), US Department of Defense (W81XWH-05-1-0265), and American Cancer Society (RP0408401) grants to L.F.P. L.F.P. is an American Cancer Society Research Professor.

REFERENCES

Aboody, K. S., Brown, A., Rainov, N. G., Bower, K. A., Liu, S., Yang, W., Small, J. E., Herrlinger, U., Ourednik, V., Black, P. M., Breakefield, X. O., and Snyder, E. Y. (2000). Neural stem cells display extensive tropism for pathology in adult brain: evidence from intracranial gliomas. *Proc. Natl. Acad. Sci. USA* **97,** 12846–12851.

Alcantara Llaguno, S., Chen, J., Kwon, C. H., Jackson, E. L., Li, Y., Burns, D. K., Alvarez-Buylla, A., and Parada, L. F. (2009a). Malignant astrocytomas originate from neural stem/progenitor cells in a somatic tumor suppressor mouse model. *Cancer Cell* **15,** 45–56.

Alcantara Llaguno, S. R., Chen, J., and Parada, L. F. (2009b). Signaling in malignant astrocytomas: role of neural stem cells and its therapeutic implications. *Clin. Cancer Res.* **15,** 7124–7129.

Altman, J. (1969). Autoradiographic and histological studies of postnatal neurogenesis. IV. Cell proliferation and migration in the anterior forebrain, with special reference to persisting neurogenesis in the olfactory bulb. *J. Comp. Neurol.* **137,** 433–457.

Altman, J., and Das, G. D. (1965). Autoradiographic and histological evidence of postnatal hippocampal neurogenesis in rats. *J. Comp. Neurol.* **124,** 319–335.

Alvarez-Buylla, A., and Lim, D. A. (2004). For the long run: maintaining germinal niches in the adult brain. *Neuron* **41,** 683–686.

Androutsellis-Theotokis, A., Leker, R. R., Soldner, F., Hoeppner, D. J., Ravin, R., Poser, S. W., Rueger, M. A., Bae, S. K., Kittappa, R., and McKay, R. D. (2006). Notch signalling regulates stem cell numbers in vitro and in vivo. *Nature* **442,** 823–826.

Bachoo, R. M., Maher, E. A., Ligon, K. L., Sharpless, N. E., Chan, S. S., You, M. J., Tang, Y., DeFrances, J., Stover, E., Weissleder, R., Rowitch, D. H., Louis, D. N., *et al.* (2002). Epidermal growth factor receptor and Ink4a/Arf: convergent mechanisms governing terminal differentiation and transformation along the neural stem cell to astrocyte axis. *Cancer Cell* **1,** 269–277.

Bao, S., Wu, Q., McLendon, R. E., Hao, Y., Shi, Q., Hjelmeland, A. B., Dewhirst, M. W., Bigner, D. D., and Rich, J. N. (2006). Glioma stem cells promote radioresistance by preferential activation of the DNA damage response. *Nature* **444,** 756–760.

Bartel, D. P. (2004). MicroRNAs: genomics, biogenesis, mechanism, and function. *Cell* **116,** 281–297.

Beachy, P. A., Karhadkar, S. S., and Berman, D. M. (2004). Tissue repair and stem cell renewal in carcinogenesis. *Nature* **432,** 324–331.

Benedetti, S., Pirola, B., Pollo, B., Magrassi, L., Bruzzone, M. G., Rigamonti, D., Galli, R., Selleri, S., Di Meco, F., De Fraja, C., Vescovi, A., Cattaneo, E., *et al.* (2000). Gene therapy of experimental brain tumors using neural progenitor cells. *Nat. Med.* **6,** 447–450.

Ben-Porath, I., Thomson, M. W., Carey, V. J., Ge, R., Bell, G. W., Regev, A., and Weinberg, R. A. (2008). An embryonic stem cell-like gene expression signature in poorly differentiated aggressive human tumors. *Nat. Genet.* **40,** 499–507.

Bleau, A. M., Hambardzumyan, D., Ozawa, T., Fomchenko, E. I., Huse, J. T., Brennan, C. W., and Holland, E. C. (2009). PTEN/PI3K/Akt pathway regulates the side population phenotype and ABCG2 activity in glioma tumor stem-like cells. *Cell Stem Cell* **4,** 226–235.

Calin, G. A., Sevignani, C., Dumitru, C. D., Hyslop, T., Noch, E., Yendamuri, S., Shimizu, M., Rattan, S., Bullrich, F., Negrini, M., and Croce, C. M. (2004). Human microRNA genes are frequently located at fragile sites and genomic regions involved in cancers. *Proc. Natl. Acad. Sci. USA* **101,** 2999–3004.

Carro, M. S., Lim, W. K., Alvarez, M. J., Bollo, R. J., Zhao, X., Snyder, E. Y., Sulman, E. P., Anne, S. L., Doetsch, F., Colman, H., Lasorella, A., Aldape, K., *et al.* (2010). The transcriptional network for mesenchymal transformation of brain tumours. *Nature* **463,** 318–325.

CBTRUS (2010). CBTRUS Statistical Report Primary Brain and Central Nervous System Tumors Diagnosed in the United States in 2004-2006.

Chan, J. A., Krichevsky, A. M., and Kosik, K. S. (2005). MicroRNA-21 is an antiapoptotic factor in human glioblastoma cells. *Cancer Res.* **65,** 6029–6033.

Cheng, L. C., Pastrana, E., Tavazoie, M., and Doetsch, F. (2009). miR-124 regulates adult neurogenesis in the subventricular zone stem cell niche. *Nat. Neurosci.* **12,** 399–408.

Ciafre, S. A., Galardi, S., Mangiola, A., Ferracin, M., Liu, C. G., Sabatino, G., Negrini, M., Maira, G., Croce, C. M., and Farace, M. G. (2005). Extensive modulation of a set of microRNAs in primary glioblastoma. *Biochem. Biophys. Res. Commun.* **334,** 1351–1358.

Clement, V., Sanchez, P., de Tribolet, N., Radovanovic, I., and Ruiz i Altaba, A. (2007). HEDGEHOG-GLI1 signaling regulates human glioma growth, cancer stem cell self-renewal, and tumorigenicity. *Curr. Biol.* **17,** 165–172.

Corsten, M. F., Miranda, R., Kasmieh, R., Krichevsky, A. M., Weissleder, R., and Shah, K. (2007). MicroRNA-21 knockdown disrupts glioma growth in vivo and displays

synergistic cytotoxicity with neural precursor cell delivered S-TRAIL in human gliomas. *Cancer Res.* **67,** 8994–9000.
Cully, M., You, H., Levine, A. J., and Mak, T. W. (2006). Beyond PTEN mutations: the PI3K pathway as an integrator of multiple inputs during tumorigenesis. *Nat. Rev. Cancer* **6,** 184–192.
Dahmane, N., Sanchez, P., Gitton, Y., Palma, V., Sun, T., Beyna, M., Weiner, H., and Ruiz i Altaba, A. (2001). The Sonic Hedgehog-Gli pathway regulates dorsal brain growth and tumorigenesis. *Development* **128,** 5201–5212.
Dalerba, P., Cho, R. W., and Clarke, M. F. (2007). Cancer stem cells: models and concepts. *Annu. Rev. Med.* **58,** 267–284.
Dang, L., White, D. W., Gross, S., Bennett, B. D., Bittinger, M. A., Driggers, E. M., Fantin, V. R., Jang, H. G., Jin, S., Keenan, M. C., Marks, K. M., Prins, R. M., *et al.* (2009). Cancer-associated IDH1 mutations produce 2-hydroxyglutarate. *Nature* **462,** 739–744.
Delaloy, C., Liu, L., Lee, J. A., Su, H., Shen, F., Yang, G. Y., Young, W. L., Ivey, K. N., and Gao, F. B. (2010). MicroRNA-9 coordinates proliferation and migration of human embryonic stem cell-derived neural progenitors. *Cell Stem Cell* **6,** 323–335.
Deng, W., Aimone, J. B., and Gage, F. H. (2010). New neurons and new memories: how does adult hippocampal neurogenesis affect learning and memory? *Nat. Rev. Neurosci.* **11,** 339–350.
Ding, H., Roncari, L., Shannon, P., Wu, X., Lau, N., Karaskova, J., Gutmann, D. H., Squire, J. A., Nagy, A., and Guha, A. (2001). Astrocyte-specific expression of activated p21-ras results in malignant astrocytoma formation in a transgenic mouse model of human gliomas. *Cancer Res.* **61,** 3826–3836.
Doetsch, F., Caille, I., Lim, D. A., Garcia-Verdugo, J. M., and Alvarez-Buylla, A. (1999). Subventricular zone astrocytes are neural stem cells in the adult mammalian brain. *Cell* **97,** 703–716.
Doetsch, F., Petreanu, L., Caille, I., Garcia-Verdugo, J. M., and Alvarez-Buylla, A. (2002). EGF converts transit-amplifying neurogenic precursors in the adult brain into multipotent stem cells. *Neuron* **36,** 1021–1034.
Dominguez, M. (2006). Interplay between Notch signaling and epigenetic silencers in cancer. *Cancer Res.* **66,** 8931–8934.
Eriksson, P. S., Perfilieva, E., Bjork-Eriksson, T., Alborn, A. M., Nordborg, C., Peterson, D. A., and Gage, F. H. (1998). Neurogenesis in the adult human hippocampus. *Nat. Med.* **4,** 1313–1317.
Fan, X., Khaki, L., Zhu, T. S., Soules, M. E., Talsma, C. E., Gul, N., Koh, C., Zhang, J., Li, Y. M., Maciaczyk, J., Nikkhah, G., Dimeco, F., *et al.* (2010). NOTCH pathway blockade depletes CD133-positive glioblastoma cells and inhibits growth of tumor neurospheres and xenografts. *Stem Cells* **28,** 5–16.
Feil, R. (2007). Conditional somatic mutagenesis in the mouse using site-specific recombinases. *Handb. Exp. Pharmacol.* 3–28.
Feil, R., Brocard, J., Mascrez, B., LeMeur, M., Metzger, D., and Chambon, P. (1996). Ligand-activated site-specific recombination in mice. *Proc. Natl. Acad. Sci. USA* **93,** 10887–10890.
Furnari, F. B., Fenton, T., Bachoo, R. M., Mukasa, A., Stommel, J. M., Stegh, A., Hahn, W. C., Ligon, K. L., Louis, D. N., Brennan, C., Chin, L., DePinho, R. A., *et al.* (2007). Malignant astrocytic glioma: genetics, biology, and paths to treatment. *Genes Dev.* **21,** 2683–2710.
Gabriely, G., Wurdinger, T., Kesari, S., Esau, C. C., Burchard, J., Linsley, P. S., and Krichevsky, A. M. (2008). MicroRNA 21 promotes glioma invasion by targeting matrix metalloproteinase regulators. *Mol. Cell. Biol.* **28,** 5369–5380.
Gage, F. H. (2000). Mammalian neural stem cells. *Science* **287,** 1433–1438.

Garzon, R., Calin, G. A., and Croce, C. M. (2009). MicroRNAs in Cancer. *Annu. Rev. Med.* **60,** 167–179.

Godlewski, J., Nowicki, M. O., Bronisz, A., Williams, S., Otsuki, A., Nuovo, G., Raychaudhury, A., Newton, H. B., Chiocca, E. A., and Lawler, S. (2008). Targeting of the Bmi-1 oncogene/stem cell renewal factor by microRNA-128 inhibits glioma proliferation and self-renewal. *Cancer Res.* **68,** 9125–9130.

Gotz, M., and Huttner, W. B. (2005). The cell biology of neurogenesis. *Nat. Rev. Mol. Cell Biol.* **6,** 777–788.

Gutmann, D. H., Rasmussen, S. A., Wolkenstein, P., MacCollin, M. M., Guha, A., Inskip, P. D., North, K. N., Poyhonen, M., Birch, P. H., and Friedman, J. M. (2002). Gliomas presenting after age 10 in individuals with neurofibromatosis type 1 (NF1). *Neurology* **59,** 759–761.

Gutmann, D. H., James, C. D., Poyhonen, M., Louis, D. N., Ferner, R., Guha, A., Hariharan, S., Viskochil, D., and Perry, A. (2003). Molecular analysis of astrocytomas presenting after age 10 in individuals with NF1. *Neurology* **61,** 1397–1400.

Han, Y. G., Spassky, N., Romaguera-Ros, M., Garcia-Verdugo, J. M., Aguilar, A., Schneider-Maunoury, S., and Alvarez-Buylla, A. (2008). Hedgehog signaling and primary cilia are required for the formation of adult neural stem cells. *Nat. Neurosci.* **11,** 277–284.

Hansen, D. V., Lui, J. H., Parker, P. R., and Kriegstein, A. R. (2010). Neurogenic radial glia in the outer subventricular zone of human neocortex. *Nature* **464,** 554–561.

Holland, E. C., Celestino, J., Dai, C., Schaefer, L., Sawaya, R. E., and Fuller, G. N. (2000). Combined activation of Ras and Akt in neural progenitors induces glioblastoma formation in mice. *Nat. Genet.* **25,** 55–57.

Huse, J. T., and Holland, E. C. (2009). Genetically engineered mouse models of brain cancer and the promise of preclinical testing. *Brain Pathol.* **19,** 132–143.

Huse, J. T., Brennan, C., Hambardzumyan, D., Wee, B., Pena, J., Rouhanifard, S. H., Sohn-Lee, C., le Sage, C., Agami, R., Tuschl, T., and Holland, E. C. (2009). The PTEN-regulating microRNA miR-26a is amplified in high-grade glioma and facilitates gliomagenesis in vivo. *Genes Dev.* **23,** 1327–1337.

Hutvagner, G., McLachlan, J., Pasquinelli, A. E., Balint, E., Tuschl, T., and Zamore, P. D. (2001). A cellular function for the RNA-interference enzyme Dicer in the maturation of the let-7 small temporal RNA. *Science* **293,** 834–838.

Ichimura, K., Ohgaki, H., Kleihues, P., and Collins, V. P. (2004). Molecular pathogenesis of astrocytic tumours. *J. Neurooncol.* **70,** 137–160.

Ikushima, H., Todo, T., Ino, Y., Takahashi, M., Miyazawa, K., and Miyazono, K. (2009). Autocrine TGF-beta signaling maintains tumorigenicity of glioma-initiating cells through Sry-related HMG-box factors. *Cell Stem Cell* **5,** 504–514.

Jacques, T. S., Swales, A., Brzozowski, M. J., Henriquez, N. V., Linehan, J. M., Mirzadeh, Z., O'Malley, C., Naumann, H., Alvarez-Buylla, A., and Brandner, S. (2010). Combinations of genetic mutations in the adult neural stem cell compartment determine brain tumour phenotypes. *EMBO J.* **29,** 222–235.

Kalani, M. Y., Cheshier, S. H., Cord, B. J., Bababeygy, S. R., Vogel, H., Weissman, I. L., Palmer, T. D., and Nusse, R. (2008). Wnt-mediated self-renewal of neural stem/progenitor cells. *Proc. Natl. Acad. Sci. USA* **105,** 16970–16975.

Kanamori, M., Kawaguchi, T., Nigro, J. M., Feuerstein, B. G., Berger, M. S., Miele, L., and Pieper, R. O. (2007). Contribution of Notch signaling activation to human glioblastoma multiforme. *J. Neurosurg.* **106,** 417–427.

Kefas, B., Godlewski, J., Comeau, L., Li, Y., Abounader, R., Hawkinson, M., Lee, J., Fine, H., Chiocca, E. A., Lawler, S., and Purow, B. (2008). microRNA-7 inhibits the epidermal growth factor receptor and the Akt pathway and is down-regulated in glioblastoma. *Cancer Res.* **68,** 3566–3572.

Kefas, B., Comeau, L., Floyd, D. H., Seleverstov, O., Godlewski, J., Schmittgen, T., Jiang, J., diPierro, C. G., Li, Y., Chiocca, E. A., Lee, J., Fine, H., *et al.* (2009). The neuronal microRNA miR-326 acts in a feedback loop with notch and has therapeutic potential against brain tumors. *J. Neurosci.* **29,** 15161–15168.

Ketting, R. F., Fischer, S. E., Bernstein, E., Sijen, T., Hannon, G. J., and Plasterk, R. H. (2001). Dicer functions in RNA interference and in synthesis of small RNA involved in developmental timing in C. elegans. *Genes Dev.* **15,** 2654–2659.

Kim, E. J., Leung, C. T., Reed, R. R., and Johnson, J. E. (2007). In vivo analysis of Ascl1 defined progenitors reveals distinct developmental dynamics during adult neurogenesis and gliogenesis. *J. Neurosci.* **27,** 12764–12774.

Kjellman, C., Olofsson, S. P., Hansson, O., Von Schantz, T., Lindvall, M., Nilsson, I., Salford, L. G., Sjogren, H. O., and Widegren, B. (2000). Expression of TGF-beta isoforms, TGF-beta receptors, and SMAD molecules at different stages of human glioma. *Int. J. Cancer* **89,** 251–258.

Kleihues, P., Louis, D. N., Scheithauer, B. W., Rorke, L. B., Reifenberger, G., Burger, P. C., and Cavenee, W. K. (2002). The WHO classification of tumors of the nervous system. *J. Neuropathol. Exp. Neurol.* **61,** 215–225 [Discussion: 226–9].

Krichevsky, A. M., and Gabriely, G. (2009). miR-21: a small multi-faceted RNA. *J. Cell. Mol. Med.* **13,** 39–53.

Krichevsky, A. M., Sonntag, K. C., Isacson, O., and Kosik, K. S. (2006). Specific microRNAs modulate embryonic stem cell-derived neurogenesis. *Stem Cells* **24,** 857–864.

Kriegstein, A., and Alvarez-Buylla, A. (2009). The glial nature of embryonic and adult neural stem cells. *Annu. Rev. Neurosci.* **32,** 149–184.

Kwon, C. H., Zhao, D., Chen, J., Alcantara, S., Li, Y., Burns, D. K., Mason, R. P., Lee, E. Y., Wu, H., and Parada, L. F. (2008). Pten haploinsufficiency accelerates formation of high-grade astrocytomas. *Cancer Res.* **68,** 3286–3294.

Lai, K., Kaspar, B. K., Gage, F. H., and Schaffer, D. V. (2003). Sonic hedgehog regulates adult neural progenitor proliferation in vitro and in vivo. *Nat. Neurosci.* **6,** 21–27.

Le, L. Q., and Parada, L. F. (2007). Tumor microenvironment and neurofibromatosis type I: connecting the GAPs. *Oncogene* **26,** 4609–4616.

Lee, R. C., Feinbaum, R. L., and Ambros, V. (1993). The C. elegans heterochronic gene lin-4 encodes small RNAs with antisense complementarity to lin-14. *Cell* **75,** 843–854.

Lee, Y., Ahn, C., Han, J., Choi, H., Kim, J., Yim, J., Lee, J., Provost, P., Radmark, O., Kim, S., and Kim, V. N. (2003). The nuclear RNase III Drosha initiates microRNA processing. *Nature* **425,** 415–419.

Li, Y., Guessous, F., Zhang, Y., Dipierro, C., Kefas, B., Johnson, E., Marcinkiewicz, L., Jiang, J., Yang, Y., Schmittgen, T. D., Lopes, B., Schiff, D., *et al.* (2009a). MicroRNA-34a inhibits glioblastoma growth by targeting multiple oncogenes. *Cancer Res.* **69,** 7569–7576.

Li, Z., Bao, S., Wu, Q., Wang, H., Eyler, C., Sathornsumetee, S., Shi, Q., Cao, Y., Lathia, J., McLendon, R. E., Hjelmeland, A. B., and Rich, J. N. (2009b). Hypoxia-inducible factors regulate tumorigenic capacity of glioma stem cells. *Cancer Cell* **15,** 501–513.

Lie, D. C., Colamarino, S. A., Song, H. J., Desire, L., Mira, H., Consiglio, A., Lein, E. S., Jessberger, S., Lansford, H., Dearie, A. R., and Gage, F. H. (2005). Wnt signalling regulates adult hippocampal neurogenesis. *Nature* **437,** 1370–1375.

Ligon, K. L., Huillard, E., Mehta, S., Kesari, S., Liu, H., Alberta, J. A., Bachoo, R. M., Kane, M., Louis, D. N., Depinho, R. A., Anderson, D. J., Stiles, C. D., *et al.* (2007). Olig2-regulated lineage-restricted pathway controls replication competence in neural stem cells and malignant glioma. *Neuron* **53,** 503–517.

Liu, X., Wang, L., Zhao, S., Ji, X., Luo, Y., and Ling, F. (2010). beta-Catenin overexpression in malignant glioma and its role in proliferation and apoptosis in glioblastma cells. *Med. Oncol.* [Epub ahead of print].

Llaguno, S. A., Chen, J., Kwon, C. H., and Parada, L. F. (2008). Neural and cancer stem cells in tumor suppressor mouse models of malignant astrocytoma. *Cold Spring Harb. Symp. Quant. Biol.* **73,** 421–426.

Louis, D. N. (2006). Molecular pathology of malignant gliomas. *Annu. Rev. Pathol.* **1,** 97–117.

Lu, J., Getz, G., Miska, E. A., Alvarez-Saavedra, E., Lamb, J., Peck, D., Sweet-Cordero, A., Ebert, B. L., Mak, R. H., Ferrando, A. A., Downing, J. R., Jacks, T., *et al.* (2005). MicroRNA expression profiles classify human cancers. *Nature* **435,** 834–838.

Maher, E. A., Furnari, F. B., Bachoo, R. M., Rowitch, D. H., Louis, D. N., Cavenee, W. K., and DePinho, R. A. (2001). Malignant glioma: genetics and biology of a grave matter. *Genes Dev.* **15,** 1311–1333.

Makeyev, E. V., Zhang, J., Carrasco, M. A., and Maniatis, T. (2007). The MicroRNA miR-124 promotes neuronal differentiation by triggering brain-specific alternative pre-mRNA splicing. *Mol. Cell* **27,** 435–448.

Martin, V., Xu, J., Pabbisetty, S. K., Alonso, M. M., Liu, D., Lee, O. H., Gumin, J., Bhat, K. P., Colman, H., Lang, F. F., Fueyo, J., and Gomez-Manzano, C. (2009). Tie2-mediated multidrug resistance in malignant gliomas is associated with upregulation of ABC transporters. *Oncogene* **28,** 2358–2363.

Marumoto, T., Tashiro, A., Friedmann-Morvinski, D., Scadeng, M., Soda, Y., Gage, F. H., and Verma, I. M. (2009). Development of a novel mouse glioma model using lentiviral vectors. *Nat. Med.* **15,** 110–116.

Merkle, F. T., Tramontin, A. D., Garcia-Verdugo, J. M., and Alvarez-Buylla, A. (2004). Radial glia give rise to adult neural stem cells in the subventricular zone. *Proc. Natl. Acad. Sci. USA* **101,** 17528–17532.

Merkle, F. T., Mirzadeh, Z., and Alvarez-Buylla, A. (2007). Mosaic organization of neural stem cells in the adult brain. *Science* **317,** 381–384.

Palma, V., Lim, D. A., Dahmane, N., Sanchez, P., Brionne, T. C., Herzberg, C. D., Gitton, Y., Carleton, A., Alvarez-Buylla, A., and Ruiz i Altaba, A. (2005). Sonic hedgehog controls stem cell behavior in the postnatal and adult brain. *Development* **132,** 335–344.

Papagiannakopoulos, T., Shapiro, A., and Kosik, K. S. (2008). MicroRNA-21 targets a network of key tumor-suppressive pathways in glioblastoma cells. *Cancer Res.* **68,** 8164–8172.

Parsons, D. W., Jones, S., Zhang, X., Lin, J. C., Leary, R. J., Angenendt, P., Mankoo, P., Carter, H., Siu, I. M., Gallia, G. L., Olivi, A., McLendon, R., *et al.* (2008). An integrated genomic analysis of human glioblastoma multiforme. *Science* **321,** 1807–1812.

Penuelas, S., Anido, J., Prieto-Sanchez, R. M., Folch, G., Barba, I., Cuartas, I., Garcia-Dorado, D., Poca, M. A., Sahuquillo, J., Baselga, J., and Seoane, J. (2009). TGF-beta increases glioma-initiating cell self-renewal through the induction of LIF in human glioblastoma. *Cancer Cell* **15,** 315–327.

Phillips, H. S., Kharbanda, S., Chen, R., Forrest, W. F., Soriano, R. H., Wu, T. D., Misra, A., Nigro, J. M., Colman, H., Soroceanu, L., Williams, P. M., Modrusan, Z., *et al.* (2006). Molecular subclasses of high-grade glioma predict prognosis, delineate a pattern of disease progression, and resemble stages in neurogenesis. *Cancer Cell* **9,** 157–173.

Piccirillo, S. G., Reynolds, B. A., Zanetti, N., Lamorte, G., Binda, E., Broggi, G., Brem, H., Olivi, A., Dimeco, F., and Vescovi, A. L. (2006). Bone morphogenetic proteins inhibit the tumorigenic potential of human brain tumour-initiating cells. *Nature* **444,** 761–765.

Pu, P., Zhang, Z., Kang, C., Jiang, R., Jia, Z., Wang, G., and Jiang, H. (2009). Downregulation of Wnt2 and beta-catenin by siRNA suppresses malignant glioma cell growth. *Cancer Gene Ther.* **16,** 351–361.

Purow, B. W., Haque, R. M., Noel, M. W., Su, Q., Burdick, M. J., Lee, J., Sundaresan, T., Pastorino, S., Park, J. K., Mikolaenko, I., Maric, D., Eberhart, C. G., *et al.* (2005). Expression of Notch-1 and its ligands, Delta-like-1 and Jagged-1, is critical for glioma cell survival and proliferation. *Cancer Res.* **65,** 2353–2363.

Rasmussen, S. A., Yang, Q., and Friedman, J. M. (2001). Mortality in neurofibromatosis 1: an analysis using U.S. death certificates. *Am. J. Hum. Genet.* **68,** 1110–1118.

Reilly, K. M., Loisel, D. A., Bronson, R. T., McLaughlin, M. E., and Jacks, T. (2000). Nf1; Trp53 mutant mice develop glioblastoma with evidence of strain-specific effects. *Nat. Genet.* **26,** 109–113.

Reya, T., Morrison, S. J., Clarke, M. F., and Weissman, I. L. (2001). Stem cells, cancer, and cancer stem cells. *Nature* **414,** 105–111.

Rich, J. N. (2007). Cancer stem cells in radiation resistance. *Cancer Res.* **67,** 8980–8984.

Sanai, N., Tramontin, A. D., Quinones-Hinojosa, A., Barbaro, N. M., Gupta, N., Kunwar, S., Lawton, M. T., McDermott, M. W., Parsa, A. T., Manuel-Garcia Verdugo, J., Berger, M. S., and Alvarez-Buylla, A. (2004). Unique astrocyte ribbon in adult human brain contains neural stem cells but lacks chain migration. *Nature* **427,** 740–744.

Sanai, N., Alvarez-Buylla, A., and Berger, M. S. (2005). Neural stem cells and the origin of gliomas. *N. Engl. J. Med.* **353,** 811–822.

Sarangi, A., Valadez, J. G., Rush, S., Abel, T. W., Thompson, R. C., and Cooper, M. K. (2009). Targeted inhibition of the Hedgehog pathway in established malignant glioma xenografts enhances survival. *Oncogene* **28,** 3468–3476.

Sareddy, G. R., Challa, S., Panigrahi, M., and Babu, P. P. (2009a). Wnt/beta-catenin/Tcf signaling pathway activation in malignant progression of rat gliomas induced by transplacental N-ethyl-N-nitrosourea exposure. *Neurochem. Res.* **34,** 1278–1288.

Sareddy, G. R., Panigrahi, M., Challa, S., Mahadevan, A., and Babu, P. P. (2009b). Activation of Wnt/beta-catenin/Tcf signaling pathway in human astrocytomas. *Neurochem. Int.* **55,** 307–317.

Schubbert, S., Shannon, K., and Bollag, G. (2007). Hyperactive Ras in developmental disorders and cancer. *Nat. Rev. Cancer* **7,** 295–308.

Sii-Felice, K., Barroca, V., Etienne, O., Riou, L., Hoffschir, F., Fouchet, P., Boussin, F. D., and Mouthon, M. A. (2008). Role of Fanconi DNA repair pathway in neural stem cell homeostasis. *Cell Cycle* **7,** 1911–1915.

Silber, J., Lim, D. A., Petritsch, C., Persson, A. I., Maunakea, A. K., Yu, M., Vandenberg, S. R., Ginzinger, D. G., James, C. D., Costello, J. F., Bergers, G., Weiss, W. A., *et al.* (2008). miR-124 and miR-137 inhibit proliferation of glioblastoma multiforme cells and induce differentiation of brain tumor stem cells. *BMC Med.* **6,** 14.

Singh, S. K., Hawkins, C., Clarke, I. D., Squire, J. A., Bayani, J., Hide, T., Henkelman, R. M., Cusimano, M. D., and Dirks, P. B. (2004). Identification of human brain tumour initiating cells. *Nature* **432,** 396–401.

Son, M. J., Woolard, K., Nam, D. H., Lee, J., and Fine, H. A. (2009). SSEA-1 is an enrichment marker for tumor-initiating cells in human glioblastoma. *Cell Stem Cell* **4,** 440–452.

Stockhausen, M., Kristoffersen, K., and Poulsen, H. S. (2010). The functional role of Notch signaling in human gliomas. *Neuro Oncol.* **12,** 199–211.

Stupp, R., Mason, W. P., van den Bent, M. J., Weller, M., Fisher, B., Taphoorn, M. J., Belanger, K., Brandes, A. A., Marosi, C., Bogdahn, U., Curschmann, J., Janzer, R. C., *et al.* (2005). Radiotherapy plus concomitant and adjuvant temozolomide for glioblastoma. *N. Engl. J. Med.* **352,** 987–996.

Stupp, R., Hegi, M. E., Mason, W. P., van den Bent, M. J., Taphoorn, M. J., Janzer, R. C., Ludwin, S. K., Allgeier, A., Fisher, B., Belanger, K., Hau, P., Brandes, A. A., *et al.* (2009). Effects of radiotherapy with concomitant and adjuvant temozolomide versus radiotherapy alone on survival in glioblastoma in a randomised phase III study: 5-year analysis of the EORTC-NCIC trial. *Lancet Oncol.* **10,** 459–466.

Suh, H., Consiglio, A., Ray, J., Sawai, T., D'Amour, K. A., and Gage, F. H. (2007). In vivo fate analysis reveals the multipotent and self-renewal capacities of Sox2+ neural stem cells in the adult hippocampus. *Cell Stem Cell* **1,** 515–528.

Suh, H., Deng, W., and Gage, F. H. (2009). Signaling in adult neurogenesis. *Annu. Rev. Cell Dev. Biol.* **25,** 253–275.

Tavazoie, M., Van der Veken, L., Silva-Vargas, V., Louissaint, M., Colonna, L., Zaidi, B., Garcia-Verdugo, J. M., and Doetsch, F. (2008). A specialized vascular niche for adult neural stem cells. *Cell Stem Cell* **3,** 279–288.

TCGA (2008). Comprehensive genomic characterization defines human glioblastoma genes and core pathways. *Nature* **455,** 1061–1068.

Tran, T. T., Uhl, M., Ma, J. Y., Janssen, L., Sriram, V., Aulwurm, S., Kerr, I., Lam, A., Webb, H. K., Kapoun, A. M., Kizer, D. E., McEnroe, G., *et al.* (2007). Inhibiting TGF-beta signaling restores immune surveillance in the SMA-560 glioma model. *Neuro Oncol.* **9,** 259–270.

Tso, C. L., Freije, W. A., Day, A., Chen, Z., Merriman, B., Perlina, A., Lee, Y., Dia, E. Q., Yoshimoto, K., Mischel, P. S., Liau, L. M., Cloughesy, T. F., *et al.* (2006). Distinct transcription profiles of primary and secondary glioblastoma subgroups. *Cancer Res.* **66,** 159–167.

Uhrbom, L., Dai, C., Celestino, J. C., Rosenblum, M. K., Fuller, G. N., and Holland, E. C. (2002). Ink4a-Arf loss cooperates with KRas activation in astrocytes and neural progenitors to generate glioblastomas of various morphologies depending on activated Akt. *Cancer Res.* **62,** 5551–5558.

Visvanathan, J., Lee, S., Lee, B., Lee, J. W., and Lee, S. K. (2007). The microRNA miR-124 antagonizes the anti-neural REST/SCP1 pathway during embryonic CNS development. *Genes Dev.* **21,** 744–749.

Wang, J. C., and Dick, J. E. (2005). Cancer stem cells: lessons from leukemia. *Trends Cell Biol.* **15,** 494–501.

Wang, J., Wakeman, T. P., Lathia, J. D., Hjelmeland, A. B., Wang, X. F., White, R. R., Rich, J. N., and Sullenger, B. A. (2010). Notch promotes radioresistance of glioma stem cells. *Stem Cells* **28,** 17–28.

Weissenberger, J., Steinbach, J. P., Malin, G., Spada, S., Rulicke, T., and Aguzzi, A. (1997). Development and malignant progression of astrocytomas in GFAP-v-src transgenic mice. *Oncogene* **14,** 2005–2013.

Xu, P., Qiu, M., Zhang, Z., Kang, C., Jiang, R., Jia, Z., Wang, G., Jiang, H., and Pu, P. (2010). The oncogenic roles of Notch1 in astrocytic gliomas in vitro and in vivo. *J. Neurooncol.* **97,** 41–51.

Yan, H., Parsons, D. W., Jin, G., McLendon, R., Rasheed, B. A., Yuan, W., Kos, I., Batinic-Haberle, I., Jones, S., Riggins, G. J., Friedman, H., Friedman, A., *et al.* (2009). IDH1 and IDH2 mutations in gliomas. *N. Engl. J. Med.* **360,** 765–773.

Yu, S. L., Chen, H. Y., Chang, G. C., Chen, C. Y., Chen, H. W., Singh, S., Cheng, C. L., Yu, C. J., Lee, Y. C., Chen, H. S., Su, T. J., Chiang, C. C., *et al.* (2008). MicroRNA signature predicts survival and relapse in lung cancer. *Cancer Cell* **13,** 48–57.

Zhao, C., Deng, W., and Gage, F. H. (2008). Mechanisms and functional implications of adult neurogenesis. *Cell* **132,** 645–660.

Zhao, S., Lin, Y., Xu, W., Jiang, W., Zha, Z., Wang, P., Yu, W., Li, Z., Gong, L., Peng, Y., Ding, J., Lei, Q., *et al.* (2009). Glioma-derived mutations in IDH1 dominantly inhibit IDH1 catalytic activity and induce HIF-1alpha. *Science* **324,** 261–265.

Zheng, H., Ying, H., Yan, H., Kimmelman, A. C., Hiller, D. J., Chen, A. J., Perry, S. R., Tonon, G., Chu, G. C., Ding, Z., Stommel, J. M., Dunn, K. L., *et al.* (2008). p53 and Pten control neural and glioma stem/progenitor cell renewal and differentiation. *Nature* **455,** 1129–1133.
Zhu, Y., and Parada, L. F. (2002). The molecular and genetic basis of neurological tumours. *Nat. Rev. Cancer* **2,** 616–626.
Zhu, Y., Guignard, F., Zhao, D., Liu, L., Burns, D. K., Mason, R. P., Messing, A., and Parada, L. F. (2005). Early inactivation of p53 tumor suppressor gene cooperating with NF1 loss induces malignant astrocytoma. *Cancer Cell* **8,** 119–130.
Zhuo, L., Theis, M., Alvarez-Maya, I., Brenner, M., Willecke, K., and Messing, A. (2001). hGFAP-cre transgenic mice for manipulation of glial and neuronal function in vivo. *Genesis* **31,** 85–94.

CHAPTER THREE

MDM2 AND MDMX IN CANCER AND DEVELOPMENT

Jean-Christophe Marine

Contents

Abstract

The p53 tumor suppressor pathway is active in cells that are subjected to stress and/or damaged, where it promotes cell cycle arrest or apoptosis. In contrast, in normal cells that are not exposed to stress signals and in tumor cells p53 is tightly kept in check or completely silenced. In most, if not all, tumor cells p53 is indeed inactivated by mutations in the p53 locus or by alternative, yet unclear, mechanisms that impinge directly or indirectly on p53 function. Recent biochemical and genetic data indicate that tumor cells hijack and enforce some of the mechanisms used by normal cells to restrain p53 function. This is best illustrated by the aberrant expression in tumor cells of MDM2 and MDMX (or MDM4), two structurally related proteins that play a critical role in maintaining p53 in an OFF state under normal conditions, but in particular in embryonic and

VIB–K.U.Leuven, Department of Molecular and Developmental Genetics, Laboratory for Molecular Cancer Biology, Leuven, Belgium

Current Topics in Developmental Biology, Volume 94
ISSN 0070-2153, DOI: 10.1016/B978-0-12-380916-2.00003-6

stem cells. These advances and their potential implications for the development of new cancer therapeutic strategies form the focus of this chapter.

1. Introduction

p53 is a sequence-specific transcription factor that activates the expression of genes that promote cell cycle arrest or cell death in response to multiple forms of cellular stress, such as DNA damage, oxidative stress, or oncogene activation (Vousden and Lu, 2002; Fig. 3.1). Among the many p53 responsive genes that mediate its biological activities are *bax* and *PUMA* (Michalak *et al.*, 2005), *p21* (el-Deiry *et al.*, 1993), *Ptprv* (Doumont *et al.*, 2005), and PAI1 (Kortlever *et al.*, 2006). Activation of p53 function is mainly due to posttranslational modifications that lead to accumulation of the ordinarily short-lived p53 protein thus increasing its transcriptional activities. Because of its growth inhibitory activities, p53 actions are kept tightly in check under most physiological conditions, including during embryonic development. This is achieved through several mechanisms including posttranslational events, occlusion of the transactivation domain through direct binding, ubiquitylation and degradation by the proteasome, and control of subcellular localization. A large body of biochemical and

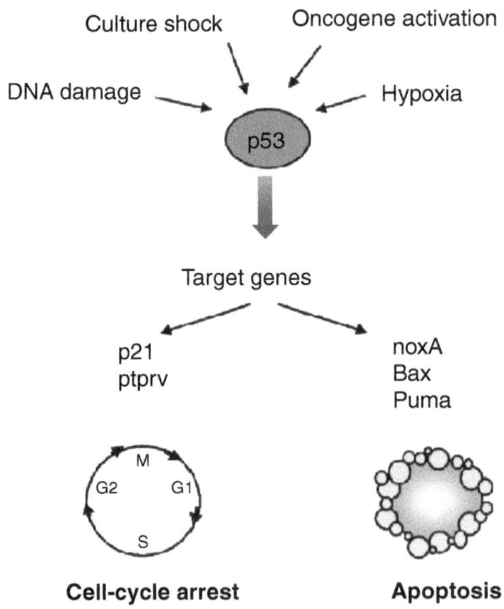

Figure 3.1 p53 as an integrator of stress. p53 protein is stabilized in response to stress and function as a transcription factor to promote either cell-cycle arrest or apoptosis.

genetic evidence indicates that the related proteins Mdm2 and Mdmx play a critical role in all these processes.

A vast majority of human cancers are associated with impaired p53 function. p53 is in fact the most frequently inactivated tumor suppressor gene in human cancer, irrespective of the tumor type, location, or patient age. Whereas half of human tumors harbor *Trp53* mutation, genetic, or functional aberrations of loci that ultimately impair p53 function is a common feature of the others. The mutual exclusivity of these widespread events underscores the central role of the p53 pathway in tumor suppression. Amplification and/or aberrant expression of MDM2 and MDMX have been observed in a number of tumors of diverse origin, especially in those that retain wild-type p53. These observations have raised the possibility that enhanced expression of MDM2 and/or MDMX could directly contribute to tumor formation via inhibition of p53 tumor suppressor function. It therefore also follows that specific MDM2 and MDMX antagonists could be used to reactivate the p53 pathway in tumors that retained wild-type p53. This strategy is best illustrated by the identification of MDMX as the first potential specific chemotherapeutic target for treating retinoblastoma (Laurie *et al.*, 2006).

This chapter provides an overview of our current knowledge on Mdm2 and Mdmx as key regulators of p53 under physiological conditions, focusing in particular on embryonic development, and as oncogenes and putative targets for cancer therapy.

2. The p53–Mdm2 Network

The murine double minute 2 gene was originally identified by virtue of its amplification in a spontaneously transformed mouse BALB/c cell line (3T3-DM) (Cahilly-Snyder *et al.*, 1987). Physical association between the MDM2 protein and the tumor suppressor protein p53 was later reported and shown to inhibit p53-mediated transcription activation (Momand *et al.*, 1992), providing a simple explanation for its transforming potential. Gene amplification of *MDM2* (also known as *HDM2*) was observed in about one third of human sarcomas that retained wild-type p53 (Oliner *et al.*, 1992), leading to the conclusion that overexpression of MDM2 is one of the molecular mechanism by which the cell can inactivate p53 in the process of tumor formation.

Genetic experiments have demonstrated the importance of the Mdm2/p53 interaction in vivo. Mdm2-deficient mice die very early in development prior to implantation, whereas mice lacking both p53 and Mdm2 are viable and indistinguishable from mice lacking only p53 (de Oca *et al.*, 1995; Jones *et al.*, 1995; Fig. 3.2). Similarly, Zdm2-deficient zebrafish embryos,

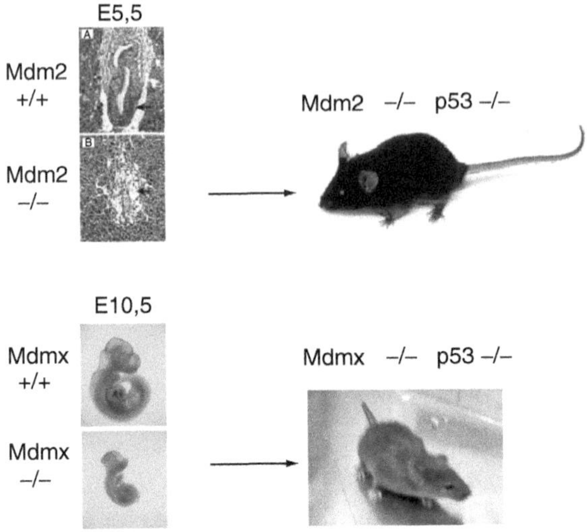

Figure 3.2 A critical role for Mdm2 and Mdmx as negative regulator of p53. Genetic ablation of Mdm2 (top panel) and Mdmx (lower panel) results in embryonic lethal phenotypes. Both of these phenotypes are rescued on a p53-deficient background. (Adapted from Montes de Oca Luna *et al.*, 1995 and Migliorini *et al.*, 2002). (See Color Insert.)

generated by injection of antisense morpholinos, exhibit widespread apoptosis, leading to developmental arrest. As in Mdm2-deficient mice, inactivation of Zp53 in Zdm2-deficient zebrafish embryos rescues this developmental defect (Langheinrich *et al.*, 2002). In addition, mice with a hypomorphic allele of *Mdm2* (equivalent to ~30% of wild-type activity) exhibit defects in thymus development, metabolism, bone marrow, and intestinal cell production (Mendrysa *et al.*, 2003). Conditional inactivation of an $Mdm2^{Lox}$ allele in cardiomyocytes (Grier *et al.*, 2006), neuronal progenitor cells (Xiong *et al.*, 2006), and smooth muscle cells (SMCs) of the Gastro-Intestinal (GI) tract (Boesten *et al.*, 2006) leads to p53-dependent cell death. Conditional expression of p53 in neuronal progenitor cells and postmitotic neurons in mice lacking *Mdm2* leads to dramatic p53 activation and cell death (Francoz *et al.*, 2006). Taken together, these observations indicate that Mdm2 is an essential p53 antagonist in the developing embryo and in mature, differentiated cells.

Mdm2 inhibits p53 function via multiple mechanisms. Mdm2 is a RING-finger-containing protein acting as an E3 ligase, essential for ubiquitylation and degradation of p53 (Haupt *et al.*, 1997; Kubbutat *et al.*, 1997). Consistently, several lines of evidence indicate that p53 undergoes constitutive degradation *in vivo*. Although in most embryonic and adult tissues p53 transcription can be measured by RT-qPCR the p53 protein is undetectable (MacCallum *et al.*, 1996). Treatment of cells with 26S proteasome inhibitors leads to accumulation of ubiquitin-p53 conjugates indicating that p53 is degraded in a ubiquitin–proteasome-dependent manner (Maki *et al.*, 1996). Mdm2 was the first p53 E3-ubiquitin ligase to be described, and induces p53 polyubiquitylation and degradation when overexpressed (Haupt *et al.*, 1997; Honda *et al.*, 1997; Kubbutat *et al.*, 1997). More recently, it was proposed that Mdm2 mediates monomeric p53 ubiquitylation on multiple lysine residues rather than polyubiquitylation (Lai *et al.*, 2001). Because chains of multiple ubiquitin molecules are necessary for efficient protein degradation, Mdm2 might not be sufficient for optimal degradation of p53. In fact, p300 was proposed to promote polyubiquitylation of p53-monoubiquitylated molecules (Grossman *et al.*, 2003). More recently it has been suggested that the ability of Mdm2 to monoubiquitinate or polyubiquitinate p53 depends on its expression level (Li *et al.*, 2003). Low levels of Mdm2 induce monoubiquitylation and nuclear export of p53, whereas high levels promote its polyubiquitylation and nuclear degradation. These distinct mechanisms might be exploited in different physiological settings. For example, Mdm2-mediated polyubiquitylation and nuclear degradation of p53 could play a critical role in suppressing p53 function during the later stages of a DNA damage response or when Mdm2 is malignantly overexpressed. In contrast, Mdm2-mediated monoubiquitylation and subsequent cytoplasmic translocation of p53 would represent an important means of p53 regulation in unstressed cells (Brooks and Gu, 2006). Importantly, decreased expression of Mdm2 in mice harboring a hypomorphic or a null allele of *mdm2* leads to activation of p53 function without concomitant increase in p53 protein levels (Mendrysa *et al.*, 2003). Thus, the reduced levels of Mdm2 in these mice (about 30% of the level in wild-type tissues) are sufficiently high to allow efficient degradation of p53. Alternatively, other p53 ubiquitin ligases might compensate for the decrease in Mdm2 expression *in vivo*. A number of other p53 ubiquitin ligases, such as Pirh2, Cop-1, Yin Yang1, and ARF/BP1, have been described and shown to function in a Mdm2-independent manner (Chen *et al.*, 2005a,b; Dornan *et al.*, 2004; Leng *et al.*, 2003; Sui *et al.*, 2004). Together, these data challenge the conventional view that Mdm2 is essential for p53 turnover *in vivo*. A direct evaluation of the physiological role of Mdm2 in the regulation of p53 stability in this setting test is unfortunately technically difficult owing to the very early embryonic lethality of Mdm2 mutants. Nevertheless, a series of recent studies using conditional alleles clearly demonstrate that p53 degradation *in vivo* occurs in a strict Mdm2-dependent manner. Conditional inactivation of *mdm2* in cardiomyocytes, in neuronal progenitor cells and terminally differentiated SMCs of the GI tract was achieved using an *mdm2* floxed allele and various specific Cre transgenic lines (Boesten *et al.*, 2006; Grier *et al.*, 2006; Xiong *et al.*, 2006). In all of these experimental settings, loss of *mdm2* leads to a dramatic accumulation of the p53 protein *in vivo*. Similar results were obtained using an alternative genetic approach (Francoz *et al.*, 2006).

Finally, additional support, even if less direct, for a key role of Mdm2 in the regulation of p53 levels *in vivo* has come from mouse mutants encoding specifically altered p53 variants. Mutation of two amino acids (L25Q and W26S), essential for the Mdm2–p53 interaction, generates a stable p53 protein which cannot be degraded (Jimenez *et al.*, 2000; Johnson *et al.*, 2005). Similarly, a p53 mutant lacking the proline-rich domain ($p53^{\Delta P}$), a region that appears to modulate Mdm2-binding and Mdm2-mediated degradation *in vitro* (Berger *et al.*, 2001; Dumaz *et al.*, 2001), exhibits a significantly shorter half-life than wild-type p53 *in vivo* (Toledo *et al.*, 2006).

Independent studies using a p53–ER^{TAM} fusion protein show that induction of p53 activity with 4-OH tamoxifen in an *Mdm2*-null background yields lethal pathologies within 5–6 days (Ringshausen *et al.*, 2006). While radiosensitive tissues such as bone marrow, thymus, spleen, small intestine, and colon were ablated due to apoptosis in these mice, cell cycle arrest was induced in the liver and testes, exemplifying the tissue-specific nature of the p53 response. Unexpectedly, this model provided evidence for Mdm2-independent degradation of p53. p53–ER stability was examined in extracts derived from spleens of these knock-in mice with and without Mdm2. Upon functional restoration of p53–ER, its levels decreased, demonstrating a functional negative feedback loop even in the absence of Mdm2. However, the delay in degradation of p53 in the absence of Mdm2 further supports the key contribution of Mdm2 to p53 degradation *in vivo*.

In summary, Mdm2 is required to maintain p53 at low levels both in proliferating progenitor cells and in, terminally differentiated, postmitotic cells. The data suggest that, even if other E3 or E4 ligases might contribute to degradation of p53, this process occurs in a strict Mdm2-dependent manner. Further germline and/or conditional knockout studies are required to determine the physiological contribution of additional ubiquitin ligases to the regulation of p53 turnover.

In addition to promoting p53 degradation, Mdm2 binds p53 in its transactivation domain. This interaction has been proposed to interfere with the recruitment of the basal transcription machinery and/or essential coactivator(s) (Thut *et al.*, 1997). Moreover, Mdm2 was also reported to promote NEDD8 conjugation to p53, a modification that inhibits its transcriptional activity (Xirodimas *et al.*, 2004). Finally, Mdm2 induces monoubiquitylation of histone H2B surrounding the p53-response elements, resulting in transcriptional repression (Minsky and Oren, 2004). Recent genetic studies, however, are not entirely consistent with a role of Mdm2 in the regulation of p53 transcriptional activity *per se* (reviewed in Marine *et al.*, 2006). Conditional inactivation of *mdm2* leads to a dramatic increase in p53 levels accompanied by an increase in transcriptional activity. Indeed loss of *mdm2* in neuronal progenitor cells and in mouse embryonic fibroblasts (MEFs) leads to enhanced transcription of p53 target genes as measured by quantitative PCR analysis (Francoz *et al.*, 2006; Xiong *et al.*, 2006).

However, when the transcriptional activity of p53 is normalized to the amount of protein present, no significant difference could be observed in cells with or without *mdm2* (Francoz *et al.*, 2006). In other words, stabilized p53 in cells lacking *mdm2* is not transcriptionally more potent than in cells expressing Mdm2. Similar conclusions were reached when mice encoding a mutant *p53* allele generated by homologous recombination that lacks the proline-rich domain ($p53^{\Delta P}$) were analysed (Toledo *et al.*, 2006). The $p53^{\Delta P}$ protein was more sensitive to Mdm2-dependent degradation and less prone to transactivation, correlating with deficient cell cycle arrest and reduced apoptotic responses. Strikingly, the compromised function of $p53^{\Delta P}$ failed to rescue *mdm2*-deficiency but fully rescued *mdmx*-deficiency (see below). Importantly, deletion of one single *mdm2* gene copy significantly increased $p53^{\Delta P}$ levels, leading to increased transactivation of p53 target genes. Notably, $p53^{\Delta P}$ did not appear more active on a per molecule basis in $p53^{\Delta P/\Delta P}mdm2^{+/-}$ cells than in $p53^{\Delta P/\Delta P}$ cells. Thus, two very different *in vivo* approaches suggest that Mdm2 does not control p53 transcriptional activity *per se*. Nevertheless, this conclusion should be taken with caution. For instance, since association with Mdm2 leads to rapid degradation of p53, it cannot be excluded that, the majority of the p53 molecules detected in Mdm2 positive cells are not physically associated with Mdm2. If this is indeed the case, then it is not surprising that transcriptional activity of p53, when normalized to the total amount of p53 present, is identical in cells with or without Mdm2. It should also be noted that these experiments were performed in cultured MEFs under nonphysiological oxygen levels. Under these conditions, MEFs accumulate significantly more DNA damage (Parrinello *et al.*, 2003), which known to modulate the levels and behavior of the Mdm2, Mdmx, and p53 proteins. Thus, until the levels of all relevant players are carefully quantified and the various complexes formed in physiological settings are precisely evaluated, this proposal has to be considered as tentative. These caveats notwithstanding, these data strongly argue that the primary physiological function of Mdm2 is to promote p53 degradation, instead of controlling its transcriptional activity.

Once activated, p53 prevents the proliferation of genetically compromised cells by regulating the expression of a battery of genes that initiate cell cycle arrest, apoptosis, and DNA repair. p53 also binds to two adjacent p53-responsive elements located within the *Mdm2* gene to promote its transcription (Barak *et al.*, 1993; Perry *et al.*, 1993). Because of the antagonistic action of Mdm2 towards p53, a negative feedback loop is established, whereby p53 transcriptionally activates Mdm2, which in turn targets p53 for degradation (Fig. 3.1). This feedback loop is likely the key mechanism for restraining p53 activity in normal cells, in the absence of stress.

3. Mdmx, an Mdm2 Relative

A clone encoding Mdmx was originally isolated from a 16-day-old whole mouse cDNA expression library during a search for novel proteins able to interact with p53 (Shvarts *et al.*, 1996). The human ortholog, MDMX, was identified a year later (Shvarts *et al.*, 1997), and the ability of both Mdmx and MDMX proteins to interact with p53 in cells was confirmed by coimmunoprecipitation experiments. Similarly to Mdm2, Mdmx overexpression inhibited p53-activated transcription (Shvarts *et al.*, 1996). cDNAs encoding MDMX were also independently identified in a yeast two-hybrid screen aimed at the identification of MDM2-associated proteins (Sharp *et al.*, 1999; Tanimura *et al.*, 1999). The physical interaction between the two related proteins was confirmed *in vivo* by coimmunoprecipitation assays. Heterooligomerization between Mdmx and Mdm2 appears to be much more stable than homooligomerization of each protein (Tanimura *et al.*, 1999). Together, these independent biochemical observations implicated Mdmx in the regulation of the p53–Mdm2 interplay.

MDMX and *MDM2* mRNAs encode structurally related proteins of 490 and 491 amino acids, respectively (Fig. 3.3). The greatest similarity between the two proteins is found at the N-terminal end, a region harboring the p53-binding domain. The amino acids required for interaction with p53 are strictly conserved in MDM2 and MDMX proteins (Shvarts *et al.*, 1996), and the same amino acids in p53 are required for both MDMX/p53 and MDM2/p53 interactions (Bottger *et al.*, 1999). Another well-conserved region common to MDMX and MDM2 is a RING-finger domain, located at the C-terminal end of each protein. The integrity of the RING-finger domain is essential for MDMX/MDM2 heterodimerization (Sharp *et al.*, 1999; Tanimura *et al.*, 1999). Both MDM2 and MDMX also contain an additional zinc finger domain. The central regions of Mdm2 and Mdmx show no significant similarity, but both regions are rich in acidic amino acids.

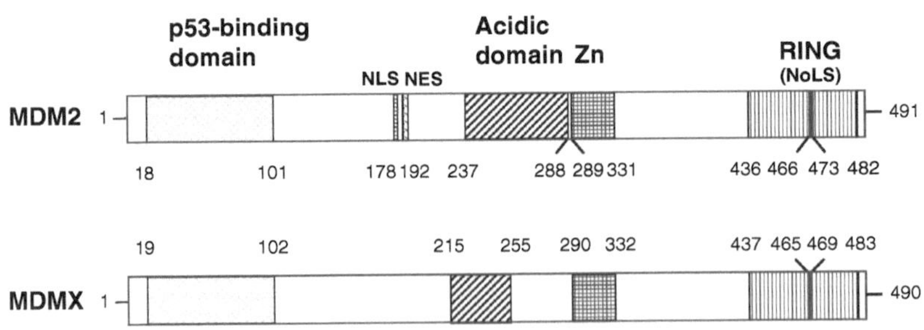

Figure 3.3 MDM2 and MDMX share a common primary structure. (NLS: Nulcear localization Signal; NES: Nuclear Export signal; NoLS: Nucleolar Localization Signal).

The posttranslational modifications of Mdmx that have been characterized to date include phosphorylation, ubiquitylation, and SUMOylation; the latter two have been shown to be important for the regulation of Mdmx stability and activity. Pan and Chen (2005) showed in transient transfection studies that Mdmx is conjugated with SUMO-1 on K254 and K379. Conversion of K254 and K379 to arginine had no effect on Mdmx function in the assays used by these authors. Jochemsen and his colleagues have extended these studies, and demonstrated that endogenous MDMX is modified by SUMO-2 on K254 and K379 (A. Jochemsen, personal communication). However, the biological relevance of these modifications remains unclear.

4. Mdmx, Another Key Regulator of p53

Because of its similarity with Mdm2 and its ability to inhibit p53-induced transcription following overexpression, Mdmx was hypothesized to act as a novel negative regulator of p53 (Migliorini *et al.*, 2002a,b; Shvarts *et al.*, 1996). Loss of function studies in mice confirmed that, similarly to Mdm2, Mdmx acts as an essential, nonredundant, negative regulator of p53 during embryonic development. Indeed, the functional relationship between Mdmx and p53 was genetically demonstrated by the observation that inactivation of p53 rescued the embryonic developmental defects in Mdmx-deficient mice (Finch *et al.*, 2002; Migliorini *et al.*, 2002a,b; Parant *et al.*, 2001; also reviewed in Marine and Jochemsen, 2004; Fig. 3.2).

Early embryonic lethality associated with Mdm2 and Mdmx-null mutations has made it difficult to assess the physiological contributions of Mdm2 and Mdmx to the regulation of p53 levels and activity. However, conditional alleles have more recently been developed that yield further insights into how and in what cell types Mdm2 and Mdmx regulate p53 (Grier *et al.*, 2002, 2006; Mendrysa *et al.*, 2003; Steinman and Jones, 2002).

To test whether Mdm2 and Mdmx are required to restrain p53 activity in a single cell type, both *Mdm2* and *Mdmx* were conditionally inactivated in neuronal progenitors (Xiong *et al.*, 2006). In addition, conditional expression of p53 was restored specifically in neuronal progenitor cells or in postmitotic cells of mice lacking *Mdm2* and/or *Mdmx* (Francoz *et al.*, 2006). Loss of *Mdmx* or Mdm2 led to distinct phenotypes but importantly, all phenotypes disappear in the absence of *p53*. These observations demonstrate that both Mdm2 and Mdmx are required to inhibit p53 activity in the same cell type and confirm the notion that Mdm2 cannot compensate for Mdmx loss *in vivo*, at least in the cell types tested. Together, the data are consistent with specific, nonoverlapping roles for Mdmx in the regulation of the Mdm2–p53 interplay.

In contrast to Mdm2, conditional inactivation of *Mdmx* in cardiomyocytes (Grier *et al.*, 2006) and SMCs of the GI tract (Boesten *et al.*, 2006) only led to minor defects in histogenesis and tissue homeostasis. Inhibition of p53 by Mdmx is likely only required in a restricted number of cell types and/or under certain physiological conditions. Availability of novel specific Mdmx-antibodies allowing detection of endogenous Mdmx in mouse tissues have recently revealed that Mdmx, in contrast to Mdm2, is preferentially expressed in brain, spleen, and thymus (de Clercq *et al.*, 2010). However, interpretation of the genetic data might be more complicated. Indeed, even in cells in which Mdmx function is critical, such as the neuronal progenitor and postmitotic cells, loss of *Mdmx* consistently led to only a moderate increase in p53 activity *in vivo*. This difference between Mdm2 and Mdmx phenotypes can be explained, at least in part, by the fact that p53 activates the transcription of *Mdm2* (Barak *et al.*, 1993; Wu *et al.*, 1993) but not of *Mdmx*. Thus, in the absence of *Mdmx*, p53 transcriptional activity is enhanced leading to the stimulation of the p53–Mdm2 negative feedback loop. In agreement, *Mdmx* loss leads to a moderate increase in Mdm2 protein levels *in vitro* and an increase in *Mdm2*-transcription *in vivo* (Francoz *et al.*, 2006; Toledo *et al.*, 2006; Xiong *et al.*, 2006). The stimulation of *Mdm2* transcription therefore complicates the interpretation of the results from *Mdmx*-deficiency. In one example, overexpression of an *Mdm2* transgene rescues the embryonic lethality associated with Mdmx-deficiency (Steinman *et al.*, 2005), indicating that high levels of Mdm2 compensate for *Mdmx* loss. Thus, as an alternative to the simplistic view of tissue-specific function for Mdmx, increased Mdm2 levels might better compensate for *Mdmx* loss in specific cell types.

Nevertheless, at the molecular level, the difference in the severity of the phenotypes observed following *Mdm2* or *Mdmx* loss is most likely due to the fact that loss of Mdm2 leads to dramatic accumulation of the p53 protein, whereas loss of *Mdmx* does not cause significant increase in p53 levels *in vivo* (see below).

The precise role of Mdmx in the control of p53 and Mdm2 stability remains unclear. Mdmx was reported to act as a ubiquitin ligase *in vitro* (Badciong and Haas, 2002) but Mdmx overexpression in cells does not lead to p53 ubiquitylation and degradation (Jackson and Berberich, 2000; Migliorini *et al.*, 2002a,b; Stad *et al.*, 2000). Transfection studies suggest that Mdmx stabilizes Mdm2, perhaps by interfering with its autoubiquitylation thereby indirectly regulating p53 stability (Gu *et al.*, 2002; Stad *et al.*, 2001). Another report, however, suggested that Mdmx stimulates not only Mdm2-mediated ubiquitylation of p53, but also Mdm2 self-ubiquitylation (Linares *et al.*, 2003). *In vivo*, p53 levels stay below the limit of detection in both Western blotting and immunohistochemistry assays, when its expression is restored in progenitor and in postmitotic neuronal cells lacking Mdmx (Francoz *et al.*, 2006). Similarly, p53 is not detected at E10.5 in the neural progenitor cells in which Mdmx is conditionally inactivated. In contrast, p53 is clearly detectable in Mdm2-deficient cells at the same stage of development (Xiong *et al.*, 2006). Loss of both *Mdm2* and *Mdmx* does not lead to any further increase in p53 levels compared to loss of *Mdm2* alone, suggesting that Mdmx does not participate in the regulation of p53 stability independently of Mdm2 (Francoz *et al.*, 2006). However, whether it does so in an Mdm2-dependent manner remains unclear.

Important insights into Mdmx function came from analyzing the hypomorphic $p53^{DP}$ mouse model (Toledo *et al.*, 2006) in which *Mdmx*-deficiency can be fully rescued. Here, the consequences of *Mdmx* loss were observed in a compromised p53 context, but did not require Cre expression. The results revealed that in the absence of Mdmx, the transactivation of *Mdm2* is to some extent stimulated, leading to slightly increased Mdm2 protein levels. These data imply that Mdmx does not significantly affect Mdm2 protein stability.

The contribution of Mdmx to the regulation of p53 transcriptional activity has also become, to some extent, clearer. The approaches described above provide the first genetic evidence that Mdmx inhibits p53 transcriptional activity independent of Mdm2. Loss of *Mdmx* indeed causes an increase in p53 activity in cultured MEFs that lack *Mdm2*, without concomitant increase in p53 levels (Francoz *et al.*, 2006). Similar conclusions were drawn from the analysis of $p53^{\Delta P}$ regulation by Mdm2 and Mdmx (Toledo *et al.*, 2006). It has been shown that interaction of Mdmx with p53 decreases the p300-mediated acetylation of p53 (Danovi *et al.*, 2004; Sabbatini and McCormick, 2002), and that endogenous p53 acetylation is increased in Mdmx knock-out cells (Danovi *et al.*, 2004). However, the exact nature of the mechanism through which Mdmx attenuates p53 transcriptional activity awaits further investigation.

The above-mentioned data suggest that Mdmx and Mdm2 cooperate to antagonize p53 protein accumulation and activity (Fig. 3.4). This key finding was further confirmed in various *in vivo* settings. Ablation of *Mdm2* and *Mdmx* specifically in the CNS leads to a phenotype that is more severe and appears earlier than the phenotype seen with loss of *Mdm2* alone (Xiong *et al.*, 2006). Similarly, the extent of p53-mediated apoptosis, upon Cre-mediated p53 expression, is significantly greater in the neuroepithelium and in postmitotic cells of mice lacking both *Mdm2* and *Mdmx* than in mice lacking *Mdm2* alone (Francoz *et al.*, 2006). Together, the data support a model in which Mdm2 and Mdmx cooperate *in vivo* to limit p53 activity, irrespectively of the proliferation/differentiation status of the cells.

Mdm2 and Mdmx have been implicated in the regulation of the stability and/or the activity of several proteins playing a key role in the control of cell proliferation such as the retinoblastoma protein pRb, the heterodimer E2F1/DP1, Numb, and Smads (Ganguli and Wasylyk, 2003; Marine and

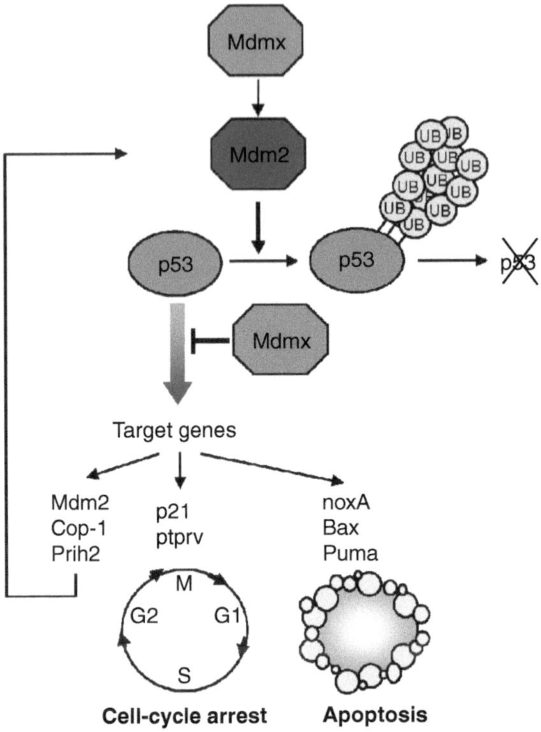

Figure 3.4 A genetic model for cooperative controls of p53 protein levels and transcriptional activity by Mdm2 and Mdmx.

Jochemsen, 2005). However, the relevance of these interactions has not been firmly established genetically. Several lines of evidence do not support p53-independent functions for regulation of Mdm2 and Mdmx under physiological conditions. These data do not exclude the possibility that supra-physiological expression levels of these two proteins affect the activity of other proteins and of p53-independent pathways. This possibility is of great interest, since both proteins are aberrantly expressed in a number of human primary tumors (see below).

5. Regulation of Mdm2-Mediated p53 Ubiquitylation

Numerous mechanisms regulate Mdm2-directed p53 ubiquitylation (reviewed in Brooks and Gu, 2006). For example, DNA damaging agents, such as ionizing radiation (IR) or UV, bring about posttranslational modifications of both Mdm2 and p53, in some cases leading to a reduction of the affinity of p53 to Mdm2. Some modifications, such as phosphorylation or acetylation, directly reduce the ability of Mdm2 to promote p53 ubiquitylation. The simplest and perhaps the most efficient way to regulate Mdm2-mediated ubiquitylation involves downregulation of its expression, for example through regulation of transcriptional rate or half-lives of the transcript and/or the protein. Interaction with other cellular factors can also play a regulatory role. The tumor suppressor ARF and oncoprotein Mdmx are the two best-studied examples (reviewed in Marine *et al.*, 2007; Sherr, 2006). Both proteins bind Mdm2 tightly and affect its cellular localization, stability, and activity through a variety of mechanisms that are yet to be fully elucidated.

All these mechanisms, which continue to grow in number, likely cooperate to impose a tight and fine-tuned regulation on p53 activity. However, since the p53–Mdm2 loop is exquisitely sensitive to a variety of small experimental parameters that are often difficult to control, such as oxygen concentrations, growth factor exposure and extent of DNA damage, great care should be taken when interpreting this flurry of data, especially those from tissue culture work. Key examples are used below to illustrate the complexity of this regulation.

In response to low dose of IR, p53 activation often leads to a transient cell cycle arrest. When the damage has been repaired, restoration of low, prestress levels of p53 must be achieved before the cell cycle can resume. Phosphatase Wip1 (also known as PPM1D) was recently identified as a key component of the p53–Mdm2 negative feedback loop (Lu *et al.*, 2007). Wip1 is a transcriptional target of p53 (Fiscella *et al.*, 1997; Rossi *et al.*, 2008) that catalyzes dephosphorylation of Mdm2 at Ser-395, resulting in Mdm2 stabilization, enhanced Mdm2–p53 binding, and subsequent ubiquitylation (Lu *et al.*, 2007). Ser-395 is a target of the DNA damage-induced kinase ATM, which participates in Mdm2 degradation in response to IR (Maya *et al.*, 2001). The data suggest that Wip1 facilitates Mdm2-mediated degradation of p53 by counteracting DNA damage-induced Mdm2 degradation. When overexpressed Wip1 may function as an oncogene; indeed, it is amplified and overexpressed in breast cancers (Yu *et al.*, 2007). The involvement of Wip1 as a key gatekeeper in the p53/Mdm2 loop is further highlighted by its role in maintaining the uniform shape of p53 pulses in response to persistent DNA damage (Batchelor *et al.*, 2008).

The hedgehog (Hh)-signaling pathway plays an important role in organogenesis during normal development but the pathway is also frequently activated in human cancers. Its role in cancer development involves, at least partially, the Mdm2–p53 hub (Abe *et al.*, 2008). Constitutively active mutants of Smoothened (Smo), a transducer of the Hh-signaling pathway, inhibit the accumulation of p53 in several human cancer cell lines by facilitating the Mdm2–p53 physical interaction and promoting p53 ubiquitylation. Hh signaling induces phosphorylation of Mdm2 on Serines 166 and 186, known activating sites for Mdm2. Importantly, mutations in Smo enhance the proliferation of MEFs, partially inhibit p53-dependent apoptosis and reduce cell growth inhibition in oncogene-expressing MEFs. Taken together, the Hh pathway seems to affect oncogenesis by enhancing Mdm2 activity and thereby inhibiting p53 tumor suppression function.

Mdm2 is also subject to other posttranslational modifications, including acetylation, ubiquitylation, and sumoylation (Meek and Knippschild, 2003). Recent genetic data suggest that SENP2, the SUMO-specific protease 2, targets Mdm2 sumoylation and thus regulates its E3 ligase activity (Chiu *et al.*, 2008). Targeted disruption of *SENP2* in mice affects the expansion of trophoblast progenitors and their maturation, a phenotype that is alleviated upon p53 downregulation. Reintroduction of SENP2 into these mutants reduces sumoylation of Mdm2, diminishes p53 levels and rescues trophoblast development. These data therefore suggest that sumoylation plays a key role in the control of Mdm2 E3 ligase activity.

Numb controls cell fate choices by antagonizing the activity of the plasma membrane receptor of the NOTCH family (Roegiers and Jan, 2004). Mdm2 had been shown to bind NUMB and promote its ubiquitylation and degradation (Colaluca *et al.*, 2008; Juven-Gershon *et al.*, 1998; Yogosawa *et al.*, 2003). Sequential coimmunoprecipitation assays further demonstrate the existence of a p53–Mdm2–Numb trimeric complex in cells, whereby Numb prevents Mdm2-mediated p53 ubiquitylation and subsequent degradation (Fig. 3.2). Purified recombinant NUMB interferes with Mdm2-dependent p53 ubiquitylation *in vitro* in ubiquitylation assays and p53 half-life, steady-state protein levels and activity are significantly reduced in NUMB knock-down cells. Consistently, NUMB expression is frequently lost in breast cancers and is associated with decreased p53 levels and increased chemoresistance (Colaluca *et al.*, 2008). Loss of NUMB expression also enhances the activity of the oncogenic receptor NOTCH. Hence a single event—loss of NUMB expression—leads to an activation of an oncogene and attenuation of the p53 tumor suppressor pathway. This alteration results in an aggressive tumor phenotype, as highlighted by the observation that NUMB-defective breast cancers display poor prognosis. The study raises a number of questions about this three-way interaction, one of which is the issue of how NUMB a cytoplasmic protein, mainly associated with biomembranes, can affect the actions of p53 and Mdm2, most of which are nuclear. Nevertheless, given the role of NUMB in binary cell fate decisions, this study suggests a role for p53 in the control of asymmetric cell division. Such a function for p53 has been suggested earlier (Sherley *et al.*, 1995) but the current study may help revisit this important issue. The proposed model predicts that inactivation of the p53–Mdm2–NUMB axis, as the result of decreased NUMB expression for instance, may cause the skewing of stem cell division towards a symmetric pattern and thus favor tumor development.

The Yin Yang 1 (YY1) transcription factor plays an essential role in development, possibly as the result of decreased p53 ubiquitylation, p53 protein accumulation and activation of its function (Sui *et al.*, 2004). Overexpression of YY1 stimulates p53 ubiquitylation and degradation, and recombinant YY1 is sufficient to induce Mdm2-mediated p53 polyubiquitylation *in vitro*. Direct physical interactions between YY1 with Mdm2 and p53 have been demonstrated suggesting that YY1 regulates p53 ubiquitylation by facilitating the p53–Mdm2 interaction. The data therefore point to YY1 as a potential cofactor for Mdm2 in the regulation of p53 homeostasis.

In contrast, the protein RYBP (RING1- and YY1-binding protein), a member of the polycomb group (PcG) decreases Mdm2-mediated p53 ubiquitylation upon binding to Mdm2 (Chen *et al.*, 2009). RYBP induces cell cycle arrest by stabilizing and activating p53. Accordingly, RYBP expression is decreased in human cancers, suggesting that it exhibits tumor suppressor activity owing to its ability to regulate the p53–Mdm2 loop.

6. Regulation of Mdm2 Stability

Mdm2 is a very short-lived protein, whose rapid degradation is controlled by ubiquitin-dependent proteolysis (Chang *et al.*, 1998). In fact, in addition to ubiquitylating p53, Mdm2 can also drive its own ubiquitylation (Honda and Yasuda, 2000). This autoubiquitylation ability is distinct from its ability to ubiquitylate p53; in fact, some forms of genotoxic damage stabilize p53 by promoting the autodegradation of Mdm2, thereby shifting the balance between the levels of the two proteins (Stommel and Wahl, 2004). Recent genetic data suggest, however, that endogenous Mdm2 does not regulate its own stability by self-ubiquitylation (Clegg *et al.*, 2008). Mdm2 steady-state levels observed in mice in which the Mdm2's RING E3 ubiquitin ligase activity was abrogated by a single point mutation were comparable to the levels in wild-type mice, implying that Mdm2 stability is controlled by another E3 ubiquitin ligase *in vivo*. The histone acetyltransferase PCAF (p300–CBP-associated factor) has recently been proposed as a putative candidate (Linares *et al.*, 2007), as knock-down of PCAF in U2OS and Hela cancer cell lines stabilized Mdm2. PCAF possesses an intrinsic ubiquitylation activity that is critical for controlling Mdm2 stability, and thus p53 function.

7. Regulation of Mdm2 Localization

Mdm2-mediated p53 ubiquitylation and degradation can also be regulated by disrupting p53 binding and sequestration of Mdm2 to a specific cellular compartment. ARF-dependent sequestration of Mdm2 to the nucleoli is one proposed mechanisms through which ARF stabilizes p53 (Weber *et al.*, 1999). Another nucleolar protein, NPM (or B23), interacts with Mdm2 and protects p53 from Mdm2-mediated degradation (Kurki *et al.*, 2004). The PML protein interacts with Mdm2 (Wei *et al.*, 2003) and, like ARF, enhances p53 stability by sequestering Mdm2 to the nucleolus (Bernardi *et al.*, 2004). Following DNA damage, both PML and Mdm2 accumulate in the nucleoli in an ARF-independent manner. The nucleolar localization of Mdm2 is impaired in PML-deficient cells, as is p53 stabilization and the induction of apoptosis. Moreover, PML physically associates with the ribosomal protein L11 and this interaction is largely responsible for the recruitment of PML to the nucleoli after DNA damage. Importantly, L11, as well as other ribosomal proteins including L5, directly interacts with Mdm2 and inhibits its E3 function (Lindstrom *et al.*, 2007). L5 and L11 may cooperate to ensure robust inhibition of the E3 activity of Mdm2, and stabilization and activation of p53 (Horn and Vousden, 2008). The interaction between the ribosomal proteins and Mdm2 may be induced under conditions of ribosomal biogenesis stress, thereby leading to an appropriate p53-dependent cellular response. Together these data provide further confirmation of a previously recognized important role for the nucleolus in the regulation of the Mdm2–p53 feedback loop (Opferman and Zambetti, 2006).

8. Other Mdm2 Binding Proteins Acting in the p53 Pathway

Knock-in mice expressing p53R175H, a mutant form of p53 frequently found in human cancers, have recently been generated and have provided *in vivo* evidence for the importance of the Mdm2-mutant p53 interaction (Terzian *et al.*, 2008). Cells from mice homozygous for the p53R172H mutation have normal levels of mutant p53. In the absence of *Mdm2*, however, p53R172H becomes stable in normal cells and leads to its gain-of-function activities *in vivo*. Thus, constitutive levels of Mdm2 regulate the normal levels of p53 whether it contains a mutation or not. The tissue-specific nature of mutant p53 stabilization indicates that either p53 is not important in some cell types or that other negative regulators of p53 take precedence over Mdm2 in specific cell types (Terzian *et al.*, 2008). However, since p53 can no longer activate transcription of the *Mdm2* the negative feedback loop cannot be established hence, upon DNA damage, mutant p53 is stabilized (Terzian *et al.*, 2008). In heterozygous mice with one mutant and one wild-type *p53* allele, the increased stability of mutant p53 in response to DNA damage functions as a dominant negative and dampens wild-type p53 activity. Proliferating cells with mutant p53 that receive DNA damage signals therefore do not need to lose the wild-type *p53* allele to accumulate additional alterations during tumor development.

Numerous transfection studies have provided existence of additional targets for Mdm2 (Ganguli and Wasylyk, 2003). Among the new putative targets are a number of proteins that act as modulators and/or downstream effectors of the p53 pathway. The JMY protein belongs to this category and functions as an essential p53 cofactor (Shikama *et al.*, 1999). Mdm2 binds JMY and catalyze its ubiquitin-dependent proteasomal degradation, thereby overcoming the ability of JMY to augment p53 response (Coutts *et al.*, 2007). Similarly, Mdm2 promotes ubiquitin-dependent proteasomal degradation of hnRNP K, another p53 cofactor (Moumen *et al.*, 2005). p53 and hnRNP K are recruited to p53-responsive promoters in a mutually dependent manner in response to DNA damage. hnRNP K protein rapidly accumulates in response to DNA damage and facilitates induction of p53-target genes and cell cycle checkpoint arrest. hnRNP K depletion strikingly impairs p53 transcriptional activity.

HIPK2 is a kinase that phosphorylates p53 at ser-46 to promote its proapoptotic activity upon severe, nonreparable, DNA damage (D'Orazi *et al.*, 2002; Hofmann *et al.*, 2002). Recent data suggest that HIPK2 undergoes Mdm2-mediated ubiquitylation and subsequent degradation (Rinaldo *et al.*, 2007). This finding suggests that p53 represses its own phosphorylation at Ser-46 as the result of its ability to induce Mdm2 expression. Thus, in cells exposed to low dose of radiation, the p53–Mdm2–HIPK2 pathway favors cell survival by inhibiting p53 proapoptotic activities.

An important target of Mdm2 is also Mdmx (see below).

9. Regulation of Mdmx Expression and Activity

Many mechanisms have been proposed, mostly involving phosphorylation, that regulate the interaction between Mdm2 and p53 or affecting the activity of Mdm2 as a p53 ubiquitin ligase (reviewed in Wahl, 2006). It is only recently that insight into Mdmx regulation after DNA damage has been obtained. In contrast to the *Mdm2* gene, no evidence has been provided so far that transcription of the *Mdmx* gene is changed upon DNA damage or mitogenic stimulation or any other type of cellular stress. Abundance of the Mdmx protein appears to be mainly regulated at the protein level and it has now been shown that Mdm2 has the ability to bind and ubiquitylate Mdmx, stimulating its proteasome-dependent degradation (de Graaf *et al.*, 2003; Kawai *et al.*, 2003; Pan and Chen, 2003). Interestingly, in normal proliferating cells, MDM2 does not play a major role in regulating MDMX stability. The MDMX protein is very stable, and knocking down MDM2 in cultured cells has little effect on the levels of MDMX (reviewed in Marine and Jochemsen, 2005). However, following DNA damage, the MDMX protein levels rapidly decline in an MDM2-dependent manner. Efficient degradation of MDMX following DNA damage requires ATM-dependent phosphorylation on S342 and S367 by Chk2 and S403 by ATM (Chen *et al.*, 2005a,b; Okamoto *et al.*, 2005; Pereg *et al.*, 2005). UV-C treatment results in Chk1-mediated phosphorylation of S367 (Jin *et al.*, 2006). MDMX phosphorylation reduces its affinity for the ubiquitin protease, HAUSP/USP7 (Meulmeester *et al.*, 2005). Expression of HAUSP is essential for maintenance of MDMX protein expression, as previously reported for MDM2 (Cummins *et al.*, 2004; Li *et al.*, 2004). MDM2 destabilization following DNA damage, as reported by Stommel and Wahl (2004), is also the result of decreased HAUSP binding, while binding of p53 to HAUSP is not affected (Meulmeester *et al.*, 2005). The destabilization of both MDMX and MDM2 is essential for proper p53 activation following DNA damage. The mechanism by which MDMX phosphorylation affects the MDMX/HAUSP interaction has not been elucidated. But loss of HAUSP binding might not be the only mechanism for MDMX protein destabilization. For example, phosphorylation of both S342 and S367 creates binding sites for a dimeric 14-3-3 molecule. Interaction of 14-3-3 with MDMX is necessary for DNA damage-induced nuclear accumulation and degradation of MDMX (LeBron *et al.*, 2006; Pereg *et al.*, 2006). It cannot be excluded, however, that 14-3-3 interaction with MDMX affects its binding to HAUSP.

Apart from the DNA damage-induced phosphorylation events, basal phosphorylation of Mdmx on Ser96 and Ser298 by CDK1/cdc34 and Casein Kinas 1-alpha, respectively, has been reported (Chen *et al.*, 2005a, b; Elias *et al.*, 2005). Ser96 is proposed to affect the regulation of Mdm2 localization, while the CK1-alphamediated phosphorylation stimulates the Mdmx-p53 interaction via an unknown mechanism. Taken together, these data indicate that Mdmx is regulated primarily by posttranslational modifications that affect its stability, subcellular localization and protein–protein interactions.

10. The Mdm2-Mdmx-p53 Network and Cancer Development

Numerous genetic experiments described above helped to define p53 as an important target of Mdm2 and Mdmx in cell survival and in development. The importance of the p53–Mdm2–Mdmx relationship to a tumor phenotype has also been demonstrated genetically. Mice with approximately 30% the levels of *Mdm2* and mice haploinsufficient for *Mdm2* are exquisitely sensitive to IR (Mendrysa *et al.*, 2003; Terzian *et al.*, 2007). These mice die within 2 weeks after treatment with a sub-lethal dose of IR. This phenotype is p53-dependent and indicates that the balance between p53 and Mdm2 levels is also critical for cell survival in response to DNA damage. More importantly, decreased levels of *Mdm2* alter the rate of tumor formation. The development of Eμ-myc-driven B-cell lymphomas and intestinal adenomas as a result of APC loss are delayed in mice with decreased levels of Mdm2 (Alt *et al.*, 2003; Mendrysa *et al.*, 2006). These data are consistent with human genetic data. Patients with *p53* mutations that also inherit an *Mdm2* polymorphism slightly increasing Mdm2 levels show a statistically significant earlier age of tumor onset (Bond *et al.*, 2004). Similarly, Mdmx haploinsufficiency delayed tumor onset, decreased tumor growth and inhibited metastasis in the well-established *TP-ras0/+* murine melanoma progression model (Terzian *et al.*, 2010). Thus even modest decrease in Mdmx levels affect tumorigenesis in mice suggesting that genetic variants of MDMX might have similar effects also in humans. Consistently, a SNP (SNP34091) in the 3′-UTR of MDMX was recently identified that creates a target site for hsa-miR-191, a microRNA that is highly expressed in normal and tumor tissues. Biochemical evidence supports specific miR-191-dependent regulation of the MDMX-C, but not MDMX-A, variant. Importantly, A/A patients have an increased risk of recurrence and tumor-related death indicating that acquisition of an illegitimate miR-191-target site causes downregulation of MDMX expression thereby significantly delaying ovarian carcinoma progression and tumor-related death (Wynendaele *et al.*, 2010). These data indicate that even modest changes in Mdm2 or Mdmx levels, as the result of polymorphisms for example, cause measurable perturbation of p53 function and eventually cancer-related phenotypical manifestations (reviewed in Whibley *et al.*, 2009).

Disruption of the p53 tumor suppressor pathway in tumors is accomplished by direct mutation of the p53 gene in about 50% of the cases. It is believed that tumors retaining wild-type p53 contain defects either in effector target genes or in the expression of p53 regulatory proteins. The above data raise the possibility that elevated Mdm2 and Mdmx expression could contribute to tumor formation via inhibition of p53 tumor suppressor function. A number of studies are consistent with this view. Overexpression of *MDM2*, which can result from gene amplification, enhanced transcription or increased translation, occurs in about a third of human sarcomas that retained wild-type p53 (Leach *et al.*, 1993). Several studies together now also implicate MDMX in tumor formation. *MDMX* was shown to be amplified/overexpressed in some gliomas (Riemenschneider *et al.*, 1999) and to be the gene that is commonly amplified in large amplicons (Riemenschneider *et al.*, 2003). Furthermore, MDMX protein is overexpressed in approximately 30% of the tumor cell lines tested, in general correlating with wild-type p53 status (Ramos *et al.*, 2001). A recent analysis of a large series of tumors indicated that *MDMX* mRNA is overexpressed in a significant percentage of several tumor types, that is, 19% of breast carcinomas, a proportion (20%) of which can be explained by gene amplification (Danovi *et al.*, 2004). The increased expression of the MDMX protein in the *MDMX*-amplified cases could be confirmed both by immunohistochemistry and by Western blotting analysis. In all samples analyzed, amplification of *MDMX* correlated with a wild-type p53 status and lack of *MDM2* amplification. The importance of enhanced MDMX expression was tested in the MCF-7 breast tumor cell line, which contains wild-type p53. Knocking down of endogenous *MDMX* increased $p21^{WAF1}$ expression without a significant increase in p53 levels and was incompatible with proliferation of MCF-7 cells, unless p53 levels were simultaneously decreased. Moreover, constitutive expression of Mdmx immortalizes MEFs in the absence of p53 mutation or loss of ARF expression (Danovi *et al.*, 2004; de Clercq *et al.*, 2010). Finally, Mdmx prevents oncogenic Ras-induced premature senescence, and Mdmx/Ras^{V12}-expressing cells are oncogenic in nude mice (Danovi *et al.*, 2004). Together these data strongly indicate that MDMX functions as an oncogene when constitutively overexpressed in human tumors as an alternative for p53 mutation.

Many tumors contain aberrantly and/or alternatively spliced MDM2 variants. The functions of these variants are still unknown, but their expression is more common in high-grade than in low-grade tumors (Bartel *et al.*, 2004). Although systematic analysis of *MDMX* splicing variants in large tumor sets is still lacking, two *MDMX* splicing variants have been identified and partly characterized. The *MDMX*-S variant essentially encodes only the p53-binding domain and a few alternative C-terminal amino acids. This variant is detected both in untransformed and transformed cells, and its expression is elevated when cells are stimulated into S-phase (Rallapalli *et al.*, 1999). Due to a higher affinity than full length MDMX for p53 and to increased nuclear localization, MDMX-S appears to be a very efficient inhibitor of p53 function (Rallapalli *et al.*, 1999, 2003). In addition, MDMX-S protein is more stable than MDMX, possibly because it can no longer interact with MDM2 and is, therefore, protected from MDM2-mediated degradation. An elevated ratio of MDMX-S/MDMX was reported in high-grade gliomas (Riemenschneider *et al.*, 2003) and analysis of soft-tissue sarcomas indicated that high MDMX-S expression levels correlates with decreased patient survival and increased risk of tumor-related death (Bartel *et al.*, 2005). Another splicing variant, MDMX211, results from splicing between the exon 2 donor site and a cryptic splice acceptor located within exon 11 (Giglio *et al.*, 2005). The resulting protein lacks the p53-binding domain but retains the RING-finger domain. Transfection and siRNA studies indicate an oncogenic activity of this protein possibly through stabilization of MDM2. Although the MDMX211 variant was identified in 2/16 analyzed nonsmall cell lung tumors, further studies are needed to find its significance in human cancer. All in all, these data show that expression of the *MDMX* gene is deregulated in a significant proportion of human tumors, indicating a role for MDMX in tumorigenesis.

11. MDMX as the First Specific Drug Target for Treating Retinoblastoma

Retinoblastoma is a rare childhood cancer of the retina that initiates *in utero* during fetal development by *RB1* gene inactivation (Dyer and Harbour, 2006). While the initiating genetic event in retinoblastoma, biallelic inactivation of the *RB1* gene, is well established, the subsequent genetic events that contribute to retinoblastoma progression have not been well characterized. In particular, the status of the p53 pathway in retinoblastoma has been a topic of considerable debate in the field. Early studies on human tumors demonstrated that the p53 gene is wild-type in retinoblastoma (Kato *et al.*, 1996) and that p53 can be activated in retinoblastoma cell lines (Nork *et al.*, 1997). However, in HPV-E7 transgenic mouse models of retinoblastoma, tumor development was greatly enhanced when p53 was inactivated (Howes *et al.*, 1994). More recently, the first knockout mouse model of retinoblastoma was developed (Zhang *et al.*, 2004) by conditionally inactivating *Rb1* in the developing retina of $p107^{-/-}$ mice. As with the HPV-E7 transgenic mouse models of retinoblastoma, simultaneous inactivation of p53 in retinal progenitor cells lacking *Rb1* and *p107* led to an aggressive invasive form of retinoblastoma in mice that more faithfully recapitulated the human disease (Dyer *et al.*, 2005). Taken together, these data suggest that p53 pathway inactivation is an important step in retinoblastoma progression, although the p53 gene itself remains intact.

Using BAC-CGH, FISH, and IHC for components of the p53 pathway we found that the *MDMX* gene shows increased copy number in 65% of human retinoblastomas and the *MDM2* gene is amplified in an additional 10% of these tumors (Laurie *et al.*, 2006). Genetic amplification of *MDMX* correlated with increased mRNA and protein expression and suppression of p53 target genes such as p21. Cell culture experiments demonstrated that MDMX regulates cell death and cell cycle exit in retinoblastoma cells in a p53-dependent manner. Moreover, ectopic expression of *MDMX* in mouse *Rb*; *p107*-deficient retinal progenitor cells led to a reduction in p53-mediated apoptosis and clonal expansion of tumor cells (Laurie *et al.*, 2006). Similar studies carried out on human fetal retinae also demonstrated that ectopic expression of MDMX rescued p53-mediated cell death as a result of activation of the $p14^{ARF}$ oncogenic stress response pathway following *RB1* gene inactivation (Laurie *et al.*, 2006). Taken together these data clearly show that the p53 pathway is suppressed in retinoblastoma cells following biallelic inactivation of *RB1* and that a majority of tumors inactivate the p53 pathway through *MDMX* gene amplification. In addition, these data show that retinoblastoma does not arise from an intrinsically death resistant cell as previously thought (Dyer *et al.*, 2005).

One of the most common ways (~65%) that the p53 pathway is inactivated in retinoblastomas is by increased MDMX expression through genetic amplification. The fact that the frequency of *MDMX* amplification is high in retinoblastoma compared to other tumor types (Danovi *et al.*, 2004) may be explained by the difference in the ability of $\mathrm{p14^{ARF}}$ to bind to MDM2 and MDMX. Biochemical studies have shown that $\mathrm{p14^{ARF}}$ can bind MDM2 but not MDMX (Wang *et al.*, 2001). Considering that $p14^{ARF}$ is directly regulated by RB1 (Aslanian *et al.*, 2004), retinal cells lacking RB1 may have a greater induction of $\mathrm{p14^{ARF}}$ than tumors that initiate with other disruptions in the Rb pathway involving p16, cyclin D1, or CDK4 (Sherr and McCormick, 2002). The biochemical data and the preferential $p14^{ARF}$ activation suggest that *MDM2* amplification would not lead to an efficient inhibition of the p53 pathway in *RB1*-deficient retinal cells. In contrast, despite high levels of $\mathrm{p14^{ARF}}$, *MDMX* amplification would be expected to efficiently silence the p53 cell death pathway in retinoblastoma because $\mathrm{p14^{ARF}}$ does not bind MDMX.

These findings not only challenge the long-standing belief that retinoblastoma is the exception to the general principle that the Rb and p53 pathways must be inactivated in cancer, but also they identify Mdmx as a specific target for chemotherapy.

12. The Mdmx–Mdm2–p53 Interplay as a Target for Therapeutic Intervention

Ample evidence suggests that transformed cells are more sensitive to p53-induced apoptosis than their normal counterparts. Restoration of p53 activity in mice causes tumor-specific cell killing or induction of senescence

(Martins *et al.*, 2006; Ventura *et al.*, 2007; Xue *et al.*, 2007). Activation of the p53 response becomes, therefore, an attractive therapeutic goal. In particular, the physical interaction between p53 and Mdm2 has recently become the target for the development of new cancer therapeutic strategies (Issaeva *et al.*, 2004; Vassilev *et al.*, 2004; Yang *et al.*, 2005).

Nutlin-3a is a small molecule that binds Mdm2 in its p53-binding domain and thereby disrupting the Mdm2–p53 interaction (Vassilev *et al.*, 2004). A family of small molecules, HLI98, also binds Mdm2, but their binding inhibits the E3 ligase activity at the carboxyl terminus (Yang *et al.*, 2005). Another small molecule—RITA (reactivation of p53 and induction of tumor cell apoptosis)—blocks the Mdm2–p53 interaction, but by direct interaction with p53 (Issaeva *et al.*, 2004). Since numerous studies in mice indicate that *Mdm2* loss leads to p53-dependent pathologies in normal tissues, the toxicity of these molecules *in vivo* will have to be assessed. However, genetic ablation of *Mdm2* is more drastic than the use of small molecule inhibitors that have a limited half-life and it is therefore possible that a therapeutic window can be found.

Since mutations in p53 abolish p53 binding to DNA but leave the Mdm2 interacting domain intact the interaction of these small molecules with mutant p53 and stabilization of the mutant p53 protein cannot be ignored. The loss of *Mdm2*, which mimics the inhibition of the Mdm2–p53 interaction, leads to a stabilization of mutant p53 and an increased incidence of metastasis as compared to mice lacking *p53* (Terzian *et al.*, 2008). The model predicts that patients with mutations in p53 will stabilize mutant p53 upon treatment with small molecule inhibitors of Mdm2, regardless of the length of small molecule administration. In tumors with p53 LOH, the stabilization of mutant p53 is likely to yield a worse outcome. In tumors that retain a wild-type *p53*, wild-type p53 may eventually overcome the stabilization of the mutant p53. Whether the presence of one mutant p53 contributes to a metastatic outcome remains unknown. However in mutant *p53* heterozygous mice treated with IR, the mutant p53 half-life is longer than that of wild-type p53; this overall decrease in p53 activity might in turn lead to the acquisition of additional mutations (Terzian *et al.*, 2008).

In light of the observations described above, small molecules inhibitors should be used in tumor cells that retain wild-type p53 whether or not Mdm2 overexpression is detected, as small differences in Mdm2 levels affect a tumor phenotype. The small molecules that bind Mdm2 may have additional efficacy as these may eliminate Mdm2-dependent p53-independent activities. On the other hand, they may stabilize Mdm2 and increase its oncogenic functions. Although this possibility has not been directly tested, HLI98, for example, was shown to stabilize both Mdm2 and p53 and result in p53-independent toxicity (Yang *et al.*, 2005).

Mdm2's numerous activities affect cell survival and tumor growth, only some of which are p53-dependent. An understanding of the Mdm2-dependent mechanisms that contribute to maintenance of tumor growth and development is crucial to making therapeutic decisions for individual patient tumors.

A recent study suggested that nutlin-3 acts primarily through MDM2 even in cells expressing high levels of MDMX (Patton *et al.*, 2006). We propose that nutlin-3 can also bind to MDMX and activate the p53 pathway. Indeed, we have shown that nutlin-3 can induce p53-mediated cell death in *Mdm2*-deficient MEFs in a manner that is dependent on MDMX (Laurie *et al.*, 2006). However, the affinity of nutlin-3 for MDMX is about 40-fold lower than for MDM2. This is consistent with structural and functional studies suggesting that MDM2 antagonists may not be optimal as MDMX antagonists (Bottger *et al.*, 1999; McCoy *et al.*, 2003). A number of recent findings have underscored the importance of inhibiting both MDMX and MDM2 in tumors with wild-type p53 because they are believed to play distinct functions: MDMX acts primarily as a transcriptional inhibitor of p53, while MDM2 regulates p53 stability (Francoz *et al.*, 2006; Toledo *et al.*, 2006). To achieve synergistic p53 induction in tumors with wild-type p53 separate MDM2 and MDMX antagonists may be required. Alternatively, if an MDM2 inhibitor such as nutlin-3 can be delivered locally (i.e., subconjunctival injections for retinoblastoma) at a high enough concentration to achieve inhibition of both MDM2 and MDMX, then a single antagonist may be sufficient. Moreover, by combining MDM2/MDMX antagonists with drugs that induce a p53 response through DNA damage (i.e., topotecan) this anti-tumor effect may be further enhanced (Fig. 3.5 Adapted from Laurie *et al.*, 2006). Therefore, retinoblastoma is not only a good model of suppression of p53-mediated cell death by *MDMX*

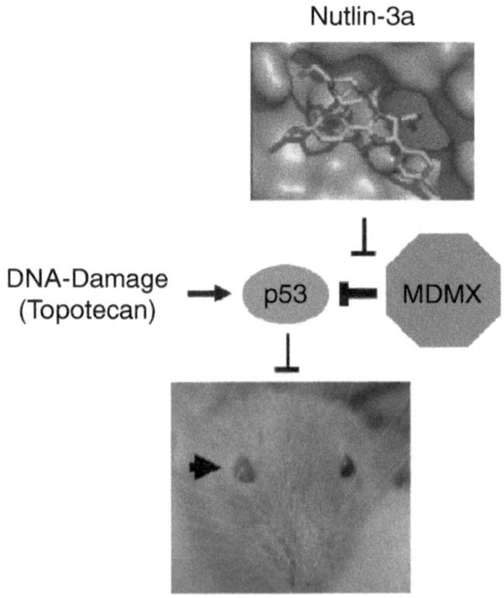

Figure 3.5 Synergistic induction of p53 tumor suppressor function in retinoblastoma by combining Mdm2/Mdmx antagonists and genotoxic drugs. (See Color Insert.)

amplification but it is an ideal system to study local delivery of chemotherapy targeted to the MDM2/MDMX-p53 pathway (Wang *et al.*, 2004).

REFERENCES

Abe, Y., Oda-Sato, E., *et al.* (2008). Hedgehog signaling overrides p53-mediated tumor suppression by activating Mdm2. *Proc. Natl. Acad. Sci. USA* **105**(12), 4838–4843.

Alt, J. R., Greiner, T. C., *et al.* (2003). Mdm2 haplo-insufficiency profoundly inhibits Myc-induced lymphomagenesis. *EMBO J.* **22**(6), 1442–1450.

Aslanian, A., Iaquinta, P. J., *et al.* (2004). Repression of the Arf tumor suppressor by E2F3 is required for normal cell cycle kinetics. *Genes Dev.* **18**(12), 1413–1422.

Badciong, J. C., and Haas, A. L. (2002). MdmX is a RING finger ubiquitin ligase capable of synergistically enhancing Mdm2 ubiquitination. *J. Biol. Chem.* **277**(51), 49668–49675.

Barak, Y., Juven, T., *et al.* (1993). mdm2 expression is induced by wild type p53 activity. *EMBO J.* **12**(2), 461–468.

Bartel, F., Harris, L. C., *et al.* (2004). MDM2 and its splice variant messenger RNAs: Expression in tumors and down-regulation using antisense oligonucleotides. *Mol. Cancer Res.* **2**(1), 29–35.

Bartel, F., Schulz, J., *et al.* (2005). Significance of HDMX-S (or MDM4) mRNA splice variant overexpression and HDMX gene amplification on primary soft tissue sarcoma prognosis. *Int. J. Cancer* **117**(3), 469–475.

Batchelor, E., Mock, C. S., *et al.* (2008). Recurrent initiation: A mechanism for triggering p53 pulses in response to DNA damage. *Mol. Cell* **30**(3), 277–289.

Berger, M., Vogt Sionov, R., *et al.* (2001). A role for the polyproline domain of p53 in its regulation by Mdm2. *J. Biol. Chem.* **276**(6), 3785–3790.

Bernardi, R., Scaglioni, P. P., *et al.* (2004). PML regulates p53 stability by sequestering Mdm2 to the nucleolus. *Nat. Cell Biol.* **6**(7), 665–672.

Boesten, L. S., Zadelaar, S. M., *et al.* (2006). Mdm2, but not Mdm4, protects terminally differentiated smooth muscle cells from p53-mediated caspase-3-independent cell death. *Cell Death Differ.* **13**(12), 2089–2098.

Bond, G. L., Hu, W., *et al.* (2004). A single nucleotide polymorphism in the MDM2 promoter attenuates the p53 tumor suppressor pathway and accelerates tumor formation in humans. *Cell* **119**(5), 591–602.

Bottger, V., Bottger, A., *et al.* (1999). Comparative study of the p53–mdm2 and p53–MDMX interfaces. *Oncogene* **18**(1), 189–199.

Brooks, C. L., and Gu, W. (2006). p53 ubiquitination: Mdm2 and beyond. *Mol. Cell* **21**(3), 307–315.

Cahilly-Snyder, L., Yang-Feng, T., *et al.* (1987). Molecular analysis and chromosomal mapping of amplified genes isolated from a transformed mouse 3T3 cell line. *Somat. Cell Mol. Genet.* **13**(3), 235–244.

Chang, Y. C., Lee, Y. S., *et al.* (1998). mdm2 and bax, downstream mediators of the p53 response, are degraded by the ubiquitin-proteasome pathway. *Cell Growth Differ.* **9**(1), 79–84.

Chen, D., Kon, N., *et al.* (2005a). ARF-BP1/Mule is a critical mediator of the ARF tumor suppressor. *Cell* **121**(7), 1071–1083.

Chen, L., Gilkes, D. M., *et al.* (2005b). ATM and Chk2-dependent phosphorylation of MDMX contribute to p53 activation after DNA damage. *EMBO J.* **24**(19), 3411–3422.

Chen, D., Zhang, J., *et al.* (2009). RYBP stabilizes p53 by modulating MDM2. *EMBO Rep.* **10**(2), 166–172.

Chiu, S. Y., Asai, N., *et al.* (2008). SUMO-specific protease 2 is essential for modulating p53–Mdm2 in development of trophoblast stem cell niches and lineages. *PLoS Biol.* **6**(12), e310.
Clegg, H. V., Itahana, K., *et al.* (2008). Unlocking the Mdm2-p53 loop: Ubiquitin is the key. *Cell Cycle* **7**(3), 287–292.
Colaluca, I. N., Tosoni, D., *et al.* (2008). NUMB controls p53 tumour suppressor activity. *Nature* **451**(7174), 76–80.
Coutts, A. S., Boulahbel, H., *et al.* (2007). Mdm2 targets the p53 transcription cofactor JMY for degradation. *EMBO Rep.* **8**(1), 84–90.
Cummins, J. M., Rago, C., *et al.* (2004). Tumour suppression: Disruption of HAUSP gene stabilizes p53. *Nature* **428**(6982), 1, p following 486.
Danovi, D., Meulmeester, E., *et al.* (2004). Amplification of Mdmx (or Mdm4) directly contributes to tumor formation by inhibiting p53 tumor suppressor activity. *Mol. Cell. Biol.* **24**(13), 5835–5843.
de Clercq, S., *et al.* (2010). Widespread overexpression of epitote tagged-MdM4 does not accelarate tumor formation in vivo. *Mol. Biol. Cell* [Epub ahead of print].
de Graaf, P., Little, N. A., *et al.* (2003). Hdmx protein stability is regulated by the ubiquitin ligase activity of Mdm2. *J. Biol. Chem.* **278**(40), 38315–38324.
de Oca, Montes, Luna, R., Wagner, D. S., *et al.* (1995). Rescue of early embryonic lethality in mdm2-deficient mice by deletion of p53. *Nature* **378**(6553), 203–206.
D'Orazi, G., Cecchinelli, B., *et al.* (2002). Homeodomain-interacting protein kinase-2 phosphorylates p53 at Ser 46 and mediates apoptosis. *Nat. Cell Biol.* **4**(1), 11–19.
Dornan, D., Wertz, I., *et al.* (2004). The ubiquitin ligase COP1 is a critical negative regulator of p53. *Nature* **429**(6987), 86–92.
Doumont, G., Martoriati, A., *et al.* (2005). G1 checkpoint failure and increased tumor susceptibility in mice lacking the novel p53 target Ptprv. *EMBO J.* **24**(17), 3093–3103.
Dumaz, N., Milne, D. M., *et al.* (2001). Critical roles for the serine 20, but not the serine 15, phosphorylation site and for the polyproline domain in regulating p53 turnover. *Biochem. J.* **359**(Pt 2), 459–464.
Dyer, M. A., and Harbour, J. W. (2006). Cellular event in tumorogenesis. *In* "Clinical ocular oncology," (A. D. Singh, B. E. Damato, A. L. Murphree, and J. D. Perry, eds.), Elsevier, London.
Dyer, M. A., Rodriguez-Galindo, C., *et al.* (2005). Use of preclinical models to improve treatment of retinoblastoma. *PLoS Med.* **2**(10), e332.
el-Deiry, W. S., Tokino, T., *et al.* (1993). WAF1, a potential mediator of p53 tumor suppression. *Cell* **75**(4), 817–825.
Elias, B., Laine, A., *et al.* (2005). Phosphorylation of MdmX by CDK2/Cdc2(p34) is required for nuclear export of Mdm2. *Oncogene* **24**(15), 2574–2579.
Finch, R. A., Donoviel, D. B., *et al.* (2002). mdmx is a negative regulator of p53 activity in vivo. *Cancer Res.* **62**(11), 3221–3225.
Fiscella, M., Zhang, H., *et al.* (1997). Wip1, a novel human protein phosphatase that is induced in response to ionizing radiation in a p53-dependent manner. *Proc. Natl. Acad. Sci. USA* **94**(12), 6048–6053.
Francoz, S., Froment, P., *et al.* (2006). Mdm4 and Mdm2 cooperate to inhibit p53 activity in proliferating and quiescent cells in vivo. *Proc. Natl. Acad. Sci. USA* **103**(9), 3232–3237.
Ganguli, G., and Wasylyk, B. (2003). p53-independent functions of MDM2. *Mol. Cancer Res.* **1**(14), 1027–1035.
Giglio, S., Mancini, F., *et al.* (2005). Identification of an aberrantly spliced form of HDMX in human tumors: A new mechanism for HDM2 stabilization. *Cancer Res.* **65**(21), 9687–9694.
Grier, J. D., Yan, W., *et al.* (2002). Conditional allele of mdm2 which encodes a p53 inhibitor. *Genesis* **32**(2), 145–147.

Grier, J. D., Xiong, S., *et al.* (2006). Tissue-specific differences of p53 inhibition by Mdm2 and Mdm4. *Mol. Cell. Biol.* **26**(1), 192–198.

Grossman, S. R., Deato, M. E., *et al.* (2003). Polyubiquitination of p53 by a ubiquitin ligase activity of p300. *Science* **300**(5617), 342–344.

Gu, J., Kawai, H., *et al.* (2002). Mutual dependence of MDM2 and MDMX in their functional inactivation of p53. *J. Biol. Chem.* **277**(22), 19251–19254.

Haupt, Y., Maya, R., *et al.* (1997). Mdm2 promotes the rapid degradation of p53. *Nature* **387** (6630), 296–299.

Hofmann, T. G., Moller, A., *et al.* (2002). Regulation of p53 activity by its interaction with homeodomain-interacting protein kinase-2. *Nat. Cell Biol.* **4**(1), 1–10.

Honda, R., and Yasuda, H. (2000). Activity of MDM2, a ubiquitin ligase, toward p53 or itself is dependent on the RING finger domain of the ligase. *Oncogene* **19**(11), 1473–1476.

Honda, R., Tanaka, H., *et al.* (1997). Oncoprotein MDM2 is a ubiquitin ligase E3 for tumor suppressor p53. *FEBS Lett.* **420**(1), 25–27.

Horn, H. F., and Vousden, K. H. (2008). Cooperation between the ribosomal proteins L5 and L11 in the p53 pathway. *Oncogene* **27**(44), 5774–5784.

Howes, K. A., Ransom, N., *et al.* (1994). Apoptosis or retinoblastoma: Alternative fates of photoreceptors expressing the HPV-16 E7 gene in the presence or absence of p53. *Genes Dev.* **8**(11), 1300–1310.

Issaeva, N., Bozko, P., *et al.* (2004). Small molecule RITA binds to p53, blocks p53-HDM-2 interaction and activates p53 function in tumors. *Nat. Med.* **10**(12), 1321–1328.

Jackson, M. W., and Berberich, S. J. (2000). MdmX protects p53 from Mdm2-mediated degradation. *Mol. Cell. Biol.* **20**(3), 1001–1007.

Jimenez, G. S., Nister, M., *et al.* (2000). A transactivation-deficient mouse model provides insights into Trp53 regulation and function. *Nat. Genet.* **26**(1), 37–43.

Jin, Y., Dai, M. S., *et al.* (2006). 14-3-3gamma binds to MDMX that is phosphorylated by UV-activated Chk1, resulting in p53 activation. *EMBO J.* **25**(6), 1207–1218.

Johnson, T. M., Hammond, E. M., *et al.* (2005). The p53QS transactivation-deficient mutant shows stress-specific apoptotic activity and induces embryonic lethality. *Nat. Genet.* **37**(2), 145–152.

Jones, S. N., Roe, A. E., *et al.* (1995). Rescue of embryonic lethality in Mdm2-deficient mice by absence of p53. *Nature* **378**(6553), 206–208.

Juven-Gershon, T., Shifman, O., *et al.* (1998). The Mdm2 oncoprotein interacts with the cell fate regulator Numb. *Mol. Cell. Biol.* **18**(7), 3974–3982.

Kato, M. V., Shimizu, T., *et al.* (1996). Loss of heterozygosity on chromosome 17 and mutation of the p53 gene in retinoblastoma. *Cancer Lett.* **106**(1), 75–82.

Kawai, H., Wiederschain, D., *et al.* (2003). DNA damage-induced MDMX degradation is mediated by MDM2. *J. Biol. Chem.* **278**(46), 45946–45953.

Kortlever, R. M., Higgins, P. J., *et al.* (2006). Plasminogen activator inhibitor-1 is a critical downstream target of p53 in the induction of replicative senescence. *Nat. Cell Biol.* **8**(8), 877–884.

Kubbutat, M. H., Jones, S. N., *et al.* (1997). Regulation of p53 stability by Mdm2. *Nature* **387**(6630), 299–303.

Kurki, S., Peltonen, K., *et al.* (2004). Nucleolar protein NPM interacts with HDM2 and protects tumor suppressor protein p53 from HDM2-mediated degradation. *Cancer Cell* **5**(5), 465–475.

Lai, Z., Ferry, K. V., *et al.* (2001). Human mdm2 mediates multiple mono-ubiquitination of p53 by a mechanism requiring enzyme isomerization. *J. Biol. Chem.* **276**(33), 31357–31367.

Langheinrich, U., Hennen, E., *et al.* (2002). Zebrafish as a model organism for the identification and characterization of drugs and genes affecting p53 signaling. *Curr. Biol.* **12**(23), 2023–2028.

Laurie, N. A., Donovan, S. L., *et al.* (2006). Inactivation of the p53 pathway in retinoblastoma. *Nature* **444**(7115), 61–66.
Leach, F. S., Tokino, T., *et al.* (1993). p53 Mutation and MDM2 amplification in human soft tissue sarcomas. *Cancer Res.* **53**(10 Suppl.), 2231–2234.
LeBron, C., Chen, L., *et al.* (2006). Regulation of MDMX nuclear import and degradation by Chk2 and 14-3-3. *EMBO J.* **25**(6), 1196–1206.
Leng, R. P., Lin, Y., *et al.* (2003). Pirh2, a p53-induced ubiquitin-protein ligase, promotes p53 degradation. *Cell* **112**(6), 779–791.
Li, M., Brooks, C. L., *et al.* (2003). Mono- versus polyubiquitination: Differential control of p53 fate by Mdm2. *Science* **302**(5652), 1972–1975.
Li, M., Brooks, C. L., *et al.* (2004). A dynamic role of HAUSP in the p53–Mdm2 pathway. *Mol. Cell* **13**(6), 879–886.
Linares, L. K., Hengstermann, A., *et al.* (2003). HdmX stimulates Hdm2-mediated ubiquitination and degradation of p53. *Proc. Natl. Acad. Sci. USA* **100**(21), 12009–12014.
Linares, L. K., Kiernan, R., *et al.* (2007). Intrinsic ubiquitination activity of PCAF controls the stability of the oncoprotein Hdm2. *Nat. Cell Biol.* **9**(3), 331–338.
Lindstrom, M. S., Deisenroth, C., *et al.* (2007). Putting a finger on growth surveillance: Insight into MDM2 zinc finger-ribosomal protein interactions. *Cell Cycle* **6**(4), 434–437.
Lu, X., Ma, O., *et al.* (2007). The Wip1 Phosphatase acts as a gatekeeper in the p53–Mdm2 autoregulatory loop. *Cancer Cell* **12**(4), 342–354.
MacCallum, D. E., Hupp, T. R., *et al.* (1996). The p53 response to ionising radiation in adult and developing murine tissues. *Oncogene* **13**(12), 2575–2587.
Maki, C. G., Huibregtse, J. M., *et al.* (1996). In vivo ubiquitination and proteasome-mediated degradation of p53(1). *Cancer Res.* **56**(11), 2649–2654.
Marine, J. C., and Jochemsen, A. G. (2004). Mdmx and Mdm2: Brothers in arms? *Cell Cycle* **3**(7), 900–904.
Marine, J. C., and Jochemsen, A. G. (2005). Mdmx as an essential regulator of p53 activity. *Biochem. Biophys. Res. Commun.* **331**(3), 750–760.
Marine, J. C., Francoz, S., *et al.* (2006). Keeping p53 in check: Essential and synergistic functions of Mdm2 and Mdm4. *Cell Death Differ.* **13**(6), 927–934.
Marine, J. C., Dyer, M. A., *et al.* (2007). MDMX: From bench to bedside. *J. Cell Sci.* **120**(Pt 3), 371–378.
Martins, C. P., Brown-Swigart, L., *et al.* (2006). Modeling the therapeutic efficacy of p53 restoration in tumors. *Cell* **127**(7), 1323–1334.
Maya, R., Balass, M., *et al.* (2001). ATM-dependent phosphorylation of Mdm2 on serine 395: Role in p53 activation by DNA damage. *Genes Dev.* **15**(9), 1067–1077.
McCoy, M. A., Gesell, J. J., *et al.* (2003). Flexible lid to the p53-binding domain of human Mdm2: Implications for p53 regulation. *Proc. Natl. Acad. Sci. USA* **100**(4), 1645–1648.
Meek, D. W., and Knippschild, U. (2003). Posttranslational modification of MDM2. *Mol. Cancer Res.* **1**(14), 1017–1026.
Mendrysa, S. M., McElwee, M. K., *et al.* (2003). mdm2 Is critical for inhibition of p53 during lymphopoiesis and the response to ionizing irradiation. *Mol. Cell. Biol.* **23**(2), 462–472.
Mendrysa, S. M., O'Leary, K. A., *et al.* (2006). Tumor suppression and normal aging in mice with constitutively high p53 activity. *Genes Dev.* **20**(1), 16–21.
Meulmeester, E., Maurice, M. M., *et al.* (2005). Loss of HAUSP-mediated deubiquitination contributes to DNA damage-induced destabilization of Hdmx and Hdm2. *Mol. Cell* **18** (5), 565–576.
Michalak, E., Villunger, A., *et al.* (2005). Death squads enlisted by the tumour suppressor p53. *Biochem. Biophys. Res. Commun.* **331**(3), 786–798.
Migliorini, D., Danovi, D., *et al.* (2002a). Hdmx recruitment into the nucleus by Hdm2 is essential for its ability to regulate p53 stability and transactivation. *J. Biol. Chem.* **277**(9), 7318–7323.

Migliorini, D., Lazzerini Denchi, E., *et al.* (2002b). Mdm4 (Mdmx) regulates p53-induced growth arrest and neuronal cell death during early embryonic mouse development. *Mol. Cell. Biol.* **22**(15), 5527–5538.
Minsky, N., and Oren, M. (2004). The RING domain of Mdm2 mediates histone ubiquitylation and transcriptional repression. *Mol. Cell* **16**(4), 631–639.
Momand, J., Zambetti, G. P., *et al.* (1992). The mdm-2 oncogene product forms a complex with the p53 protein and inhibits p53-mediated transactivation. *Cell* **69**(7), 1237–1245.
Moumen, A., Masterson, P., *et al.* (2005). hnRNP K: An HDM2 target and transcriptional coactivator of p53 in response to DNA damage. *Cell* **123**(6), 1065–1078.
Nork, T. M., Poulsen, G. L., *et al.* (1997). p53 regulates apoptosis in human retinoblastoma. *Arch. Ophthalmol.* **115**(2), 213–219.
Okamoto, K., Kashima, K., *et al.* (2005). DNA damage-induced phosphorylation of MdmX at serine 367 activates p53 by targeting MdmX for Mdm2-dependent degradation. *Mol. Cell. Biol.* **25**(21), 9608–9620.
Oliner, J. D., Kinzler, K. W., *et al.* (1992). Amplification of a gene encoding a p53-associated protein in human sarcomas. *Nature* **358**(6381), 80–83.
Opferman, J. T., and Zambetti, G. P. (2006). Translational research? Ribosome integrity and a new p53 tumor suppressor checkpoint. *Cell Death Differ.* **13**(6), 898–901.
Pan, Y., and Chen, J. (2003). MDM2 promotes ubiquitination and degradation of MDMX. *Mol. Cell. Biol.* **23**(15), 5113–5121.
Pan, Y., and Chen, J. (2005). Modification of MDMX by sumoylation. *Biochem. Biophys. Res. Commun.* **332**(3), 702–709.
Parant, J., Chavez-Reyes, A., *et al.* (2001). Rescue of embryonic lethality in Mdm4-null mice by loss of Trp53 suggests a nonoverlapping pathway with MDM2 to regulate p53. *Nat. Genet.* **29**(1), 92–95.
Parrinello, S., Samper, E., *et al.* (2003). Oxygen sensitivity severely limits the replicative lifespan of murine fibroblasts. *Nat. Cell Biol.* **5**(8), 741–747.
Patton, J. T., Mayo, L. D., *et al.* (2006). Levels of HdmX expression dictate the sensitivity of normal and transformed cells to Nutlin-3. *Cancer Res.* **66**(6), 3169–3176.
Pereg, Y., Shkedy, D., *et al.* (2005). Phosphorylation of Hdmx mediates its Hdm2- and ATM-dependent degradation in response to DNA damage. *Proc. Natl. Acad. Sci. USA* **102**(14), 5056–5061.
Pereg, Y., Lam, S., *et al.* (2006). Differential roles of ATM- and Chk2-mediated phosphorylations of Hdmx in response to DNA damage. *Mol. Cell. Biol.* **26**(18), 6819–6831.
Perry, M. E., Piette, J., *et al.* (1993). The mdm-2 gene is induced in response to UV light in a p53-dependent manner. *Proc. Natl. Acad. Sci. USA* **90**(24), 11623–11627.
Rallapalli, R., Strachan, G., *et al.* (1999). A novel MDMX transcript expressed in a variety of transformed cell lines encodes a truncated protein with potent p53 repressive activity. *J. Biol. Chem.* **274**(12), 8299–8308.
Rallapalli, R., Strachan, G., *et al.* (2003). Identification of a domain within MDMX-S that is responsible for its high affinity interaction with p53 and high-level expression in mammalian cells. *J. Cell. Biochem.* **89**(3), 563–575.
Ramos, Y. F., Stad, R., *et al.* (2001). Aberrant expression of HDMX proteins in tumor cells correlates with wild-type p53. *Cancer Res.* **61**(5), 1839–1842.
Riemenschneider, M. J., Buschges, R., *et al.* (1999). Amplification and overexpression of the MDM4 (MDMX) gene from 1q32 in a subset of malignant gliomas without TP53 mutation or MDM2 amplification. *Cancer Res.* **59**(24), 6091–6096.
Riemenschneider, M. J., Knobbe, C. B., *et al.* (2003). Refined mapping of 1q32 amplicons in malignant gliomas confirms MDM4 as the main amplification target. *Int. J. Cancer* **104**(6), 752–757.
Rinaldo, C., Prodosmo, A., *et al.* (2007). MDM2-regulated degradation of HIPK2 prevents p53Ser46 phosphorylation and DNA damage-induced apoptosis. *Mol. Cell* **25**(5), 739–750.

Ringshausen, I., O'Shea, C. C., *et al.* (2006). Mdm2 is critically and continuously required to suppress lethal p53 activity in vivo. *Cancer Cell* **10**(6), 501–514.
Roegiers, F., and Jan, Y. N. (2004). Asymmetric cell division. *Curr. Opin. Cell Biol.* **16**(2), 195–205.
Rossi, M., Demidov, O. N., *et al.* (2008). Induction of PPM1D following DNA-damaging treatments through a conserved p53 response element coincides with a shift in the use of transcription initiation sites. *Nucleic Acids Res.* **36**(22), 7168–7180.
Sabbatini, P., and McCormick, F. (2002). MDMX inhibits the p300/CBP-mediated acetylation of p53. *DNA Cell Biol.* **21**(7), 519–525.
Sharp, D. A., Kratowicz, S. A., *et al.* (1999). Stabilization of the MDM2 oncoprotein by interaction with the structurally related MDMX protein. *J. Biol. Chem.* **274**(53), 38189–38196.
Sherley, J. L., Stadler, P. B., *et al.* (1995). Expression of the wild-type p53 antioncogene induces guanine nucleotide-dependent stem cell division kinetics. *Proc. Natl. Acad. Sci. USA* **92**(1), 136–140.
Sherr, C. J. (2006). Divorcing ARF and p53: An unsettled case. *Nat. Rev. Cancer* **6**(9), 663–673.
Sherr, C. J., and McCormick, F. (2002). The RB and p53 pathways in cancer. *Cancer Cell* **2**(2), 103–112.
Shikama, N., Lee, C. W., *et al.* (1999). A novel cofactor for p300 that regulates the p53 response. *Mol. Cell* **4**(3), 365–376.
Shvarts, A., Steegenga, W. T., *et al.* (1996). MDMX: A novel p53-binding protein with some functional properties of MDM2. *EMBO J.* **15**(19), 5349–5357.
Shvarts, A., Bazuine, M., *et al.* (1997). Isolation and identification of the human homolog of a new p53-binding protein, Mdmx. *Genomics* **43**(1), 34–42.
Stad, R., Ramos, Y. F., *et al.* (2000). Hdmx stabilizes Mdm2 and p53. *J. Biol. Chem.* **275**(36), 28039–28044.
Stad, R., Little, N. A., *et al.* (2001). Mdmx stabilizes p53 and Mdm2 via two distinct mechanisms. *EMBO Rep.* **2**(11), 1029–1034.
Steinman, H. A., and Jones, S. N. (2002). Generation of an Mdm2 conditional allele in mice. *Genesis* **32**(2), 142–144.
Steinman, H. A., Hoover, K. M., *et al.* (2005). Rescue of Mdm4-deficient mice by Mdm2 reveals functional overlap of Mdm2 and Mdm4 in development. *Oncogene* **24**(53), 7935–7940.
Stommel, J. M., and Wahl, G. M. (2004). Accelerated MDM2 auto-degradation induced by DNA-damage kinases is required for p53 activation. *EMBO J.* **23**(7), 1547–1556.
Sui, G., Affar, B., *et al.* (2004). Yin Yang 1 is a negative regulator of p53. *Cell* **117**(7), 859–872.
Tanimura, S., Ohtsuka, S., *et al.* (1999). MDM2 interacts with MDMX through their RING finger domains. *FEBS Lett.* **447**(1), 5–9.
Terzian, T., Suh, Y. A., *et al.* (2008). The inherent instability of mutant p53 is alleviated by Mdm2 or p16INK4a loss. *Genes Dev.* **22**(10), 1337–1344.
Terzian, T., Torchia, E. C., *et al.* (2010). p53 prevents progression of nevi to melanoma predominantly through cell cycle regulation. *Pigment Cell Melanoma Res.* [Sept 2010 Epub ahead of printing].
Terzian, T., Wang, Y., Van Pelt, C. S., Box, N. F., Travis, E. L., and Lozano, G. (2007). Haploinsufficiency of Mdm2 and Mdm4 in Tumorigenesis and Development. *Mol. Cell Biol.* **27**(15), 5479–5485.
Thut, C. J., Goodrich, J. A., *et al.* (1997). Repression of p53-mediated transcription by MDM2: A dual mechanism. *Genes Dev.* **11**(15), 1974–1986.
Toledo, F., Krummel, K. A., *et al.* (2006). A mouse p53 mutant lacking the proline-rich domain rescues Mdm4 deficiency and provides insight into the Mdm2-Mdm4-p53 regulatory network. *Cancer Cell* **9**(4), 273–285.

Vassilev, L. T., Vu, B. T., *et al.* (2004). In vivo activation of the p53 pathway by small-molecule antagonists of MDM2. *Science* **303**(5659), 844–848.

Ventura, A., Kirsch, D. G., *et al.* (2007). Restoration of p53 function leads to tumour regression in vivo. *Nature* **445**(7128), 661–665.

Vousden, K. H., and Lu, X. (2002). Live or let die: The cell's response to p53. *Nat. Rev. Cancer* **2**(8), 594–604.

Wahl, G. M. (2006). Mouse bites dogma: How mouse models are changing our views of how P53 is regulated in vivo. *Cell Death Differ.* **13**(6), 973–983.

Wang, X., Arooz, T., *et al.* (2001). MDM2 and MDMX can interact differently with ARF and members of the p53 family. *FEBS Lett.* **490**(3), 202–208.

Wang, X., Taplick, J., *et al.* (2004). Inhibition of p53 degradation by Mdm2 acetylation. *FEBS Lett.* **561**(1–3), 195–201.

Weber, J. D., Taylor, L. J., *et al.* (1999). Nucleolar Arf sequesters Mdm2 and activates p53. *Nat. Cell Biol.* **1**(1), 20–26.

Wei, X., Yu, Z. K., *et al.* (2003). Physical and functional interactions between PML and MDM2. *J. Biol. Chem.* **278**(31), 29288–29297.

Whibley, C., Pharoah, P. D., *et al.* (2009). p53 polymorphisms: Cancer implications. *Nat. Rev. Cancer* **9**(2), 95–107.

Wu, X., Bayle, J. H., *et al.* (1993). The p53-mdm-2 autoregulatory feedback loop. *Genes Dev.* **7**(7A), 1126–1132.

Wynendaele, J., Böhnke, A., Leucci, E., Nielsen, S. J., Lambertz, I., Hammer, S., Sbrzesny, N., Kubitza, D., Wolf, A., Gradhand, E., Balschun, K., Braicu, I., *et al.* (2010). An illegitimate microRNA target site within the 3' UTR of MDM4 affects ovarian cancer progression and chemosensitivity. *Cancer Res.* **70**(23), 9641–9649.

Xiong, S., Van Pelt, C. S., *et al.* (2006). Synergistic roles of Mdm2 and Mdm4 for p53 inhibition in central nervous system development. *Proc. Natl. Acad. Sci. USA* **103**(9), 3226–3231.

Xirodimas, D. P., Saville, M. K., *et al.* (2004). Mdm2-mediated NEDD8 conjugation of p53 inhibits its transcriptional activity. *Cell* **118**(1), 83–97.

Xue, W., Zender, L., *et al.* (2007). Senescence and tumour clearance is triggered by p53 restoration in murine liver carcinomas. *Nature* **445**(7128), 656–660.

Yang, Y., Ludwig, R. L., *et al.* (2005). Small molecule inhibitors of HDM2 ubiquitin ligase activity stabilize and activate p53 in cells. *Cancer Cell* **7**(6), 547–559.

Yogosawa, S., Miyauchi, Y., *et al.* (2003). Mammalian Numb is a target protein of Mdm2, ubiquitin ligase. *Biochem. Biophys. Res. Commun.* **302**(4), 869–872.

Yu, E., Ahn, Y. S., *et al.* (2007). Overexpression of the wip1 gene abrogates the p38 MAPK/p53/Wip1 pathway and silences p16 expression in human breast cancers. *Breast Cancer Res. Treat.* **101**(3), 269–278.

Zhang, J., Schweers, B., *et al.* (2004). The first knockout mouse model of retinoblastoma. *Cell Cycle* **3**(7), 952–959.

CHAPTER FOUR

The Connections Between Neural Crest Development and Neuroblastoma

Manrong Jiang,* Jennifer Stanke,*,† *and* Jill M. Lahti*

Contents

* Department of Genetics and Tumor Cell Biology, St. Jude Children's Research Hospital, Memphis, Tennessee, USA
† Department of Developmental Neurobiology, St. Jude Children's Research Hospital, Memphis, Tennessee, USA

Current Topics in Developmental Biology, Volume 94
ISSN 0070-2153, DOI: 10.1016/B978-0-12-380916-2.00004-8

Abstract

Neuroblastoma (NB), the most common extracranial solid tumor in childhood, is an extremely heterogeneous disease both biologically and clinically. Although significant progress has been made in identifying molecular and genetic markers for NB, this disease remains an enigmatic challenge. Since NB is thought to be an embryonal tumor that is derived from precursor cells of the peripheral (sympathetic) nervous system, understanding the development of normal sympathetic nervous system may highlight abnormal events that contribute to NB initiation. Therefore, this review focuses on the development of the peripheral trunk neural crest, the current understanding of how developmental factors may contribute to NB and on recent advances in the identification of important genetic lesions and signaling pathways involved in NB tumorigenesis and metastasis. Finally, we discuss how future advances in identification of molecular alterations in NB may lead to more effective, less toxic therapies, and improve the prognosis for NB patients.

1. Clinical and Biological Characteristics of Neuroblastoma (NB)

NB is the most common extracranial solid tumor in childhood, accounting for approximately 7–10% of pediatric cancers and 15% of all pediatric cancer deaths in patients less than 15 years old (Brodeur, 2003; Maris *et al.*, 2007; Schor, 1999). NB is an extremely heterogeneous disease both biologically and clinically (Brodeur, 2003; Evans *et al.*, 1971; Maris *et al.*, 2007). NB is thought to be an embryonal tumor that is derived from precursor cells of the peripheral (sympathetic) nervous system (Brodeur, 2003; Grimmer and Weiss, 2006; Nakagawara and Ohira, 2004). The tumor can arise anywhere along the sympathetic chain but is most frequently in the adrenal medulla and paraspinal ganglia (Fig. 4.1; Johnsen *et al.*, 2009; Nakagawara and Ohira, 2004).

Historically, NB is clinically classified into five different stages (1–4 and 4S) according to the International Neuroblastoma Staging System (INSS; Brodeur, 2003; Maris *et al.*, 2007; Schor, 1999; van Noesel and Versteeg, 2004). Early stage NB tumors (i.e., stages 1, 2) do not metastasize to bone or bone marrow and are treatable with chemotherapeutic drugs and irradiation. Advanced-stage NB tumors (stages 3 and 4) are highly metastatic and usually

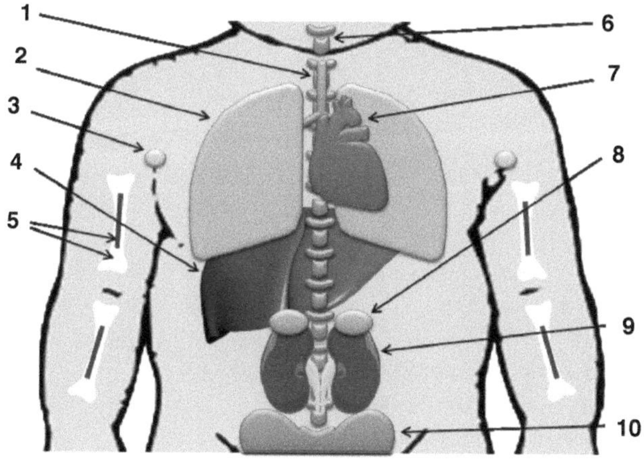

Figure 4.1 Neuroblastoma localization. Neuroblastoma primary tumors derive from precursor cells of the peripheral (sympathetic) nervous system and can arise anywhere along the sympathetic chain, most frequently in the adrenal gland (position 8 as shown). Neuroblastoma may also develop from spinal cord of neck (position 6) and pelvis (position 10). Neuroblastomas mainly metastasize to lymph nodes (position 3), liver (position 4), bone and bone marrow (position 5), and also spread to central nervous system (position 1) and lungs (position 2) in infants.

respond positively to initial treatment. However, they often become resistant to chemotherapy and irradiation. A fifth stage of NB tumors (stage 4S) undergoes spontaneous regression with minimum treatment or even without medical intervention.

Ninety percent of the children with this disease are diagnosed before the age of 5 years and in those patients older than 1 year, 75% of the patients present with stage 3 or 4 metastatic diseases (Brodeur, 2003; Maris *et al.*, 2007). Metastatic disease remains a major clinical challenge in the treatment of NB since greater than 50% of all NB patients are diagnosed with metastatic diseases (Maris, 2005; Maris *et al.*, 2007). In contrast, infants with this disease tend to be at lower stages (stage 1, 2, and 4s) and to have a better prognosis (Brodeur, 2003; Maris, 2005; Maris *et al.*, 2007; van Noesel and Versteeg, 2004).

In addition to classification by stage, NB tumors are also classified into three risk groups (low, intermediate, and high risk) according to age, MYCN status, and histology (Table 4.1). A new system for tumor staging has recently been implemented by the International Neuroblastoma Risk Group Staging System (INRGSS; Cohn *et al.*, 2009; Monclair *et al.*, 2009). This system, which is based on a combination clinical and imaging data, classifies patients as L1 (localized disease without imaging risk factors), L2 (localized disease with imaging risk factors), M (metastatic tumors), and Ms

Table 4.1 Neuroblastoma risk stratification based on age, MYCN status and histology

INSS stage	Age	MYCN status	Histology	Risk group	3-year survival rate
1	0–21 years	Any	Any	Low	>90%
2	0–21 years	Non-Amp	Any	Low	70–90% in this stage
	1–21 years	Amp	Favorable	Low	
	1–21 years	Amp	Unfavorable	High	
3	<1 year	Non-Amp	Any	Intermediate	30–50% in this stage
	1–21 years	Non-Amp	Favorable	Intermediate	
	0–21 years	Amp	Any	High	
	0–21 years	Non-Amp	Unfavorable	High	
4	<1 year	Non-Amp	Any	Intermediate	<30% in this stage
	0–21 years	Amp	Any	High	
4S	<1 year	Non-Amp	Favorable	Low	50–80% in this stage
	<1 year	Non-Amp	Unfavorable	Intermediate	
	<1 year	Amp	Any	High	

INSS, International Neuroblastoma Staging System; Amp, amplified; Non-Amp, not amplified.

(metastatic disease with metastasis only in skin, liver, and/or bone marrow; Cohn *et al.*, 2009; Monclair *et al.*, 2009). NB tumors are also graded in terms of histology using the International Pathology Classification System (INPC) which is based on the Shimada histology grading system (Shimada *et al.*, 1999). This system distinguishes good and poor prognosis tumors based on the degree of differentiation, the Schwannian stromal content, the mitotic-karyorrhexis index (MKI), and the age at diagnosis (Shimada *et al.*, 1999). Unfavorable tumors tend to include undifferentiated tumors with high MKI of any age, poorly differentiated tumors or intermediate MKI in patients older than 18 months, and differentiated tumor or low MKI in patients 5 years old or older. In contrast all other cases, including those with ganglioneuroma or tumor with regions of ganglioneuroma intermixed, have a good prognosis.

In the remainder of this review, we will discuss the development of the peripheral neural crest with a focus on how developmental factors may contribute to NB tumorigenesis and metastasis, and highlight the current understanding of other genetic changes related to NB and their importance in NB diagnosis and treatment.

2. Neural Development and NB

2.1. Neural crest contribution to sympathetic ganglia and adrenal gland

The majority of NB tumors appear to arise from neural crest-derived cells in the abdomen adjacent to the aorta in the region of the kidney or in the medullary region of the adrenal gland (Brodeur, 2003; Maris *et al.*, 2007). Thus, NB is a sympaticoadrenal lineage neural crest-derived tumor. The neural crest arises from the dorsal region of the closing neural tube beneath the ectoderm (Le Dourin and Kalcheim, 1999). This transient population of cells produces multipotential progenitor cells that give rise to the peripheral nervous system, the enteric nervous system, pigment cells, Schwann cells, adrenal medullary cells, and cells of the craniofacial skeleton (Le Dourin and Kalcheim, 1999). This process is regulated by both extrinsic and intrinsic factors. The Hedgehog and Wnt signaling pathways are especially crucial for proper neural crest development (Dupin *et al.*, 2007; Le Dourin and Kalcheim, 1999; Morales *et al.*, 2005). Lineage studies in the developing embryo have shown that neural crest cells within the trunk region generate multiple neural crest derivatives such as melanocytes, Schwann cells, glia, and neurons of the dorsal root ganglia (Fontaine-Perus *et al.*, 1982; Lallier and Bronner-Fraser, 1988; Teillet *et al.*, 1987; Weston, 1963). A subset of these trunk crest cells, commonly referred to as the sympathoadrenal lineage, contributes to the sympathetic ganglia and medullary region of the adrenal gland (Anderson and Axel, 1986; Anderson *et al.*, 1991). This lineage of cells is thought to be the origin of NB (Brodeur, 2003; Maris *et al.*, 2007). However, given the fact that NB can develop anywhere along the sympathetic axis, it is likely that NB can also arise from earlier crest derivatives, before development of the sympathethoadreanal lineage but after the initial fate specification (Brodeur, 2003; Maris *et al.*, 2007). This could contribute to the heterogeneous histology and pathology of NB.

Neural crest cells develop in response to extracellular signals (Bronner-Fraser, 1993). The signals that induce formation of the crest (BMP/Shh) appear to be similar along the dorsal/ventral axis of the embryo. In contrast, different factors appear to confer cell fate along the anterior/posterior axis of the embryo. Heterotopic crest cell transplantation studies indicate that the positional identity of the cells is based on their location during development, rather than the characteristics of cells in original locations (Bronner-Fraser *et al.*, 1980; Ruffins *et al.*, 1998). Although NB can arise anywhere along the developing sympathetic axis, the majority of cases arise in the abdomen (65%), frequently in the adrenal medulla, where the sympathoadrenal lineage is specified; while others are found in the paraspinal sympathetic ganglia in places such as the neck (5%), chest (20%), and pelvis (5%) (Maris *et al.*, 2007). Therefore, it is reasonable to postulate that the positional identity of the cell along the anterior posterior axis of the embryo, and factors that specify this region, likely contribute to the oncogenic potential of the crest derivatives in this region. As such, developmental signals fating the proper development, migratory pathways, and regulated cell death will be examined here in the context of NB.

2.2. Neural crest migratory pathways

Neural crest-derived cells are highly migratory. Shortly after induction the crest cells undergo an epithelial to mesenchymal transition (EMT). This EMT transition results in acquisition of enhanced migratory abilities and decreased requirements for intercellular contact which allows the neural crest cells to leave the dorsal neural tube (Le Dourin and Kalcheim, 1999). They then migrate either between the dermatome and epidermis in a dorsolateral pathway or delaminate from the neural tube via a ventrolateral pathway (Le Dourin and Kalcheim, 1999). Importantly, a similar EMT transition may also play a role in NB metastasis as described below in Section 6.1.

Neural crest migration pathways are determined by signal from the mesoderm which develops prior to the arrival of the crest cells. Trunk region neural crest cells either migrate ventrolaterally and remain in the sclerotome form the doral root ganglia or continue migrating to a more ventral position to form sympathetic ganglia. N-myc, an oncogene which plays a role in aggressive NB, appears to be required for the migration, survival and/or differentiation of cells that migrate to the dorsal aorta since N-myc deficient mouse embryos have decreased numbers of mature cells in the both the dorsal root ganglia and sympathetic ganglia (Charron *et al.*, 1992; Sawai *et al.*, 1993; Stanton *et al.*, 1992). Hedgehog, Wnt, and additional growth factors all play a role in inducing N-myc expression in the neural crest (Grimmer and Weiss, 2006). More detailed information on the role of N-myc in NB is present below in Sections 3.2.1 and 5.2.3.

The most complete studies on neural crest cell migration and development have been carried out in birds. These studies revealed that the avian trunk crest, which is located between somites 6 and the tail, gives rise to sympathetic neurons and that a subset of these (between somites 18 and 24) contribute to the adrenal medulla (Le Dourin and Kalcheim, 1999). The crest cells that contribute to the adrenal medulla, the site where the majority of NB tumors are found, follow the ventrolateral migratory pathway. Upon arrival to the proper target tissue, the cells then undergo final differentiation, and regulated cell death.

2.3. Sympathoadrenal lineage development

The sympathoadrenal lineage is thought to derive from a common progenitor that aggregates at the dorsal aorta after migrating from the crest utilizing the ventral pathway (Anderson, 1993). During the migration to the dorsal aorta, the crest cells encounter signals from the somites, the ventral neural tube, and the notochord. Local signals from the dorsal aorta, such as BMPS, specify the future differentiation of the crest cell as either a catecholaminergic/adrenal chromaffin cell or sympathetic neuron (Ernsberger *et al.*, 2005; Howard *et al.*, 2000; Reissmann *et al.*, 1996; Schneider *et al.*, 1999; Shah *et al.*, 1996). The sympathoadrenal lineage is specified by a tightly regulated set of transcription factors. Trunk crest cells can first be identified by the expression of Sox10 (Betters *et al.*, 2010; Huber *et al.*, 2008). As the crest cells migrate along the ventral pathway, they are exposed to BMP signals (e.g., BMP2, 4, and 7) that induce the expression of Mash1, a helix-loop-helix transcription factor expressed throughout the autonomic progeny (Huber, 2006). Shortly thereafter Phox2b expression occurs in sympathoadrenal lineage cells. Phox2b is required for the maintenance of Mash1 and recent evidence has shown that temporal difference in expression of these two factors may separate the sympathoadrenal lineage into separate sympathetic and adrenal lineages earlier in development than thought (Huber, 2006). Mash 1 induces the expression of Phox2a which is required for the production of the biosynthetic enzymes, dopamine beta-hydroxylase (DBH) and tyrosine hydroxylase (TH) in noradrenergic cells. Phox2b also appears to be important in the development of NB since PHOX2b mutations have been found in a subset of familial NB patients (see Section 3.1.1). Shortly after the migrating cells reach the dorsal aorta, they begin to acquire their respective sympathetic and adrenal cell fates and undergo a secondary migration to the presumptive prevertebral ganglia, the adrenal medulla, and secondary sympathetic ganglia where they complete their differentiation (Fig. 4.2; Huber, 2006; Le Dourin and Kalcheim, 1999; Morales *et al.*, 2005).

Although the complete mechanisms for proper chromaffin cell development are unknown, evidence indicates a highly intrinsic program. In animals that lack an adrenal cortex, chromaffin cells migrate to the suprarenal region, downregulate neuronal markers and contain large chromaffin granules. However, the generation of proper numbers of chromaffin cells and the expression of secretogranin II and PNMT requires intact glucocorticoid signaling. The human adrenal gland is remodeled throughout fetal development, infancy, and into adulthood in a process that is greatly affected by perturbations in levels of IGF II, FGF, and epidermal growth factor (EGF) levels. Importantly, as discussed below several of these growth factors are involved in cellular proliferation and signaling in NB (Section 4.2).

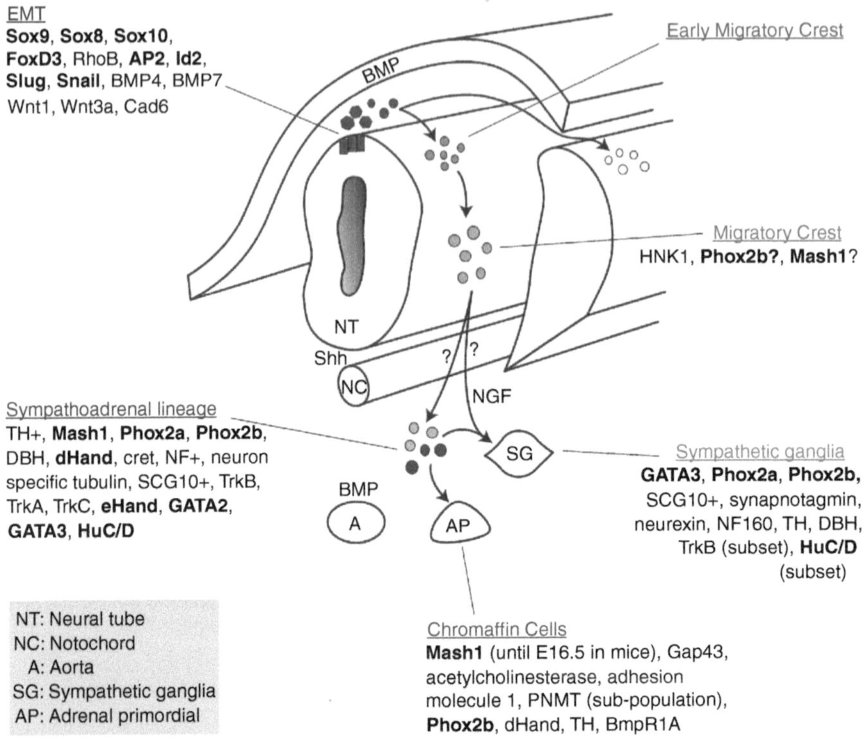

Figure 4.2 General schema of the development of chromaffin cells and sympathetic ganglia. Cells at the dorsal region of the neural tube undergo EMT (red population), delaminate from the neural tube (orange), and migrate ventrally to the aorta (green) where they are commonly referred to as the sympathoadrenal progenitors (blue and purple). From the aortic region, the cells then migrate to the developing adrenal gland (AP) to become chromaffin cells or differentiate to become sympathetic ganglia (SG). As cells begin to differentiate as sympathetic ganglia they upregulate neural markers while chromaffin cells upregulate proteins found in the adrenal gland. Recent studies suggest that the chromaffin cell and sympathetic ganglia may come from divergent lineages rather than a common sympathoadrenal lineage. The question marks between the migrating crest and the sympathoadrenal progenitors address this possibility. A more detailed discussion on the temporal expression of the transcription factors in the sympathoadreanal lineage (including developmental stages) can be found in (Howard *et al.*, 2000) and details about the distinct chromaffin and sympathetic lineages can be found in (Ernsberger *et al.*, 2005). Additional of neurotrophins and their receptors can be found in (Straub *et al.*, 2007). Transcription Factors are shown in bold and factors implicated in neuroblastoma have been underlined. Abbreviations: NT, Neural Tube; NC, notochord; A, Aorta; SG, sympathetic ganglia; AP, adrenal primordial. (See Color Insert.)

3. Genetic Lesions in NB

3.1. Familial genetic lesions

Hereditary NB is both rare and heterogeneous, accounting for less than 5% of all NBs (Maris *et al.*, 2002). In addition to the known hereditary mutations that are described below, hereditary NB predisposition loci have been mapped to chromosomes 16p12–13 and 4p16 indicating other familiar predisposition mutations may exist, but no genes have been shown to be inactivated or mutated in these regions, to date (Maris *et al.*, 2002; Perri *et al.*, 2002).

3.1.1. Phox2b

Germline mutations in the paired-like homeobox 2B (*PHOX2b*) gene on chromosome 4p13 are the first predisposition mutations identified in NB (Mosse *et al.*, 2004; Trochet *et al.*, 2004). As mentioned earlier, Phox2b, as well as to MASH1, are expressed early in the developing sympathoadrenal progenitors (Alenina *et al.*, 2006; Nakagawara, 2004; Nakagawara and Ohira, 2004). Shortly after expression of MASH1 and Phox2b in the sympathoadrenal lineage, Hand2, Phox2a, and GATA2/3 appear (Alenina *et al.*, 2006; Nakagawara, 2004; Nakagawara and Ohira, 2004). Phox2b has also been shown to be essential for the expression of the glial family ligand tyrosine kinase coreceptor RET (REarranged during Transfection) and for the specification of noradrenergic fates, particularly the biosynthetic enzymes TH and DBH (Alenina *et al.*, 2006; Nakagawara, 2004; Nakagawara and Ohira, 2004).

NB patients with PHOX2b mutations also have familial disorders of the neural crest such as Hirschsprung's disease (HSCR) and congenital hypoventilation syndrome (CCHS; Mosse *et al.*, 2004; Trochet *et al.*, 2004). It is unclear that the mutations in PHOX2b found in familiar NB result in gain or loss of function, although many PHOX2b mutations stabilize the Phox2b protein and decrease or eliminate the ability of Phox2b to transactivate the DBH promoter (Raabe *et al.*, 2008; Trochet *et al.*, 2005). The findings that Phox2b is necessary for the differentiation of autonomic neurons and overexpression of Phox2b inhibits proliferation in neuron progenitors and cell lines suggests Phox2b is a tumor suppressor (Raabe *et al.*, 2008; Trochet *et al.*, 2005, 2009). However, the absence of tumors with loss of heterozygosity (LOH) or mutation in second allele suggests gain-of-function, dominant-negative effect, or haploinsufficiency (Benailly *et al.*, 2003; Bourdeaut *et al.*, 2005).

3.1.2. Anaplastic lymphoma kinase (ALK)

ALK is a member of receptor tyrosine kinases (RTK) and was first identified as a part of the fusion gene nucleophosmin (NMP)–ALK in anaplastic large cell lymphoma via chromosome translocation of t(2;5)(p23;q25)

(Morris *et al.*, 1994, 1997). ALK is thought to play a role in the normal development of the central and peripheral nervous system since ALK mRNA is expressed throughout the nervous system in mouse and rat, but is not present in normal hematopoietic cells (Degoutin *et al.*, 2009; Hurley *et al.*, 2006; Iwahara *et al.*, 1997; Morris *et al.*, 1997; Vernersson *et al.*, 2006). More detailed studies in chick embryos have shown a similar pattern of ALK expression in the developing central nervous system in which ALK localizes primarily to the spinal motor neuron, the sympathetic ganglia, and the dorsal root ganglia. In mice, expression of ALK in the nervous system decreases after birth but is maintained at low levels in adults. Similar patterns of expression are observed in humans although additional ALK transcripts of differing size, most likely due to alternative splicing, have been observed in colon, prostrate, testis, small intestine, and brain of adults (Iwahara *et al.*, 1997; Palmer *et al.*, 2009).

Full-length ALK protein is comprised of an extracellular region and an intracellular region containing a RTK domain, linked by a transmembrane (TM)-spanning segment, whereas the NMP–ALK fusion protein generated as a result of the t(2;5)(p23;125) translocation contains the N-terminal of NMP and C-terminal kinase domain of ALK. Translocation of the *ALK* gene is also found in other tumors, such as inflammatory myofibroblastic tumor (IMT), and nonsmall cell lung cancer (NSCLC), but not in NB (Palmer *et al.*, 2009). Overexpression of wild-type ALK has also been observed in thyroid carcinoma, breast cancer, NB, melanoma, small cell lung carcinoma, glioblastoma, astrocytoma, retinoblastoma, Ewing sarcoma, and rhabdomyosarcomas NB (Cheng and Ott, 2010; Mosse *et al.*, 2009; Palmer *et al.*, 2009).

During 2008, at least five papers described ALK point mutations in 8–12% of all NB patient (both hereditary and sporadic) and some NB cell lines as well (Caren *et al.*, 2008; Chen *et al.*, 2008; George *et al.*, 2008; Janoueix-Lerosey *et al.*, 2008; Mosse *et al.*, 2008). With one exception, all the point mutations identified to date occur in the kinase domain and result in the constitutive activation of ALK. Two of these activating ALK mutants were able to transform NIH3T3 fibroblasts and induce tumor formation in nude mice (Chen *et al.*, 2008). In addition, knockdown of ALK or small molecular ALK inhibitors could reduce cell proliferation and induce apoptosis (George *et al.*, 2008; Janoueix-Lerosey *et al.*, 2008). Amplification of the *ALK* gene and/or overexpression of the ALK protein is seen in as many as 77% of all NB tumors (Passoni *et al.*, 2009) suggesting that overexpression of the ALK protein may also contribute to NB.

The downstream effects of ALK in NB need to be defined. Current data suggest that ALK may function through the Shc and MAP kinase pathways (Motegi *et al.*, 2004; Osajima-Hakomori *et al.*, 2005; Souttou *et al.*, 2001). More recent studies also suggest that activation of ALK enhances RAP1 activity via interaction with C3G, a Crk-binding protein and Crk-like protein (CRKL), and that this complex contributes to NB tumor cell growth and neurite outgrowth (Schonherr *et al.*, 2010).

3.2. Chromosome gain and oncogene activation

Many genetic abnormalities have been identified in nonfamilial NB tumors, including amplification of the MYCN proto-oncogene (25–33% of patients) and consistent areas of chromosomal deletion and rearrangement that result in loss of 1p36 (25–35%), 11q23 (35–45%), and 14q23 (16–27%), as well as unbalanced gain of 17q22 (~50%) (Table 4.2; Brodeur, 2003; Maris *et al.*, 2007; Schor, 1999). In contrast, known tumor suppressor genes (TSGs) such as $p16^{INK4a}$, pRb, p53, and $p14^{ARF}$ are not frequently deleted or mutated in NB, although the nuclear localization of the $p16^{INK4a}$ and p53 proteins has been reported to be altered in some tumor cell lines (Brodeur, 2003; Maris *et al.*, 2007; Schor, 1999; Teitz *et al.*, 2001; van Noesel and Versteeg, 2004). Many of these abnormalities are powerful prognostic markers and are highly related to clinical outcome. For example, amplification of MYCN in NB patients is correlated with chromosome 1p36 LOH. NB tumors which harbor 1p36 LOH and MYCN amplification are usually advanced-stage (stages 3 and 4) aggressive tumors that are frequently metastatic and generally respond poorly to chemotherapy/irradiation (Brodeur, 2003; Maris *et al.*, 2007). In the recent years, clinical trials are increasingly based on the tumor genetic characteristics.

3.2.1. Amplification of MYCN and the 2p24 locus

In 1983, Schwab *et al.* found that a novel myc homolog gene was amplified in several NB cell lines and one NB tumor (Schwab *et al.*, 1983). Later, several papers termed this gene as *MYCN* based on homology to c-myc and expression pattern in the developing nervous system, and identified its location at chromosome 2p24 (Kohl *et al.*, 1983; Schwab *et al.*, 1984). Additional studies have shown that N-myc protein is a nuclear phosphoprotein that is a member of the myc family of helix-loop-helix transcription factors (Pelengaris *et al.*, 2002). Amplification of the *MYCN* gene in patient tumors ranges from 10-fold to more than 500-fold, although the majority of tumors exhibit 50- to 100-fold *MYCN* gene amplification levels. The amplified DNA typically contains a large region of chromosome 2 ranging from 100 kb to 1 Mb which includes the entire *MYCN* gene and varying amounts of adjacent DNA. Although other genes may be coamplified with *MYCN*, *MYCN* is only consistent amplified gene from this region (Reiter and Brodeur, 1996, 1998).

MYCN amplification is rarely observed on chromosome 2p24 in primary tumors but is found to be at homogeneously staining regions (HSRs) on different chromosomes or, more frequently, as double minutes (DMs; which are small fragments of extrachromosomal DNA; Emanuel *et al.*, 1985; Schwab *et al.*, 1984). During cell culture, the amplification unit frequently integrates into chromosomes to become HSRs. The reason for

Table 4.2 Frequent chromosomal region abnormalities in neuroblastoma

Chromosomal region	Status	Frequency	Relation with MYCN amplification	Involved genes	Clinical group
1p36	Loss	25–35%	Correlation	*CHD5*, *miR-34a*, *KIF1Bβ*	Unfavorable
11q23	Loss	35–45%	Inversed correlation	*TSLC1/IGSF4*	Unfavorable
14q23	Loss	16–27%	Inversed correlation		All groups
17q22	Gain	~50%	Correlation	*Survivin*, *NM23A*	Unfavorable

the differences in the location of the amplicon in primary tumors and cultured cells remains unclear.

Amplification of MYCN is highly associated with aggressive NB tumors and poor outcome. Although the entire role of MYCN in NB is still being uncovered, amplification of the *MYCN* gene is usually accompanied by overexpression of the N-myc protein. Studies on N-myc regulation suggest that the transcription factor and signaling pathways responsible for the upregulation of N-myc are dependent on cell type (Hurlin, 2005). These factors include IL-7 and Pax-5, NF-κB in pre-B cells, and insulin-like growth factors I and II (IGFI and IGFII) in NB cells (Strieder and Lutz, 2003). In contrast, N-*myc* transcription is repressed by retinoic acid (RA) in association with E2F binding, nerve growth factor (NGF) binding to TrkA receptor, the iron chelator deferoxamine mesylate, and transforming growth factor-beta 1 (TGF-β1; Strieder and Lutz, 2003; Wada *et al.*, 1992).

Myc proteins form heterodimers with the Max protein. These heterodimers bind to E-box elements (CACGTG) to activate transcription. However, Myc–Max dimers can also associate with other transcription factors such as Miz-1 and Smad and bind to Inr (initiator) element to repress transcription. Max can also form homodimers or heterodimers with Mad to compete or suppress Myc–Max binding (Pelengaris *et al.*, 2002; Thompson, 1998; Fig. 4.3). The targets of Myc–Max are involved in various cellular processes, including cell growth, proliferation, loss of differentiation, and apoptosis (Adhikary and Eilers, 2005; Pelengaris *et al.*, 2002; Thompson, 1998), and include proteins such as MASH1 and important molecules in the normal development of sympathocoadrenal lineage cells, such as the multidrug resistance protein 1 (MRP1), α-prothymosin, telomerase, Id2, MCM7; leukemia inhibitory factor, activin A, Pax-3, and MDM2 (Breit and Schwab, 1989; Haber *et al.*, 1999; Harris *et al.*, 2002; Hatzi *et al.*, 2002; Lasorella *et al.*, 2002; Mac *et al.*, 2000; Pelengaris *et al.*, 2002; Shohet *et al.*, 2002; Slack *et al.*, 2005). Many other putative N-myc targets with E-boxes in or near the promoter have also been identified although studies are still ongoing to determine which E-boxes actually bind Myc. Overexpression of N-myc is also reported to influence the expression of IL-6, NDRG1, MHC class I genes, and integrins (Chambery and Mohseni-Zadeh, 1999; Lutz *et al.*, 1996; Mac *et al.*, 2000), although the mechanisms responsible for these effects are unknown.

The transgenic mouse model demonstrates that MYCN overexpression is an initial event in NB tumorigenesis. In this model, overexpression of the human MYCN is driven by the rat TH promoter, which is expressed in migrating cells of the neural crest early in development (Banerjee *et al.*, 1992), causes the formation of NB tumors in transgenic mice (Weiss *et al.*, 1997). These tumors recapitulate most of the histological and pathological aspects of the human disease, including tumor localization, positive staining for neuronal markers, and gains and losses of chromosomes in regions

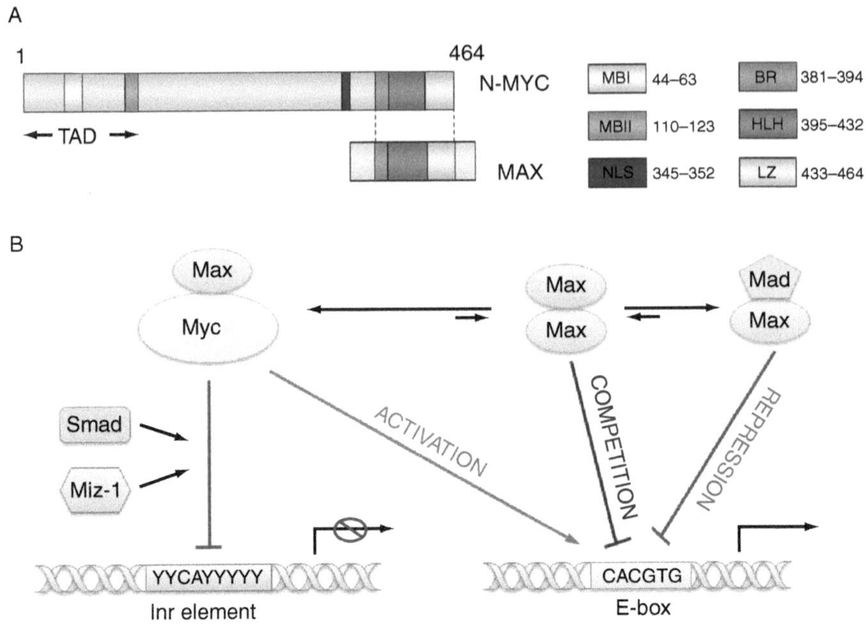

Figure 4.3 The structure of N-myc protein and transcription regulation by N-myc. (A) The structure of N-myc protein. The N-terminal transactivation domain (TAD) contains two conserved Myc box I and II (MBI and MBII), which are essential for DNA binding. The C-terminal domain (CTD) harbors basic region (BR), helix-loop-helix (HLH) motif, and leucine zipper (LZ) for dimerization with Max. There is a nuclear localization signal (NLS) before CTD. (B) The model for the transcription regulation by N-myc. Myc–Max heterodimer may bind to E-box element (CACGTG) to activate transcription, however, Myc–Max dimer can associate with other transcription factors such as Miz-1, Smad, and bind to Inr (initiator, weak consensus) element to repress transcription. Max can also form homodimers or heterodimers with Mad to compete or suppress Myc–Max binding to E-box. (See Color Insert.)

syntenic with those observed in human NB (Weiss *et al.*, 1997). However, other factors are also likely to be involved in the early stages of tumor formation since amplification of the MYCN oncogene occurs in only about one-third of NBs. In addition, the tumors in these transgenic mice rarely exhibit significant metastasis despite the presence of high levels of N-myc protein suggesting that other genetic alterations and/or epigenetic changes are needed for tumor formation and metastasis. These and other studies suggest that N-myc regulates neural progenitor cell proliferation, nuclear size and differentiation (Knoepfler *et al.*, 2002). Importantly, studies using chick/quail chimera reveal that overexpression of N-myc in the early neural crest induces premature ventral migration of neural crest cells and promotes the differentiation of these cells (Wakamatsu *et al.*, 1997). In addition, other studies have shown that high level N-myc mRNA or protein expression in

NB cells accelerate cell cycle progression (Lutz *et al.*, 1996), and that overexpression of N-myc in postmitotic sympathetic neurons causes quiescent cells to reenter the cell cycle and enhances the survival of these cells upon NGF withdrawal (Wartiovaara *et al.*, 2002).

The high level expression of N-myc in NB is also consistent with the hypothesis that NB arises during development. N-myc is normally expressed at the beginning of the preimplantation stage of development. As the embryo develops, N-myc expression is observed in the central nervous system and neural crest. By embryonic day 9.5, relatively high levels of N-myc are observed in the fetal brain, kidney, and in the neural crest and early stage migrating neural crest cells (Zimmerman *et al.*, 1986). During later stages of neural crest migration N-myc expression is only observed in cells undergoing neuronal migration. Even in these cells, N-myc expression is gradually downregulated as cells differentiate and become quiescent (Lee *et al.*, 1984; Zimmerman *et al.*, 1986). These data are consistent with studies using N-myc knockout mice that demonstrate that the loss of N-myc results in embryonic death at day 10.5 of gestation due to defects in the nervous system, limb, heart, liver, lung, gut, mesonephros, and genital ridge (Sawai *et al.*, 1991).

3.2.2. Gain of 17q

Gain of chromosome 17q was first identified by G-banded cytogenetic analyses in early 1980s (Gilbert *et al.*, 1984). However, researchers paid little attention to these observations since their interests focused on the genetic abnormalities of MYCN amplification and 1p LOH at that time. In the middle 1990s, NB scientists realized the importance of 17q abnormalities since FISH technology indicated that translocation of this chromosome arm occurs in about half of the NB primary tumors. The translocation results in unbalanced gain of one to three copies of 17q (Brodeur, 2003; Maris *et al.*, 2007; Schor, 1999). Although the breakpoint of 17q varies, the frequent gain of regions from 17q22 suggests that increased dosage of one or more genes from this region may confer a selective survival advantage for NB tumor cells. It is estimated that as much as 20 Mb of the 17q chromosome fragment, which could include more than 200 genes, can be translocated in NB tumors. Therefore, it is difficult to identify the genes responsible for the selective advantages (Brodeur, 2003; Maris *et al.*, 2007; Schor, 1999). Several genes in this area are considered good candidate oncogenes or tumor suppressors based on correlations between expression levels and unbalanced gain of 17q. These include survivin, NM23A, and PPM1D (Godfried *et al.*, 2002; Islam *et al.*, 2000; Saito-Ohara *et al.*, 2003). Notably, survivin is a member of apoptosis inhibiting protein family and is frequently overexpressed in many tumor types, including NB where expression has been correlated with late stage disease and poor prognosis;

whereas NM23A is metastasis-related gene and PPM1D a phosphatase that suppresses stress induced apoptosis (Almgren *et al.*, 2004; Bown *et al.*, 1999).

Unbalanced gain of 17q correlates with other chromosomal deletions. The most frequent deletion site is the short arm of chromosome 1, followed by 11q. At least 30 translocation sites on 20 different chromosomes have been identified in various patient samples and cell lines (Bown *et al.*, 1999; Lastowska *et al.*, 1997; Meddeb *et al.*, 1996). Nevertheless, NB tumors harboring unbalanced gain of 17q exhibit a more aggressive phenotype and a poorer prognosis than those without this abnormality.

3.2.3. Amplification and chromosome gain of other loci

In addition to the amplification of *MYCN* gene, several other regions of gene amplifications have been identified in small groups of NB cases. These include amplification of the *MDM2* gene at 12q13, the *DDX1* gene at 2p24, the *MYCL* gene at 1p32, and unidentified DNA from chromosome 2p22 and 2p13 (Corvi *et al.*, 1995a,b; Jinbo *et al.*, 1989; Van *et al.*, 1995). The *MDM2* gene was initially found to be amplified in three NB cell lines and one primary tumor (Corvi *et al.*, 1995a). Like the *MYCN* gene amplification, the MDM2 amplification unit first developed within DMs and then integrates into a different chromosome to form HSRs (Corvi *et al.*, 1995b). The *DDX1* gene, which encodes a RNA helicase, was found to be coamplified with MYCN in 4/6 NB cell lines and 6/16 tumors with MYCN amplification; however, DDX1 amplification was not found without MYCN amplification (George *et al.*, 1996). One paper also indicated that *MYCL* gene is coamplified with MYCN in NB cell lines. MYCL, another member of myc gene family, is frequently overexpressed in small cell lung carcinoma (Jinbo *et al.*, 1989). In addition to gain of 17q, other chromosome gains have been identified on 1q, 4q, 5q, 6p, 7q, 18q using comparative genomic hybridization (CGH) methodology, although their biological and clinical significance remain unclear (Hirai *et al.*, 1999; Lastowska *et al.*, 1997; Meltzer *et al.*, 1996; Takita *et al.*, 2000; Vandesompele *et al.*, 1998).

3.3. Chromosome loss and tumor supressor genes (TSGs)

In addition to mutation, gene amplification and increased chromosome copy number, NB tumors also experience loss of genetic material and deletion of putative TSGs.

3.3.1. LOH of chromosome 1p and CHD5, miR-34, KIF1Bβ

Loss of the short arm of chromosome 1 occurs in about 25–35% NB tumors. 1p LOH is correlated with amplification of MYCN in NB patients. As mentioned above, loss of 1p correlates with and may be a result of unbalanced gain of 17q, however, the exact mechanism that is responsible for these two events is not clear. The importance of 1p LOH is highlighted by studies showing that transferring chromosome 1p material into human NB cells *in vitro* led to differentiation and suppression of tumorigenicity (Bader *et al.*, 1991). In search of potential TSGs that reside in this region, extensive efforts have been made to identify the smallest region of overlap (SRO) that would include the TSG candidate. These studies delineate the 1p36.1–36.3 as the SRO. Several candidate genes, such as p73, reside in this area. However, further studies failed to demonstrate a correlation between p73 loss and NB development (Ichimiya *et al.*, 1999).

Whereas patients with 1p36 abnormalities without MYCN amplification have been identified, the reverse situation virtually never occurs suggesting either that 1p36 LOH provides a permissive environment for MYCN amplification or that tumors with these two associated genetic defects have a high degree of genomic instability (Brodeur, 2003). NB tumors with 1p36 LOH and MYCN amplification are usually aggressive tumors that are frequently metastatic and generally resistant to chemotherapy/irradiation. Although the chromosomal regions described above are known to be important in NB, the TSGs that reside within these regions have not been definitively identified.

Recent studies have identified three new putative tumor suppressors on chromosome 1p36: the chromodomain helicase DNA-binding domain 5 (CHD5), microRNA-34a (mir-34a), and the kinesin superfamily protein 1B beta (KIF1Bβ; Bagchi *et al.*, 2007; Munirajan *et al.*, 2008; Welch *et al.*, 2007). All three of these proteins have affects on cell growth. For example, Bagchi *et al.* demonstrated that the effects of CHD5 on cell growth were dependent on p53 and that CDH5 positively regulates p53 via effects on $p19^{ARF}$ expression. Thus, overexpression of CHD5 results in enhanced apoptosis and senescence, increased p53 and $p19^{ARF}$ levels, and sequestration of MDM2, the negative regulator of p53, by $p19^{ARF}$. Conversely, cells lacking CHD5 exhibit decreased p16 and $p19^{ARF}$ expression. This decrease in $p19^{ARF}$ was mirrored by a decrease in p53 levels and enhanced cellular proliferation. Thus, CHD5 appears to function as a tumor suppressor that controls proliferation, apoptosis, and senescence via effects on the $p19^{ARF}$/p53 pathway. These effects are most likely due to changes in the accessibility of the p16/$p19^{ARF}$ gene locus resulting from the chromatin remodeling function of CHD5 (Bagchi *et al.*, 2007).

In addition to CHD5, Chen *et al.* found that mir-34a was expressed at very low levels in unfavorable primary tumors and NB cell lines. This group further showed that introduction of this microRNA (miRNA) into cell lines resulted in decreased cell proliferation and caspase-dependent apoptosis. They also found that mir-34a directly targeted the E2F3 mRNA and repressed its expression (Chen and Stallings, 2007; Welch *et al.*, 2007). E2F3 is a transcription factor that induces the expression of many genes with roles in cellular proliferation.

Finally, overexpression of KIF1Bβ induced cell death while decreased KIF1Bβ levels correlated with cell proliferation and enhanced tumor formation in nude mice, suggesting that KIF1Bβ is also a potential TSG candidate (Munirajan *et al.*, 2008). Kaelin's group also found that KIF1Bβ is a downstream target of prolyl hydroxylase EglN3 and induced apoptosis in neuronal progenitor cells or NB cells when NGF is limited. In addition, they identified missense mutations of KIF1Bβ in inherited NBs and pheochromocytomas (Schlisio *et al.*, 2008), supporting the hypothesis that KIF1Bβ is a potential TSG candidate.

3.3.2. Loss of 11q and TSLC1

Loss of the long arm of chromosome 11 has been identified in 35–45% NB primary tumors with a single copy *MYCN* gene. Two large patient studies analyzed 295 NB primary tumors. These studies found loss of 11q in 44% cases, and common regions of LOH located at 11q23, suggesting there are TSGs residing in this area (Guo *et al.*, 1999; Maris and Matthay, 1999). Loss of 11q correlated with adverse clinical features including late stage disease, older age of disease onset and unfavorable histology, although it is strikingly inversely correlated with MYNC amplification and 1p loss. Therefore, 11q loss is a useful and important marker in determining the clinical prognosis for those advanced-stage tumors without MYCN amplification. Transfer of chromosome 11 induced differentiation in NB cell lines supporting the importance of loss of 11q in tumorigenesis (Bader *et al.*, 1991).

One putative tumor suppressor, the *IGSF4* (immunoglobulin superfamily 4) gene, was first localized to the common 11q23 LOH region in 1999 (Gomyo *et al.*, 1999). This gene which is also known as TSLC1/CADM1 (Tumor suppressor in lung cancer 1/cell adhesion molecule 1), is considered as a potential TSG for lung cancers. A recent CGH study which examined 236 primary tumor samples found TSLC1 LOH locus in 35% tumors. Importantly, the level of TSLC1 expression correlated with tumor stage, histological classification, MYCN and TrkA expression levels. Reduced expression of TSLC1 was found in unfavorable tumors. Further, introduction of TSLC1 decreased cell proliferation in NB cell lines (Ando *et al.*, 2008). These results indicated that TSLC1 is a good NB tumor suppressor candidate. Interestingly, a recent study indicates that expression of both KIF1Bβ and TSLC1 is controlled by the polycomb protein Bmi1, whose expression is regulated by N-myc (Ochiai *et al.*, 2010).

3.3.3. LOH of 14q

Loss of the long arm of chromosome 14 is also commonly found in NB primary tumors (~16–27% of the patients). LOH on chromosome 14q was first identified in 1989 using a polymorphic DNA marker which detected allelic deletion at specific 14q23 loci (Suzuki *et al.*, 1989). LOH analysis of 14q in a large number of primary tumors using 11 polymorphic DNA markers found 14q LOH in 83 of 372 tumors (22%) (Thompson *et al.*, 2001). 14q LOH was highly correlated with 11q loss and had an inverse relationship with 1p loss and MYCN amplification (Thompson *et al.*, 2001). However, LOH for 14q was present in tumors from all clinical stages, suggesting this abnormality may be a universal early event during tumor development.

In addition, to the genetic changes described above, there have also been reports of LOH and/or allelic imbalance at chromosome arms 2q, 3p, 4p, 9p, and 19q (Caron, 1996; Ejeskar *et al.*, 1998; Marshall *et al.*, 1997; Mora *et al.*, 2001; Takita *et al.*, 2001), however, the significance of these genetic changes is not clear.

4. The Role of Neurotrophins and Growth Factors in the Development of the Sympathetic Nervous System and in NB

As the neural crest cells migrate to the aorta but prior to reaching the adrenal medulla they begin to express TH which in turn controls the expression of other enzymes needed for catecholamine biosynthesis as described above in Section 2.3. Since NB appears to arise from cells that are transformed at various times during this migration, the majority of NB tumors secrete catecholamines. Indeed, the presence of high levels of catecholamines in patient urine samples is used as one of the diagnostic criteria for the disease (LaBrosse *et al.*, 1976). Based on this data, Sawada began to screen the urine of 6-month-old infants for increased catecholamine metabolites from 1984 and found that the incidence of *in situ* NB was much higher than the number of sporadic cases that had been observed previously (Sawada, 1992). These data agree with a previous hypothesis of Beckwith and Perrin who postulated that during the development of sympathetic neurons the incidence of *in situ* NB is higher than the incidence of sporadic cases (Beckwith and Perrin, 1963). Most of these *in situ* NBs spontaneously regress as the child ages (Brodeur, 2003; Maris, 2005; van Noesel and Versteeg, 2004), suggesting they are resolved using normal developmental programs. Developmental studies and studies from knockout mice suggest the TrkA is crucial for the development of many sympathetic lineage cell types (Brodeur *et al.*, 2009). This is consistent with data indicating the NGF is required for the differentiation and survival of many sympathetic lineage cells (Nakagawara, 2004; Nakagawara and Ohira, 2004). In addition to NGF and TRKs several other growth factors also play roles in the development of the sympathoadrenal lineage cells. These include EGF which is expressed in neural crest cells and is thought to contribute to the formation of neuron and melanocytes at later point during neural crest migration (Erickson and Turley, 1987), vascular endothelial growth factor (VEGF) which is expressed by surrounding cells (McLennan *et al.*, 2010), and IGFI and IGFII which is expressed in the neural crest, the dorsal root, the sympathetic ganglia, and the adrenal medulla (Coppola *et al.*, 2009; D'Ercole *et al.*, 1996).

4.1. Neurotrophin receptors

Although the steps in the transformation of sympathetic neuroblasts to NB cells is not clear, extensive evidence suggests that neurotrophin receptors are involved in NB tumorigenesis and in development of the nervous system and sympaticoadrenal lineage cells. The neurotrophin receptors TrkA, TrkB, and TrkC, are encoded by the *NTRK1, 2, 3* genes, respectively. Upon binding to their ligands, NGF, brain-derived neurotrophic factor (BDNF), neurotrophin-3, respectively, these receptors regulate proliferation, survival, and differentiation in normal neuronal cells (Brodeur *et al.*, 2009; Maris and Matthay, 1999; Maris *et al.*, 2007; Straub *et al.*, 2007). These receptors all associate with p75, a low affinity receptor that may enhance the binding of ligand to the Trk proteins or alter the function of the Trk receptors (Brodeur *et al.*, 2009; Maris and Matthay, 1999; Maris *et al.*, 2007; Straub *et al.*, 2007). High levels of TrkA, in association with very low/no N-myc expression, are detected in favorable NB tumors which often spontaneous regress (Nakagawara, 1993; Nakagawara *et al.*, 1992). These favorable NB tumors cells usually express a small amount of NGF as do some of the surrounding cells (Nakagawara, 1993). Cells that express the most NGF are thought to undergo differentiation, while those that express less NGF undergo apoptosis (Nakagawara, 1993). TrkA expression is dramatically decreased in MYCN amplified NB tumors (Nakagawara *et al.*, 1992). Thus, the TrkA/NGF pathway could play an important role in determining the ability of these favorable NB tumors to differentiate or to regress in response to the microenvironment. While in general TrkA expression is correlated with a good prognosis, a novel TrkA splice variant has been found in advanced-stage tumors with adverse biological features (Tacconelli *et al.*, 2004). This TrkA isoform is constitutively active and promotes cell survival and angiogenesis independently of NGF expression. NGF signaling may also be linked to IGFII expression since a study in which SHSY5Y cells were transfected with TrkA found increased expression of IGFII in response to NGF binding to the transfected receptor (Kim *et al.*, 1999).

In contrast, TrkB is preferentially expressed in clinically unfavorable NB tumors and expression of TrkB strongly correlates with MYCN amplification (Nakagawara *et al.*, 1994). The TrkB ligand, BDNF, is also highly expressed in these tumors. Coexpression of ligand and receptor may form an autocrine loop to enhance survival, metastasis, and drug resistance (Douma *et al.*, 2004; Nakagawara, 1994). Interestingly, a truncated form of TrkB which lacks tyrosine kinase activity is expressed in some favorable NB tumors (Ho *et al.*, 2002).

Finally, TrkC is commonly expressed in favorable NB tumors. These tumors also express very limited amount of the TrkC ligand neurotrophin-3 and coexpress TrkA (Svensson *et al.*, 1997; Yamashiro *et al.*, 1996).

4.2. Other growth factors and growth factor receptors

In addition to NGF and BDNF several other growth factors also play roles in the development of the sympathoadrenal lineage cells and NB tumorigenesis. These include EGF, VEGF, and insulin-like IGFI and IGFII. EGF receptor 1 (EGFR1) expression is found on both primary NB tumors and tumor-derived cell lines (Ho *et al.*, 2005). Binding of EGF to EGFR1 causes receptor autophosphorylation and increases proliferation via effects on the MAPK and PI3K/AKT pathways (Henson and Gibson, 2006). Exogenous VEGF also stimulates the PI3K/AKT pathway and increases expression of survivin, an antiapoptotic gene in NB cells (Beierle *et al.*, 2005). Endocrine-derived VEGF has also been shown to play a role in the proliferation and differentiation of neural crest cells during development and to promote NB cell growth. IGF1 receptors (IGF1Rs) are expressed in the majority of NB primary tumors (Martin *et al.*, 1992). This receptor binds both IGFI and IGFII suggesting that these growth factors are important for NB tumorigenesis. Expression of IGF1R activates the PI3K/AKT and MAPK pathways and enhances cellular proliferation, cell survival, migration, and invasion, and induces chemotherapeutic resistance and reduced response to other apoptotic stimuli (Valentinis and Baserga, 2001). IGFII has been reported to be upregulated upon TRKA activation and downregulated upon TRKA overexpression suggesting potential feedback loops between these proteins (Kim *et al.*, 1999). In addition, IGF1R is transcriptionally activated by N-myc and in turn high IGF1R levels induce N-myc protein and mRNA expression suggesting the presence of an amplification loop that enhances the ability of these proteins to promote tumorigenesis (Chambery and Mohseni-Zadeh, 1999). Inhibition of IGF1R signaling has also been shown to increase N-myc phosphorylation by GSK-3β which inactivates N-myc and enhances N-myc turnover resulting in decreased cell growth both in culture and in mice model systems (Coulter *et al.*, 2009). IGF1R has also been shown to enhance NB metastasis to bone most likely due to its ability to enhance migration and invasion and to the presence of IGF ligand in bone marrow (van Golen *et al.*, 2006). In addition, IGF increases cellular survival under hypoxic conditions via increased expression of hypoxia-inducing factor 1α (HIF1α) and VEGF (Treins *et al.*, 2005).

5. Programmed Cell Death (PCD; Apoptosis) in Development of NB

5.1. The role of cell death in development

Another important process during development of the peripheral nervous system is PCD, also known as apoptosis. This process is used during development to eliminate redundant cells, control cell number, and for remodeling and repair. Cell death also occurs in the developing peripheral nervous system in response to loss of essential growth factors and cytokines (De Zio *et al.*, 2005). Neural crest development therefore is a balance between proliferation, cell death, migration, and differentiation. Errors in any of these processes may leave the cell more prone to transformation and potentially increase tumorigenesis. This is especially true of pediatric cancers, such as NB, since these cancers develop during normal development. Indeed, stage 4S NB spontaneously regresses with little to no intervention (Blaschke *et al.*, 1998; De Zio *et al.*, 2005; Johnsen *et al.*, 2009). Tumors of this stage are present in younger children and the prognosis is good. It is unknown why these tumors suddenly die or cease to grow, although it is thought that apoptosis is likely to be involved in the disappearance of these tumors (Brodeur, 2003; Maris *et al.*, 2007; Schor, 1999). Several hypotheses have been suggested to explain the regression of these tumors: (1) these tumors are dependent on growth factors or other proteins that are present in low levels and that once a tumor reaches a certain size the factors are depleted and the tumor undergoes apoptosis, (2) as the child develops, the tumor is recognized by the immune system and destroyed, and/or (3) that the cells become responsive to the environmental cues and developmental regulatory apoptotic pathways as they mature which triggers apoptosis. Though any of these possibilities may be the case, there is sparse literature investigating apoptotic pathways in the developing trunk crest as well as the peripheral nervous system. Although it is known that crest cells from rhombomeres 3 and 5 in the chicken and mammals undergo PCD (Kulesa *et al.*, 2004; Morales *et al.*, 2005) and that expression of Snail in the early migratory crest cells protects the migratory crest cells form apoptosis (Vega *et al.*, 2004), a more detailed investigation regarding specific cell death programs in crest cells contributing to the developing sympathoadrenal lineage requires more characterization.

5.2. The role of apoptosis genes in NB

As mentioned above, apoptosis or PCD in multicellular organisms is a tightly regulated process required for normal growth, development, and cellular specialization (Danial and Korsmeyer, 2004; Hengartner, 2000). Defective expression of proteins and aberrant function of constituents in the apoptotic cascade have been implicated in oncogenesis, tumor progression, and treatment resistance (Fesik, 2005; Kaufmann and Vaux, 2003; Reed and Tomaselli, 2000). There are two major apoptotic pathways in mammalian cells: the death receptor (or extrinsic) pathway and the mitochondrial (or intrinsic) pathway. Caspases (cysteine proteases) are the central components of the apoptotic machinery. At least 14 distinct caspases have been identified in mammals and are grouped into three categories (nonapoptotic, initiators, and effectors; Danial and Korsmeyer, 2004; Shi, 2004). Importantly for this review, alterations in caspase-8 expression have been observed in NB.

5.2.1. Caspase-8

Human caspase-8 (also known as FLICE) is encoded by the *CASP8* gene which is located on chromosome 2q33–2q34 (Grenet *et al.*, 1999). Caspase-8 exists as a monomer in the cell and requires dimerization or oligomerization for its activation (Boatright and Salvesen, 2003; Salvesen and Abrams, 2004). Subsequent cleavage events, although not essential for activity, further stabilize the activated protein (Boatright and Salvesen, 2003; Boatright *et al.*, 2003; Salvesen and Abrams, 2004). Activated caspase-8 is an important initiator caspase in the death receptor-mediated apoptotic pathway. The death receptor pathway is triggered by members of the death receptor superfamily (FasR, TNFRI, DR5, etc.; Hengartner, 2000). Binding of the ligand (e.g., FasL) to its receptor (FasR) induces the formation of the death-inducing signaling complex (DISC) containing death receptors, adaptor proteins and procaspase-8 (Danial and Korsmeyer, 2004; Hengartner, 2000). Procaspase-8 then dimerizes, resulting in activation and cleavage (Danial and Korsmeyer, 2004; Hengartner, 2000; Shi, 2004). The active enzyme subsequently cleaves downstream effector caspases, resulting in their activation and leading to apoptosis (Danial and Korsmeyer, 2004; Hengartner, 2000; Shi, 2004). Active caspase-8 can also cleave Bid, which is a Bcl2 family member. Cleaved Bid, termed t-Bid, translocates to the mitochondria and promotes cytochrome c release, thereby activating the mitochondrial pathway (Danial and Korsmeyer, 2004; Hengartner, 2000; Shi, 2004).

Our laboratory first found that caspase-8 is deleted or more commonly, silenced in most NB cell lines and patient tumor samples. We further identified a region within the *CASP8* gene which is methylated (Teitz *et al.*, 2000). This finding has subsequently been verified by several groups (Fulda and Debatin, 2006; Fulda *et al.*, 2001; Teitz *et al.*, 2000; Yang *et al.*, 2007). Several additional genes with roles in tumorigenesis are also hypermethylated in NB including the PCDHB gene family, BLU, TSP-1, RASSFIA, TIG1, HIN-1, DcR1, DcR2, DcR4, etc. (Abe *et al.*, 2005; Astuti *et al.*, 2001; van Noesel *et al.*, 2002; Yang *et al.*, 2003, 2007). Among these, CASP8, PCDHB, BLU, DcR2, and HIN-1 are found to be associated with high-risk factors and poor outcome in NB patients (Abe *et al.*, 2007; Yang *et al.*, 2007). Methylation of this region of the *CASP8* gene has been correlated with decreased caspase-8 protein expression in many (Fulda *et al.*, 2001; Hopkins-Donaldson *et al.*, 2000; Teitz *et al.*, 2000), but not all studies (Banelli *et al.*, 2002; van Noesel *et al.*, 2002). However, even in studies where methylation has not been correlated with expression, decreased caspase-8 protein levels are observed in 50–70% of NB patient tumors. Since our initial discovery in NB, loss of caspase-8 expression has been found in many other tumors, including peripheral neuroectodermal tumors, medulloblastoma, glioblastoma multiforme, rhabdomyosarcoma, retinoblastoma, small cell lung carcinoma, and Wilms tumors (Gonzalez-Gomez *et al.*, 2004; Harada *et al.*, 2002; Hopkins-Donaldson *et al.*, 2003). Caspase-8 mutations in adult gastric and colorectal tumors have also been reported (Kim *et al.*, 2003; Soung *et al.*, 2005). Finally, loss of caspase-8 expression has been correlated with poor prognosis in medulloblastoma (Pingoud-Meier *et al.*, 2003) and with relapse and aggressive metastatic disease in glioblastoma multiforme (Martinez *et al.*, 2007) A role for caspase-8 in tumorigenesis is also supported by recent studies demonstrating increased transformation of caspase-8 null SV40 immortalized mouse embryo fibroblasts (Krelin *et al.*, 2008). In addition, we have demonstrated that the loss of caspase-8 expression has biological significance in NB cell metastasis (Lahti *et al.*, 2006; Stupack *et al.*, 2006; Teitz *et al.*, 2006). Loss of caspase-8 increases metastasis by blocking integrin-mediated death, a caspase-8-dependent process (Lahti *et al.*, 2006; Stupack *et al.*, 2006; Teitz *et al.*, 2006). The role of caspase-8 in metastasis will be discussed in more detail in the Section 6.2 describing proteins that play a role in NB metastasis.

Multiple groups have shown that caspase-8 deficient NB cells are resistant to death receptor signals and some chemotherapeutic drugs. Importantly, these defects can be corrected by reexpression of caspase-8 via demethylating agents, IFN-γ treatment and caspase-8 expressing retroviral vectors. However, in a few instances secondary defects in death receptor proteins and/or signaling pathway or increased expression of FLIP prevented restoration of the extrinsic cell death pathway (Fulda *et al.*, 2001; Johnsen *et al.*, 2004; Muhlethaler-Mottet *et al.*, 2003; Teitz *et al.*, 2000; Tekautz *et al.*, 2006; Yang *et al.*, 2003).

5.2.2. Bcl-2 family

Another group of proteins that play a role in NB tumorigenesis and metastasis are the Bcl-2 family members. This family of proteins plays a central role in the intrinsic apoptotic pathway. There are more than 20 genes of Bcl-2 family identified in mammals, including antiapoptotic members (Bcl-2, Bcl-XL, Bcl-W, Bcl-G, Bfl1, Mcl-1) and proapoptotic members (Bcl-XS, Bcl-B, Bax, Bak, Bad, Bid, Bik, Bok, Bim, Puma, Noxa, Nix, Nip3, Hrk, Mtd). They usually regulate apoptosis by controlling mitochondrial outer membrane permeabilization (MOMP) to promote or prevent the release of cytochrome c (Green and Kroemer, 2004; Johnsen *et al.*, 2009; Reed, 2006).

The antiapoptotic members, such as Bcl-2, Bcl-XL, Mcl-1, are highly expressed in neural progenitor cells, indicating their protection role in neuron development. Some proapoptotic regulators of the intrinsic pathway, such as Bid and caspase-9, are expressed preferentially in favorable tumors, whereas antiapoptotic regulators such as Mcl-1 were expressed at high levels in unfavorable tumors (Abel *et al.*, 2005), suggesting an imbalance between anti- and proapoptotic factors in NB tumors with favorable or unfavorable biology. Bcl-2 is highly expressed in the majority of NB cell lines and primary tumors and inversely correlates with the degree of cell differentiation (Ramani and Lu, 1994). Although conflicting data exists about the correlation between Bcl-2 levels and MYCN amplification or poor prognosis (Castle *et al.*, 1993; Ikegaki *et al.*, 1995; Mejia *et al.*, 1998), high Bcl-2 level is considered as an important reason for chemoresistance and transfection of Bcl-2 or Bcl-XL into NB cells conferred drug resistance (Dole *et al.*, 1994; Dole *et al.*, 1995).

5.2.3. Regulation of apoptosis-related genes by MYCN

Although the *Myc* gene was originally identified as an oncogene, it is involved in various cellular processes, including cell growth, proliferation, loss of differentiation, and apoptosis (Adhikary and Eilers, 2005; Pelengaris *et al.*, 2002; Thompson, 1998). N-myc has been found to sensitize NB cells to death receptor induced apoptosis in the absence of cytokines, growth factors or other conditions of stress (Cui *et al.*, 2005; Fulda *et al.*, 1999; Lutz *et al.*, 1998). Cell death in response to MYCN results in caspase-8 cleavage which can be blocked using the caspase inhibitor zVAD-fmk (Cui *et al.*, 2005; Fulda *et al.*, 1999). Upregulation of p53, a direct target of MYCN, is an important mechanism of sensitization of the cell to apoptosis by N-myc (Chen *et al.*, 2010). These data suggest that resistance of NB cells with MYCN amplification to chemotherapy requires additional dysfunction in apoptotic pathways, such as silencing of caspase-8 or inhibiting the p53 pathway. Currently, there are conflicting data on a possible direct relationship between N-myc and caspase-8. Genome wide studies looking for E-box binding sites have identified caspase-8 as a target of both N-myc and c-myc (Fernandez *et al.*, 2003; Perini *et al.*, 2005). Perini and coworkers further demonstrated that the E-box sites in the caspase-8 promoter are methylated in NB preventing binding of N-myc/Max heterodimers. Furthermore, when these NB cells were treated with the demethylating agent 5-azacytidine, N-myc/Max binding was restored and caspase-8 expression was increased (Perini *et al.*, 2005). In contrast to this data which suggests a direct relationship, Fulda and coworkers failed to observe any changes in caspase-8 expression upon overexpression of N-myc or after downregulation of N-myc with antisense RNA; however, since these authors did not examine the methylation state of the E-box sequences in caspase-8, this study is somewhat difficult to interpret (Fulda *et al.*, 1999). In addition to data suggesting a possible direct relationship between N-myc and caspase-8, several studies have reported that myc expression promotes activation or priming of the mitochondrial pathway which increases the sensitivity of cells to death receptor signaling. Once the pathway is activated, caspase-8 cleavage of Bid provides a further amplification loop (Klefstrom *et al.*, 2002; Nieminen *et al.*, 2007). Additionally others have shown that the death receptor DR5 contains two E-box sequences that bind N-myc resulting in increased expression of this receptor and further augmentation of the extrinsic pathway (Cui *et al.*, 2005). Some proapoptotic Bcl-2 family members such as Bax and puma are transactivated by MYC, which play an important role in switching p53 downstream effects from G1 arrest to apoptosis (Seoane *et al.*, 2002).

Although p53 is a transcriptional target of N-myc, N-myc also inhibits its function through increased levels of p53 negative regulators, such as MDM2 and TWIST. MDM2, the essential inhibitor of p53, is a direct target of N-myc and inhibition of N-myc results in decreased MDM2, stabilized p53, and apoptosis. TWIST, a transcription factor of bHLH family with antiapoptotic activity, is frequently overexpressed in MYCN amplified NB (Valsesia-Wittmann *et al.*, 2004). Furthermore, TWIST overexpression inhibits ARF/p53 preventing apoptosis in response to N-myc overexpression (Valsesia-Wittmann *et al.*, 2004).

6. The Role of Epithelial to Mesenchymal Transition (EMT) in Development and Metastasis

6.1. EMT in development

Although greater than 50% of all NB patients present with metastatic disease, little is known of the process and the mechanism. Microarray mRNA expression analysis studies comparing highly metastatic human NBs (stage 4) to nonmetastatic human tumors (stage 1 and 2) have provided information on some of the proteins that are involved in NB metastasis (Scaruffi *et al.*, 2005). Of note, transcripts encoding proteins related to the developmental program of the epithelial to mesenchymal transition or EMT, are expressed at higher levels in metastatic human NB (Scaruffi *et al.*, 2005). As mentioned above, early in development, as the neural folds join to form the neural tube, the epidermal ectoderm expresses BMP signals and the floorplate of the neural tube produces Shh. These gradients interact and result in the formation of the neural crest at the dorsal aspect of the neural tube, between the epidermal ectoderm and the dorsal neural tube. The epithelial-like cells of the former neural folds undergo the EMT program. EMT is characterized by: (1) lose of epithelial morphology, (2) downregulation of junctional complexes (E-cadherin, cytokeratin, occludins, and claudins), (3) upregulation of intracellular migratory proteins (RhoB), (4) increased expression of matrix modulators (collagenase, matrilysin, urokinase, heparanase, matrix metalloproteinases—MMP), and (5) upregulation of matrix recognition molecules (N-cadherin; Thiery *et al.*, 2009). Metastatic cells exhibit many of these features and evidence is accumulating that several regulators of EMT are misregulated in NB (Shimono *et al.*, 2000; Valsesia-Wittmann *et al.*, 2004; Vitali *et al.*, 2008).

The transcription factors Snail, Twist, and SIP1/ZEB2 are considered to be master regulators of the EMT transition. Snail is induced primarily by TGFB signaling and directly represses E-cadherin by binding of the E-boxes in the E-cadherin promoter (Kang and Massague, 2004). The repression of E-cadherin is a hallmark of the induction of the EMT program, allowing for the motility of the crest cells. Similarly Twist, a basic helix-loop-helix transcription factor developmentally regulated via the NF-κB pathway, represses E-cadherin expression via the E-boxes within the E-cadherin promoter, although this interaction has not been shown to be direct (Sosic and Olson, 2003). Snail also induces the expression of ZEB factors (inhibitors of E-cadherin expression) which are involved in an EMT regulation loop with miRNA-200 family (Bracken *et al.*, 2008). The miRNA-200 family inhibits the expression of the ZEB family that relieves repression of E-cadherin (Bracken *et al.*, 2008).

6.2. Metastasis-related genes

Although greater than 50% of all NB patients present with metastatic disease, little is known of the process and the mechanism. The data from our group and collaborators first indicated that loss of caspase-8 increases metastasis by blocking integrin-mediated death. Loss of caspase-8 facilitates survival in foreign environments both during development and during metastasis. This data may also explain the observation that at least 50% of NB patients present with metastatic disease (Lahti *et al.*, 2006; Stupack *et al.*, 2006; Teitz *et al.*, 2006).

Another protein that is involved in NB metastasis is CD44, a cell surface glycoprotein that plays a role in cell adhesion and metastasis. CD44 is highly expressed in colon tumors where it affects tumor invasion, however, CD44 expression variable in NB tumors (Shtivelman and Bishop, 1991; Tanabe *et al.*, 1993). There are conflicting data about the correlation between CD44 expression and MYCN amplification, however, high expression of CD44 always correlates with favorable tumors and with the presence of more differentiated cells (Combaret *et al.*, 1996; Munchar *et al.*, 2003). The observation that CD44 is only expressed in favorable tumors is consistent with its functions as a metastasis inhibitor (Munchar *et al.*, 2003; Shtivelman and Bishop, 1991).

The Nm23A protein (nucleoside diphosphate kinase A) encoded by *NM23-H1* gene mapped to 17q22 locus is highly expressed in unfavorable NB tumors. A point mutation (S120G) was identified in a subset of NB tumors with elevated NM23A expression (Chang *et al.*, 1996). In addition, NM23A promotes NB metastasis in the xenograft NB animal model and S120G mutation reduced cell adhesion and increased cell migration

(Almgren *et al.*, 2004). In contrast, NM23A expression is very low in melanoma, breast, colon tumors and NM23A appears to serve as metastasis suppressor in these tumors (Bown *et al.*, 1999). The reason for the opposite expression pattern may be due to the different regulatory mechanisms in specific tissues (Okabe-Kado *et al.*, 2005).

The matrix metalloproteinases (MMPs), such as MMP-2 and MMP-9, are highly expressed in advanced-stage NB tumors. MMP-2 and MMP-9 promote cell invasion and metastasis by degrading extracellular matrix including type IV, V, VII, and X collagens as well as fibronectin. However, there is no correlation between these MMPs and MYCN status (Ribatti *et al.*, 2004; Sugiura *et al.*, 1998).

Twist1, which is originally identified as a key inducer of mesoderm formation in Drosophila, is found to play an essential role during metastasis in many types of tumors. First evidence of Twist1 involvement in metastasis came from human breast cancers. High levels of Twist1 are associated with invasive carcinoma and loss of EMT. Suppression of Twist1 inhibits the ability of cancer cells to metastasize, while overexpression of Twist1 increases cell motility and decreases cell adhesion (Yang *et al.*, 2004). Further study demonstrated Twist1 may promote metastasis through direct induction of microRNA-10b, which inhibits homeobox D10, resulting in upregulation of a well-characterized prometastatic gene, RHOC (Ma *et al.*, 2007). Later studies found that Twist1 also promotes migration and invasion in bladder and prostate cancer, hepatocellular carcinoma and colorectal cancer (Matsuo *et al.*, 2009; McConkey *et al.*, 2009; Valdes-Mora and Gomez Test, 2009). Twist1 is also overexpressed in N-myc amplified NB, where it inhibits N-myc induced apoptosis by inhibiting the p53 pathway in part via downregulation of $p19^{ARF}$ (Valsesia-Wittmann *et al.*, 2004).

7. The Role of miRNA in Development and NB

miRNAs are endogenous small noncoding RNAs of ~22 nucleotides in length that negatively regulate gene expression by mRNA cleavage or translational repression of the target mRNA (Bartel, 2004). miRNAs play an important role in regulating most cellular processes, and contribute to the process of tumorigenesis and metastasis (Zhang *et al.*, 2010). miRNA expression profiles have been correlated with prognosis, differentiation, and apoptosis in NB tumors (Chen and Stallings, 2007), suggesting that miRNAs could function as TSGs or oncogenes in NB. As mentioned above, miR-34a, which is located at 1p36, is a good TSG candidate (Welch *et al.*, 2007). Mir-34a is a direct target of p53 and knockdown of mir-34a reduced p53-dependent apoptosis (He *et al.*, 2007). In addition, one study found that N-myc is a direct target of mir-34a (Wei *et al.*, 2008), although this data is contrary to a previous study (Welch *et al.*, 2007).

The discrepancy between these two studies may reflect the slightly different systems that were used for these two studies.

Recent data showed that seven miRNAs including miR-17 to -92 cluster were induced by N-myc *in vitro*, and their high expression correlates with MYCN amplification in tumors (Schulte *et al.*, 2008), suggesting this cluster is a potential oncogene in NB. This cluster, which is transcribed as a polycistronic transcript containing miR-17, -18, -19, -20, and -92, was first found to be a target of c-myc, a potential oncogene in B-cell lymphoma (He *et al.*, 2005). The miR-17 to -92 cluster is also overexpressed in lung cancers where it has been shown to enhance cell proliferation (Hayashita *et al.*, 2005). miRNAs also mediate tumor metastasis, for example, miR-10b was shown to initiate tumor invasion and metastasis in breast cancer (Ma *et al.*, 2007).

One very recent paper compared miRNA expression patterns between primary and metastatic NB tumors and found significant changes of 54 miRNAs in metastatic samples, among which 35 miRNAs were upregulated and 19 miRNAs were downregulated (Guo *et al.*, 2010). Some miRNAs, such as miR-10b, miR-29a/b, miR-335, which are known to promote metastasis were upregulated in metastatic NB tumors. In contrast, miR-7, miR-338-3p, and the let-7 family were the three of the top 10 downregulated miRNAs in the metastatic group. These miRNAs have been shown to play antimetastatic roles in other tumors (Zhang *et al.*, 2010). The authors also analyzed the predicted miRNA targets of these miRNAs and found that many of these targets are related to metastasis. For instance, both caspase-8 and integrin beta1 are the predicted targets of miR-29a and miR-29b, miRNAs that promote metastasis in breast cancer (Gebeshuber *et al.*, 2009). However, the actual function of these identified miRNAs in metastasis needs to be investigated.

8. Other Important Genes in NB

8.1. Telomerase

Telomerase is a specialized ribonucleoprotein polymerase that synthesizes the TTAGGG telomeric repeats found at the end of chromosomes to maintain the length of the telomere. This enzyme is expressed in germ line cells but not in the majority of somatic cells. Thus, telomeres in somatic cells undergo progressive shortening and eventually lose the ability to protect chromosome ends, resulting in cell senescence and/or death. Increased telomerase expression, which results in unlimited cell replication and repression of cell senescence, is found in many tumors (Bodnar *et al.*, 1998; Hahn *et al.*, 1999). Telomerase activity was detected in most NBs, but not in ganglioneuromas or normal adrenal tissue (Hiyama *et al.*, 1997). In addition, high telomerase activity usually correlated with MYCN amplification and poor outcome, suggesting telomerase could be a prognostic marker for poor survival (Hiyama *et al.*, 1997; Ohali *et al.*, 2006; Reynolds *et al.*, 1997).

8.2. MDR1 and MRP gene family

Most NBs exhibit a strong initial response to chemotherapy followed by the appearance of multidrug resistance (MDR) in more than half of cases with remission/remaining tumor (Keshelava *et al.*, 2001; Tweddle *et al.*, 2003; Xue *et al.*, 2007). Although p53 seems to have an important role, the MDR1 gene (multidrug resistance gene 1) and the MRP (multidrug transporter MDR-associated protein) gene family also contribute to this resistance. MDR1 expression was found to be increased after treatment of NB tumors (Bourhis *et al.*, 1989; Chan *et al.*, 1991). In one report, MDR1 expression was inversely correlated with MYCN amplification and poor outcome (Chan *et al.*, 1991), while other groups did not find this correlation (Dhooge *et al.*, 1997; Kutlik *et al.*, 2002). Likewise, some studies found that expression of MRP gene family members such as MRP1, MRP4, strongly correlated with MYCN expression and chemoresistance in NBs (Haber *et al.*, 2006; Norris *et al.*, 1996, 2005), while other studies failed to find a correlation (De Cremoux *et al.*, 2007; Goto *et al.*, 2000).

8.3. GD2 and Bmi-1

Several other genes are also abnormally expressed in NB tumors such as GD2 and Bmi-1. The ganglioside GD2 is a glycolipid that is most commonly expressed in the majority of NB tumors, thus representing a good diagnostic marker and therapeutic target. Indeed recent studies using an anti-GD2 antibodies in association with GMC-SF and IL2 and RA results in a significant increase in event free survival (2-year estimates of 66% ± 5% vs. 46% ± 5%), and preliminary data suggest in overall survival (86% ± 4% vs. 75% ± 5% at 2 years $p = 0.0223$) (Yu *et al.*, 2009). Bmi-1, a polycomb ring finger gene, was found to be strongly expressed in primary NB tumors in 2006 and very recent data indicated that it is a *MYCN* target gene that promotes tumorigenesis through inhibition of two potential TSGs (KIF1Bβ and TSCL1) in NB cells (Nowak *et al.*, 2006; Ochiai *et al.*, 2010).

9. Clinical Treatment Overview

Current treatment for NB consists of surgery, chemotherapy, radiation, and biotherapy. The clinical strategy usually depends on a patient's risk stratification (Table 4.1). For examples, exposure to chemotherapy is generally limited for low risk group patients, whereas, high-risk group patients are treated with multiagent chemotherapy to reduce the overall burden of the disease before the surgical removal of the primary tumor (Haase *et al.*, 1999; Park *et al.*, 2008).

9.1. Low risk group and intermediate risk group

Low risk group encompasses diseases at stages 1, 2, and 4S with favorable characteristics (Brodeur, 2003; Castleberry *et al.*, 1997; Matthay, 1995). Most low risk group patients at stages 1 or 2 are successfully treated with surgery alone and complete resection of the tumor is the goal. The chances of these tumors recurring or progressing to advanced-stage NB are very low and chemotherapy is reserved for those patients with recurrences. NBs at stage 4S without MYCN amplification almost always spontaneously regresses and these patients can be safely observed without any treatment. The survival rate of patients of this group is greater than 95% (Alvarado *et al.*, 2000; Park *et al.*, 2008; Simon *et al.*, 2004).

Intermediate risk group patients include stage 3 patients of any age with favorable features, stage 4 infants with favorable features, and 4S patients without MYCN amplification and with unfavorable histology (Brodeur, 2003; Castleberry *et al.*, 1997; Matthay, 1995). For intermediate risk group patients, surgery, and moderate dose of multiagent chemotherapy are the basic therapeutic strategies. Sometimes radiation is also used to remove the residual tumors. Aggressive chemotherapy is utilized for patients who respond poorly to initial treatment or experience recurrence. Current treatments for intermediate risk group patients have a cure rate about 70–90% (Matthay *et al.*, 1998; Park *et al.*, 2008; Schmidt *et al.*, 2000).

9.2. High-risk group patients

High-risk group consist of patients with unfavorable biological features at stages 2, 3, 4, and those with MYCN amplification at stage 4S (Brodeur, 2003; Castleberry *et al.*, 1997; Matthay, 1995). The standard clinical strategy for this group is comprised of 4 steps: initial induction chemotherapy, local control, consolidation, and biology therapy (Haase *et al.*, 1999; Park *et al.*, 2008). Initial induction chemotherapy consists of combinations of chemotherapeutic agents such as cisplatin, etoposide, doxorubicin, cyclophosphamide, topectan/ironotecan, and vincristine. Following the completion of induction therapy, local control is achieved by aggressive surgical resection of the primary tumor followed by external beam radiation. After that, consolidation is provided by high-dose chemotherapy and autologous hematopoietic stem cell rescue, using stem cells that are prepared during the induction therapy. The purpose of consolidation therapy is to eliminate any remaining tumors. Consolidation therapy typically employs agents such as carboplatin, etoposide, and melphalan. Sometime focal radiotherapy is also applied to the primary tumor sites. After recovery from the consolidation, patients receive biological therapy to eradicate minimal residual disease because relapse is a frequent occurrence after autologous transplantation in this group of patients. The most common used biological agent is *cis*-retinoic acid (*cis*-RA), which is a noncytotoxic differentiation inducer. The use of monoclonal antibodies against tumor-specific antigens, such as GD2, provides an alternative and promising therapy to eliminate minimal residual tumor cells as described above (Matthay *et al.*, 1999; Pearson *et al.*, 2008; Zage *et al.*, 2008).

Even with these complicated and aggressive treatments, the overall cure rate for high-risk patients is only about 30% during the last two decades (Matthay *et al.*, 1999; Pearson *et al.*, 2008; Zage *et al.*, 2008). Therefore, the development of new agents and methods to more effectively treat these patients is underway. These therapeutic approaches include immune-therapy aimed at NB-specific antigens, targeted delivery of radioactive molecules to NB, or new chemotherapy such as retinoids for inducing differentiation, tyrosine kinase inhibitors targeted at ALK, demethylating agents and HDAC inhibitors for inducing reexpression of caspase-8 or other epigenetically silenced genes (Kelleher and McDermott, 2010; Schor, 2009; Wagner and Danks, 2009; Witt *et al.*, 2009).

10. Conclusion

The investigation and identification of genomic abnormalities and gene expression changes has improved the understanding of the molecular basis of biological and clinical characteristics of NBs (summary in Tables 4.2 and 4.3). Although great progress has been made in recent 20 years, much work needs to be done to identify tumor-specific targets for therapy. Also, deeper understanding of the development of normal sympathetic nervous system will help us find the important abnormal events that initiate NB development. In conclusion, more precise identification of molecular alterations should allow more effective and less toxic therapies with improved cure rate.

ACKNOWLEDGMENTS

We apologize to all of our colleagues whose work was either not cited or was cited in a review article. We thank the members of the Lahti lab, especially Judith Hyle, for their input. This work was funded by NIH grants R01 CA067938 to JML, Comprehensive Cancer Center Support Grant P01 CA021765 and the American Syrian Lebanese Associated Charities ALSAC.

Table 4.3 Summary of genetic changes in neuroblastoma

Gene symbol	Gene name	Chromosomal locus	Gene function	Gene alterations	References
ALK	Anaplastic lymphoma receptor tyrosine kinase	2p23	Receptor tyrosine kinase	Mutation/ amplification	Caren *et al.* (2008), Chen *et al.* (2008), George *et al.* (2008), Janoueix-Lerosey *et al.* (2008), Mosse *et al.* (2008), Passoni *et al.* (2009)
Bcl2	B-cell CLL/lymphoma 2	18q21.3	Apoptosis suppression	High expression	Abel *et al.* (2005), Ramani and Lu (1994), Castle *et al.* (1993), Ikegaki *et al.* (1995), Mejia *et al.* (1998)
Bmi1	BMI1 polycomb ring finger oncogene	10p11.23	Oncogene	Overexpression	Ochiai *et al.* (2010), Nowak *et al.* (2006)
Casp8	Caspase 8	2q33–q34	Apoptosis/ metastasis	Silence/deletion	Teitz *et al.* (2000, 2006), Yang *et al.* (2007), Fulda *et al.* (2001), Fulda and Debatin (2006), Lahti *et al.* (2006), Stupack *et al.* (2006)
CD44	CD44 molecule	11p13	Integrin/ antimetastasis	Expression in favorable tumors	Munchar *et al.* (2003), Combaret *et al.* (1996)
CHD5	Chromodomain helicase DNA-binding protein 5	1p36.31	Helicase/tumor suppressor	Deletion/low expression	Bagchi *et al.* (2007)
DDX1	ATP-dependent RNA helicase DDX1	2p24	RNA helicase/ Oncogene	Amplification	George *et al.* (1996)
B4GALNT1	GM2/GD2 synthase	12q13.3	Ganglioside	High expression	Yu *et al.* (2009)

(continued)

Table 4.3 *(continued)*

Gene symbol	Gene name	Chromosomal locus	Gene function	Gene alterations	References
KIF1b	Kinesin family member 1beta	1p36.2	Kinesin/tumor suppressor	Deletion/low expression	Munirajan *et al.* (2008), Schlisio *et al.* (2008)
MDM2	Mdm2 p53-binding protein homolog	12q14.3–q15	Oncogene	Amplification	Corvi *et al.* (1995a)
ABCB1	Multidrug resistance protein 1 (MRP1)	7q21.12	Multidrug resistance	High expression	Bourhis *et al.* (1989), Chan *et al.* (1991), Dhooge *et al.* (1997), Kutlik *et al.* (2002)
Mir34a	MicroRNA 34a	1p36.22	Micro RNA/ tumor suppressor	Deletion/low expression	Welch *et al.* (2007), Chen and Stallings (2007)
Mir17-92	MicroRNA cluster 17-92 (mir17, 18, 19, 20, 92)	13q31.3	Micro RNA/ Oncogene	High expression	Schulte *et al.* (2008)
MMP2, MMP	Matrix metallopeptidase 2	16q13–q21	Proteinase/ metastasis	High expression	Sugiura *et al.* (1998), Ribatti *et al.* (2004)
MMP9	Matrix metallopeptidase 9	20q11.2–q13.1	Proteinase/ metastasis	High expression	(Sugiura *et al.* (1998), Ribatti *et al.* (2004))
ABCC1	Multidrug resistance-associated protein 1	16p13.1	Multidrug resistance	High expression	Norris *et al.* (1996, 2005), Haber *et al.* (2006), Goto *et al.* (2000), De Cremoux *et al.* (2007)
MYCL	Myc-related gene from lung cancer	1p34.2	Oncogene	Amplification	Jinbo *et al.* (1989)
MYCN	Neuroblastoma MYC oncogene	2p24.1	Oncogene	Amplification	Schwab *et al.* (1983, 1984), Kohl *et al.* (1983), Pelengaris *et al.* (2002), Reiter and Brodeur (1996, 1998)

NME1	Metastasis inhibition factor NM23	17q22	Nucleoside kinase/ antimetastasis	Overexpression	Godfried *et al.* (2002), Almgren *et al.* (2004), Chang *et al.* (1996)
PHOX2B	Paired-like homeobox 2b	4p13	Neuron development	Mutation	Trochet *et al.* (2004), Mosse *et al.* (2004)
TERT	Telomerase reverse transcriptase	5p15.33	Telomere maintenance	High expression	Hiyama *et al.* (1997), Reynolds *et al.* (1997), Ohali *et al.* (2006)
NTRK1	Neurotrophic tyrosine kinase receptor 1 (TrKA)	1q21–q22	Receptor tyrosine kinase	Inverse correlation with Mycn	Nakagawara *et al.* (1992), Nakagawara (1993), Tacconelli *et al.* (2004), Kim *et al.* (1999)
NTRK2	Neurotrophic tyrosine kinase receptor 2 (TrKB)	9q22.1	Receptor tyrosine kinase	Strong correlation with Mycn	Nakagawara *et al.* (1994), Nakagawara (1994), Douma *et al.* (2004), Ho *et al.* (2002)
NTRK3	Neurotrophic tyrosine kinase receptor 3 (TrKC)	15q25	Receptor tyrosine kinase	Coexpression with TrkA	Yamashiro *et al.* (1996), Svensson *et al.* (1997)
CADM1	Tumor suppressor in lung cancer 1 (TSLC1)	11q23.2	Cell adhesion/ Tumor suppressor	Deletion/low expression	Gomyo *et al.* (1999), Ando *et al.* (2008)
TWIST1	Twist homolog 1	7p21.2	Apoptosis/ metastasis	High expression	Valsesia-Wittmann *et al.* (2004)

REFERENCES

Abe, M., Ohira, M., Kaneda, A., Yagi, Y., Yamamoto, S., Kitano, Y., Takato, T., Nakagawara, A., and Ushijima, T. (2005). CpG island methylator phenotype is a strong determinant of poor prognosis in neuroblastomas. *Cancer Res.* **65,** 828–834.

Abe, M., Westermann, F., Nakagawara, A., Takato, T., Schwab, M., and Ushijima, T. (2007). Marked and independent prognostic significance of the CpG island methylator phenotype in neuroblastomas. *Cancer Lett.* **247,** 253–258.

Abel, F., Sjoberg, R. M., Nilsson, S., Kogner, P., and Martinsson, T. (2005). Imbalance of the mitochondrial pro- and anti-apoptotic mediators in neuroblastoma tumours with unfavourable biology. *Eur. J. Cancer* **41,** 635–646.

Adhikary, S., and Eilers, M. (2005). Transcriptional regulation and transformation by Myc proteins. *Nat. Rev. Mol. Cell Biol.* **6,** 635–645.

Alenina, N., Bashammakh, S., and Bader, M. (2006). Specification and differentiation of serotonergic neurons. *Stem Cell Rev.* **2,** 5–10.

Almgren, M. A., Henriksson, K. C., Fujimoto, J., and Chang, C. L. (2004). Nucleoside diphosphate kinase A/nm23-H1 promotes metastasis of NB69-derived human neuroblastoma. *Mol. Cancer Res.* **2,** 387–394.

Alvarado, C. S., London, W. B., Look, A. T., Brodeur, G. M., Altmiller, D. H., Thorner, P. S., Joshi, V. V., Rowe, S. T., Nash, M. B., Smith, E. I., Castleberry, R. P., and Cohn, S. L. (2000). Natural history and biology of stage A neuroblastoma: A Pediatric Oncology Group Study. *J. Pediatr. Hematol. Oncol.* **22,** 197–205.

Anderson, D. J. (1993). Molecular control of cell fate in the neural crest: The sympathoadrenal lineage. *Annu. Rev. Neurosci.* **16,** 129–158.

Anderson, D. J., and Axel, R. (1986). A bipotential neuroendocrine precursor whose choice of cell fate is determined by NGF and glucocorticoids. *Cell* **47,** 1079–1090.

Anderson, D. J., Carnahan, J. F., Michelsohn, A., and Patterson, P. H. (1991). Antibody markers identify a common progenitor to sympathetic neurons and chromaffin cells in vivo and reveal the timing of commitment to neuronal differentiation in the sympathoadrenal lineage. *J. Neurosci.* **11,** 3507–3519.

Ando, K., Ohira, M., Ozaki, T., Nakagawa, A., Akazawa, K., Suenaga, Y., Nakamura, Y., Koda, T., Kamijo, T., Murakami, Y., and Nakagawara, A. (2008). Expression of TSLC1, a candidate tumor suppressor gene mapped to chromosome 11q23, is downregulated in unfavorable neuroblastoma without promoter hypermethylation. *Int. J. Cancer* **123,** 2087–2094.

Astuti, D., Agathanggelou, A., Honorio, S., Dallol, A., Martinsson, T., Kogner, P., Cummins, C., Neumann, H. P., Voutilainen, R., Dahia, P., Eng, C., Maher, E. R., *et al.* (2001). RASSF1A promoter region CpG island hypermethylation in phaeochromocytomas and neuroblastoma tumours. *Oncogene* **20,** 7573–7577.

Bader, S. A., Fasching, C., Brodeur, G. M., and Stanbridge, E. J. (1991). Dissociation of suppression of tumorigenicity and differentiation in vitro effected by transfer of single human chromosomes into human neuroblastoma cells. *Cell Growth Differ.* **2,** 245–255.

Bagchi, A., Papazoglu, C., Wu, Y., Capurso, D., Brodt, M., Francis, D., Bredel, M., Vogel, H., and Mills, A. A. (2007). CHD5 is a tumor suppressor at human 1p36. *Cell* **128,** 459–475.

Banelli, B., Casciano, I., Croce, M., di Vinci, A., Gelvi, I., Pagnan, G., Brignole, C., Allemanni, G., Ferrini, S., Ponzoni, M., and Romani, M. (2002). Expression and methylation of CASP8 in neuroblastoma: Identification of a promoter region. *Nat. Med.* **8,** 1333–1335.

Banerjee, S. A., Hoppe, P., Brilliant, M., and Chikaraishi, D. M. (1992). 5' flanking sequences of the rat tyrosine hydroxylase gene target accurate tissue-specific, developmental, and transsynaptic expression in transgenic mice. *J. Neurosci.* **12,** 4460–4467.

Bartel, D. P. (2004). MicroRNAs: Genomics, biogenesis, mechanism, and function. *Cell* **116,** 281–297.

Beckwith, J. B., and Perrin, E. V. (1963). In situ neuroblastomas: A contribution to the natural history of neural crest tumors. *Am. J. Pathol.* **43,** 1089–1104.

Beierle, E. A., Nagaram, A., Dai, W., Iyengar, M., and Chen, M. K. (2005). VEGF-mediated survivin expression in neuroblastoma cells. *J. Surg. Res.* **127,** 21–28.

Benailly, H. K., Lapierre, J. M., Laudier, B., Amiel, J., Attie, T., De Blois, M. C., Vekemans, M., and Romana, S. P. (2003). PMX2B, a new candidate gene for Hirschsprung's disease. *Clin. Genet.* **64,** 204–209.

Betters, E., Liu, Y., Kjaeldgaard, A., and Sundstrom, E.Garcia-Castro, M. (2010). Analysis of early human neural crest development. *Dev. Biol.*

Blaschke, A. J., Weiner, J. A., and Chun, J. (1998). Programmed cell death is a universal feature of embryonic and postnatal neuroproliferative regions throughout the central nervous system. *J. Comp. Neurol.* **396,** 39–50.

Boatright, K. M., and Salvesen, G. S. (2003). Mechanisms of caspase activation. *Curr. Opin. Cell Biol.* **15,** 725–731.

Boatright, K. M., Renatus, M., Scott, F. L., Sperandio, S., Shin, H., Pedersen, I. M., Ricci, J. E., Edris, W. A., Sutherlin, D. P., Green, D. R., and Salvesen, G. S. (2003). A unified model for apical caspase activation. *Mol. Cell* **11,** 529–541.

Bodnar, A. G., Ouellette, M., Frolkis, M., Holt, S. E., Chiu, C. P., Morin, G. B., Harley, C. B., Shay, J. W., Lichtsteiner, S., and Wright, W. E. (1998). Extension of life-span by introduction of telomerase into normal human cells. *Science* **279,** 349–352.

Bourdeaut, F., Trochet, D., Janoueix-Lerosey, I., Ribeiro, A., Deville, A., Coz, C., Michiels, J. F., Lyonnet, S., Amiel, J., and Delattre, O. (2005). Germline mutations of the paired-like homeobox 2B (PHOX2B) gene in neuroblastoma. *Cancer Lett.* **228,** 51–58.

Bourhis, J., Benard, J., Hartmann, O., Boccon-Gibod, L., Lemerle, J., and Riou, G. (1989). Correlation of MDR1 gene expression with chemotherapy in neuroblastoma. *J. Natl. Cancer Inst.* **81,** 1401–1405.

Bown, N., Cotterill, S., Lastowska, M., O'Neill, S., Pearson, A. D., Plantaz, D., Meddeb, M., Danglot, G., Brinkschmidt, C., Christiansen, H., Laureys, G., Speleman, F., *et al.* (1999). Gain of chromosome arm 17q and adverse outcome in patients with neuroblastoma. *N Engl J. Med.* **340,** 1954–1961.

Bracken, C. P., Gregory, P. A., Kolesnikoff, N., Bert, A. G., Wang, J., Shannon, M. F., and Goodall, G. J. (2008). A double-negative feedback loop between ZEB1-SIP1 and the microRNA-200 family regulates epithelial-mesenchymal transition. *Cancer Res.* **68,** 7846–7854.

Breit, S., and Schwab, M. (1989). Suppression of MYC by high expression of NMYC in human neuroblastoma cells. *J. Neurosci. Res.* **24,** 21–28.

Brodeur, G. M. (2003). Neuroblastoma: Biological insights into a clinical enigma. *Nat. Rev. Cancer* **3,** 203–216.

Brodeur, G. M., Minturn, J. E., Ho, R., Simpson, A. M., Iyer, R., Varela, C. R., Light, J. E., Kolla, V., and Evans, A. E. (2009). Trk receptor expression and inhibition in neuroblastomas. *Clin. Cancer Res.* **15,** 3244–3250.

Bronner-Fraser, M. (1993). Neural crest cell migration in the developing embryo. *Trends Cell Biol.* **3,** 392–397.

Bronner-Fraser, M., Sieber-Blum, M., and Cohen, A. M. (1980). Clonal analysis of the avian neural crest: Migration and maturation of mixed neural crest clones injected into host chicken embryos. *J. Comp. Neurol.* **193,** 423–434.

Caren, H., Abel, F., Kogner, P., and Martinsson, T. (2008). High incidence of DNA mutations and gene amplifications of the ALK gene in advanced sporadic neuroblastoma tumours. *Biochem. J.* **416,** 153–159.

Caron, H., van, S. P., Buschman, R., Pereira do, T. R., Maes, P., Beks, L., de, K. J., Voute, P. A., Vergnaud, G., Westerveld, A., Slater, R., and Versteeg, R. (1996). Allelic loss of the short arm of chromosome 4 in neuroblastoma suggests a novel tumour suppressor gene locus. *Hum. Genet.* **97,** 834–837.

Castle, V. P., Heidelberger, K. P., Bromberg, J., Ou, X., Dole, M., and Nunez, G. (1993). Expression of the apoptosis-suppressing protein bcl-2, in neuroblastoma is associated with unfavorable histology and N-myc amplification. *Am. J. Pathol.* **143,** 1543–1550.

Castleberry, R. P., Pritchard, J., Ambros, P., Berthold, F., Brodeur, G. M., Castel, V., Cohn, S. L., De, B. B., Cks-Mireaux, C., Frappaz, D., Haase, G. M., Haber, M., *et al.* (1997). The International Neuroblastoma Risk Groups (INRG): A preliminary repor. *Eur. J. Cancer* **33,** 2113–2116.

Chambery, D., Mohseni-Zadeh, S., de, G. B., and Babajko, S. (1999). N-myc regulation of type I insulin-like growth factor receptor in a human neuroblastoma cell line. *Cancer Res.* **59,** 2898–2902.

Chan, H. S., Haddad, G., Thorner, P. S., DeBoer, G., Lin, Y. P., Ondrusek, N., Yeger, H., and Ling, V. (1991). P-glycoprotein expression as a predictor of the outcome of therapy for neuroblastoma. *N Engl J. Med.* **325,** 1608–1614.

Chang, C. L., Strahler, J. R., Thoraval, D. H., Qian, M. G., Hinderer, R., and Hanash, S. M. (1996). A nucleoside diphosphate kinase A (nm23–H1) serine 120–> glycine substitution in advanced stage neuroblastoma affects enzyme stability and alters protein-protein interaction. *Oncogene* **12,** 659–667.

Charron, J., Malynn, B. A., Fisher, P., Stewart, V., Jeannotte, L., Goff, S. P., Robertson, E. J., and Alt, F. W. (1992). Embryonic lethality in mice homozygous for a targeted disruption of the N-myc gene. *Genes Dev.* **6,** 2248–2257.

Chen, Y., and Stallings, R. L. (2007). Differential patterns of microRNA expression in neuroblastoma are correlated with prognosis, differentiation, and apoptosis. *Cancer Res.* **67,** 976–983.

Chen, Y., Takita, J., Choi, Y. L., Kato, M., Ohira, M., Sanada, M., Wang, L., Soda, M., Kikuchi, A., Igarashi, T., Nakagawara, A., Hayashi, Y., *et al.* (2008). Oncogenic mutations of ALK kinase in neuroblastoma. *Nature* **455,** 971–974.

Chen, L., Iraci, N., Gherardi, S., Gamble, L. D., Wood, K. M., Perini, G., Lunec, J., and Tweddle, D. A. (2010). p53 is a direct transcriptional target of MYCN in neuroblastoma. *Cancer Res.* **70,** 1377–1388.

Cheng, M., and Ott, G. R. (2010). Anaplastic lymphoma kinase as a therapeutic target in anaplastic large cell lymphoma, non-small cell lung cancer and neuroblastoma. *Anticancer Agents Med. Chem.* **10,** 236–249.

Cohn, S. L., Pearson, A. D., London, W. B., Monclair, T., Ambros, P. F., Brodeur, G. M., Faldum, A., Hero, B., Iehara, T., Machin, D., Mosseri, V., Simon, T., *et al.* (2009). The International Neuroblastoma Risk Group (INRG) classification system: An INRG Task Force report. *J. Clin. Oncol.* **27,** 289–297.

Combaret, V., Gross, N., Lasset, C., Frappaz, D., Peruisseau, G., Philip, T., Beck, D., and Favrot, M. C. (1996). Clinical relevance of CD44 cell-surface expression and N-myc gene amplification in a multicentric analysis of 121 pediatric neuroblastomas. *J. Clin. Oncol.* **14,** 25–34.

Coppola, D., Ouban, A., and Gilbert-Barness, E. (2009). Expression of the insulin-like growth factor receptor 1 during human embryogenesis. *Fetal Pediatr. Pathol.* **28,** 47–54.

Corvi, R., Savelyeva, L., Amler, L., Handgretinger, R., and Schwab, M. (1995a). Cytogenetic evolution of MYCN and MDM2 amplification in the neuroblastoma LS tumour and its cell line. *Eur. J. Cancer* **31A,** 520–523.

Corvi, R., Savelyeva, L., and Schwab, M. (1995b). Duplication of N-MYC at its resident site 2p24 may be a mechanism of activation alternative to amplification in human neuroblastoma cells. *Cancer Res.* **55,** 3471–3474.

Coulter, D. W., Wilkie, M. B., and Moats-Staats, B. M. (2009). Inhibition of IGF-I receptor signaling in combination with rapamycin or temsirolimus increases MYC-N phosphorylation. *Anticancer Res.* **29,** 1943–1949.

Cui, H., Li, T., and Ding, H. F. (2005). Linking of N-Myc to death receptor machinery in neuroblastoma cells. *J. Biol. Chem.* **280,** 9474–9481.

Danial, N. N., and Korsmeyer, S. J. (2004). Cell death: Critical control points. *Cell* **116,** 205–219.

De Zio, D., Giunta, L., Corvaro, M., Ferraro, E., and Cecconi, F. (2005). Expanding roles of programmed cell death in mammalian neurodevelopment. *Semin. Cell Dev. Biol.* **16,** 281–294.

De Cremoux, P., Jourdan-Da-Silva, N., Couturier, J., Tran-Perennou, C., Schleiermacher, G., Fehlbaum, P., Doz, F., Mosseri, V., Delattre, O., Klijanienko, J., Vielh, P., and Michon, J. (2007). Role of chemotherapy resistance genes in outcome of neuroblastoma. *Pediatr. Blood Cancer* **48,** 311–317.

Degoutin, J., Brunet-de, C. N., Cifuentes-Diaz, C., and Vigny, M. (2009). ALK (Anaplastic Lymphoma Kinase) expression in DRG neurons and its involvement in neuron-Schwann cells interaction. *Eur. J. Neurosci.* **29,** 275–286.

D'Ercole, A. J., Ye, P., Calikoglu, A. S., and Gutierrez-Ospina, G. (1996). The role of the insulin-like growth factors in the central nervous system. *Mol. Neurobiol.* **13,** 227–255.

Dhooge, C. R., De Moerloose, B. M., Benoit, Y. C., and Van, R. N. (1997). Philippe, and G. G. Laureys, Expression of the MDR1 gene product P-glycoprotein in childhood neuroblastoma. *Cancer* **80,** 1250–1257.

Dole, M., Nunez, G., Merchant, A. K., Maybaum, J., Rode, C. K., Bloch, C. A., and Castle, V. P. (1994). Bcl-2 inhibits chemotherapy-induced apoptosis in neuroblastoma. *Cancer Res.* **54,** 3253–3259.

Dole, M. G., Jasty, R., Cooper, M. J., Thompson, C. B., Nunez, G., and Castle, V. P. (1995). Bcl-xL is expressed in neuroblastoma cells and modulates chemotherapy-induced apoptosis. *Cancer Res.* **55,** 2576–2582.

Douma, S., Van, L. T., Zevenhoven, J., Meuwissen, R., Van, G. E., and Peeper, D. S. (2004). Suppression of anoikis and induction of metastasis by the neurotrophic receptor TrkB. *Nature* **430,** 1034–1039.

Dupin, E., Calloni, G., Real, C., Goncalves-Trentin, A., and Le Douarin, N. M. (2007). Neural crest progenitors and stem cells. *C. R. Biol.* **330,** 521–529.

Ejeskar, K., Aburatani, H., Abrahamsson, J., Kogner, P., and Martinsson, T. (1998). Loss of heterozygosity of 3p markers in neuroblastoma tumours implicate a tumour-suppressor locus distal to the FHIT gene. *Br. J. Cancer* **77,** 1787–1791.

Emanuel, B. S., Balaban, G., Boyd, J. P., Grossman, A., Negishi, M., Parmiter, A., and Glick, M. C. (1985). N-myc amplification in multiple homogeneously staining regions in two human neuroblastomas. *Proc. Natl. Acad. Sci. USA* **82,** 3736–3740.

Erickson, C. A., and Turley, E. A. (1987). The effects of epidermal growth factor on neural crest cells in tissue culture. *Exp. Cell Res.* **169,** 267–279.

Ernsberger, U., Esposito, L., Partimo, S., Huber, K., Franke, A., Bixby, J. L., Kalcheim, C., and Unsicker, K. (2005). Expression of neuronal markers suggests heterogeneity of chick sympathoadrenal cells prior to invasion of the adrenal anlagen. *Cell Tissue Res.* **319,** 1–13.

Evans, A. E., D'Angio, G. J., and Randolph, J. (1971). A proposed staging for children with neuroblastoma. Children's cancer study group A. *Cancer* **27,** 374–378.

Fernandez, P. C., Frank, S. R., Wang, L., Schroeder, M., Liu, S., Greene, J., Cocito, A., and Amati, B. (2003). Genomic targets of the human c-Myc protein. *Genes Dev.* **17,** 1115–1129.

Fesik, S. W. (2005). Promoting apoptosis as a strategy for cancer drug discovery. *Nat. Rev. Cancer* **5,** 876–885.

Fontaine-Perus, J. C., Chanconie, M., and Le Douarin, N. M. (1982). Differentiation of peptidergic neurones in quail-chick chimaeric embryos. *Cell Differ.* **11,** 183–193.

Fulda, S., and Debatin, K. M. (2006). 5-Aza-2'-deoxycytidine and IFN-gamma cooperate to sensitize for TRAIL-induced apoptosis by upregulating caspase-8. *Oncogene* **25,** 5125–5133.

Fulda, S., Lutz, W., Schwab, M., and Debatin, K. M. (1999). MycN sensitizes neuroblastoma cells for drug-induced apoptosis. *Oncogene* **18,** 1479–1486.

Fulda, S., Kufer, M. U., Meyer, E., van Valen, F., Dockhorn-Dworniczak, B., and Debatin, K. M. (2001). Sensitization for death receptor- or drug-induced apoptosis by re-expression of caspase-8 through demethylation or gene transfer. *Oncogene* **20,** 5865–5877.

Gebeshuber, C. A., Zatloukal, K., and Martinez, J. (2009). miR-29a suppresses tristetraprolin, which is a regulator of epithelial polarity and metastasis. *EMBO Rep.* **10,** 400–405.

George, R. E., Kenyon, R. M., McGuckin, A. G., Malcolm, A. J., Pearson, A. D., and Lunec, J. (1996). Investigation of co-amplification of the candidate genes ornithine decarboxylase, ribonucleotide reductase, syndecan-1 and a DEAD box gene, DDX1, with N-myc in neuroblastoma. United Kingdom Children's Cancer Study Group. *Oncogene* **12,** 1583–1587.

George, R. E., Sanda, T., Hanna, M., Frohling, S., Luther, W., Zhang, J., Ahn, Y., Zhou, W., London, W. B., McGrady, P., Xue, L., Zozulya, S., *et al.* (2008). Activating mutations in ALK provide a therapeutic target in neuroblastoma. *Nature* **455,** 975–978.

Gilbert, F., Feder, M., Balaban, G., Brangman, D., Lurie, D. K., Podolsky, R., Rinaldt, V., Vinikoor, N., and Weisband, J. (1984). Human neuroblastomas and abnormalities of chromosomes 1 and 17. *Cancer Res.* **44,** 5444–5449.

Godfried, M. B., Veenstra, M., Sluis, P., Boon, K., Asperen, R., Hermus, M. C., Schaik, B. D., Voute, T. P., Schwab, M., Versteeg, R., and Caron, H. N. (2002). The N-myc and c-myc downstream pathways include the chromosome 17q genes nm23-H1 and nm23-H2. *Oncogene* **21,** 2097–2101.

Gomyo, H., Arai, Y., Tanigami, A., Murakami, Y., Hattori, M., Hosoda, F., Arai, K., Aikawa, Y., Tsuda, H., Hirohashi, S., Asakawa, S., Shimizu, N., *et al.* (1999). A 2-Mb sequence-ready contig map and a novel immunoglobulin superfamily gene IGSF4 in the LOH region of chromosome 11q23.2. *Genomics* **62,** 139–146.

Gonzalez-Gomez, P., Bello, M. J., Inda, M. M., Alonso, M. E., Arjona, D., Aminoso, C., Lopez-Marin, I., de Campos, J. M., Sarasa, J. L., Castresana, J. S., and Rey, J. A. (2004). Deletion and aberrant CpG island methylation of Caspase 8 gene in medulloblastoma. *Oncol. Rep.* **12,** 663–666.

Goto, H., Keshelava, N., Matthay, K. K., Lukens, J. N., Gerbing, R. B., Stram, D. O., Seeger, R. C., and Reynolds, C. P. (2000). Multidrug resistance-associated protein 1 (MRP1) expression in neuroblastoma cell lines and primary tumors. *Med. Pediatr. Oncol.* **35,** 619–622.

Green, D. R., and Kroemer, G. (2004). The pathophysiology of mitochondrial cell death. *Science* **305,** 626–629.

Grenet, J., Teitz, T., Wei, T., Valentine, V., and Kidd, V. J. (1999). Structure and chromosome localization of the human CASP8 gene. *Gene* **226,** 225–232.

Grimmer, M. R., and Weiss, W. A. (2006). Childhood tumors of the nervous system as disorders of normal development. *Curr. Opin. Pediatr.* **18,** 634–638.

Guo, C., White, P. S., Weiss, M. J., Hogarty, M. D., Thompson, P. M., Stram, D. O., Gerbing, R., Matthay, K. K., Seeger, R. C., Brodeur, G. M., and Maris, J. M. (1999). Allelic deletion at 11q23 is common in MYCN single copy neuroblastomas. *Oncogene* **18,** 4948–4957.

Guo, J., Dong, Q., Fang, Z., Chen, X., Lu, H., Wang, K., Yin, Y., Cai, X., Zhao, N., Chen, J., Zen, K., Zhang, J., *et al.* (2010). Identification of miRNAs that are associated with tumor metastasis in neuroblastoma. *Cancer Biol. Ther.* **9,** 446–452.

Haase, G. M., Perez, C., and Atkinson, J. B. (1999). Current aspects of biology, risk assessment, and treatment of neuroblastoma. *Semin. Surg. Oncol.* **16,** 91–104.

Haber, M., Bordow, S. B., Gilbert, J., Madafiglio, J., Kavallaris, M., Marshall, G. M., Mechetner, E. B., Fruehauf, J. P., Tee, L., Cohn, S. L., Salwen, H., Schmidt, M. L., *et al.* (1999). Altered expression of the MYCN oncogene modulates MRP gene expression and response to cytotoxic drugs in neuroblastoma cells. *Oncogene* **18,** 2777–2782.

Haber, M., Smith, J., Bordow, S. B., Flemming, C., Cohn, S. L., London, W. B., Marshall, G. M., and Norris, M. D. (2006). Association of high-level MRP1 expression with poor clinical outcome in a large prospective study of primary neuroblastoma. *J. Clin. Oncol.* **24,** 1546–1553.

Hahn, W. C., Stewart, S. A., Brooks, M. W., York, S. G., Eaton, E., Kurachi, A., Beijersbergen, R. L., Knoll, J. H., Meyerson, M., and Weinberg, R. A. (1999). Inhibition of telomerase limits the growth of human cancer cells. *Nat. Med.* **5,** 1164–1170.

Harada, K., Toyooka, S., Shivapurkar, N., Maitra, A., Reddy, J. L., Matta, H., Miyajima, K., Timmons, C. F., Tomlinson, G. E., Mastrangelo, D., Hay, R. J., Chaudhary, P. M., *et al.* (2002). Deregulation of caspase 8 and 10 expression in pediatric tumors and cell lines. *Cancer Res.* **62,** 5897–5901.

Harris, R. G., White, E., Phillips, E. S., and Lillycrop, K. A. (2002). The expression of the developmentally regulated proto-oncogene Pax-3 is modulated by N-Myc. *J. Biol. Chem.* **277,** 34815–34825.

Hatzi, E., Murphy, C., Zoephel, A., Ahorn, H., Tontsch, U., Bamberger, A. M., Yamauchi-Takihara, K., Schweigerer, L., and Fotsis, T. (2002). N-myc oncogene overexpression down-regulates leukemia inhibitory factor in neuroblastoma. *Eur. J. Biochem.* **269,** 3732–3741.

Hayashita, Y., Osada, H., Tatematsu, Y., Yamada, H., Yanagisawa, K., Tomida, S., Yatabe, Y., Kawahara, K., Sekido, Y., and Takahashi, T. (2005). A polycistronic microRNA cluster, miR-17-92, is overexpressed in human lung cancers and enhances cell proliferation. *Cancer Res.* **65,** 9628–9632.

He, L., Thomson, J. M., Hemann, M. T., Hernando-Monge, E., Mu, D., Goodson, S., Powers, S., Cordon-Cardo, C., Lowe, S. W., Hannon, G. J., and Hammond, S. M. (2005). A microRNA polycistron as a potential human oncogene. *Nature* **435,** 828–833.

He, L., He, X., Lim, L. P., de, S. E., Xuan, Z., Liang, Y., Xue, W., Zender, L., Magnus, J., Ridzon, D., Jackson, A. L., Linsley, P. S., *et al.* (2007). A microRNA component of the p53 tumour suppressor network. *Nature* **447,** 1130–1134.

Hengartner, M. O. (2000). The biochemistry of apoptosis. *Nature* **407,** 770–776.

Henson, E. S., and Gibson, S. B. (2006). Surviving cell death through epidermal growth factor (EGF) signal transduction pathways: Implications for cancer therapy. *Cell. Signal.* **18,** 2089–2097.

Hirai, M., Yoshida, S., Kashiwagi, H., Kawamura, T., Ishikawa, T., Kaneko, M., Ohkawa, H., Nakagawara, A., Miwa, M., and Uchida, K. (1999). 1q23 gain is associated with progressive neuroblastoma resistant to aggressive treatment. *Genes Chromosom. Cancer* **25,** 261–269.

Hiyama, E., Hiyama, K., Ohtsu, K., Yamaoka, H., Ichikawa, T., Shay, J. W., and Yokoyama, T. (1997). Telomerase activity in neuroblastoma: Is it a prognostic indicator of clinical behaviour? *Eur. J. Cancer* **33,** 1932–1936.

Ho, R., Eggert, A., Hishiki, T., Minturn, J. E., Ikegaki, N., Foster, P., Camoratto, A. M., Evans, A. E., and Brodeur, G. M. (2002). Resistance to chemotherapy mediated by TrkB in neuroblastomas. *Cancer Res.* **62,** 6462–6466.

Ho, R., Minturn, J. E., Hishiki, T., Zhao, H., Wang, Q., Cnaan, A., Maris, J., Evans, A. E., and Brodeur, G. M. (2005). Proliferation of human neuroblastomas mediated by the epidermal growth factor receptor. *Cancer Res.* **65,** 9868–9875.

Hopkins-Donaldson, S., Bodmer, J. L., Bourloud, K. B., Brognara, C. B., Tschopp, J., and Gross, N. (2000). Loss of caspase-8 expression in neuroblastoma is related to malignancy and resistance to TRAIL-induced apoptosis. *Med. Pediatr. Oncol.* **35,** 608–611.

Hopkins-Donaldson, S., Ziegler, A., Kurtz, S., Bigosch, C., Kandioler, D., Ludwig, C., Zangemeister-Wittke, U., and Stahel, R. (2003). Silencing of death receptor and caspase-8 expression in small cell lung carcinoma cell lines and tumors by DNA methylation. *Cell Death Differ.* **10,** 356–364.

Howard, M. J., Stanke, M., Schneider, C., Wu, X., and Rohrer, H. (2000). The transcription factor dHAND is a downstream effector of BMPs in sympathetic neuron specification. *Development* **127,** 4073–4081.

Huber, K. (2006). The sympathoadrenal cell lineage: Specification, diversification, and new perspectives. *Dev. Biol.* **298,** 335–343.

Huber, K., Franke, A., Bruhl, B., Krispin, S., Ernsberger, U., Schober, A., von Bohlen und, H. O., Rohrer, H., Kalcheim, C., and Unsicker, K. (2008). Persistent expression of BMP-4 in embryonic chick adrenal cortical cells and its role in chromaffin cell development. *Neural Dev.* **3,** 28.

Hurley, S. P., Clary, D. O., Copie, V., and Lefcort, F. (2006). Anaplastic lymphoma kinase is dynamically expressed on subsets of motor neurons and in the peripheral nervous system. *J. Comp. Neurol.* **495,** 202–212.

Hurlin, P. J. (2005). N-Myc functions in transcription and development. *Birth Defects Res. C Embryo Today* **75,** 340–352.

Ichimiya, S., Nimura, Y., Kageyama, H., Takada, N., Sunahara, M., Shishikura, T., Nakamura, Y., Sakiyama, S., Seki, N., Ohira, M., Kaneko, Y., McKeon, F., *et al.* (1999). p73 at chromosome 1p36.3 is lost in advanced stage neuroblastoma but its mutation is infrequent. *Oncogene* **18,** 1061–1066.

Ikegaki, N., Katsumata, M., Tsujimoto, Y., Nakagawara, A., and Brodeur, G. M. (1995). Relationship between bcl-2 and myc gene expression in human neuroblastoma. *Cancer Lett.* **91,** 161–168.

Islam, A., Kageyama, H., Takada, N., Kawamoto, T., Takayasu, H., Isogai, E., Ohira, M., Hashizume, K., Kobayashi, H., Kaneko, Y., and Nakagawara, A. (2000). High expression of Survivin, mapped to 17q25, is significantly associated with poor prognostic factors and promotes cell survival in human neuroblastoma. *Oncogene* **19,** 617–623.

Iwahara, T., Fujimoto, J., Wen, D., Cupples, R., Bucay, N., Arakawa, T., Mori, S., Ratzkin, B., and Yamamoto, T. (1997). Molecular characterization of ALK, a receptor tyrosine kinase expressed specifically in the nervous system. *Oncogene* **14,** 439–449.

Janoueix-Lerosey, I., Lequin, D., Brugieres, L., Ribeiro, A., De, P. L., Combaret, V., Raynal, V., Puisieux, A., Schleiermacher, G., Pierron, G., Valteau-Couanet, D., Frebourg, T., *et al.* (2008). Somatic and germline activating mutations of the ALK kinase receptor in neuroblastoma. *Nature* **455,** 967–970.

Jinbo, T., Iwamura, Y., Kaneko, M., and Sawaguchi, S. (1989). Coamplification of the L-myc and N-myc oncogenes in a neuroblastoma cell line. *Jpn J. Cancer Res.* **80,** 299–301.

Johnsen, J. I., Pettersen, I., Ponthan, F., Sveinbjornsson, B., Flaegstad, T., and Kogner, P. (2004). Synergistic induction of apoptosis in neuroblastoma cells using a combination of cytostatic drugs with interferon-gamma and TRAIL. *Int. J. Oncol.* **25,** 1849–1857.

Johnsen, J. I., Kogner, P., Albihn, A., and Henriksson, M. A. (2009). Embryonal neural tumours and cell death. *Apoptosis* **14,** 424–438.

Kang, Y., and Massague, J. (2004). Epithelial-mesenchymal transitions: Twist in development and metastasis. *Cell* **118,** 277–279.

Kaufmann, S. H., and Vaux, D. L. (2003). Alterations in the apoptotic machinery and their potential role in anticancer drug resistance. *Oncogene* **22,** 7414–7430.

Kelleher, F. C., and McDermott, R. (2010). The emerging pathogenic and therapeutic importance of the anaplastic lymphoma kinase gene. *Eur. J. Cancer* **46,** 2357–2368.

Keshelava, N., Zuo, J. J., Chen, P., Waidyaratne, S. N., Luna, M. C., Gomer, C. J., Triche, T. J., and Reynolds, C. P. (2001). Loss of p53 function confers high-level multidrug resistance in neuroblastoma cell lines. *Cancer Res.* **61,** 6185–6193.

Kim, C. J., Matsuo, T., Lee, K. H., and Thiele, C. J. (1999). Up-regulation of insulin-like growth factor-II expression is a feature of TrkA but not TrkB activation in SH-SY5Y neuroblastoma cells. *Am. J. Pathol.* **155,** 1661–1670.

Kim, H. S., Lee, J. W., Soung, Y. H., Park, W. S., Kim, S. Y., Lee, J. H., Park, J. Y., Cho, Y. G., Kim, C. J., Jeong, S. W., Nam, S. W., Kim, S. H., *et al.* (2003). Inactivating mutations of caspase-8 gene in colorectal carcinomas. *Gastroenterology* **125,** 708–715.

Klefstrom, J., Verschuren, E. W., and Evan, G. (2002). c-Myc augments the apoptotic activity of cytosolic death receptor signaling proteins by engaging the mitochondrial apoptotic pathway. *J. Biol. Chem.* **277,** 43224–43232.

Knoepfler, P. S., Cheng, P. F., and Eisenman, R. N. (2002). N-myc is essential during neurogenesis for the rapid expansion of progenitor cell populations and the inhibition of neuronal differentiation. *Genes Dev.* **16,** 2699–2712.

Kohl, N. E., Kanda, N., Schreck, R. R., Bruns, G., Latt, S. A., Gilbert, F., and Alt, F. W. (1983). Transposition and amplification of oncogene-related sequences in human neuroblastomas. *Cell* **35,** 359–367.

Krelin, Y., Zhang, L., Kang, T. B., Appel, E., Kovalenko, A., and Wallach, D. (2008). Caspase-8 deficiency facilitates cellular transformation in vitro. *Cell Death Differ.* **15,** 1350–1355.

Kulesa, P., Ellies, D. L., and Trainor, P. A. (2004). Comparative analysis of neural crest cell death, migration, and function during vertebrate embryogenesis. *Dev. Dyn.* **229,** 14–29.

Kutlik, M. T., Ayhan, A., Gogus, S., Yalcin, B., Caglar, M., and Buyukpamukcu, M. (2002). Glutathione S-transferase and P-glycoprotein expressions in neuroblastoma. *Pediatr. Hematol. Oncol.* **19,** 337–345.

LaBrosse, E. H., Comoy, E., Bohuon, C., Zucker, J. M., and Schweisguth, O. (1976). Catecholamine metabolism in neuroblastoma. *J. Natl. Cancer Inst.* **57,** 633–638.

Lahti, J. M., Teitz, T., and Stupack, D. G. (2006). Does integrin-mediated cell death confer tissue tropism in metastasis? *Cancer Res.* **66,** 5981–5984.

Lallier, T. E., and Bronner-Fraser, M. (1988). A spatial and temporal analysis of dorsal root and sympathetic ganglion formation in the avian embryo. *Dev. Biol.* **127,** 99–112.

Lasorella, A., Boldrini, R., Dominici, C., Donfrancesco, A., Yokota, Y., Inserra, A., and Iavarone, A. (2002). Id2 is critical for cellular proliferation and is the oncogenic effector of N-myc in human neuroblastoma. *Cancer Res.* **62,** 301–306.

Lastowska, M., Nacheva, E., McGuckin, A., Curtis, A., Grace, C., Pearson, A., and Bown, N. (1997). Comparative genomic hybridization study of primary neuroblastoma tumors. United Kingdom Children's Cancer Study Group, Genes Chromosomes. *Cancer* **18,** 162–169.

Le Dourin, N. M., and Kalcheim, G. (1999). *The Neural Crest* 2nd edn. Cambridge University Press.

Lee, W. H., Murphree, A. L., and Benedict, W. F. (1984). Expression and amplification of the N-myc gene in primary retinoblastoma. *Nature* **309,** 458–460.

Lutz, W., Stohr, M., Schurmann, J., Wenzel, A., Lohr, A., and Schwab, M. (1996). Conditional expression of N-myc in human neuroblastoma cells increases expression of alpha-prothymosin and ornithine decarboxylase and accelerates progression into S-phase early after mitogenic stimulation of quiescent cells. *Oncogene* **13,** 803–812.

Lutz, W., Fulda, S., Jeremias, I., Debatin, K. M., and Schwab, M. (1998). MycN and IFNgamma cooperate in apoptosis of human neuroblastoma cells. *Oncogene* **17,** 339–346.

Ma, L., Teruya-Feldstein, J., and Weinberg, R. A. (2007). Tumour invasion and metastasis initiated by microRNA-10b in breast cancer. *Nature* **449,** 682–688.

Mac, S. M., D'Cunha, C. A., and Farnham, P. J. (2000). Direct recruitment of N-myc to target gene promoters. *Mol. Carcinog.* **29,** 76–86.

Maris, J. M. (2005). The biologic basis for neuroblastoma heterogeneity and risk stratification. *Curr. Opin. Pediatr.* **17,** 7–13.

Maris, J. M., and Matthay, K. K. (1999). Molecular biology of neuroblastoma. *J. Clin. Oncol.* **17,** 2264–2279.

Maris, J. M., Weiss, M. J., Mosse, Y., Hii, G., Guo, C., White, P. S., Hogarty, M. D., Mirensky, T., Brodeur, G. M., Rebbeck, T. R., Urbanek, M., and Shusterman, S. (2002). Evidence for a hereditary neuroblastoma predisposition locus at chromosome 16p12-13. *Cancer Res.* **62,** 6651–6658.

Maris, J. M., Hogarty, M. D., Bagatell, R., and Cohn, S. L. (2007). Neuroblastoma. *Lancet* **369,** 2106–2120.

Marshall, B., Isidro, G., Martins, A. G., and Boavida, M. G. (1997). Loss of heterozygosity at chromosome 9p21 in primary neuroblastomas: Evidence for two deleted regions. *Cancer Genet. Cytogenet.* **96,** 134–139.

Martin, D. M., Yee, D., Carlson, R. O., and Feldman, E. L. (1992). Gene expression of the insulin-like growth factors and their receptors in human neuroblastoma cell lines. *Brain Res. Mol. Brain Res.* **15,** 241–246.

Martinez, R., Setien, F., Voelter, C., Casado, S., Quesada, M. P., Schackert, G., and Esteller, M. (2007). CpG island promoter hypermethylation of the pro-apoptotic gene caspase-8 is a common hallmark of relapsed glioblastoma multiforme. *Carcinogenesis* **28,** 1264–1268.

Matsuo, N., Shiraha, H., Fujikawa, T., Takaoka, N., Ueda, N., Tanaka, S., Nishina, S., Nakanishi, Y., Uemura, M., Takaki, A., Nakamura, S., Kobayashi, Y., *et al.* (2009). Twist expression promotes migration and invasion in hepatocellular carcinoma. *BMC Cancer* **9,** 240.

Matthay, K. K. (1995). Neuroblastoma: A clinical challenge and biologic puzzle. *CA Cancer J. Clin.* **45,** 179–192.

Matthay, K. K., Perez, C., Seeger, R. C., Brodeur, G. M., Shimada, H., Atkinson, J. B., Black, C. T., Gerbing, R., Haase, G. M., Stram, D. O., Swift, P., and Lukens, J. N. (1998). Successful treatment of stage III neuroblastoma based on prospective biologic staging: A Children's Cancer Group study. *J. Clin. Oncol.* **16,** 1256–1264.

Matthay, K. K., Villablanca, J. G., Seeger, R. C., Stram, D. O., Harris, R. E., Ramsay, N. K., Swift, P., Shimada, H., Black, C. T., Brodeur, G. M., Gerbing, R. B., and Reynolds, C. P. (1999). Treatment of high-risk neuroblastoma with intensive chemotherapy, radiotherapy, autologous bone marrow transplantation, and 13-*cis*-retinoic acid. Children's Cancer Group. *N Engl J. Med.* **341,** 1165–1173.

McConkey, D. J., Choi, W., Marquis, L., Martin, F., Williams, M. B., Shah, J., Svatek, R., Das, A., Adam, L., Kamat, A., Siefker-Radtke, A., and Dinney, C. (2009). Role of epithelial-to-mesenchymal transition (EMT) in drug sensitivity and metastasis in bladder cancer. *Cancer Metastasis Rev.* **28,** 335–344.

McLennan, R., Teddy, J. M., Kasemeier-Kulesa, J. C., Romine, M. H., and Kulesa, P. M. (2010). Vascular endothelial growth factor (VEGF) regulates cranial neural crest migration in vivo. *Dev. Biol.* **339,** 114–125.

Meddeb, M., Danglot, G., Chudoba, I., Venuat, A. M., Benard, J., vet-Loiseau, H., Vasseur, B., Le, P. D., Terrier-Lacombe, M. J., Hartmann, O., and Bernheim, A. (1996). Additional copies of a 25 Mb chromosomal region originating from 17q23.1-17qter are present in 90% of high-grade neuroblastomas, Genes Chromosomes. *Cancer* **17,** 156–165.

Mejia, M. C., Navarro, S., Pellin, A., Castel, V., and Llombart-Bosch, A. (1998). Study of bcl-2 protein expression and the apoptosis phenomenon in neuroblastoma. *Anticancer Res.* **18,** 801–806.

Meltzer, S. J., O'Doherty, S. P., Frantz, C. N., Smolinski, K., Yin, J., Cantor, A. B., Liu, J., Valentine, M., Brodeur, G. M., and Berg, P. E. (1996). Allelic imbalance on chromosome 5q predicts long-term survival in neuroblastoma. *Br. J. Cancer* **74,** 1855–1861.

Monclair, T., Brodeur, G. M., Ambros, P. F., Brisse, H. J., Cecchetto, G., Holmes, K., Kaneko, M., London, W. B., Matthay, K. K., Nuchtern, J. G., von, S. D., Simon, T., *et al.* (2009). The International Neuroblastoma Risk Group (INRG) staging system: An INRG Task Force report. *J. Clin. Oncol.* **27,** 298–303.

Mora, J., Cheung, N. K., Chen, L., Qin, J., and Gerald, W. (2001). Loss of heterozygosity at 19q13.3 is associated with locally aggressive neuroblastoma. *Clin. Cancer Res.* **7,** 1358–1361.

Morales, A. V., Barbas, J. A., and Nieto, M. A. (2005). How to become neural crest: From segregation to delamination. *Semin. Cell Dev. Biol.* **16,** 655–662.

Morris, S. W., Kirstein, M. N., Valentine, M. B., Dittmer, K. G., Shapiro, D. N., Saltman, D. L., and Look, A. T. (1994). Fusion of a kinase gene, ALK, to a nucleolar protein gene, NPM, in non-Hodgkin's lymphoma. *Science* **263,** 1281–1284.

Morris, S. W., Naeve, C., Mathew, P., James, P. L., Kirstein, M. N., Cui, X., and Witte, D. P. (1997). ALK, the chromosome 2 gene locus altered by the t(2;5) in non-Hodgkin's lymphoma, encodes a novel neural receptor tyrosine kinase that is highly related to leukocyte tyrosine kinase (LTK). *Oncogene* **14,** 2175–2188.

Mosse, Y. P., Laudenslager, M., Khazi, D., Carlisle, A. J., Winter, C. L., Rappaport, E., and Maris, J. M. (2004). Germline PHOX2B mutation in hereditary neuroblastoma. *Am. J. Hum. Genet.* **75,** 727–730.

Mosse, Y. P., Laudenslager, M., Longo, L., Cole, K. A., Wood, A., Attiyeh, E. F., Laquaglia, M. J., Sennett, R., Lynch, J. E., Perri, P., Laureys, G., Speleman, F., *et al.* (2008). Identification of ALK as a major familial neuroblastoma predisposition gene. *Nature* **455,** 930–935.

Mosse, Y. P., Wood, A., and Maris, J. M. (2009). Inhibition of ALK signaling for cancer therapy. *Clin. Cancer Res.* **15,** 5609–5614.

Motegi, A., Fujimoto, J., Kotani, M., Sakuraba, H., and Yamamoto, T. (2004). ALK receptor tyrosine kinase promotes cell growth and neurite outgrowth. *J. Cell Sci.* **117,** 3319–3329.

Muhlethaler-Mottet, A., Balmas, K., Auderset, K., Joseph, J. M., and Gross, N. (2003). Restoration of TRAIL-induced apoptosis in a caspase-8-deficient neuroblastoma cell line by stable re-expression of caspase-8. *Ann. NY Acad. Sci.* **1010,** 195–199.

Munchar, M. J., Sharifah, N. A., Jamal, R., and Looi, L. M. (2003). CD44s expression correlated with the International Neuroblastoma Pathology Classification (Shimada system) for neuroblastic tumours. *Pathology* **35,** 125–129.

Munirajan, A. K., Ando, K., Mukai, A., Takahashi, M., Suenaga, Y., Ohira, M., Koda, T., Hirota, T., Ozaki, T., and Nakagawara, A. (2008). KIF1Bbeta functions as a haploinsufficient tumor suppressor gene mapped to chromosome 1p36.2 by inducing apoptotic cell death. *J. Biol. Chem.* **283,** 24426–24434.

Nakagawara, A. (2004). Neural crest development and neuroblastoma: The genetic and biological link. *Prog. Brain Res.* **146,** 233–242.

Nakagawara, A., and Ohira, M. (2004). Comprehensive genomics linking between neural development and cancer: Neuroblastoma as a model. *Cancer Lett.* **204,** 213–224.

Nakagawara, A., Arima, M., Azar, C. G., Scavarda, N. J., and Brodeur, G. M. (1992). Inverse relationship between trk expression and N-myc amplification in human neuroblastomas. *Cancer Res.* **52,** 1364–1368.

Nakagawara, A., rima-Nakagawara, M., Scavarda, N. J., Azar, C. G., Cantor, A. B., and Brodeur, G. M. (1993). Association between high levels of expression of the TRK gene and favorable outcome in human neuroblastoma. *N Engl J. Med.* **328,** 847–854.

Nakagawara, A., Azar, C. G., Scavarda, N. J., and Brodeur, G. M. (1994a). Expression and function of TRK-B and BDNF in human neuroblastomas. *Mol. Cell. Biol.* **14,** 759–767.

Nakagawara, A., rima-Nakagawara, M., Azar, C. G., Scavarda, N. J., and Brodeur, G. M. (1994b). Clinical significance of expression of neurotrophic factors and their receptors in neuroblastoma. *Prog. Clin. Biol. Res.* **385,** 155–161.

Nieminen, A. I., Partanen, J. I., and Klefstrom, J. (2007). c-Myc blazing a trail of death: Coupling of the mitochondrial and death receptor apoptosis pathways by c-Myc. *Cell Cycle* **6,** 2464–2472.

Norris, M. D., Bordow, S. B., Marshall, G. M., Haber, P. S., Cohn, S. L., and Haber, M. (1996). Expression of the gene for multidrug-resistance-associated protein and outcome in patients with neuroblastoma. *N Engl J. Med.* **334,** 231–238.

Norris, M. D., Smith, J., Tanabe, K., Tobin, P., Flemming, C., Scheffer, G. L., Wielinga, P., Cohn, S. L., London, W. B., Marshall, G. M., Allen, J. D., and Haber, M. (2005). Expression of multidrug transporter MRP4/ABCC4 is a marker of poor prognosis in neuroblastoma and confers resistance to irinotecan in vitro. *Mol. Cancer Ther.* **4,** 547–553.

Nowak, K., Kerl, K., Fehr, D., Kramps, C., Gessner, C., Killmer, K., Samans, B., Berwanger, B., Christiansen, H., and Lutz, W. (2006). BMI1 is a target gene of E2F-1 and is strongly expressed in primary neuroblastomas. *Nucleic Acids Res.* **34,** 1745–1754.

Ochiai, H., Takenobu, H., Nakagawa, A., Yamaguchi, Y., Kimura, M., Ohira, M., Okimoto, Y., Fujimura, Y., Koseki, H., Kohno, Y., Nakagawara, A., and Kamijo, T. (2010). Bmi1 is a MYCN target gene that regulates tumorigenesis through repression of KIF1Bbeta and TSLC1 in neuroblastoma. *Oncogene* **29,** 2681–2690.

Ohali, A., Avigad, S., Ash, S., Goshen, Y., Luria, D., Feinmesser, M., Zaizov, R., and Yaniv, I. (2006). Telomere length is a prognostic factor in neuroblastoma. *Cancer* **107,** 1391–1399.

Okabe-Kado, J., Kasukabe, T., Honma, Y., Hanada, R., Nakagawara, A., and Kaneko, Y. (2005). Clinical significance of serum NM23-H1 protein in neuroblastoma. *Cancer Sci.* **96,** 653–660.

Osajima-Hakomori, Y., Miyake, I., Ohira, M., Nakagawara, A., Nakagawa, A., and Sakai, R. (2005). Biological role of anaplastic lymphoma kinase in neuroblastoma. *Am. J. Pathol.* **167,** 213–222.

Palmer, R. H., Vernersson, E., Grabbe, C., and Hallberg, B. (2009). Anaplastic lymphoma kinase: Signalling in development and disease. *Biochem. J.* **420,** 345–361.

Park, J. R., Eggert, A., and Caron, H. (2008). Neuroblastoma: Biology, prognosis, and treatment. *Pediatr. Clin. North Am.* **55,** 97–120, (x).

Passoni, L., Longo, L., Collini, P., Coluccia, A. M., Bozzi, F., Podda, M., Gregorio, A., Gambini, C., Garaventa, A., Pistoia, V., Del, G. F., Tonini, G. P., *et al.* (2009). Mutation-independent anaplastic lymphoma kinase overexpression in poor prognosis neuroblastoma patients. *Cancer Res.* **69,** 7338–7346.

Pearson, A. D., Pinkerton, C. R., Lewis, I. J., Imeson, J., Ellershaw, C., and Machin, D. (2008). High-dose rapid and standard induction chemotherapy for patients aged over 1 year with stage 4 neuroblastoma: A randomised trial. *Lancet Oncol.* **9,** 247–256.

Pelengaris, S., Khan, M., and Evan, G. (2002). c-MYC: More than just a matter of life and death. *Nat. Rev. Cancer* **2,** 764–776.

Perini, G., Diolaiti, D., Porro, A., and Della, V. G. (2005). In vivo transcriptional regulation of N-Myc target genes is controlled by E-box methylation. *Proc. Natl. Acad. Sci. USA* **102,** 12117–12122.

Perri, P., Longo, L., Cusano, R., McConville, C. M., Rees, S. A., Devoto, M., Conte, M., Ferrara, G. B., Seri, M., Romeo, G., and Tonini, G. P. (2002). Weak linkage at 4p16 to predisposition for human neuroblastoma. *Oncogene* **21,** 8356–8360.

Pingoud-Meier, C., Lang, D., Janss, A. J., Rorke, L. B., Phillips, P. C., Shalaby, T., and Grotzer, M. A. (2003). Loss of caspase-8 protein expression correlates with unfavorable survival outcome in childhood medulloblastoma. *Clin. Cancer Res.* **9,** 6401–6409.

Raabe, E. H., Laudenslager, M., Winter, C., Wasserman, N., Cole, K., LaQuaglia, M., Maris, D. J., Mosse, Y. P., and Maris, J. M. (2008). Prevalence and functional consequence of PHOX2B mutations in neuroblastoma. *Oncogene* **27,** 469–476.

Ramani, P., and Lu, Q. L. (1994). Expression of bcl-2 gene product in neuroblastoma. *J. Pathol.* **172,** 273–278.

Reed, J. C. (2006). Proapoptotic multidomain Bcl-2/Bax-family proteins: Mechanisms, physiological roles, and therapeutic opportunities. *Cell Death Differ.* **13,** 1378–1386.

Reed, J. C., and Tomaselli, K. J. (2000). Drug discovery opportunities from apoptosis research. *Curr. Opin. Biotechnol.* **11,** 586–592.

Reissmann, E., Ernsberger, U., Francis-West, P. H., Rueger, D., Brickell, P. M., and Rohrer, H. (1996). Involvement of bone morphogenetic protein-4 and bone morphogenetic protein-7 in the differentiation of the adrenergic phenotype in developing sympathetic neurons. *Development* **122,** 2079–2088.

Reiter, J. L., and Brodeur, G. M. (1996). High-resolution mapping of a 130-kb core region of the MYCN amplicon in neuroblastomas. *Genomics* **32,** 97–103.

Reiter, J. L., and Brodeur, G. M. (1998). MYCN is the only highly expressed gene from the core amplified domain in human neuroblastomas. *Genes Chromosom. Cancer* **23,** 134–140.

Reynolds, C. P., Zuo, J. J., Kim, N. W., Wang, H., Lukens, J. N., Matthay, K. K., and Seeger, R. C. (1997). Telomerase expression in primary neuroblastomas. *Eur. J. Cancer* **33,** 1929–1931.

Ribatti, D., Marimpietri, D., Pastorino, F., Brignole, C., Nico, B., Vacca, A., and Ponzoni, M. (2004). Angiogenesis in neuroblastoma. *Ann. NY Acad. Sci.* **1028,** 133–142.

Ruffins, S., Artinger, K. B., and Bronner-Fraser, M. (1998). Early migrating neural crest cells can form ventral neural tube derivatives when challenged by transplantation. *Dev. Biol.* **203,** 295–304.

Saito-Ohara, F., Imoto, I., Inoue, J., Hosoi, H., Nakagawara, A., Sugimoto, T., and Inazawa, J. (2003). PPM1D is a potential target for 17q gain in neuroblastoma. *Cancer Res.* **63,** 1876–1883.

Salvesen, G. S., and Abrams, J. M. (2004). Caspase activation—Stepping on the gas or releasing the brakes? Lessons from humans and flies. *Oncogene* **23,** 2774–2784.

Sawada, T. (1992). Past and future of neuroblastoma screening in Japan. *Am. J. Pediatr. Hematol. Oncol.* **14,** 320–326.

Sawai, S., Shimono, A., Hanaoka, K., and Kondoh, H. (1991). Embryonic lethality resulting from disruption of both N-myc alleles in mouse zygotes. *New Biol.* **3,** 861–869.

Sawai, S., Shimono, A., Wakamatsu, Y., Palmes, C., Hanaoka, K., and Kondoh, H. (1993). Defects of embryonic organogenesis resulting from targeted disruption of the N-myc gene in the mouse. *Development* **117,** 1445–1455.

Scaruffi, P., Valent, A., Schramm, A., Astrahantseff, K., Eggert, A., and Tonini, G. P. (2005). Application of microarray-based technology to neuroblastoma. *Cancer Lett.* **228,** 13–20.

Schlisio, S., Kenchappa, R. S., Vredeveld, L. C., George, R. E., Stewart, R., Greulich, H., Shahriari, K., Nguyen, N. V., Pigny, P., Dahia, P. L., Pomeroy, S. L., Maris, J. M., *et al.* (2008). The kinesin KIF1Bbeta acts downstream from EglN3 to induce apoptosis and is a potential 1p36 tumor suppressor. *Genes Dev.* **22,** 884–893.

Schmidt, M. L., Lukens, J. N., Seeger, R. C., Brodeur, G. M., Shimada, H., Gerbing, R. B., Stram, D. O., Perez, C., Haase, G. M., and Matthay, K. K. (2000). Biologic factors determine prognosis in infants with stage IV neuroblastoma: A prospective Children's Cancer Group study. *J. Clin. Oncol.* **18,** 1260–1268.

Schneider, C., Wicht, H., Enderich, J., Wegner, M., and Rohrer, H. (1999). Bone morphogenetic proteins are required in vivo for the generation of sympathetic neurons. *Neuron* **24,** 861–870.

Schonherr, C., Yang, H. L., Vigny, M., Palmer, R. H., and Hallberg, B. (2010). Anaplastic lymphoma kinase activates the small GTPase Rap1 via the Rap1-specific GEF C3G in both neuroblastoma and PC12 cells. *Oncogene* **29,** 2817–2830.

Schor, N. F. (1999). Neuroblastoma as a neurobiological disease. *J. Neurooncol.* **41,** 159–166.

Schor, N. F. (2009). New approaches to pharmacotherapy of tumors of the nervous system during childhood and adolescence. *Pharmacol. Ther.* **122,** 44–55.

Schulte, J. H., Horn, S., Otto, T., Samans, B., Heukamp, L. C., Eilers, U. C., Krause, M., Astrahantseff, K., Klein-Hitpass, L., Buettner, R., Schramm, A., Christiansen, H., *et al.* (2008). MYCN regulates oncogenic MicroRNAs in neuroblastoma. *Int. J. Cancer* **122,** 699–704.

Schwab, M., Alitalo, K., Klempnauer, K. H., Varmus, H. E., Bishop, J. M., Gilbert, F., Brodeur, G., Goldstein, M., and Trent, J. (1983). Amplified DNA with limited homology to myc cellular oncogene is shared by human neuroblastoma cell lines and a neuroblastoma tumour. *Nature* **305,** 245–248.

Schwab, M., Varmus, H. E., Bishop, J. M., Grzeschik, K. H., Naylor, S. L., Sakaguchi, A. Y., Brodeur, G., and Trent, J. (1984). Chromosome localization in normal human cells and neuroblastomas of a gene related to c-myc. *Nature* **308,** 288–291.

Seoane, J., Le, H. V., and Massague, J. (2002). Myc suppression of the p21(Cip1) Cdk inhibitor influences the outcome of the p53 response to DNA damage. *Nature* **419,** 729–734.

Shah, N. M., Groves, A. K., and Anderson, D. J. (1996). Alternative neural crest cell fates are instructively promoted by TGFbeta superfamily members. *Cell* **85,** 331–343.

Shi, Y. (2004). Caspase activation: Revisiting the induced proximity model. *Cell* **117,** 855–858.

Shimada, H., Ambros, I. M., Dehner, L. P., Hata, J., Joshi, V. V., Roald, B., Stram, D. O., Gerbing, R. B., Lukens, J. N., Matthay, K. K., and Castleberry, R. P. (1999). The International Neuroblastoma Pathology Classification (the Shimada system). *Cancer* **86,** 364–372.

Shimono, R., Matsubara, S., Takamatsu, H., Fukushige, T., and Ozawa, M. (2000). The expression of cadherins in human neuroblastoma cell lines and clinical tumors. *Anticancer Res.* **20,** 917–923.

Shohet, J. M., Hicks, M. J., Plon, S. E., Burlingame, S. M., Stuart, S., Chen, S. Y., Brenner, M. K., and Nuchtern, J. G. (2002). Minichromosome maintenance protein MCM7 is a direct target of the MYCN transcription factor in neuroblastoma. *Cancer Res.* **62,** 1123–1128.

Shtivelman, E., and Bishop, J. M. (1991). Expression of CD44 is repressed in neuroblastoma cells. *Mol. Cell. Biol.* **11,** 5446–5453.

Simon, T., Spitz, R., Faldum, A., Hero, B., and Berthold, F. (2004). New definition of low-risk neuroblastoma using stage, age, and 1p and MYCN status. *J. Pediatr. Hematol. Oncol.* **26,** 791–796.

Slack, A., Chen, Z., Tonelli, R., Pule, M., Hunt, L., Pession, A., and Shohet, J. M. (2005). The p53 regulatory gene MDM2 is a direct transcriptional target of MYCN in neuroblastoma. *Proc. Natl. Acad. Sci. USA* **102,** 731–736.

Sosic, D., and Olson, E. N. (2003). A new twist on twist–modulation of the NF-kappa B pathway. *Cell Cycle* **2,** 76–78.

Soung, Y. H., Lee, J. W., Kim, S. Y., Jang, J., Park, Y. G., Park, W. S., Nam, S. W., Lee, J. Y., Yoo, N. J., and Lee, S. H. (2005). CASPASE-8 gene is inactivated by somatic mutations in gastric carcinomas. *Cancer Res.* **65,** 815–821.

Souttou, B., Carvalho, N. B., Raulais, D., and Vigny, M. (2001). Activation of anaplastic lymphoma kinase receptor tyrosine kinase induces neuronal differentiation through the mitogen-activated protein kinase pathway. *J. Biol. Chem.* **276,** 9526–9531.

Stanton, B. R., Perkins, A. S., Tessarollo, L., Sassoon, D. A., and Parada, L. F. (1992). Loss of N-myc function results in embryonic lethality and failure of the epithelial component of the embryo to develop. *Genes Dev.* **6,** 2235–2247.

Straub, J. A., Sholler, G. L., and Nishi, R. (2007). Embryonic sympathoblasts transiently express TrkB in vivo and proliferate in response to brain-derived neurotrophic factor in vitro. *BMC Dev. Biol.* **7,** 10.

Strieder, V., and Lutz, W. (2003). E2F proteins regulate MYCN expression in neuroblastomas. *J. Biol. Chem.* **278,** 2983–2989.

Stupack, D. G., Teitz, T., Potter, M. D., Mikolon, D., Houghton, P. J., Kidd, V. J., Lahti, J. M., and Cheresh, D. A. (2006). Potentiation of neuroblastoma metastasis by loss of caspase-8. *Nature* **439,** 95–99.

Sugiura, Y., Shimada, H., Seeger, R. C., Laug, W. E., and DeClerck, Y. A. (1998). Matrix metalloproteinases-2 and -9 are expressed in human neuroblastoma: Contribution of stromal cells to their production and correlation with metastasis. *Cancer Res.* **58,** 2209–2216.

Suzuki, T., Yokota, J., Mugishima, H., Okabe, I., Ookuni, M., Sugimura, T., and Terada, M. (1989). Frequent loss of heterozygosity on chromosome 14q in neuroblastoma. *Cancer Res.* **49,** 1095–1098.

Svensson, T., Ryden, M., Schilling, F. H., Dominici, C., Sehgal, R., Ibanez, C. F., and Kogner, P. (1997). Coexpression of mRNA for the full-length neurotrophin receptor trk-C and trk-A in favourable neuroblastoma. *Eur. J. Cancer* **33,** 2058–2063.

Tacconelli, A., Farina, A. R., Cappabianca, L., Desantis, G., Tessitore, A., Vetuschi, A., Sferra, R., Rucci, N., Argenti, B., Screpanti, I., Gulino, A., and Mackay, A. R. (2004). TrkA alternative splicing: A regulated tumor-promoting switch in human neuroblastoma. *Cancer Cell* **6,** 347–360.

Takita, J., Hayashi, Y., Takei, K., Yamaguchi, N., Hanada, R., Yamamoto, K., and Yokota, J. (2000). Allelic imbalance on chromosome 18 in neuroblastoma. *Eur. J. Cancer* **36,** 508–513.

Takita, J., Yang, H. W., Chen, Y. Y., Hanada, R., Yamamoto, K., Teitz, T., Kidd, V., and Hayashi, Y. (2001). Allelic imbalance on chromosome 2q and alterations of the caspase 8 gene in neuroblastoma. *Oncogene* **20,** 4424–4432.

Tanabe, K. K., Ellis, L. M., and Saya, H. (1993). Expression of CD44R1 adhesion molecule in colon carcinomas and metastases. *Lancet* **341,** 725–726.

Teillet, M. A., Kalcheim, C., and Le Douarin, N. M. (1987). Formation of the dorsal root ganglia in the avian embryo: Segmental origin and migratory behavior of neural crest progenitor cells. *Dev. Biol.* **120,** 329–347.

Teitz, T., Wei, T., Valentine, M. B., Vanin, E. F., Grenet, J., Valentine, V. A., Behm, F. G., Look, A. T., Lahti, J. M., and Kidd, V. J. (2000). Caspase 8 is deleted or silenced preferentially in childhood neuroblastomas with amplification of MYCN. *Nat. Med.* **6,** 529–535.

Teitz, T., Lahti, J. M., and Kidd, V. J. (2001). Aggressive childhood neuroblastomas do not express caspase-8: An important component of programmed cell death. *J. Mol. Med.* **79,** 428–436.

Teitz, T., Stupack, D. G., and Lahti, J. M. (2006). Halting neuroblastoma metastasis by controlling integrin-mediated death. *Cell Cycle* **5,** 681–685.

Tekautz, T. M., Zhu, K., Grenet, J., Kaushal, D., Kidd, V. J., and Lahti, J. M. (2006). Evaluation of IFN-gamma effects on apoptosis and gene expression in neuroblastoma–preclinical studies. *Biochim. Biophys. Acta* **1763,** 1000–1010.

Thiery, J. P., Acloque, H., Huang, R. Y., and Nieto, M. A. (2009). Epithelial-mesenchymal transitions in development and disease. *Cell* **139,** 871–890.

Thompson, E. B. (1998). The many roles of c-Myc in apoptosis. *Annu. Rev. Physiol.* **60,** 575–600.

Thompson, P. M., Seifried, B. A., Kyemba, S. K., Jensen, S. J., Guo, C., Maris, J. M., Brodeur, G. M., Stram, D. O., Seeger, R. C., Gerbing, R., Matthay, K. K., Matise, T. C., *et al.* (2001). Loss of heterozygosity for chromosome 14q in neuroblastoma. *Med. Pediatr. Oncol.* **36,** 28–31.

Treins, C., Giorgetti-Peraldi, S., Murdaca, J., Monthouel-Kartmann, M. N., and Van, O. E. (2005). Regulation of hypoxia-inducible factor (HIF)-1 activity and expression of HIF hydroxylases in response to insulin-like growth factor I. *Mol. Endocrinol.* **19,** 1304–1317.

Trochet, D., Bourdeaut, F., Janoueix-Lerosey, I., Deville, A., de, P. L., Schleiermacher, G., Coze, C., Philip, N., Frebourg, T., Munnich, A., Lyonnet, S., Delattre, O., *et al.* (2004). Germline mutations of the paired-like homeobox 2B (PHOX2B) gene in neuroblastoma. *Am. J. Hum. Genet.* **74,** 761–764.

Trochet, D., Hong, S. J., Lim, J. K., Brunet, J. F., Munnich, A., Kim, K. S., Lyonnet, S., Goridis, C., and Amiel, J. (2005). Molecular consequences of PHOX2B missense, frameshift and alanine expansion mutations leading to autonomic dysfunction. *Hum. Mol. Genet.* **14,** 3697–3708.

Trochet, D., Mathieu, Y., Pontual, L., Savarirayan, R., Munnich, A., Brunet, J. F., Lyonnet, S., Goridis, C., and Amiel, J. (2009). In vitro studies of non poly alanine PHOX2B mutations argue against a loss-of-function mechanism for congenital central hypoventilation. *Hum. Mutat.* **30,** E421–E431.

Tweddle, D. A., Pearson, A. D., Haber, M., Norris, M. D., Xue, C., Flemming, C., and Lunec, J. (2003). The p53 pathway and its inactivation in neuroblastoma. *Cancer Lett.* **197,** 93–98.

Valdes-Mora, F., Gomez Del, P. T., Bandres, E., Cejas, P., Ramirez De, M. A., Perez-Palacios, R., Gallego-Ortega, D., Garcia-Cabezas, M. A., Casado, E., Larrauri, J., Nistal, M., Gonzalez-Baron, M., *et al.* (2009). TWIST1 overexpression is associated with nodal invasion and male sex in primary colorectal cancer. *Ann. Surg. Oncol.* **16,** 78–87.

Valentinis, B., and Baserga, R. (2001). IGF-I receptor signalling in transformation and differentiation. *Mol. Pathol.* **54,** 133–137.

Valsesia-Wittmann, S., Magdeleine, M., Dupasquier, S., Garin, E., Jallas, A. C., Combaret, V., Krause, A., Leissner, P., and Puisieux, A. (2004). Oncogenic cooperation between H-Twist and N-Myc overrides failsafe programs in cancer cells. *Cancer Cell* **6,** 625–630.

van Golen, C. M., Schwab, T. S., Kim, B., Soules, M. E., Su, O. S., Fung, K., van Golen, K. L., and Feldman, E. L. (2006). Insulin-like growth factor-I receptor expression regulates neuroblastoma metastasis to bone. *Cancer Res.* **66,** 6570–6578.

van Noesel, M. M., and Versteeg, R. (2004). Pediatric neuroblastomas: Genetic and epigenetic 'danse macabre'. *Gene* **325,** 1–15.

van Noesel, M. M., van Bezouw, S., Salomons, G. S., Voute, P. A., Pieters, R., Baylin, S. B., Herman, J. G., and Versteeg, R. (2002). Tumor-specific down-regulation of the tumor necrosis factor-related apoptosis-inducing ligand decoy receptors DcR1 and DcR2 is associated with dense promoter hypermethylation. *Cancer Res.* **62,** 2157–2161.

Van, R. N., Forus, A., Myklebost, O., Cheng, N. C., Versteeg, R., and Speleman, F. (1995). Identification of two distinct chromosome 12-derived amplification units in neuroblastoma cell line NGP. *Cancer Genet. Cytogenet.* **82,** 151–154.

Vandesompele, J., Van, R. N., Van, G. M., Laureys, G., Ambros, P., Heimann, P., Devalck, C., Schuuring, E., Brock, P., Otten, J., Gyselinck, J., De, P. A., *et al.* (1998). Genetic heterogeneity of neuroblastoma studied by comparative genomic hybridization. *Genes Chromosom. Cancer* **23,** 141–152.

Vega, S., Morales, A. V., Ocana, O. H., Valdes, F., Fabregat, I., and Nieto, M. A. (2004). Snail blocks the cell cycle and confers resistance to cell death. *Genes Dev.* **18,** 1131–1143.

Vernersson, E., Khoo, N. K., Henriksson, M. L., Roos, G., Palmer, R. H., and Hallberg, B. (2006). Characterization of the expression of the ALK receptor tyrosine kinase in mice. *Gene Expr. Patterns* **6,** 448–461.

Vitali, R., Mancini, C., Cesi, V., Tanno, B., Mancuso, M., Bossi, G., Zhang, Y., Martinez, R. V., Calabretta, B., Dominici, C., and Raschella, G. (2008). Slug (SNAI2) down-regulation by RNA interference facilitates apoptosis and inhibits invasive growth in neuroblastoma preclinical models. *Clin. Cancer Res.* **14,** 4622–4630.

Wada, R. K., Seeger, R. C., Reynolds, C. P., Alloggiamento, T., Yamashiro, J. M., Ruland, C., Black, A. C., and Rosenblatt, J. D. (1992). Cell type-specific expression and negative regulation by retinoic acid of the human N-myc promoter in neuroblastoma cells. *Oncogene* **7,** 711–717.

Wagner, L. M., and Danks, M. K. (2009). New therapeutic targets for the treatment of high-risk neuroblastoma. *J. Cell. Biochem.* **107,** 46–57.

Wakamatsu, Y., Watanabe, Y., Nakamura, H., and Kondoh, H. (1997). Regulation of the neural crest cell fate by N-myc: Promotion of ventral migration and neuronal differentiation. *Development* **124,** 1953–1962.

Wartiovaara, K., Barnabe-Heider, F., Miller, F. D., and Kaplan, D. R. (2002). N-myc promotes survival and induces S-phase entry of postmitotic sympathetic neurons. *J. Neurosci.* **22,** 815–824.

Wei, J. S., Song, Y. K., Durinck, S., Chen, Q. R., Cheuk, A. T., Tsang, P., Zhang, Q., Thiele, C. J., Slack, A., Shohet, J., and Khan, J. (2008). The MYCN oncogene is a direct target of miR-34a. *Oncogene* **27,** 5204–5213.

Weiss, W. A., Aldape, K., Mohapatra, G., Feuerstein, B. G., and Bishop, J. M. (1997). Targeted expression of MYCN causes neuroblastoma in transgenic mice. *EMBO J.* **16,** 2985–2995.

Welch, C., Chen, Y., and Stallings, R. L. (2007). MicroRNA-34a functions as a potential tumor suppressor by inducing apoptosis in neuroblastoma cells. *Oncogene* **26,** 5017–5022.

WESTON, J. A. (1963). A radioautographic analysis of the migration and localization of trunk neural crest cells in the chick. *Dev. Biol.* **6,** 279–310.

Witt, O., Deubzer, H. E., Lodrini, M., Milde, T., and Oehme, I. (2009). Targeting histone deacetylases in neuroblastoma. *Curr. Pharm. Des.* **15,** 436–447.

Xue, C., Haber, M., Flemming, C., Marshall, G. M., Lock, R. B., MacKenzie, K. L., Gurova, K. V., Norris, M. D., and Gudkov, A. V. (2007). p53 determines multidrug sensitivity of childhood neuroblastoma. *Cancer Res.* **67,** 10351–10360.

Yamashiro, D. J., Nakagawara, A., Ikegaki, N., Liu, X. G., and Brodeur, G. M. (1996). Expression of TrkC in favorable human neuroblastomas. *Oncogene* **12,** 37–41.

Yang, Q. W., Liu, S., Tian, Y., Salwen, H. R., Chlenski, A., Weinstein, J., and Cohn, S. L. (2003). Methylation-associated silencing of the thrombospondin-1 gene in human neuroblastoma. *Cancer Res.* **63,** 6299–6310.

Yang, J., Mani, S. A., Donaher, J. L., Ramaswamy, S., Itzykson, R. A., Come, C., Savagner, P., Gitelman, I., Richardson, A., and Weinberg, R. A. (2004). Twist, a master regulator of morphogenesis, plays an essential role in tumor metastasis. *Cell* **117,** 927–939.

Yang, Q., Kiernan, C. M., Tian, Y., Salwen, H. R., Chlenski, A., Brumback, B. A., London, W. B., and Cohn, S. L. (2007). Methylation of CASP8, DCR2, and HIN-1 in neuroblastoma is associated with poor outcome. *Clin. Cancer Res.* **13,** 3191–3197.

Yu, A. L., Gilman, A. L., Ozkaynak, M. F., London, W. B., Kreissman, S., Chen, H. X., Matthay, K. K., Cohn, S. L., Maris, J. M., and Sondel, P. (2009). A phase III randomized trial of the chimeric anti-GD2 antibody ch14.18 with GM-CSF and IL2 as immunotherapy following dose intensive chemotherapy for high-risk neuroblastoma: Children's Oncology Group (COG) study ANBL0032. *J. Clin. Oncol.* **27,** 15s, (suppl; abstr 10067z).

Zage, P. E., Kletzel, M., Murray, K., Marcus, R., Castleberry, R., Zhang, Y., London, W. B., and Kretschmar, C. (2008). Outcomes of the POG 9340/9341/9342 trials for children with high-risk neuroblastoma: A report from the Children's Oncology Group. *Pediatr. Blood Cancer* **51,** 747–753.

Zhang, H., Li, Y., and Lai, M. (2010). The microRNA network and tumor metastasis. *Oncogene* **29,** 937–948.

Zimmerman, K. A., Yancopoulos, G. D., Collum, R. G., Smith, R. K., Kohl, N. E., Denis, K. A., Nau, M. M., Witte, O. N., Toran-Allerand, D., Gee, C. E., *et al.* (1986). Differential expression of myc family genes during murine development. *Nature* **319,** 780–783.

CHAPTER FIVE

RB1, Development, and Cancer

Meenalakshmi Chinnam *and* David W. Goodrich

Contents

Abstract

The *RB1* gene is the first tumor suppressor gene identified whose mutational inactivation is the cause of a human cancer, the pediatric cancer retinoblastoma. The 25 years of research since its discovery has not only illuminated a general role for *RB1* in human cancer, but also its critical importance in normal development. Understanding the molecular function of the *RB1* encoded protein, pRb, is a long-standing goal that promises to inform our understanding of cancer, its relationship to normal development, and possible therapeutic strategies to combat this disease. Achieving this goal has been difficult, complicated by the complexity of pRb and related proteins. The goal of this review is to explore the hypothesis that, at its core, the molecular function of pRb is to dynamically regulate the location-specific assembly or disassembly of protein complexes on the DNA in response to the output of various signaling pathways. These protein complexes participate in a variety of molecular processes relevant to DNA including gene transcription, DNA replication, DNA repair, and mitosis. Through regulation of these processes, *RB1* plays a uniquely prominent role in normal development and cancer.

Department of Pharmacology and Therapeutics, Roswell Park Cancer Institute, Buffalo, New York, USA

Current Topics in Developmental Biology, Volume 94
ISSN 0070-2153, DOI: 10.1016/B978-0-12-380916-2.00005-X

1. Introduction

If there is one lesson to be learned from the preceding three decades of cancer research, it is that there are a large variety of genetic, epigenetic, and chromosomal changes that accumulate in cancer cells. If only a small number of these changes actually drive malignant behavior, the possible combinations of contributing molecular alterations are dizzying. This molecular heterogeneity is perhaps why even the most successful therapeutic regimens fail in a significant fraction of patients, bedeviling efforts to understand and effectively treat cancer. How can this molecular complexity be distilled into important general principles that would advance understanding and predict successful therapeutic strategies?

Confronted by complexity, scientists have traditionally sought refuge in simple model systems. Simple systems are easier to study, and the general principles gleaned are often relevant to the more complicated cases. In the world of human cancer research, perhaps the simplest "model" system is the pediatric eye cancer retinoblastoma. A subset of retinoblastoma cases are clearly hereditary, with susceptibility transmitted as a simple autosomal dominant trait. The remaining cases arise sporadically. Hereditary cases are diagnosed earlier and exhibit tumors in both eyes while sporadic cases are diagnosed later and typically only occur in one eye. Mathematical modeling of the time to diagnosis suggests hereditary retinoblastoma as a one hit phenomenon while sporadic retinoblastoma is a two-hit phenomenon (Knudson, 1971). This now famous "two-hit hypothesis" suggests that genetic mutation of both alleles of a single gene is the cause (Comings, 1973). The molecular cloning of the *RB1* gene has verified the essential features of this hypothesis (Friend *et al.*, 1986; Fung *et al.*, 1987; Lee *et al.*, 1987). Mutational inactivation of both *RB1* alleles is necessary and rate limiting, but likely not sufficient, for the genesis of retinoblastoma. The molecular etiology of retinoblastoma, therefore, is perhaps the simplest among all human cancers. In addition to its causative role in retinoblastoma, deregulation of the molecular pathway in which *RB1* functions occurs with substantial frequency in virtually every other type of cancer where it has been examined (Hanahan and Weinberg, 2000; Sherr and McCormick, 2002). Thus, the possibility of learning important general principles relevant to cancer from a detailed molecular understanding of *RB1* has long been its siren song.

The relative genetic simplicity of retinoblastoma, however, belies the significant functional complexity of the *RB1* encoded protein (pRb). The first cellular function described for pRb, and the most thoroughly studied, is as a negative regulator of the cell cycle (Goodrich *et al.*, 1991). Loss of pRb-mediated cell cycle control is frequently observed in cancer (Malumbres and Barbacid, 2001; Weinberg, 1995). As cancer is a disease of abnormal cell proliferation, it makes intuitive sense that the key function underlying pRb-mediated tumor suppression is cell cycle regulation. However, pRb loss also has profound effects on many other cellular processes relevant to cancer including differentiation (Korenjak *et al.*, 2005; McClellan *et al.*, 2007b; Nguyen and McCance, 2005), survival (Chau *et al.*, 2003; Delston *et al.*, 2006; Hallstrom *et al.*, 2009), senescence (Ben-Porath *et al.*, 2005; Liu *et al.*, 2004), and genome stability (Knudsen *et al.*, 2006) to name a few. This functional complexity is mirrored by the variety of molecular interactions involving pRb. *RB1* protein interacts with a large and steadily growing list of cellular proteins, and an even greater number of genes. Deriving satisfying general cancer principles from the study of pRb thus remains elusive.

The goal of this review is to explore the emerging hypothesis that the core molecular function of pRb is to dynamically regulate the location-specific assembly or disassembly of protein complexes on the DNA. In essence, pRb serves as an adaptor, physically linking sequence-specific DNA binding proteins with other proteins that influence chromatin and DNA in various ways. The pRb protein interactions central to this function are regulated by post-translational modification, and these modifications represent the output of different signaling pathways. *RB1* protein thus integrates the output of signaling pathways and translates them into genome location-specific changes in protein complexes that influence a variety of molecular processes relevant to DNA including gene transcription, DNA replication, DNA repair, and mitosis. Through its regulation of these important processes, *RB1* plays a uniquely prominent role in normal development and cancer.

2. The Pocket Proteins

RB1 is the founding member of a small gene family that also includes the *RBl1* (p107) and *RBl2* (p130) genes. Of these three family members, *RB1* is clearly the focus of attention because of its well-documented function as a tumor suppressor gene. Despite the fact that they share partially overlapping functions, *RBl1* and *RBl2* are rarely mutated in human cancer. Most of the experiments we review here are focused on *RB1*. For more coverage of the similarities and differences among the different pocket proteins, the reader is referred to comprehensive reviews on the gene family (Classon and Dyson, 2001; Claudio *et al.*, 2002).

The proteins encoded by these three genes are structurally related with their defining feature being the "pocket," a structural domain required for their established cellular and molecular functions. The pocket is composed of two highly conserved and well-ordered subdomains separated by a less

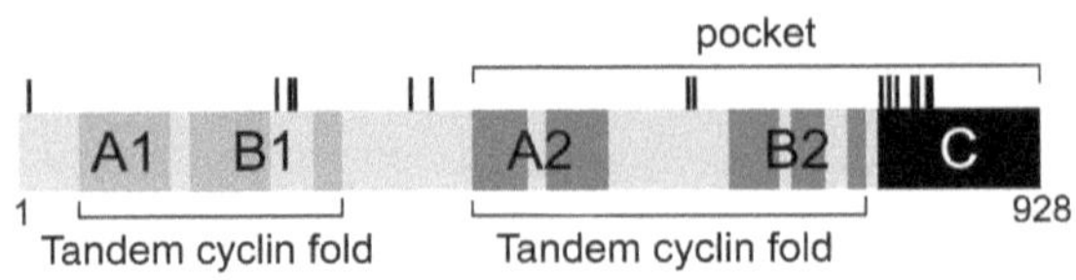

Figure 5.1 Structural organization of the *Rb1* encoded protein. The schematic identifies major structural and functional elements including the "pocket," the dual tandem cyclin folds (A1 + B1, A2 + B2), the unstructured tail (C), and threonine and serine residues, marked by vertical lines, whose phosphorylation regulates intermolecular and intramolecular protein interactions. The colored regions mark the positions of the five alpha helices that are characteristic of cyclin folds. Not shown are additional structural elements (i.e., alpha helices) that contribute to overall pRb structure. The schematic is approximately to scale.

structured spacer and followed by a similarly unstructured tail (Fig. 5.1). The structure of the subdomains has been solved by X-ray crystallography (Kim and Cho, 1997; Kim *et al.*, 2001; Lee, 2002; Lee *et al.*, 1998; Liu and Marmorstein, 2007; Xiao *et al.*, 2003). Each subdomain bears similarity to the cyclin box fold (Kim and Cho, 1997). Thus, the structure of the pocket includes tandem cyclin box folds. Cyclin folds are well-characterized structures involved in mediating protein–protein interactions. The unstructured carboxy terminal tail of pRb also contributes to intermolecular protein interactions and can make intramolecular contacts with the tandem cyclin fold (Dick and Dyson, 2003; Rubin *et al.*, 2005; Welch and Wang, 1993). These intramolecular contacts compete for access to a tandem cyclin fold protein interaction surface, thereby regulating intermolecular protein interactions. Interestingly, a similar tandem cyclin fold structure is present in the amino terminal half of pRb (Hassler *et al.*, 2007). It too mediates intermolecular interactions with cellular proteins as well as intramolecular interactions with the pocket tandem cyclin fold. The amino terminal tandem cyclin fold thus adds additional protein interaction surfaces through which pRb can function and be regulated.

Threading analysis suggests that all members of the *RB1* family have an analogous dual tandem cyclin fold structure (Hassler *et al.*, 2007). However, it is clear that p107 and p130 (47% identity) are more closely related to each other than either one is to pRb (<21% identity). The greatest degree of similarity is within the pocket tandem cyclin fold (30–40% identity). The amino terminal tandem cyclin fold and the unstructured spacers and carboxy terminal tails are not as well conserved (10–20% identity). Differences in the dual tandem cyclin folds as well as the unstructured regions that make intramolecular contacts with the tandem cyclin folds conceivably could account for some of the functional differences between the pocket proteins. However, why pRb is often mutated in human cancer while p107 and p130 are not is an unresolved question.

It is clear from structure/function studies conducted over the past two decades that both the tandem cyclin folds and the unstructured tail that comprise the pocket are required for activity in cell culture based assays that have been used to define various aspects of pRb cellular function, including assays relevant to the cell cycle, differentiation, and apoptosis. The amino terminal tandem cyclin fold domain of pRb is much less studied, likely because it is generally not required for activity in such assays. However, there is evidence that the amino terminal tandem cyclin fold is required for normal function *in vivo*, although this conclusion is controversial (Goodrich, 2003). A small number of *Rb1* missense point mutants have been identified from retinoblastoma patients that specifically affect the amino terminal tandem cyclin fold while sparing the carboxy terminal pocket. There are also a number of cellular proteins that interact specifically with the amino terminal tandem cyclin fold, some of which also play a role in functions regulated by pRb. It seems likely that additional amino terminal tandem cyclin fold interacting proteins will be discovered as more study is devoted to this domain. Finally, the amino terminal tandem cyclin fold is conserved among pocket protein orthologues in different species (Hassler *et al.*, 2007).

The multiple protein interaction surfaces available within the overall pRb structure suggest that it can interact with more than one protein at a time, making it ideally suited to serve as an adaptor. Function as an adaptor molecule is also in keeping with one of the defining characteristics of the pocket proteins, their ability to interact with a large variety of cellular and viral proteins (Goodrich, 2006; Morris and Dyson, 2001). Over 150 different pRb protein interactions have been documented in the literature to date, although the biological relevance of most of them has not been validated. The proteins that p107 and p130 interact with have not been well characterized. Nonetheless, it is clear they can interact with some, but not all, of the proteins that interact with pRb. A significant majority of the proteins that interact with pRb are involved, in some way, in the regulation of RNA transcription and gene expression. Other protein interaction partners have roles in different processes involving DNA including DNA replication, chromosome condensation, and DNA repair. Thus, it is likely that the pocket protein adaptor is a tool used in a number of biological contexts where location-specific changes in protein complex assembly or activity are important.

3. The pRb/E2F Cell Cycle Switch

The hypothesis that pRb functions as an adaptor is based on the paradigm of its well-characterized interaction with the sequence-specific DNA binding E2F transcription factors and the role of this complex in regulating cell proliferation. *E2F1* is the founding member of the E2F gene family (Helin *et al.*, 1992; Kaelin *et al.*, 1992; Shan *et al.*, 1992); the family is currently comprised of eight structurally related transcription factors, five of which (E2F1-5) interact with the pocket proteins (Chen *et al.*, 2009a). E2Fs have traditionally been subclassified as either transcriptional activators or repressors. Of the E2Fs that interact with pocket proteins, E2F1-3 are activators. They share a winged-helix DNA binding domain, a leucine zipper, a transcriptional activation domain, and marked box domains that specify binding to their obligate heterodimerization partners, the DP proteins (*TFDP1-3*). Within the transcriptional activation domain is a region required for pocket protein interaction. *E2F1-3* proteins are most commonly found in complex with pRb, although p107 and p130 can bind E2F1-3 under certain conditions (Lee *et al.*, 2002). *E2F4-5* proteins are classified as transcriptional repressors. They have an overall structure similar to E2F1-3, but lack a nuclear localization signal and instead contain a nuclear export signal. Localization to the nucleus appears to depend on heterodimerization with DP proteins and interaction with pocket proteins. E2F4-5 are most frequently found in complex with p107 and p130, although pRb can also be found in complex with E2F4. The subdivision of E2F1-5 into activators and repressors is based largely on *in vitro* studies, reporter gene assays for example, and may not always reflect their actual role *in vivo*.

When bound to E2Fs, pocket proteins are able to negatively regulate transcription of E2F target genes by at least two different mechanisms. The transcriptional activation domain of E2F1 interacts with general transcription factors and histone acetylases that are necessary for transcriptional activation (Emili and Ingles, 1995; Hagemeier *et al.*, 1993; Lang *et al.*, 2001). Due to their overlapping interaction surfaces within the transcriptional activation domain, pRb competes with these factors for binding E2F thereby blocking transcriptional activation (Pearson and Greenblatt, 1997). Pocket proteins can also bind a host of factors that influence chromatin structure including histone deacetylases, histone methyltransferases, histone demethylases, DNA methyltransferases, and histone remodeling complexes (Table 5.1). In some cases such as the histone deacetylases, binding to pRb appears to be important for recruiting these factors to particular sites on the DNA (Brehm *et al.*, 1998; Luo *et al.*, 1998; Muzumdar *et al.*, 2007). In other cases like the histone H4K20 methyltransferases SUV420H1 and SUV420H2, interaction with pRb appears to stabilize their localization to a specific location rather than to specify it (Gonzalo *et al.*, 2005). pRb can also displace chromatin modifying factors from promoters (Benevolenskaya *et al.*, 2005). For a comprehensive review of pRb interactions with such chromatin modifying factors, see Frolov and Dyson (2004). The typical net effect of pRb at these E2F bound promoters is alteration to a more closed chromatin structure that represses gene transcription. E2Fs regulate many

Table 5.1 pRb interacting gene products involved in chromatin regulation[a]

Gene[b]	Biochemical activity; role in chromatin biology	References
HDAC1, 2, 3	Histone deacetylase; transcriptional repression	Brehm *et al.* (1998), Lai *et al.* (1999, 2001), Luo *et al.* (1998), Magnaghi-Jaulin *et al.* (1998)
SIRT1	NAD-dependent protein deacetylase; transcriptional repression	Wong and Weber (2007)
ARID4A (RBP1)	Component of Sin3-histone deacetylase complexes; imprinting, transcriptional repression	Defeo-Jones *et al.* (1991), Fattaey *et al.* (1993)
PA2G4 (EBP1)	RNA binding protein, component of HDAC containing complexes; transcriptional repression	Xia *et al.* (2001)
EP300	Histone acetyltransferase; transcriptional activation	Chan *et al.* (2001)
KAT5 (TIP60)	Histone acetyltransferase; transcriptional activation	Leduc *et al.* (2006)
SUV420H1, 2	Histone H4 methyltransferase; organization of heterochromatin	Gonzalo *et al.* (2005), Isaac *et al.* (2006)
SUV39H1	Histone H3 methyltransferase; organization of heterochromatin, transcriptional repression	Nielsen *et al.* (2001), Vandel *et al.* (2001)
RBBP5	Component of the MLL1/MLL histone H3 methyltransferase complex; transcriptional activation	Saijo *et al.*, (1995)
KDM1A (LSD1)	Histone H3 demethylase; transcriptional repression or activation	Chau *et al.* (2008)
KDM5A (RBP2)	Histone H3 demethylase; transcriptional repression or activation	Benevolenskaya *et al.* (2005), Defeo-Jones *et al.* (1991), Fattaey *et al.* (1993)
DNMT1	DNA methyltransferase; transcriptional silencing	Robertson *et al.* (2000)
CBX1, 3, 5 (HP1)	Methyl-lysine binding protein; organization of heterochromatin, transcriptional repression	Nielsen *et al.* (2001)
MORF4L1 (MRG15)	Methyl-lysine binding protein; component of histone acetylase and deacetylase complexes, transcriptional regulation	Leung *et al.* (2001)

(*continued*)

Table 5.1 (*continued*)

Gene[b]	Biochemical activity; role in chromatin biology	References
L3MBTL	Methyl-lysine binding protein; nucleosome compaction	Trojer *et al.* (2007)
RBBP7 (RpAp46)	Histone binding protein; component of multiple chromatin assembly, remodeling, and transcriptional repressor complexes	Qian and Lee (1995)
RBBP4 (RpAp48)	Histone binding protein; component of multiple chromatin assembly, remodeling, and transcriptional repressor complexes	Qian *et al.* (1993)
SMARCA4 (BRG1)	ATP-dependent helicase, component of the SWI/SNF chromatin remodeling complex	Dunaief *et al.* (1994)
SMARCA2 (BRM)	ATP-dependent helicase, component of the SWI/SNF chromatin remodeling complex	Strober *et al.* (1996), Trouche *et al.* (1997)
NCAPD3	Subunit of the condensin II complex, organization of chromosome condensation and cohesion during mitosis	Longworth *et al.* (2008), Manning *et al.* (2010)
SMC4	Part of the ATPase subunit of condensin complexes, chromosome condensation during mitosis	Longworth *et al.* (2008)
RBBP8 (CtIP)	Interacts with CTBP/polycomb transcriptional repressor complexes	Dahiya *et al.* (2001), Meloni *et al.* (1999)

[a] This list is restricted to proteins where evidence supports direct binding to pRb. Additional proteins with roles in chromatin biology interact with pRb indirectly as part of multiprotein complexes. Other chromatin modifying complexes appear to interact primarily with other members of the pocket protein family (i.e., DREAM and LINC complexes).
[b] Common alias in parenthesis.

genes important for the cell cycle (Ishida *et al.*, 2001; Ren *et al.*, 2002) and are required for a normal cell cycle in some cells (Humbert *et al.*, 2000; Wu *et al.*, 2001). Thus by both blocking E2F transcriptional activation and actively silencing expression of E2F target genes, pRb restrains cell proliferation.

Cyclin-dependent kinases (CDKs) can phosphorylate pRb *in vitro*. The phosphorylation status of pRb in cells correlates with the cyclic activity of CDKs during the cell cycle, and pRb is phosphorylated on a dozen or more sites that match the consensus target sites for proline directed kinases like CDKs. These observations suggest that CDKs are physiological kinases for pRb. Phosphorylation sites are scattered throughout the length of pRb, but invariably cluster in the unstructured regions rather than the cyclin folds (Fig. 5.1). The overall organization of the phosphorylation sites is similar among the different pocket proteins. While less is known about the phosphorylation of p107 and p130, regulation is likely to be analogous to that of pRb (Leng *et al.*, 2002). Phosphorylation inhibits pRb/E2F complex formation (Dynlacht *et al.*, 1994) and pRb-mediated cell cycle arrest activity (Connell-Crowley *et al.*, 1997). How phosphorylation regulates other myriad pRb protein interactions is not well known, but most of the interactions studied to date are favored when pRb is relatively unphosphorylated.

High affinity pRb/E2F interaction involves two major contacts. One is between the transcriptional activation domain of E2F and the pocket tandem cyclin fold of pRb. This contact is modulated by two clusters of phosphorylation sites, S608/S612 and T356/T373. S608/S612 is located within the unstructured spacer between the pocket dual cyclin folds, and phosphorylation of these residues promotes an intramolecular interaction between the spacer and the dual cyclin fold that inhibits access to the E2F binding surface (Burke *et al.*, 2010). T356/T373 is located within the linker between the amino terminal and pocket tandem cyclin folds. Phosphorylation of these residues induces an intramolecular interaction between the two tandem cyclin folds, also interfering with the interaction between pRb and the E2F transcriptional activation domain. The other major contact is between the flexible carboxy terminal tail of pRb and the marked box domains of the E2F/DP heterodimers. This contact is also modulated by two clusters of phosphorylation sites, S788/S795 and T821/T826. S788/S795 phosphorylation directly destabilizes one set of interactions while T821/T826 phosphorylation induces an intramolecular interaction between the tail and the dual cyclin fold that indirectly destabilizes the remaining interaction (Rubin *et al.*, 2005).

All cell cycle CDKs that have been tested are able to phosphorylate pRb *in vitro*, and they do so with some site specificity (Connell-Crowley *et al.*, 1997; Takaki *et al.*, 2005; Zarkowska and Mittnacht, 1997). Other proline directed serine/threonine kinases can also phosphorylate pRb including the p38 stress activated kinase 1 (Hou *et al.*, 2002; Nath *et al.*, 2003), ERK1/2

(Guo *et al.*, 2005), AMP-activated protein kinase (Dasgupta *et al.*, 2009), protein kinase C beta (Suzuma *et al.*, 2002), Aurora B (Nair *et al.*, 2009), Chk1/2 (Inoue *et al.*, 2007), and Raf-1 (Dasgupta *et al.*, 2004). The particular subsets of sites phosphorylated by these kinases are overlapping, but typically different. While some phosphorylation sites appear to be critical for regulation (Connell-Crowley *et al.*, 1997; Knudsen and Wang, 1996), how phosphorylation at individual sites or combinations of sites regulates pRb protein interactions and function has not been clearly defined (Barrientes *et al.*, 2000; Brown *et al.*, 1999; Knudsen and Wang, 1997). While dephosphorylation of pRb may be dynamic and occur throughout the cell cycle, pRb is converted to a largely unphosphorylated form as cells transit mitosis (Durfee *et al.*, 1993; Ludlow *et al.*, 1993).

These observations define the pRb-mediated cell cycle switch that currently serves as the paradigm for *RB1* tumor suppressor function (Fig. 5.2). In its unphosphorylated active state, pRb binds to E2F at the promoters of E2F regulated cell proliferation genes. *RB1* protein prevents assembly of protein complexes required for transcriptional transactivation and instead assembles alternative complexes that alter local chromatin structure to silence gene expression. Mitogenic signaling pathways impinge on pRb by altering its phosphorylation status. These phosphorylation events alter pRb protein interactions resulting in derepression and activation of E2F target cell cycle genes, including cyclins. CDK activity is enhanced by increased cyclin levels, and increased CDK activity augments pRb phosphorylation forming a positive feedback loop. This positive feedback amplifies the mitogenic signals, inactivates the pRb-mediated block to cell proliferation, and commits cells to a round of cell division. The switch is also subject to negative feedback control as the *RB1* gene itself is subject to transcriptional control by pRb/E2F (Burkhart *et al.*, 2010). Increases in E2F transcriptional transactivation increase *RB1* expression which, in turn, dampens E2F activity. Positive and negative feedback control makes the pRb/E2F switch bistable (Yao *et al.*, 2008). Negative feedback keeps the initial threshold of mitogenic stimulation necessary to derepress and activate transcription of E2F target genes high. Positive feedback through activation of CDK activity lowers the threshold of mitogenic stimulation required to maintain the switch in an "on" position. Thus, the switch tends to exist in either an off or on state, spending less time in intermediate states. As cells transit mitosis, pRb is dephosphorylated. This resets the switch, making mitogenic stimulation necessary again for continued cell proliferation.

It is important to note that relatively normal cell cycles occur in the complete absence of the pocket protein family (Sage *et al.*, 2000). Rather than execution of the cell cycle, their role is in regulating entry into and transit through different phases of the cell cycle. Thus, pocket proteins translate the output of signaling pathways into binary go/no go decisions on whether to enter and traverse the cell cycle. This switch is important in

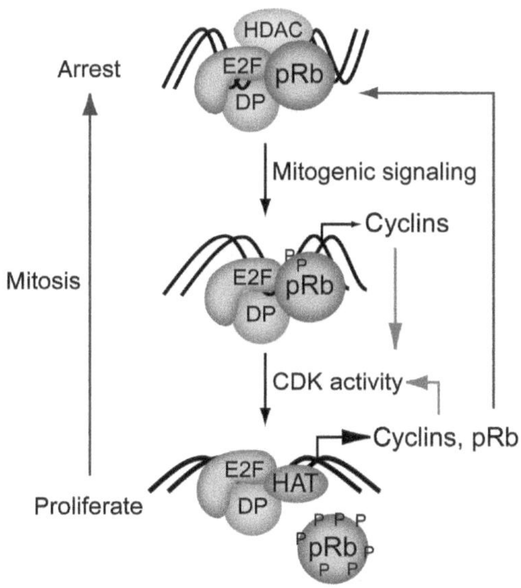

Figure 5.2 The pRb/E2F cell proliferation switch. The pRb/E2F switch is bistable, existing primarily on one of two states, either permissive or nonpermissive for cell proliferation. Bistability is established by both positive (red lines) and negative (blue line) feedback loops (see text). In its unphosphorylated state, pRb complexes with E2F transcription factors at gene promoters, blocks the ability of E2F to recruit transcriptional coactivators like histone acetylases (HAT), and recruits corepressors like histone deacetylases (HDAC). These coactivators and corepressors alter chromatin to a more open or condensed states, respectively. Mitogenic kinases phosphorylate pRb, altering its ability to interact with E2F and transcriptional corepressors. The resulting transcriptional derepression of genes like cyclins increases cyclin dependent kinase activity (CDK). Increased CDK activity, in turn, augments pRb phosphorylation thus creating a positive feedback loop that amplifies the mitogenic signal and triggers commitment to a round of cell division. Negative feedback is provided by the fact that the *Rb1* promoter is itself regulated by E2F. As cells transit mitosis, pRb is dephosphorylated, resetting the switch to a nonproliferative state. (See Color Insert.)

many biological contexts that require coordination with the cell cycle including DNA damage responses, cellular differentiation, cell death signaling, contact inhibition, and tissue morphogenesis among others. Loss of pRb alters the dynamics of this switch, making the conditions necessary for cell proliferation in these different contexts more permissive. Permissive cell proliferation is presumably a major reason why pRb loss has profound effects on tumorigenesis.

While the pRb/E2F cell cycle switch is an intuitively satisfying explanation for *RB1* mediated tumor suppression, there is evidence that the switch is more versatile and can be used to regulate gene expression in biological contexts unrelated to cell cycle control. For example, compound loss of *E2F3* rescues a cell cycle independent defect in neuronal migration observed in *Rb1* null mice, presumably because the pRb/E2F switch regulates genes like neogenin with known roles in neuronal migration (McClellan *et al.*, 2007a). Similarly, compound loss of *E2F3a* rescues a cell cycle-independent defect in the differentiation of retinal starburst amacrine cells that occurs in *Rb1*-deficient mice (Chen *et al.*, 2007). *Rb1* loss also causes a defect in retinal rod cell differentiation that appears to be unrelated to the cell cycle (Zhang *et al.*, 2004). Interestingly, the pRb/E2F switch is also used to regulate metabolic responses. pRb/E2F regulates insulin secretion by pancreatic β cells through modulation of *KIR6.2* expression, a gene that encodes a K+ ATP channel that controls membrane polarization (Annicotte *et al.*, 2009). CDK4 is activated in response to glucose in these cells, resulting in pRb phosphorylation, liberation of pE2F, and increased *KIR6.2* transcription. This feedback loop involving the pRb/E2F switch thus contributes to blood glucose homeostasis. Such examples likely underrepresent the significance of cell cycle independent functions of the pRb/E2F switch as it is often difficult to disentangle direct effects from the indirect effects caused by deregulated cell proliferation. Thus, it appears the pRb/E2F switch has been co-opted to regulate gene expression for a number of different purposes. How the pRb/E2F complexes are specifically directed to regulate unique subsets of genes in these different biological contexts remains unknown.

4. Evolutionary Conservation of the Switch

E2F and pocket proteins are conserved through evolution, with orthologues present in vertebrates, invertebrates, plants, and unicellular algae. The invertebrate orthologues most thoroughly studied are from *Drosophila melanogaster* and *Caenorhabditis elegans* where the reduced number of E2F and pocket protein family members has facilitated analysis. Studies of these model organisms have not only verified the evolutionary conservation of the pocket protein cell cycle switch, but have also yielded new potential roles for the switch in DNA replication and cell fate commitment.

Drosophila contains two pocket protein family members (*RBF1-2*), one activating E2F (*dE2F1*), one repressor E2F (*dE2F2*), and one DP partner for E2F (*dDP*) (van den Heuvel and Dyson, 2008). Activating *dE2F1* is essential for normal cell proliferation in flies (Duronio *et al.*, 1995), foreshadowing an analogous cell proliferation requirement for activating *E2F1-3* observed in mouse fibroblasts (Wu *et al.*, 2001). *RBF1* is required for embryogenesis in *Drosophila*, and *RBF1* deficient cells exhibit deregulated *dE2F1* target gene expression, loss of normal cell cycle control, and increased apoptosis (Du and Dyson, 1999). These phenotypes are broadly similar to those observed in mice lacking *Rb1* (Clarke *et al.*, 1992; Jacks *et al.*, 1992; Lee *et al.*, 1992).

Phenotypes observed in *RBF1* deficient flies can be rescued by mutations in *dE2F1*, in particular mutations that specifically compromise its ability to transactivate transcription (Du, 2000). This observation suggests that the ability of pRb to block E2F transactivation, rather than its ability to actively silence gene expression, is paramount for control of cell proliferation in *Drosophila*. There is also evidence from studies in *Drosophila* that the pRb/E2F switch is responsive to signaling pathways. For example, Notch signaling promotes release of dE2F1 from RBF1, expression of dE2F1 cell cycle target genes, and cell proliferation (Baonza and Freeman, 2005).

C. elegans contains a single *RB1* like orthologue (*lin-35*), a single DP gene (*DPL-1*), and three possible E2F-like genes (*EFL-1*, *EFL-2*, *F49E12.6*). In worms, the pRb/E2F switch appears to have a less prominent role in the regulation of cell proliferation. *Lin-35* mutant worms have subtle cell cycle defects (Grishok and Sharp, 2005; Ouellet and Roy, 2007), and the canonical pRb/E2F cell cycle switch is only revealed upon combining multiple mutations in switch components (Boxem and van den Heuvel, 2001, 2002; Fay *et al.*, 2002; Park and Krause, 1999). Rather, studies in *C. elegans* have emphasized a role for the pRb/E2F switch in cell fate commitment. Mutations of *lin-35*, *dpl-1*, or *efl-1* all have similar phenotypes (Ceol and Horvitz, 2001; Page *et al.*, 2001) suggesting, they function as a unit to repress gene transcription (Kirienko and Fay, 2007). Thus in worms, it appears the ability of the pRb/E2F switch to actively silence gene expression may be most important for the cell fate commitment phenotypes observed. Of particular interest is that pRb/E2F may be involved in specifying the distinction between germline and somatic gene expression patterns. Germline-specific expression patterns in worms are maintained by a number of different mechanisms including epigenetic chromatin modifications and RNA interference (Bender *et al.*, 2006; Fong *et al.*, 2002; Ketting *et al.*, 1999; Sijen and Plasterk, 2003). Mutation of *lin-35* causes a soma to germline transformation in expression pattern that is mediated, at least in part, by chromatin modifying factors like the MES-4 histone methyltransferase (Wang *et al.*, 2005). In *C. elegans*, the pRb/E2F switch appears to facilitate a soma associated gene expression pattern by maintaining repressive chromatin at germline-specific genes. Loss of this switch presumably leads to a relaxed chromatin environment permissive for transcription of these germline genes.

The essential features of the pRb/E2F cell cycle switch, including its role in regulating chromatin structure and gene expression, are also conserved in plants (Sanchez Mde *et al.*, 2008; Shen, 2002). *Arabidopsis thaliana* contains a single *RB1* related gene (*RBR*). As in animals, studies in *Arabidopsis* have reinforced the notion that pocket proteins have context dependent roles in regulating both cell fate commitment and cell proliferation. RNA interference-induced knockdown of *RBR* expression causes a defect in the ability of stem cells to commit to a differentiating lineage

(Borghi *et al.*, 2010; Wildwater *et al.*, 2005). This initially results in expansion of the stem cell pool, but ultimately compromises stem cell maintenance causing an arrest of plant development. Increased expression of *RBR* has the opposite effect, decreasing the number of multipotent stem cells due to over commitment to differentiating lineages. While the phenotype of these plants is associated with disruption of the normal patterns of cell proliferation, the effects of *RBR* on stem cells appear to be primarily in cell fate commitment rather than deregulated cell proliferation. Loss of *RBR* also compromises gametogenesis (Chen *et al.*, 2009b; Ebel *et al.*, 2004; Ingouff *et al.*, 2006). In this case, however, defects are attributed primarily to deregulated cell proliferation.

A comparative analysis of the pRb and pE2F function in evolutionarily divergent species reinforces the hypothesis that the pRb/E2F switch functions to control assembly and disassembly of protein complexes at specific regions of the genome. The switch has been adapted for various context dependent uses. A primary function is in the control of gene expression, and the genes it regulates can directly influence cell proliferation, differentiation, migration, and metabolic homeostasis, to name a few.

5. Variations on the Switch

The complexity of the pRb/E2F switch is considerable given that the above discussion has summarized a small subset of the literature published on the topic. This complexity brings to mind the principle that evolution often builds upon a good thing. Once a useful molecular tool is built to satisfy a particular need, it is often reused to solve new problems in different biological contexts. By creating variations on the basic tool, its repertoire of possible functions can be enlarged. As an adaptor, pRb is ideally suited for functional diversity. Increasing the number of sequence-specific DNA binding factors it interacts with will increase the variety of locations where it can be recruited. By increasing the variety of proteins, it interacts with at those locations, the possible biological effects can be expanded and more finely tuned. The possible combinatorial permutations increase functional diversity further still.

RB1 protein can interact with a number of sequence-specific DNA binding transcription factors beyond E2F including MyoD (Gu *et al.*, 1993), C/EBPs (Chen *et al.*, 1996b), NF-IL6 (Chen *et al.*, 1996a), ATF2 (Kim *et al.*, 1992), Pax8 (Miccadei *et al.*, 2005), CBFA1/RUNX2 (Thomas *et al.*, 2001), Tbx2 (Vance *et al.*, 2010), nuclear hormone receptors (Batsche *et al.*, 2005; Lu and Danielsen, 1998; Singh *et al.*, 1995), and GATA1(Kadri *et al.*, 2009). In a number of cases, loss of *RB1* compromises transcriptional activation and differentiation that is dependent on such transcription factors. Thus in contrast to the effects of the canonical pRb/E2F switch in blocking gene expression, pRb can also function as a transcriptional coactivator for lineage-specific transcription factors. For example, pRb augments differentiation into muscle cells by binding muscle-specific transcription factors like MyoD and augmenting their ability to drive expression of genes required for muscle differentiation and function (Gu *et al.*, 1993; Novitch *et al.*, 1996, 1999; Schneider *et al.*, 1994; Zacksenhaus *et al.*, 1996). Likewise, *RB1* can function as a coactivator for the CBFA1/RUNX2 and the C/EBP transcription factors that participate in osteogenic and adipogenic differentiation, respectively (Chen *et al.*, 1996b; Thomas *et al.*, 2001). Importantly, the differentiation defects that result appear to be unrelated to cell proliferation. For example, adipogenesis defects are independent of cell proliferation since compound loss of *E2F4* rescues cell proliferation defects, but does not rescue differentiation defects (Landsberg *et al.*, 2003).

It seems likely that additional interactions between pRb and tissue-specific transcription factors necessary for differentiation may exist. *Rb1* deficiency *in vivo* affects the differentiation of a relatively large number of distinct tissues and cell types. *Rb1* null mice die in midgestation (Clarke *et al.*, 1992; Jacks *et al.*, 1992; Lee *et al.*, 1992) with defects in the eye lens, brain, peripheral nervous system, muscle, liver, placenta, and the hematopoietic system (Iavarone *et al.*, 2004; Novitch *et al.*, 1996; Spike *et al.*, 2004; Tsai *et al.*, 1998; Wu *et al.*, 2003; Zacksenhaus *et al.*, 1996). Conditional *Rb1* loss in adult mice affects the differentiation of additional cell types and tissues including the epidermis (Balsitis *et al.*, 2003; Ruiz *et al.*, 2004), melanocytes (Yu *et al.*, 2003), hair cells (Sage *et al.*, 2005), prostate (Maddison *et al.*, 2004), lung (Wikenheiser-Brokamp, 2004), cerebellum (Marino *et al.*, 2003), pituitary (Vooijs *et al.*, 1998), small intestine (Guo *et al.*, 2009), and retina (Chen *et al.*, 2004; MacPherson *et al.*, 2004; Zhang *et al.*, 2004) among others. The mechanisms responsible for these defects have not been completely defined. While the effects of deregulated E2F and cell proliferation probably contribute, it is possible that disruption of pRb interactions with currently unidentified tissue-specific transcription factors also plays a part in these phenotypes.

RB1 function is thus extended by interaction with additional sequence-specific transcription factors that recruit pRb to additional genome locations in different biological contexts. The observation that pRb can function as a transcriptional coactivator indicates that the net effect of pRb recruitment is not always to repress transcription. It is currently not clear how pRb functions as a transcriptional coactivator. One possible mechanism is that pRb can bind and block the activity of differentiation inhibitors like EID-1 (MacLellan *et al.*, 2000; Miyake *et al.*, 2000) and Id2 (Iavarone *et al.*, 1994, 2004). EID-1 binds and inhibits p300 histone acetylase activity that normally functions as a coactivator for transcription factors like MyoD. When bound by pRb, EID-1 is targeted for proteolytic degradation. In this case, pRb inhibits an inhibitor of a transcriptional coactivator. In a similar way, pRb also inhibits the Id2 differentiation inhibitor. Id2 functions as a dominant negative helix-loop-helix protein that heterodimerizes with basic helix-loop-helix transcription factors. Heterodimerization with Id2 prevents these transcription factors from binding their E box response elements on DNA, thus preventing transcriptional activation. Locally recruited pRb can compete for binding Id2, thus shifting the equilibrium away from the inhibitory Id2/bHLH interaction. Thus, pRb recruitment may disassemble inhibitory interactions that prevent transcriptional activation.

It is also conceivable that pRb may coactivate transcription by altering chromatin structure. While many repressive chromatin modifying factors are recruited or potentiated by pRb, it is also possible that pRb may oppose their activity. RBP2 is a histone demethylase specific for di- and trimethylated H3K4. Trimethylated H3K4 typically marks chromatin that is permissive for transcription (Christensen *et al.*, 2007; Iwase *et al.*, 2007; Klose *et al.*, 2007). Removal of this methyl mark by RBP2 is expected to tip the balance in favor of repressive chromatin. Consistent with this, knockdown of RBP2 expression facilitates the transcription of genes necessary for differentiation in some contexts (Benevolenskaya *et al.*, 2005). Interaction with pRb displaces RBP2 from promoters of genes such as osteocalcin, thus facilitating transcriptional activation. While pRb interaction opposes transcriptional repression by RBP2 at some promoters, they actually work together to activate transcription at other promoters. For example, binding of pRb and RBP2 to the *BRD2* and *BRD8* promoters coincides with increased expression of both native and reporter genes (Benevolenskaya *et al.*, 2005). The mechanisms underlying this observed transcriptional activation are not currently known.

Such observations make clear that it is currently difficult to predict what the net effect of pRb will be on gene expression. While pRb recruitment typically results in transcriptional repression, there clearly are exceptions where pRb is required for robust transcriptional activation. Given that the preponderance of studies to date examines the effects of pRb/E2F interaction at cell cycle related genes, examples where pRb functions as a transcriptional activator may be underrepresented in the published literature. It is not well understood why pRb functions to repress transcription at some locations, but functions as a transcriptional activator at others. What is becoming clear is that the effects of pRb recruitment to a particular location are highly context dependent, probably reflecting the diversity of proteins present within the local environment. Current efforts to map the genome location of pocket proteins, transcription factors, chromatin modifying factors, and histone marks will likely inform these issues, especially when correlated with gene expression profiling.

6. *Rb1* Function Beyond Transcription

Studies in *Drosophila* have provided some of the earliest clues that the pRb/E2F switch, or variations of the switch, may be used for purposes beyond transcriptional regulation. While it is clear that the switch can influence DNA replication indirectly through the regulation of gene expression in a number of biological contexts, the effects of pRb/E2F loss appear to be too rapid to be accounted for by changes in gene expression. A more direct role for pRb/E2F in DNA replication is suggested by the observation that *Drosophila* RBF and dE2f are found in complex with the origin recognition complex (DmORC) during chorion gene amplification (Bosco *et al.*, 2001). Elimination of the RBF/dE2F complex by *RBF* or *dE2F* mutation increases DNA amplification and genomic replication without detectable effects on E2F target gene expression (Bosco *et al.*, 2001; Royzman *et al.*, 1999). Thus, pRb/E2F appears to directly inhibit the activity of DmORC.

While it is not well understood how pRb/E2F inhibits DmORC, some possibilities have been suggested. DNA replication initiation during *Drosophila* chorion gene amplification is associated with acetylation of histones at the origin (Hartl *et al.*, 2007). Chromatin structure is known to regulate replication origin activity in *Drosophila* (Aggarwal and Calvi, 2004). Loss of RBF results in the persistence of acetylated histones and increased DNA replication, perhaps due to lack of histone deacetylase activity normally recruited by pRb. This suggests that the pRb/E2F adaptor is used in this case to alter chromatin structure for regulation of DNA replication rather than gene transcription. Other possible mechanisms have been suggested by mammalian systems. For example, pRb interacts with replication factors such as MCM7 and replication factor C, blocking activation of the prereplication complex (Mukherjee *et al.*, 2010; Pennaneach *et al.*, 2001; Sterner *et al.*, 1998). *RB1* protein has also been suggested to impede the loading of the DNA replication processivity factor PCNA (Angus *et al.*, 2004). Consistent with these findings, pRb can inhibit DNA replication *in vitro*, and this pRb-mediated activity is regulated by CDK phosphorylation (Gladden and Diehl, 2003; Sterner *et al.*, 1998). Thus, targeting pRb to replication origins may impede assembly or activation of protein complexes necessary for initiation and elongation of DNA replication.

Convincing evidence has recently been accumulating that pRb may also play a direct role in mitosis. Loss of pRb has long been known to compromise genome stability, and one component of this instability is reflected in the appearance of aneuploidy. Chromosome losses occur with a frequency that is three orders of magnitude higher in embryonic stem cells lacking pRb compared to wild-type cells (Zheng *et al.*, 2002). Acute loss of pRb in human or mouse primary fibroblasts causes supernumerary centrosomes, aneuploidy, and micronuclei (Amato *et al.*, 2009; Iovino *et al.*, 2006). Both loss of normal spindle assembly checkpoint control and defects in centrosome homeostasis can contribute to such effects. The pRb/E2F switch does regulate a number of genes involved in these processes (Ishida *et al.*, 2001; Ren *et al.*, 2002), and deregulated expression of these genes does contribute to chromosome instability observed upon pRb loss. For example, the mitotic checkpoint gene *MAD2* arrests cells in metaphase by blocking activation of the anaphase promoting complex that is required for destruction of the chromosome cohesion regulator Securin. *MAD2* expression is controlled by the pRb/E2F switch, and *MAD2* overexpression resulting from pRb loss leads to aneuploidy and tumorigenesis (Hernando *et al.*, 2004; Sotillo *et al.*, 2007).

While transcriptional effects of the pRb/E2F switch are clearly important for chromosome stability, pRb also plays a direct, nontranscriptional role in chromosome structure and mitosis. Loss of *Drosophila* RBF causes a defect in normal chromosome condensation in mitotic cells; chromosomes in these animals contain regions of condensed chromatin interspersed with regions of uncondensed chromatin. Both genetic and biochemical data indicate that this effect is mediated by direct interaction of pRb with the condensin II complex subunit CAP-D3 (Longworth *et al.*, 2008). RBF appears to be required for optimal loading of CAP-D3 onto chromosomes. This function appears to be specific for CAP-D3 as other condensin subunits localize normally to chromosomes in the absence of RBF. As the two condensin complexes load to distinct regions of chromatin, the specificity of the pRb/CAP-D3 interaction may explain the region-specific chromatin condensation defect.

This pRb/CAP-D3 interaction is also observed in human and rodent cells (Longworth *et al.*, 2008; Manning *et al.*, 2010; van Harn *et al.*, 2010). RNAi mediated silencing of *Rb1* in immortalized rat pigment epithelial cells causes a number of subtle mitotic defects including delayed progression through mitosis, an increased intercentromeric distance, premature loss of sister chromatid cohesion, a defect in chromosome congression, and lagging chromosomes in anaphase. These defects are associated with a deficiency in the loading of cohesin and CAP-D3 to centromeric locations. The net effect of these defects is to cause chromosome missegregation at a frequency comparable to that observed in human tumor cell lines, suggesting that chromosome instability upon pRb loss may contribute to tumorigenesis.

Many chromatin regulators contain the so-called "LXCXE" amino acid motif that binds pRb through an interaction surface in the pocket tandem cyclin fold. Alanine substitutions at three critical amino acids in this pRb interaction surface disrupt interactions with LXCXE containing proteins, and a knock-in *Rb1* allele containing these mutations has been created in the mouse (Isaac *et al.*, 2006). Mice homozygous for the mutant allele are viable without pronounced cell cycle or differentiation defects. However, cells from these mice have subtle defects in chromatin structure, particularly a reduction in trimethylated H4K20 at pericentric heterochromatin. The mutant pRb is unable to bind CAP-D3, and mutant cells show defects in chromosome loading of CAP-D3 and other condensin II components. Because of these defects, mutant cells take longer to complete mitosis and exhibit aneuploidy. Importantly, the mutant *Rb1* allele exacerbates tumorigenesis in *Trp53* deficient mice (Coschi *et al.*, 2010). The median time to death is reduced in these double mutant mice, compared to mice deficient only for *Trp53*, and tumors from these mice exhibit increased numbers of chromosomal changes.

These observations reinforce the hypothesis that pRb has direct functions in organizing chromosome structure during mitosis, and these functions are independent of the canonical pRb/E2F switch. Importantly, loss of these functions in mitosis can contribute directly to tumorigenesis. While these studies demonstrate that pRb plays a role in recruiting condensin II complexes, other mechanisms of regulating chromosome condensation during mitosis exist (Takemoto *et al.*, 2009). Further, it is clear that pocket proteins are not strictly required for mitosis, but rather contribute to the fidelity of the process. It is currently unknown how pRb is located to particular regions of chromosomes during mitosis to assist in recruitment of condensin complexes, or whether pRb has additional effects on the function of these complexes once recruited to the chromosome. A possibly relevant observation is that pRb is also known to directly interact with the targeting subunits of E3 ubiquitin ligases like SCF^{skp2} and APC^{cdh1} (Binne *et al.*, 2007; Ji *et al.*, 2004) and such interactions are important for tumorigenesis (Wang *et al.*, 2010). Regulated protein degradation is required for proper execution of mitosis. Thus, pRb can recruit biochemical activities that may be relevant to mitosis.

Most of the examples reviewed here have emphasized the effects of pRb on local chromatin structure and gene expression. The effects of pRb deficiency on chromosome structure during mitosis suggest that it may also contribute to the organization of chromatin domains over larger regions (Longworth and Dyson, 2009). The emerging roles of pRb in cellular senescence are consistent with this possibility. Cellular senescence is a physiological response to a variety of stresses that result in long-term or permanent cessation of cell proliferation. It provides an important barrier to tumorigenesis (Prieur and Peeper, 2008). *RB1* and *TP53* represent two key tumor suppressor gene pathways that are required for enforcing this senescence response (Courtois-Cox *et al.*, 2008). It is not obvious whether pRb is essential for the initiation or maintenance of senescence, or both. Loss of all three pocket protein family members is sufficient to trigger cell cycle reentry in senescent mouse embryonic fibroblasts (Sage *et al.*, 2003), suggesting they are at least required for maintenance of the senescent state. Ectopic expression of pRb in some cell lines can induce a senescence response, suggesting pRb may also be involved in initiation (Hinds *et al.*, 1992; Huang *et al.*, 1988; Templeton *et al.*, 1991). The molecular mechanisms underlying pRb's role in cellular senescence are not completely defined, but the pRb/E2F switch appears to be involved. The pRb/E2F switch regulates a subset E2F target genes, particularly those involved in DNA replication, specifically in the context of cellular senescence (Chicas *et al.*, 2010). The regulation of these genes is unique to pRb among the pocket proteins.

Of note is that in order to enforce the stability of the nonproliferative state, cellular senescence is accompanied by global reorganization of chromatin domains. In particular, senescence is associated with the appearance of microscopically visible foci of heterochromatin called senescence associated heterochromatin foci (SAHF; Narita *et al.*, 2003). These foci are highly enriched for trimethylated H3K9, the methyl-lysine binding proteins HP1, the histone variant macroH2A, chromatin regulators HIRA and ASF1a, and high mobility group A proteins (Funayama *et al.*, 2006; Narita *et al.*, 2006; Zhang *et al.*, 2005). These factors participate in the formation of the transcriptionally repressive facultative heterochromatin at SAHF. A number of the genes within SAHF are E2F target genes, and both pRb and marks of heterochromatin can be found at these genes. As discussed above, pRb can interact with a number of chromatin modifying factors involved in SAHF formation, including the H3K9 methylase SUV39H1, HP1, and histone deacetylases (Nielsen *et al.*, 2001; Vaute *et al.*, 2002).

These observations have led to the hypothesis that pRb can specify the regional organization of chromatin into SAHF during senescence. Recruitment of pRb to specific genomic location potentiates H3K9 methylation that attracts HP1. HP1 recruitment is augmented by local histone deacetylases. HP1 facilitates the formation of local heterochromatin as well as heterochromatin spreading (Verschure *et al.*, 2005). This hypothesis is supported by evidence indicating that loss of pRb changes the dynamics of HP1 localization, the sensitivity of chromatin to Micrococcal nuclease digestion, and changes in histone H1 phosphorylation (Herrera *et al.*, 1996; Siddiqui *et al.*, 2007). These changes are consistent with a more open chromatin structure in the absence of pRb. Thus, pRb may participate in the organization of chromatin structure in contexts as diverse as cellular senescence and mitosis.

7. *Rb1*, Stem Cells, and the Cell of Tumor Origin

How does pRb loss initiate retinoblastoma? Given the diversity of pRb function reviewed, the answer is certainly not obvious. As much of this functional diversity is context dependent, it is difficult to learn how pRb loss initiates tumorigenesis without knowing the type of cell it is lost in. Thus, it is useful to know the cell of tumor origin. Cancer is thought to be a clonal disease in an evolutionary sense. It arises from a single ancestor cell that undergoes rounds of genetic and epigenetic alteration followed by selection, eventually evolving into a malignant cancer. Thus, identifying the ancestor cell of origin for a particular type of cancer is important as it allows identification of the initiating genetic mutations and how those mutations affect the cells most relevant to tumor initiation.

In the case of retinoblastoma, the initiating genetic mutation is known. Despite the simple molecular etiology of retinoblastoma, however, the cell of retinoblastoma origin has remained controversial for more than a hundred years since the initial attempts at assignment based on similarity to photoreceptors (Wintersteiner, 1897). The crux of the problem is that the cell of origin is inferred from the morphological, molecular, and functional properties of the presenting tumors, and retinoblastoma exhibits features of multiple cell types. Recent work has provided support for the hypothesis that a photoreceptor is the cell of origin (Xu *et al.*, 2009), but other work has suggested that an interneuron (Johnson *et al.*, 2007; Laurie *et al.*, 2006) or a progenitor cell is the tumor originating cell (Kyritsis *et al.*, 1984). This is not just due to the mixture of multiple cell types within a heterogeneous tumor, as a recent unpublished examination of human and mouse retinoblastoma suggests individual retinoblastoma cells simultaneously exhibit molecular, cellular, and neurochemical features of multiple cell types (M. Dyer, personal communication). The cells express markers that are specific for progenitor, interneuron, and photoreceptor cells. The expression of such markers is normally mutually exclusive. This suggests it may not be possible to infer the retinoblastoma cell of origin from examination of retinoblastoma cells as they take on the identity of multiple cell lineages. What can be inferred, however, is that retinoblastoma cells have lost the ability to establish or maintain differentiated gene expression patterns associated with these distinct cell lineages. Thus we are left to guess the cell of retinoblastoma origin, but this guess is informed by the knowledge that it will most likely be a cell where pRb loss has the greatest potential to disrupt cell fate commitment and elaboration of lineage specifying gene expression programs.

We have reviewed evidence that supports the hypothesis that the core function of pRb is to serve as a signaling pathway responsive adaptor that regulates the assembly or activity of protein complexes at specific locations in the genome. Most typically, this results in alteration of local and regional chromatin structure to a more closed confirmation that is repressive for gene transcription. Based on this, we suggest that a relatively uncommitted stem or progenitor cell is the most likely candidate for the retinoblastoma cell of origin. Loss of pRb is expected to cause a loosening of local and regional chromatin making it more permissive for transcription. We surmise that stem or progenitor cells are more sensitive to such effects as they exhibit a neutral chromatin structure at important lineage specifying genes. This neutral chromatin exhibits histone modifications associated with both permissive and repressive chromatin suggesting it is poised for either transcriptional activation or repression in accordance with cell fate commitment (Hemberger *et al.*, 2009; Mohn and Schubeler, 2009; Sang *et al.*, 2009). Loss of pRb in such cells is expected to tip the balance toward transcriptional activation by preventing stable transcriptional repression. This would potentially allow simultaneous expression of genes that are normally restricted to distinct lineages. Loss of pRb in more committed, differentiated cells would be expected to have less effect on gene expression patterns as permissive or repressive chromatin states are reinforced by multiple redundant mechanisms, some of which may not be impacted by pRb loss. Thus the relative plasticity of stem or progenitor cells, as reflected in the programming of their chromatin structure, may make them particularly sensitive to pRb loss (Fig. 5.3).

This is not to say that loss of pRb cannot initiate retinoblastoma in lineage committed cells. Indeed, mouse models of retinoblastoma have been created that comprise interneurons with a striking degree of morphological, molecular, and functional differentiation (Ajioka *et al.*, 2007). Rather, we suggest that the probability of transformation decreases with increasing lineage commitment and differentiation. Of course, the lower probability of transformation may be offset by larger numbers of lineage committed cells. One potentially salient observation is that retinoblastoma is never initiated in adults, even if they have inherited an *RB1* mutation. The penetrance of hereditary retinoblastoma is very high, but not 100%, such that if a person inheriting a mutant *RB1* allele does not get retinoblastoma as a child, they will never get it. This suggests the cell of retinoblastoma origin exists only during a narrow window of retinal development, past which the cells change or differentiate in such a way as to make them resistant to the effects of *RB1* loss. This argues against the candidacy of cell types that persist into adulthood for the cell of retinoblastoma origin. While we favor this interpretation, there may be something unique about the pediatric retinal microenvironment that facilitates the genesis of retinoblastoma. In this case, the cell of tumor origin persists into adulthood, but the microenvironment cannot support neoplastic transformation and tumorigenesis.

Beyond retinoblastoma, small cell lung cancer (SCLC) exhibits the highest rate of *RB1* mutation in human cancers, up to 90% (Wistuba *et al.*, 2001). Hereditary retinoblastoma patients do exhibit an increased risk of lung cancer (Marees *et al.*, 2008). Thus *RB1* mutation may be an initiating event in small cell lung carcinoma. Consistent with this notion, compound loss of pRb and p53 in the mouse lung generates tumors with strong resemblance to human SCLC (Meuwissen *et al.*, 2003). SCLC expresses genes characteristic of the neuroendocrine lineage like *ASCL1*,

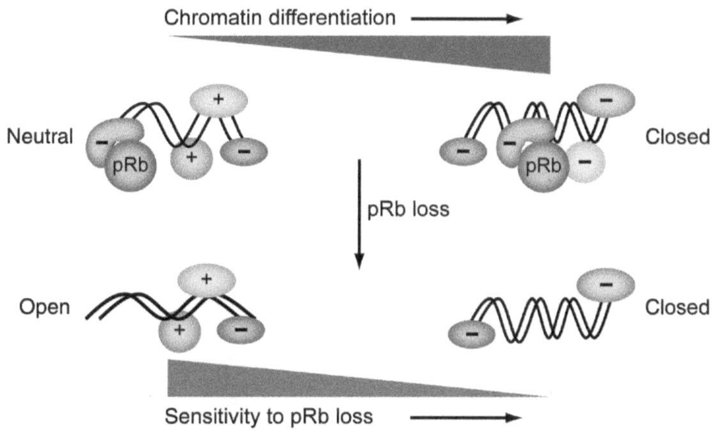

Figure 5.3 Stem/progenitor cells and sensitivity to pRb loss. The figure summarizes the hypothesis that there exists an inverse correlation between differentiation and sensitivity to pRb loss. Stem/progenitor cells are relatively undifferentiated and have a neutral chromatin state at key lineage specifying regulatory genes. Chromatin at such genes is characterized by the simultaneous presence of protein complexes and chromatin marks characteristic of both open (shapes marked +) and closed chromatin (shapes marked −). pRb is typically associated with factors that create a closed chromatin state repressive for transcription (although there are notable exceptions). Loss of pRb in such cells is expected to have a large effect on gene expression as it will tip the balance in favor of gene expression. In more differentiated cells, chromatin is more rigidly committed to an open or closed state by multiple, redundant mechanisms. Loss of pRb in such cells is expected to have less effect on the expression of stably repressed genes as there are redundant mechanisms in place to maintain the closed chromatin state. (See Color Insert.)

synaptophysin, and *NCAM1*. They also express markers of stem/progenitor cells (Meuwissen *et al.*, 2003; Moreira *et al.*, 2010). Interestingly, *Rb1* loss is required for this unique SCLC phenotype as mouse lung tumors retaining pRb are phenotypically adenocarcinomas. Thus in more than one type of human tissue, stem or progenitor cells may be particularly sensitive to loss of pRb in the context of tumorigenesis.

There is a fair amount of current evidence that supports the hypothesis that stem or progenitor cells are uniquely susceptible to pRb loss. As reviewed above, studies in plants and invertebrates have demonstrated that pRb deficiency causes defects in stem cell pool size and cell fate commitment. There are hints that similar effects occur upon *Rb1* loss in vertebrate stem and progenitor cells. For example, *Rb1* null mouse embryonic stem cells show defects in differentiation *in vitro* (Papadimou *et al.*, 2005). Loss of pRb in mouse trophoblast stem cells causes an abnormal expansion of this stem cell pool (Wenzel *et al.*, 2007). Such defects are not observed when pRb is ablated in more differentiated trophoblast derivatives. Loss of pRb causes stem or progenitor cell-related defects in a number of additional mouse tissues including the skin (Ruiz *et al.*, 2004), bone (Gutierrez *et al.*, 2008), small intestine (Guo *et al.*, 2009), and blood (Macleod, 2008). However, the effects of pRb loss on stem/progenitor cells are likely context dependent. *Rb1* status does not apparently affect the rescue of differentiating prostate tissue by transplantation of stem cell enriched embryonic urogenital sinus (Day *et al.*, 2002). Other pocket protein family members, like p107, may also serve the predominant role in regulating stem cells in some biological contexts (Vanderluit *et al.*, 2004).

There is also evidence that stem or progenitor cells are particularly sensitive to pRb loss in the context of neoplastic transformation. In the mouse, the effects of conditional *Rb1* deletion have been examined in a number of different tissues. Typically, this is not sufficient for tumorigenesis, but tumors often do occur when *Rb1* deletion is combined with other mutations, commonly deletion of *Trp53*. For example, conditional deletion of floxed *Rb1* and *Trp53* alleles using a probasin promoter based Cre transgene causes metastatic prostate cancer (Zhou *et al.*, 2006). Analogous to retinoblastoma, individual prostate tumor cells express markers of multiple lineages, in this case, both luminal epithelial and neuroendocrine lineages. The probasin promoter is androgen responsive and maximally expressed in differentiated prostate epithelial cells in postpubescent males. Despite frequent deletion of *Rb1* and *Trp53* and neoplastic changes in these differentiated epithelial cells, lethal tumors first arise in an anatomical region enriched for prostatic stem cells (Zhou *et al.*, 2007). Early neoplastic lesions in this region coexpress stem cell, epithelial, and neuroendocrine markers, as do a small number of normal cells that exist there. These observations suggest that prostatic stem or progenitor cells are the cell of origin for the first lethal tumors, and thus that they are more sensitive to neoplastic transformation upon pRb loss than differentiated epithelial cells. In a similar situation, deletion of *Rb1* and *Trp53* in the skin causes squamous cell carcinoma that first arises in the hair follicle, an anatomical location enriched with epidermal stem cells (Martinez-Cruz *et al.*, 2008). Early neoplastic lesions in the hair follicle express stem cell markers. It has also been observed that deletion of *Rb1* and *Trp53* in brain stem cells, but not mature astrocytes, causes brain tumors (Jacques *et al.*, 2010). The resulting tumors have a relatively undifferentiated neuroectodermal phenotype. Mouse models of osteosarcoma (Berman *et al.*, 2008) and gastrointestinal tract cancer (Kucherlapati *et al.*, 2008) have provided additional evidence consistent with the involvement of stem, progenitor, or immature precursor cells. These observations reinforce the notion that there may be an inverse relationship between the degree of cellular differentiation and susceptibility to loss of pRb.

8. *RB1* AND TUMOR PROGRESSION

RB1 is clearly important for the initiation of retinoblastoma and likely for other tumors like SCLC. Mouse models have demonstrated that *Rb1* deletion can contribute to the initiation of tumorigenesis in a wide variety of tissues. For many common human malignancies, however, it is clear that *RB1* mutation is a rather late event in tumor progression (Burkhart and Sage, 2008). This indicates that *RB1* mutation does not typically contribute to tumor initiation in some cancers. Assuming late *RB1* mutation is a driver of malignant behavior, why is it selected for late, rather than early, in tumor progression?

One possibility is that a cell type susceptible to tumor initiation upon pRb loss does not exist in a particular tissue, or may be too rare to support a significant probability of tumorigenesis. In such tissues, tumorigenesis is initiated in an *RB1* resistant cell by other genetic alterations. Once the cell is transformed, however, it may become sensitive to the effects of pRb loss, thus providing selection for acquisition of *RB1* mutations that further drive tumor progression. A common feature of tumorigenesis is progressive loss of differentiated features. If this is accompanied by increased plasticity in the programming of chromatin structure, then loss of pRb may be expected to have greater effect on gene expression patterns in such cells than in the original cells from which transformation is initiated. This possibility has not been tested as the cell types susceptible to transformation upon pRb loss have not been well characterized in most tissues, and because mouse models that test the effects of pRb loss in preexisting neoplastic lesions currently do not exist. However, pRb has been implicated in a number of processes that are particularly relevant to the progression of preexisting tumors. For example, autophagy is induced in hypoxic tumors. The pRb/E2F switch regulates a number of genes important for hypoxia induced autophagy (Polager *et al.*, 2008; Tracy *et al.*, 2007). The pRb/E2F switch also regulates genes involved in angiogenesis (Gabellini *et al.*, 2006). Loss of pRb could thus alleviate hypoxic stress as tumors grow and also influence the responses to that stress. We have noted above that pRb functions to enforce cellular senescence and the fidelity of mitosis. Increased chromosome instability and loss of senescence responses in the absence of pRb will also facilitate tumor progression.

Another possibility is that there may be selection against *RB1* mutation early in tumorigenesis that is ultimately relaxed later in tumorigenesis. This would suggest a paradoxical oncogenic role for pRb in early stages of tumorigenesis in some tissues. Surprisingly, there is experimental precedent for such an oncogenic role for pRb. E2F1 activity is a well-known mediator of apoptotic cell death in response to pRb loss (Tsai *et al.*, 1998), and the regulation of E2F1-mediated apoptosis may be a normal physiological function of pRb during the DNA damage response (Inoue *et al.*, 2007; Markham *et al.*, 2006). E2F1-mediated apoptosis appears to be regulated by a unique interaction surface in the pRb carboxy terminal tail (Dick and Dyson, 2003). Thus in some contexts, pRb-mediated suppression of apoptosis may facilitate the survival of neoplastic cells, particularly since they are prone to DNA damage (Halazonetis *et al.*, 2008). Consistent with this, constitutive activation of pRb in the mouse mammary gland initiates tumorigenesis (Jiang and Zacksenhaus, 2002). Further, pRb appears to be required for transformation by oncogenes in some cases (Williams *et al.*, 2006).

9. Summary

Here, we have reviewed the hypothesis that the core function of pRb is as a signaling pathway responsive adaptor that facilitates the assembly or disassembly of location-specific protein complexes on the DNA. *RB1* protein interacts with a number of different sequence-specific DNA binding factors that determine its location. The presence of pRb affects the composition or activity of protein complexes at those locations in a variety of ways. It can recruit new proteins to the location or stabilize their occupancy. The primary example of this is the recruitment of chromatin modifying factors like histone deacetylases. Recruitment of such factors results in the alteration of chromatin structure, typically to a closed state that is repressive for transcription. *RB1* protein can also displace proteins from a location, or inhibit the activity of a protein at that location. In this way, pRb can function as a transcriptional coactivator by blocking the effects of transcriptional inhibitors. In sum, pRb is a major regulator of chromatin structure that establishes gene expression patterns necessary for cell cycle control, cell fate commitment, cellular differentiation, and development. This function is presumed to be a key reason why *RB1* mutation has such profound effects on normal development and cancer.

While the function of pRb as a programmer of chromatin structure is a major theme, another theme is the diversity of pRb function. The pRb adaptor is a tool that has been co-opted for use in many disparate biological contexts relevant to DNA metabolism. For example, the pRb adaptor is not only used to control local chromatin structure at individual genes, but it has also been adapted for use in the regional organization of chromatin structure that is important for mitosis and cellular senescence. The pRb adaptor has also been used to influence DNA replication and DNA repair, either indirectly through modulation of chromatin or directly by modulating the protein complexes that carry out these biochemical processes. As an adaptor, the range of its potential functions is quite large given the number of DNA binding factors it interacts with, the number of DNA modifying proteins it interacts with, the number of signaling pathways that impinge upon these interactions, and the potential combinatorial permutations.

An important implication that follows from this functional diversity is that the consequences of pRb loss will be highly context dependent. As an adaptor, pRb function will be largely defined by the subset of proteins available for interaction in a particular cell or subcellular location, and the signals impinging on that cell which influence pRb posttranslational modification. The state of chromatin in a cell at the time of pRb loss will also heavily influence the resulting effects. Chromatin structure is regulated by multiple and redundant mechanisms. As cells become committed to a particular differentiated cell type, their chromatin tends to become progressively fixed to stably maintain the necessary gene expression pattern. Thus, the impact of pRb loss on gene expression is likely to be greater in a cell with a more plastic chromatin state. The diverse, context-dependent functions of pRb suggest that there may not be a lone universal mechanism of pRb-mediated tumor suppression. How pRb loss impacts tumorigenesis is likely to differ in different tissues.

An important motivation for studying *RB1* is that it will result in effective new approaches for the diagnosis and treatment of a range of human cancer. This has not been realized as quickly as once hoped, possibly because of the large diversity of pRb function. However, with an increasing understanding of the core pRb functions, the context dependent variations on these functions, and the increasingly sophisticated tools available to study these functions, this goal remains in sight. Another key motivation for studying *RB1* is the possibility of discerning important general cancer principles from the study of a case with simple molecular etiology. From one perspective, the dizzying functional diversity of pRb renders this a false hope. From another perspective, what we have learned about *RB1* is echoed in important principles emerging in the field of cancer research. Both the developing fields of cancer stem cells and cancer epigenetics (Gupta *et al.*, 2009; Maenhaut *et al.*, 2010; Sharma *et al.*, 2010) are directly relevant to *RB1*. Perhaps, we are beginning to see the forest for the trees.

ACKNOWLEDGMENTS

We apologize to our colleagues whose work we were unable to cite due to constraints of space and topic. We thank Drs. Jennifer Black and Adam Karpf for critical reading of the chapter. We thank Dr. Mike Dyer for discussion of unpublished data. We are grateful to numerous colleagues in the *RB1* field and members of the Goodrich lab for stimulating conversations. This work is supported by a grant from the National Cancer Institute (CA70292).

REFERENCES

Aggarwal, B. D., and Calvi, B. R. (2004). Chromatin regulates origin activity in *Drosophila* follicle cells. *Nature* **430,** 372–376.

Ajioka, I., Martins, R. A., Bayazitov, I. T., Donovan, S., Johnson, D. A., Frase, S., Cicero, S. A., Boyd, K., Zakharenko, S. S., and Dyer, M. A. (2007). Differentiated horizontal interneurons clonally expand to form metastatic retinoblastoma in mice. *Cell* **131,** 378–390.

Amato, A., Lentini, L., Schillaci, T., Iovino, F., and Di Leonardo, A. (2009). RNAi mediated acute depletion of retinoblastoma protein (pRb) promotes aneuploidy in human primary cells via micronuclei formation. *BMC Cell Biol.* **10,** 79.

Angus, S. P., Mayhew, C. N., Solomon, D. A., Braden, W. A., Markey, M. P., Okuno, Y., Cardoso, M. C., Gilbert, D. M., and Knudsen, E. S. (2004). RB reversibly inhibits DNA replication via two temporally distinct mechanisms. *Mol. Cell. Biol.* **24,** 5404–5420.

Annicotte, J. S., Blanchet, E., Chavey, C., Iankova, I., Costes, S., Assou, S., Teyssier, J., Dalle, S., Sardet, C., and Fajas, L. (2009). The CDK4-pRB-E2F1 pathway controls insulin secretion. *Nat. Cell Biol.* **11,** 1017–1023.

Balsitis, S. J., Sage, J., Duensing, S., Munger, K., Jacks, T., and Lambert, P. F. (2003). Recapitulation of the effects of the human papillomavirus type 16 E7 oncogene on mouse epithelium by somatic Rb deletion and detection of pRb-independent effects of E7 in vivo. *Mol. Cell. Biol.* **23,** 9094–9103.

Baonza, A., and Freeman, M. (2005). Control of cell proliferation in the *Drosophila* eye by Notch signaling. *Dev. Cell* **8,** 529–539.

Barrientes, S., Cooke, C., and Goodrich, D. W. (2000). Glutamic acid mutagenesis of retinoblastoma protein phosphorylation sites has diverse effects on function. *Oncogene* **19,** 562–570.

Batsche, E., Moschopoulos, P., Desroches, J., Bilodeau, S., and Drouin, J. (2005). Retinoblastoma and the related pocket protein p107 act as coactivators of neuroD1 to enhance gene transcription. *J. Biol. Chem.* **280,** 16088–16095.

Bender, L. B., Suh, J., Carroll, C. R., Fong, Y., Fingerman, I. M., Briggs, S. D., Cao, R., Zhang, Y., Reinke, V., and Strome, S. (2006). MES-4: An autosome-associated histone methyltransferase that participates in silencing the X chromosomes in the *C. elegans* germ line. *Development* **133,** 3907–3917.

Benevolenskaya, E. V., Murray, H. L., Branton, P., Young, R. A., and Kaelin, W. G. (2005). Binding of pRB to the PHD protein RBP2 promotes cellular differentiation. *Mol. Cell* **18,** 623–635.

Ben-Porath, I., Weinberg, R. A., Ben-Porath, I., and Weinberg, R. A. (2005). The signals and pathways activating cellular senescence. *Int. J. Biochem. Cell Biol.* **37,** 961–976.

Berman, S. D., Calo, E., Landman, A. S., Danielian, P. S., Miller, E. S., West, J. C., Fonhoue, B. D., Caron, A., Bronson, R., Bouxsein, M. L., Mukherjee, S., and Lees, J. A. (2008). Metastatic osteosarcoma induced by inactivation of Rb and p53 in the osteoblast lineage. *Proc. Natl. Acad. Sci. USA* **105,** 11851–11856.

Binne, U. K., Classon, M. K., Dick, F. A., Wei, W., Rape, M., Kaelin, W. G., Naar, A. M., and Dyson, N. J. (2007). Retinoblastoma protein and anaphase-promoting complex physically interact and functionally cooperate during cell-cycle exit. *Nat. Cell Biol.* **9,** 225–232.

Borghi, L., Gutzat, R., Futterer, J., Laizet, Y.h., Hennig, L., and Gruissem, W. (2010). Arabidopsis retinoblastoma-related is required for stem cell maintenance, cell differentiation, and lateral organ production. *Plant Cell* **22,** 1792–1811.

Bosco, G., Du, W., and Orr-Weaver, T. L. (2001). DNA replication control through interaction of E2F-RB and the origin recognition complex. *Nat. Cell Biol.* **3,** 289–295.

Boxem, M., and van den Heuvel, S. (2001). lin-35 Rb and cki-1 Cip/Kip cooperate in developmental regulation of G1 progression in *C. elegans*. *Development* **128,** 4349–4359.
Boxem, M., and van den Heuvel, S. (2002). *C. elegans* class B synthetic multivulva genes act in G(1) regulation. *Curr. Biol.* **12,** 906–911.
Brehm, A., Miska, E. A., McCance, D. J., Reid, J. L., Bannister, A. J., and Kouzarides, T. (1998). Retinoblastoma protein recruits histone deacetylase to repress transcription. *Nature* **391,** 597–601.
Brown, V. D., Phillips, R. A., and Gallie, B. L. (1999). Cumulative effect of phosphorylation of pRB on regulation of E2F activity. *Mol. Cell. Biol.* **19,** 3246–3256.
Burke, J. R., Deshong, A. J., Pelton, J. G., and Rubin, S. M. (2010). Phosphorylation-induced conformational changes in the retinoblastoma protein inhibit E2F transactivation domain binding. *J. Biol. Chem.* **285,** 16286–16293.
Burkhart, D. L., and Sage, J. (2008). Cellular mechanisms of tumour suppression by the retinoblastoma gene. *Nat. Rev. Cancer* **8,** 671–682.
Burkhart, D. L., Ngai, L. K., Roake, C. M., Viatour, P., Thangavel, C., Ho, V. M., Knudsen, E. S., and Sage, J. (2010). Regulation of RB transcription in vivo by RB family members. *Mol. Cell. Biol.* **30,** 1729–1745.
Ceol, C. J., and Horvitz, H. R. (2001). dpl-1 DP and efl-1 E2F act with lin-35 Rb to antagonize Ras signaling in *C. elegans* vulval development. *Mol. Cell* **7,** 461–473.
Chan, H. M., Krstic-Demonacos, M., Smith, L., Demonacos, C., and La Thangue, N. B. (2001). Acetylation control of the retinoblastoma tumour-suppressor protein. *Nat. Cell Biol.* **3,** 667–674.
Chau, B. N., Wang, J. Y., Chau, B. N., and Wang, J. Y. J. (2003). Coordinated regulation of life and death by RB. *Nat. Rev. Cancer* **3,** 130–138.
Chau, C. M., Deng, Z., Kang, H., and Lieberman, P. M. (2008). Cell cycle association of the retinoblastoma protein Rb and the histone demethylase LSD1 with the Epstein-Barr virus latency promoter Cp. *J. Virol.* **82,** 3428–3437.
Chen, P. L., Riley, D. J., Chen-Kiang, S., and Lee, W. H. (1996a). Retinoblastoma protein directly interacts with and activates the transcription factor NF-IL6. *Proc. Natl. Acad. Sci. USA* **93,** 465–469.
Chen, P. L., Riley, D. J., Chen, Y., and Lee, W. H. (1996b). Retinoblastoma protein positively regulates terminal adipocyte differentiation through direct interaction with C/EBPs. *Genes Dev.* **10,** 2794–2804.
Chen, D., Livne-bar, I., Vanderluit, J. L., Slack, R. S., Agochiya, M., and Bremner, R. (2004). Cell-specific effects of RB or RB/p107 loss on retinal development implicate an intrinsically death-resistant cell-of-origin in retinoblastoma. *Cancer Cell* **5,** 539–551.
Chen, D., Opavsky, R., Pacal, M., Tanimoto, N., Wenzel, P., Seeliger, M. W., Leone, G., and Bremner, R. (2007). Rb-mediated neuronal differentiation through cell-cycle-independent regulation of E2f3a. *PLoS Biol.* **5,** e179.
Chen, H. Z., Tsai, S. Y., and Leone, G. (2009a). Emerging roles of E2Fs in cancer: An exit from cell cycle control. *Nat. Rev. Cancer* **9,** 785–797.
Chen, Z., Hafidh, S., Poh, S. H., Twell, D., and Berger, F. (2009b). Proliferation and cell fate establishment during Arabidopsis male gametogenesis depends on the Retinoblastoma protein. *Proc. Natl. Acad. Sci. USA* **106,** 7257–7262.
Chicas, A., Wang, X., Zhang, C., McCurrach, M., Zhao, Z., Mert, O., Dickins, R. A., Narita, M., Zhang, M., and Lowe, S. W. (2010). Dissecting the unique role of the retinoblastoma tumor suppressor during cellular senescence. *Cancer Cell* **17,** 376–387.
Christensen, J., Agger, K., Cloos, P. A., Pasini, D., Rose, S., Sennels, L., Rappsilber, J., Hansen, K. H., Salcini, A. E., and Helin, K. (2007). RBP2 belongs to a family of demethylases, specific for tri- and dimethylated lysine 4 on histone 3. *Cell* **128,** 1063–1076.

Clarke, A. R., Maandag, E. R., van Roon, M., van der Lugt, N. M., van der Valk, M., Hooper, M. L., Berns, A., and te Riele, H. (1992). Requirement for a functional Rb-1 gene in murine development. *Nature* **359,** 328–330.
Classon, M., and Dyson, N. (2001). p107 and p130: Versatile proteins with interesting pockets. *Exp. Cell Res.* **264,** 135–147.
Claudio, P. P., Tonini, T., and Giordano, A. (2002). The retinoblastoma family: Twins or distant cousins? *Genome Biol.* **3**(reviews 3012).
Comings, D. E. (1973). A general theory of carcinogenesis. *Proc. Natl. Acad. Sci. USA* **70,** 3324–3328.
Connell-Crowley, L., Harper, J. W., and Goodrich, D. W. (1997). Cyclin D1/Cdk4 regulates retinoblastoma protein-mediated cell cycle arrest by site-specific phosphorylation. *Mol. Biol. Cell* **8,** 287–301.
Coschi, C. H., Martens, A. L., Ritchie, K., Francis, S. M., Chakrabarti, S., Berube, N. G., and Dick, F. A. (2010). Mitotic chromosome condensation mediated by the retinoblastoma protein is tumor-suppressive. *Genes Dev.* **24,** 1351–1363.
Courtois-Cox, S., Jones, S. L., and Cichowski, K. (2008). Many roads lead to oncogene-induced senescence. *Oncogene* **27,** 2801–2809.
Dahiya, A., Wong, S., Gonzalo, S., Gavin, M., and Dean, D. C. (2001). Linking the Rb and polycomb pathways. *Mol. Cell* **8,** 557–568.
Dasgupta, P., Sun, J., Wang, S., Fusaro, G., Betts, V., Padmanabhan, J., Sebti, S. M., and Chellappan, S. P. (2004). Disruption of the Rb–Raf-1 interaction inhibits tumor growth and angiogenesis. *Mol. Cell. Biol.* **24,** 9527–9541.
Dasgupta, B., Milbrandt, J., Dasgupta, B., and Milbrandt, J. (2009). AMP-activated protein kinase phosphorylates retinoblastoma protein to control mammalian brain development. *Dev. Cell* **16,** 256–270.
Day, K. C., McCabe, M. T., Zhao, X., Wang, Y., Davis, J. N., Phillips, J., Von Geldern, M., Ried, T., KuKuruga, M. A., Cunha, G. R., Hayward, S. W., and Day, M. L. (2002). Rescue of embryonic epithelium reveals that the homozygous deletion of the retinoblastoma gene confers growth factor independence and immortality but does not influence epithelial differentiation or tissue morphogenesis. *J. Biol. Chem.* **277,** 44475–44484.
Defeo-Jones, D., Huang, P. S., Jones, R. E., Haskell, K. M., Vuocolo, G. A., Hanobik, M. G., Huber, H. E., and Oliff, A. (1991). Cloning of cDNAs for cellular proteins that bind to the retinoblastoma gene product. *Nature* **352,** 251–254.
Delston, R. B., Harbour, J. W., Delston, R. B., and Harbour, J. W. (2006). Rb at the interface between cell cycle and apoptotic decisions. *Curr. Mol. Med.* **6,** 713–718.
Dick, F. A., and Dyson, N. (2003). pRB contains an E2F1-specific binding domain that allows E2F1-induced apoptosis to be regulated separately from other E2F activities. *Mol. Cell* **12,** 639–649.
Du, W. (2000). Suppression of the rbf null mutants by a de2f1 allele that lacks transactivation domain. *Development* **127,** 367–379.
Du, W., and Dyson, N. (1999). The role of RBF in the introduction of G1 regulation during *Drosophila* embryogenesis. *EMBO J.* **18,** 916–925.
Dunaief, J. L., Strober, B. E., Guha, S., Khavari, P. A., Alin, K., Luban, J., Begemann, M., Crabtree, G. R., and Goff, S. P. (1994). The retinoblastoma protein and BRG1 form a complex and cooperate to induce cell cycle arrest. *Cell* **79,** 119–130.
Durfee, T., Becherer, K., Chen, P. L., Yeh, S. H., Yang, Y., Kilburn, A. E., Lee, W. H., and Elledge, S. J. (1993). The retinoblastoma protein associates with the protein phosphatase type 1 catalytic subunit. *Genes Dev.* **7,** 555–569.
Duronio, R. J., O'Farrell, P. H., Xie, J. E., Brook, A., and Dyson, N. (1995). The transcription factor E2F is required for S phase during *Drosophila* embryogenesis. *Genes Dev.* **9,** 1445–1455.

Dynlacht, B. D., Flores, O., Lees, J. A., and Harlow, E. (1994). Differential regulation of E2F transactivation by cyclin/cdk2 complexes. *Genes Dev.* **8,** 1772–1786.

Ebel, C., Mariconti, L., and Gruissem, W. (2004). Plant retinoblastoma homologues control nuclear proliferation in the female gametophyte. *Nature* **429,** 776–780.

Emili, A., and Ingles, C. J. (1995). Promoter-dependent photocross-linking of the acidic transcriptional activator E2F-1 to the TATA-binding protein. *J. Biol. Chem.* **270,** 13674–13680.

Fattaey, A. R., Helin, K., Dembski, M. S., Dyson, N., Harlow, E., Vuocolo, G. A., Hanobik, M. G., Haskell, K. M., Oliff, A., Defeo-Jones, D., *et al.* (1993). Characterization of the retinoblastoma binding proteins RBP1 and RBP2. *Oncogene* **8,** 3149–3156.

Fay, D. S., Keenan, S., and Han, M. (2002). fzr-1 and lin-35/Rb function redundantly to control cell proliferation in *C. elegans* as revealed by a nonbiased synthetic screen. *Genes Dev.* **16,** 503–517.

Fong, Y., Bender, L., Wang, W., and Strome, S. (2002). Regulation of the different chromatin states of autosomes and X chromosomes in the germ line of *C. elegans. Science* **296,** 2235–2238.

Friend, S. H., Bernards, R., Rogelj, S., Weinberg, R. A., Rapaport, J. M., Albert, D. M., and Dryja, T. P. (1986). A human DNA segment with properties of the gene that predisposes to retinoblastoma and osteosarcoma. *Nature* **323,** 643–646.

Frolov, M. V., and Dyson, N. J. (2004). Molecular mechanisms of E2F-dependent activation and pRB-mediated repression. *J. Cell Sci.* **117,** 2173–2181.

Funayama, R., Saito, M., Tanobe, H., and Ishikawa, F. (2006). Loss of linker histone H1 in cellular senescence. *J. Cell Biol.* **175,** 869–880.

Fung, Y. K., Murphree, A. L., T'Ang, A., Qian, J., Hinrichs, S. H., and Benedict, W. F. (1987). Structural evidence for the authenticity of the human retinoblastoma gene. *Science* **236,** 1657–1661.

Gabellini, C., Del Bufalo, D., and Zupi, G. (2006). Involvement of RB gene family in tumor angiogenesis. *Oncogene* **25,** 5326–5332.

Gladden, A. B., and Diehl, J. A. (2003). The cyclin D1-dependent kinase associates with the pre-replication complex and modulates RB.MCM7 binding. *J. Biol. Chem.* **278,** 9754–9760.

Gonzalo, S., Garcia-Cao, M., Fraga, M. F., Schotta, G., Peters, A. H., Cotter, S. E., Eguia, R., Dean, D. C., Esteller, M., Jenuwein, T., and Blasco, M. A. (2005). Role of the RB1 family in stabilizing histone methylation at constitutive heterochromatin. *Nat. Cell Biol.* **7,** 420–428.

Goodrich, D. W. (2003). How the other half lives, the amino-terminal domain of the retinoblastoma tumor suppressor protein. *J. Cell. Physiol.* **197,** 169–180.

Goodrich, D. W. (2006). The retinoblastoma tumor-suppressor gene, the exception that proves the rule. *Oncogene* **25,** 5233–5243.

Goodrich, D. W., Wang, N. P., Qian, Y. W., Lee, E. Y., and Lee, W. H. (1991). The retinoblastoma gene product regulates progression through the G1 phase of the cell cycle. *Cell* **67,** 293–302.

Grishok, A., and Sharp, P. A. (2005). Negative regulation of nuclear divisions in *Caenorhabditis elegans* by retinoblastoma and RNA interference-related genes. *Proc. Natl. Acad. Sci. USA* **102,** 17360–17365.

Gu, W., Schneider, J. W., Condorelli, G., Kaushal, S., Mahdavi, V., and Nadal-Ginard, B. (1993). Interaction of myogenic factors and the retinoblastoma protein mediates muscle cell commitment and differentiation. *Cell* **72,** 309–324.

Guo, J., Sheng, G., and Warner, B. W. (2005). Epidermal growth factor-induced rapid retinoblastoma phosphorylation at Ser780 and Ser795 is mediated by ERK1/2 in small intestine epithelial cells. *J. Biol. Chem.* **280,** 35992–35998.

Guo, J., Longshore, S., Nair, R., and Warner, B. W. (2009). Retinoblastoma protein (pRb), but not p107 or p130, is required for maintenance of enterocyte quiescence and differentiation in small intestine. *J. Biol. Chem.* **284,** 134–140.

Gupta, P. B., Chaffer, C. L., and Weinberg, R. A. (2009). Cancer stem cells: Mirage or reality? *Nat. Med.* **15,** 1010–1012.

Gutierrez, G. M., Kong, E., Sabbagh, Y., Brown, N. E., Lee, J. S., Demay, M. B., Thomas, D. M., and Hinds, P. W. (2008). Impaired bone development and increased mesenchymal progenitor cells in calvaria of RB1−/− mice. *Proc. Natl. Acad. Sci. USA* **105,** 18402–18407.

Hagemeier, C., Cook, A., and Kouzarides, T. (1993). The retinoblastoma protein binds E2F residues required for activation in vivo and TBP binding in vitro. *Nucleic Acids Res.* **21,** 4998–5004.

Halazonetis, T. D., Gorgoulis, V. G., and Bartek, J. (2008). An oncogene-induced DNA damage model for cancer development. *Science* **319,** 1352–1355.

Hallstrom, T. C., Nevins, J. R., Hallstrom, T. C., and Nevins, J. R. (2009). Balancing the decision of cell proliferation and cell fate. *Cell Cycle* **8,** 532–535.

Hanahan, D., and Weinberg, R. A. (2000). The hallmarks of cancer. *Cell* **100,** 57–70.

Hartl, T., Boswell, C., Orr-Weaver, T. L., and Bosco, G. (2007). Developmentally regulated histone modifications in *Drosophila* follicle cells: Initiation of gene amplification is associated with histone H3 and H4 hyperacetylation and H1 phosphorylation. *Chromosoma* **116,** 197–214.

Hassler, M., Singh, S., Yue, W. W., Luczynski, M., Lakbir, R., Sanchez-Sanchez, F., Bader, T., Pearl, L. H., and Mittnacht, S. (2007). Crystal structure of the retinoblastoma protein N domain provides insight into tumor suppression, ligand interaction, and holoprotein architecture. *Mol. Cell* **28,** 371–385.

Helin, K., Lees, J. A., Vidal, M., Dyson, N., Harlow, E., and Fattaey, A. (1992). A cDNA encoding a pRB-binding protein with properties of the transcription factor E2F. *Cell* **70,** 337–350.

Hemberger, M., Dean, W., and Reik, W. (2009). Epigenetic dynamics of stem cells and cell lineage commitment: Digging Waddington's canal. *Nat. Rev. Mol. Cell Biol.* **10,** 526–537.

Hernando, E., Nahle, Z., Juan, G., Diaz-Rodriguez, E., Alaminos, M., Hemann, M., Michel, L., Mittal, V., Gerald, W., Benezra, R., Lowe, S. W., and Cordon-Cardo, C. (2004). Rb inactivation promotes genomic instability by uncoupling cell cycle progression from mitotic control. *Nature* **430,** 797–802.

Herrera, R. E., Chen, F., and Weinberg, R. A. (1996). Increased histone H1 phosphorylation and relaxed chromatin structure in Rb-deficient fibroblasts. *Proc. Natl. Acad. Sci. USA* **93,** 11510–11515.

Hinds, P. W., Mittnacht, S., Dulic, V., Arnold, A., Reed, S. I., and Weinberg, R. A. (1992). Regulation of retinoblastoma protein functions by ectopic expression of human cyclins. *Cell* **70,** 993–1006.

Hou, S. T., Xie, X., Baggley, A., Park, D. S., Chen, G., and Walker, T. (2002). Activation of the Rb/E2F1 pathway by the nonproliferative p38 MAPK during Fas (APO1/CD95)-mediated neuronal apoptosis. *J. Biol. Chem.* **277,** 48764–48770.

Huang, H. J., Yee, J. K., Shew, J. Y., Chen, P. L., Bookstein, R., Friedmann, T., Lee, E. Y., and Lee, W. H. (1988). Suppression of the neoplastic phenotype by replacement of the RB gene in human cancer cells. *Science* **242,** 1563–1566.

Humbert, P. O., Verona, R., Trimarchi, J. M., Rogers, C., Dandapani, S., and Lees, J. A. (2000). E2f3 is critical for normal cellular proliferation. *Genes Dev.* **14,** 690–703.

Iavarone, A., Garg, P., Lasorella, A., Hsu, J., and Israel, M. A. (1994). The helix-loop-helix protein Id-2 enhances cell proliferation and binds to the retinoblastoma protein. *Genes Dev.* **8,** 1270–1284.

Iavarone, A., King, E. R., Dai, X. M., Leone, G., Stanley, E. R., and Lasorella, A. (2004). Retinoblastoma promotes definitive erythropoiesis by repressing Id2 in fetal liver macrophages. *Nature* **432,** 1040–1045.

Ingouff, M., Jullien, P. E., and Berger, F. (2006). The female gametophyte and the endosperm control cell proliferation and differentiation of the seed coat in Arabidopsis. *Plant Cell* **18,** 3491–3501.

Inoue, Y., Kitagawa, M., and Taya, Y. (2007). Phosphorylation of pRB at Ser612 by Chk1/2 leads to a complex between pRB and E2F-1 after DNA damage. *EMBO J.* **26,** 2083–2093.

Iovino, F., Lentini, L., Amato, A., and Di Leonardo, A. (2006). RB acute loss induces centrosome amplification and aneuploidy in murine primary fibroblasts. *Mol. Cancer* **5,** 38.

Isaac, C. E., Francis, S. M., Martens, A. L., Julian, L. M., Seifried, L. A., Erdmann, N., Binne, U. K., Harrington, L., Sicinski, P., Berube, N. G., Dyson, N. J., and Dick, F. A. (2006). The retinoblastoma protein regulates pericentric heterochromatin. *Mol. Cell. Biol.* **26,** 3659–3671.

Ishida, S., Huang, E., Zuzan, H., Spang, R., Leone, G., West, M., and Nevins, J. R. (2001). Role for E2F in control of both DNA replication and mitotic functions as revealed from DNA microarray analysis. *Mol. Cell. Biol.* **21,** 4684–4699.

Iwase, S., Lan, F., Bayliss, P., de la Torre-Ubieta, L., Huarte, M., Qi, H. H., Whetstine, J. R., Bonni, A., Roberts, T. M., and Shi, Y. (2007). The X-linked mental retardation gene SMCX/JARID1C defines a family of histone H3 lysine 4 demethylases. *Cell* **128,** 1077–1088.

Jacks, T., Fazeli, A., Schmitt, E. M., Bronson, R. T., Goodell, M. A., and Weinberg, R. A. (1992). Effects of an Rb mutation in the mouse. *Nature* **359,** 295–300.

Jacques, T. S., Swales, A., Brzozowski, M. J., Henriquez, N. V., Linehan, J. M., Mirzadeh, Z., Malley, C. O., Naumann, H., Alvarez-Buylla, A., and Brandner, S. (2010). Combinations of genetic mutations in the adult neural stem cell compartment determine brain tumour phenotypes. *Embo J.* **29,** 222–235.

Ji, P., Jiang, H., Rekhtman, K., Bloom, J., Ichetovkin, M., Pagano, M., and Zhu, L. (2004). An Rb-Skp2-p27 pathway mediates acute cell cycle inhibition by Rb and is retained in a partial-penetrance Rb mutant. *Mol. Cell* **16,** 47–58.

Jiang, Z., and Zacksenhaus, E. (2002). Activation of retinoblastoma protein in mammary gland leads to ductal growth suppression, precocious differentiation, and adenocarcinoma. *J. Cell Biol.* **156,** 185–198.

Johnson, D. A., Zhang, J., Frase, S., Wilson, M., Rodriguez-Galindo, C., and Dyer, M. A. (2007). Neuronal differentiation and synaptogenesis in retinoblastoma. *Cancer Res.* **67,** 2701–2711.

Kadri, Z., Shimizu, R., Ohneda, O., Maouche-Chretien, L., Gisselbrecht, S., Yamamoto, M., Romeo, P. H., Leboulch, P., and Chretien, S. (2009). Direct binding of pRb/E2F-2 to GATA-1 regulates maturation and terminal cell division during erythropoiesis. *PLoS Biol.* **7,** e1000123.

Kaelin, W. G., Jr., Krek, W., Sellers, W. R., DeCaprio, J. A., Ajchenbaum, F., Fuchs, C. S., Chittenden, T., Li, Y., Farnham, P. J., Blanar, M. A., *et al.* (1992). Expression cloning of a cDNA encoding a retinoblastoma-binding protein with E2F-like properties. *Cell* **70,** 351–364.

Ketting, R. F., Haverkamp, T. H., van Luenen, H. G., and Plasterk, R. H. (1999). Mut-7 of *C. elegans,* required for transposon silencing and RNA interference, is a homolog of Werner syndrome helicase and RNaseD. *Cell* **99,** 133–141.

Kim, H. Y., and Cho, Y. (1997). Structural similarity between the pocket region of retinoblastoma tumour suppressor and the cyclin-box. *Nat. Struct. Biol.* **4,** 390–395.

Kim, S. J., Wagner, S., Liu, F., O'Reilly, M. A., Robbins, P. D., and Green, M. R. (1992). Retinoblastoma gene product activates expression of the human TGF-beta 2 gene through transcription factor ATF-2. *Nature* **358,** 331–334.

Kim, H. Y., Ahn, B. Y., and Cho, Y. (2001). Structural basis for the inactivation of retinoblastoma tumor suppressor by SV40 large T antigen. *EMBO J.* **20,** 295–304.

Kirienko, N. V., and Fay, D. S. (2007). Transcriptome profiling of the *C. elegans* Rb ortholog reveals diverse developmental roles. *Dev. Biol.* **305,** 674–684.

Klose, R. J., Yan, Q., Tothova, Z., Yamane, K., Erdjument-Bromage, H., Tempst, P., Gilliland, D. G., Zhang, Y., and Kaelin, W. G., Jr. (2007). The retinoblastoma binding protein RBP2 is an H3K4 demethylase. *Cell* **128,** 889–900.

Knudsen, E. S., and Wang, J. Y. J. (1996). Differential regulation of retinoblastoma protein function by specific Cdk phosphorylation sites. *J. Biol. Chem.* **271,** 8313–8320.

Knudsen, E. S., and Wang, J. Y. J. (1997). Dual mechanisms for the inhibition of E2F binding to RB by cyclin-dependent kinase-mediated RB phosphorylation. *Mol. Cell. Biol.* **17,** 5771–5783.

Knudsen, E. S., Sexton, C. R., Mayhew, C. N., Knudsen, E. S., Sexton, C. R., and Mayhew, C. N. (2006). Role of the retinoblastoma tumor suppressor in the maintenance of genome integrity. *Curr. Mol. Med.* **6,** 749–757.

Knudson, A. G., Jr. (1971). Mutation and cancer: Statistical study of retinoblastoma. *Proc. Natl. Acad. Sci. USA* **68,** 820–823.

Korenjak, M., Brehm, A., Korenjak, M., and Brehm, A. (2005). E2F-Rb complexes regulating transcription of genes important for differentiation and development. *Curr. Opin. Genet. Dev.* **15,** 520–527.

Kucherlapati, M. H., Yang, K., Fan, K., Kuraguchi, M., Sonkin, D., Rosulek, A., Lipkin, M., Bronson, R. T., Aronow, B. J., and Kucherlapati, R. (2008). Loss of Rb1 in the gastrointestinal tract of Apc1638N mice promotes tumors of the cecum and proximal colon. *Proc. Natl. Acad. Sci. USA* **105,** 15493–15498.

Kyritsis, A. P., Tsokos, M., Triche, T. J., and Chader, G. J. (1984). Retinoblastoma—origin from a primitive neuroectodermal cell? *Nature* **307,** 471–473.

Lai, A., Lee, J. M., Yang, W. M., DeCaprio, J. A., Kaelin, W. G., Jr., Seto, E., and Branton, P. E. (1999). RBP1 recruits both histone deacetylase-dependent and -independent repression activities to retinoblastoma family proteins. *Mol. Cell. Biol.* **19,** 6632–6641.

Lai, A., Kennedy, B. K., Barbie, D. A., Bertos, N. R., Yang, X. J., Theberge, M. C., Tsai, S. C., Seto, E., Zhang, Y., Kuzmichev, A., Lane, W. S., Reinberg, D., *et al.* (2001). RBP1 recruits the mSIN3-histone deacetylase complex to the pocket of retinoblastoma tumor suppressor family proteins found in limited discrete regions of the nucleus at growth arrest. *Mol. Cell. Biol.* **21,** 2918–2932.

Landsberg, R. L., Sero, J. E., Danielian, P. S., Yuan, T. L., Lee, E. Y., and Lees, J. A. (2003). The role of E2F4 in adipogenesis is independent of its cell cycle regulatory activity. *Proc. Natl. Acad. Sci. USA* **100,** 2456–2461.

Lang, S. E., McMahon, S. B., Cole, M. D., and Hearing, P. (2001). E2F transcriptional activation requires TRRAP and GCN5 cofactors. *J. Biol. Chem.* **276,** 32627–32634.

Laurie, N. A., Donovan, S. L., Shih, C.-S., Zhang, J., Mills, N., Fuller, C., Teunisse, A., Lam, S., Ramos, Y., Mohan, A., Johnson, D., Wilson, M., *et al.* (2006). Inactivation of the p53 pathway in retinoblastoma. *Nature* **444,** 61–66.

Leduc, C., Claverie, P., Eymin, B., Col, E., Khochbin, S., Brambilla, E., and Gazzeri, S. (2006). p14ARF promotes RB accumulation through inhibition of its Tip60-dependent acetylation. *Oncogene* **25,** 4147–4154.

Lee, C. C. J. H. (2002). Structural basis for the recognition of the E2F transactivation domain by the retinoblastoma tumor suppressor. *Genes Dev.* **16,** 3199–3212.

Lee, W. H., Bookstein, R., Hong, F., Young, L. J., Shew, J. Y., and Lee, E. Y. H. P. (1987). Human retinoblastoma susceptibility gene: Cloning, identification, and sequence. *Science* **235,** 1394–1399.

Lee, E. Y., Chang, C. Y., Hu, N., Wang, Y. C., Lai, C. C., Herrup, K., and Lee, W. H. (1992). Mice deficient for Rb are nonviable and show defects in neurogenesis and haematopoiesis. *Nature* **359,** 288–294.

Lee, J. O., Russo, A. A., and Pavletich, N. P. (1998). Structure of the retinoblastoma tumour-suppressor pocket domain bound to a peptide from HPV E7. *Nature* **391,** 859–865.

Lee, E. Y., Cam, H., Ziebold, U., Rayman, J. B., Lees, J. A., and Dynlacht, B. D. (2002). E2F4 loss suppresses tumorigenesis in Rb mutant mice. *Cancer Cell* **2,** 463–472.

Leng, X., Noble, M., Adams, P. D., Qin, J., and Harper, J. W. (2002). Reversal of growth suppression by p107 via direct phosphorylation by cyclin D1/cyclin-dependent kinase 4. *Mol. Cell. Biol.* **22,** 2242–2254.

Leung, J. K., Berube, N., Venable, S., Ahmed, S., Timchenko, N., and Pereira-Smith, O. M. (2001). MRG15 activates the B-myb promoter through formation of a nuclear complex with the retinoblastoma protein and the novel protein PAM14. *J. Biol. Chem.* **276,** 39171–39178.

Liu, X., and Marmorstein, R. (2007). Structure of the retinoblastoma protein bound to adenovirus E1A reveals the molecular basis for viral oncoprotein inactivation of a tumor suppressor. *Genes Dev.* **21,** 2711–2716.

Liu, H., Dibling, B., Spike, B., Dirlam, A., and Macleod, K. (2004). New roles for the RB tumor suppressor protein. *Curr. Opin. Genet. Dev.* **14,** 55–64.

Longworth, M. S., and Dyson, N. J. (2009). pRb, a local chromatin organizer with global possibilities. *Chromosoma* **119,** 1–11.

Longworth, M. S., Herr, A., Ji, J. Y., and Dyson, N. J. (2008). RBF1 promotes chromatin condensation through a conserved interaction with the Condensin II protein dCAP-D3. *Genes Dev.* **22,** 1011–1024.

Lu, J., and Danielsen, M. (1998). Differential regulation of androgen and glucocorticoid receptors by retinoblastoma protein. *J. Biol. Chem.* **273,** 31528–31533.

Ludlow, J. W., Glendening, C. L., Livingston, D. M., and DeCaprio, J. A. (1993). Specific enzymatic dephosphorylation of the retinoblastoma protein. *Mol. Cell. Biol.* **13,** 367–372.

Luo, R. X., Postigo, A. A., and Dean, D. C. (1998). Rb interacts with histone deacetylase to repress transcription. *Cell* **92,** 463–473.

MacLellan, W. R., Xiao, G., Abdellatif, M., and Schneider, M. D. (2000). A novel Rb- and p300-binding protein inhibits transactivation by MyoD. *Mol. Cell. Biol.* **20,** 8903–8915.

Macleod, K. F. (2008). The role of the RB tumour suppressor pathway in oxidative stress responses in the haematopoietic system. *Nat. Rev. Cancer* **8,** 769–781.

MacPherson, D., Sage, J., Kim, T., Ho, D., McLaughlin, M. E., and Jacks, T. (2004). Cell type-specific effects of Rb deletion in the murine retina. *Genes Dev.* **18,** 1681–1694.

Maddison, L. A., Sutherland, B. W., Barrios, R. J., and Greenberg, N. M. (2004). Conditional deletion of Rb causes early stage prostate cancer. *Cancer Res.* **64,** 6018–6025.

Maenhaut, C., Dumont, J. E., Roger, P. P., and van Staveren, W. C. (2010). Cancer stem cells: A reality, a myth, a fuzzy concept or a misnomer? An analysis. *Carcinogenesis* **31,** 149–158.

Magnaghi-Jaulin, L., Groisman, R., Naguibneva, I., Robin, P., Lorain, S., LeVillain, J. P., Troalen, F., Trouche, D., and Harel-Bellan, A. (1998). Retinoblastoma protein represses transcription by recruiting a histone deacetylase. *Nature* **391,** 601–605.

Malumbres, M., and Barbacid, M. (2001). To cycle or not to cycle: A critical decision in cancer. *Nat. Rev. Cancer* **1,** 222–231.

Manning, A. L., Longworth, M. S., and Dyson, N. J. (2010). Loss of pRB causes centromere dysfunction and chromosomal instability. *Genes Dev.* **24,** 1364–1376.
Marees, T., Moll, A. C., Imhof, S. M., de Boer, M. R., Ringens, P. J., and van Leeuwen, F. E. (2008). Risk of second malignancies in survivors of retinoblastoma: More than 40 years of follow-up. *J. Natl. Cancer Inst.* **100,** 1771–1779.
Marino, S., Hoogervoorst, D., Brandner, S., and Berns, A. (2003). Rb and p107 are required for normal cerebellar development and granule cell survival but not for Purkinje cell persistence. *Development* **130,** 3359–3368.
Markham, D., Munro, S., Soloway, J., O'Connor, D. P., and La Thangue, N. B. (2006). DNA-damage-responsive acetylation of pRb regulates binding to E2F-1. *EMBO Rep.* **7,** 192–198.
Martinez-Cruz, A. B., Santos, M., Lara, M. F., Segrelles, C., Ruiz, S., Moral, M., Lorz, C., Garcia-Escudero, R., and Paramio, J. M. (2008). Spontaneous squamous cell carcinoma induced by the somatic inactivation of retinoblastoma and Trp53 tumor suppressors. *Cancer Res.* **68,** 683–692.
McClellan, K. A., Ruzhynsky, V. A., Douda, D. N., Vanderluit, J. L., Ferguson, K. L., Chen, D., Bremner, R., Park, D. S., Leone, G., and Slack, R. S. (2007a). Unique requirement for Rb/E2F3 in neuronal migration: Evidence for cell cycle-independent functions. *Mol. Cell. Biol.* **27,** 4825–4843.
McClellan, K. A., Slack, R. S., McClellan, K. A., and Slack, R. S. (2007b). Specific in vivo roles for E2Fs in differentiation and development. *Cell Cycle* **6,** 2917–2927.
Meloni, A. R., Smith, E. J., and Nevins, J. R. (1999). A mechanism for Rb/p130-mediated transcription repression involving recruitment of the CtBP corepressor. *Proc. Natl. Acad. Sci. USA* **96,** 9574–9579.
Meuwissen, R., Linn, S. C., Linnoila, R. I., Zevenhoven, J., Mooi, W. J., and Berns, A. (2003). Induction of small cell lung cancer by somatic inactivation of both Trp53 and Rb1 in a conditional mouse model. *Cancer Cell* **4,** 181–189.
Miccadei, S., Provenzano, C., Mojzisek, M., Giorgio Natali, P., and Civitareale, D. (2005). Retinoblastoma protein acts as Pax 8 transcriptional coactivator. *Oncogene* **24,** 6993–7001.
Miyake, S., Sellers, W. R., Safran, M., Li, X., Zhao, W., Grossman, S. R., Gan, J., DeCaprio, J. A., Adams, P. D., and Kaelin, W. G., Jr. (2000). Cells degrade a novel inhibitor of differentiation with E1A-like properties upon exiting the cell cycle. *Mol. Cell. Biol.* **20,** 8889–8902.
Mohn, F., and Schubeler, D. (2009). Genetics and epigenetics: Stability and plasticity during cellular differentiation. *Trends Genet.* **25,** 129–136.
Moreira, A. L., Gonen, M., Rekhtman, N., and Downey, R. J. (2010). Progenitor stem cell marker expression by pulmonary carcinomas. *Mod. Pathol.* **23,** 889–895.
Morris, E. J., and Dyson, N. J. (2001). Retinoblastoma protein partners. *Adv. Cancer Res.* **82,** 1–54.
Mukherjee, P., Winter, S. L., Alexandrow, M. G., Mukherjee, P., Winter, S. L., and Alexandrow, M. G. (2010). Cell cycle arrest by transforming growth factor beta1 near G1/S is mediated by acute abrogation of prereplication complex activation involving an Rb–MCM interaction. *Mol. Cell. Biol.* **30,** 845–856.
Muzumdar, M. D., Tasic, B., Miyamichi, K., Li, L., and Luo, L. (2007). A global double-fluorescent Cre reporter mouse. *Genesis* **45,** 593–605.
Nair, J. S., Ho, A. L., Tse, A. N., Coward, J., Cheema, H., Ambrosini, G., Keen, N., Schwartz, G. K., Nair, J. S., Ho, A. L., Tse, A. N., Coward, J., *et al.* (2009). Aurora B kinase regulates the postmitotic endoreduplication checkpoint via phosphorylation of the retinoblastoma protein at serine 780. *Mol. Biol. Cell* **20,** 2218–2228.

Narita, M., Nunez, S., Heard, E., Narita, M., Lin, A. W., Hearn, S. A., Spector, D. L., Hannon, G. J., and Lowe, S. W. (2003). Rb-mediated heterochromatin formation and silencing of E2F target genes during cellular senescence. *Cell* **113,** 703–716.

Narita, M., Narita, M., Krizhanovsky, V., Nunez, S., Chicas, A., Hearn, S. A., Myers, M. P., and Lowe, S. W. (2006). A novel role for high-mobility group a proteins in cellular senescence and heterochromatin formation. *Cell* **126,** 503–514.

Nath, N., Wang, S., Betts, V., Knudsen, E., and Chellappan, S. (2003). Apoptotic and mitogenic stimuli inactivate Rb by differential utilization of p38 and cyclin-dependent kinases. *Oncogene* **22,** 5986–5994.

Nguyen, D. X., and McCance, D. J. (2005). Role of the retinoblastoma tumor suppressor protein in cellular differentiation. *J. Cell. Biochem.* **94,** 870–879.

Nielsen, S. J., Schneider, R., Bauer, U. M., Bannister, A. J., Morrison, A., O'Carroll, D., Firestein, R., Cleary, M., Jenuwein, T., Herrera, R. E., and Kouzarides, T. (2001). Rb targets histone H3 methylation and HP1 to promoters. *Nature* **412,** 561–565.

Novitch, B. G., Mulligan, G. J., Jacks, T., and Lassar, A. B. (1996). Skeletal muscle cells lacking the retinoblastoma protein display defects in muscle gene expression and accumulate in S and G2 phases of the cell cycle. *J. Cell Biol.* **135,** 441–456.

Novitch, B. G., Spicer, D. B., Kim, P. S., Cheung, W. L., and Lassar, A. B. (1999). pRb is required for MEF2-dependent gene expression as well as cell-cycle arrest during skeletal muscle differentiation. *Curr. Biol.* **9,** 449–459.

Ouellet, J., and Roy, R. (2007). The lin-35/Rb and RNAi pathways cooperate to regulate a key cell cycle transition in *C. elegans*. *BMC Dev. Biol.* **7,** 38.

Page, B. D., Guedes, S., Waring, D., and Priess, J. R. (2001). The *C. elegans* E2F- and DP-related proteins are required for embryonic asymmetry and negatively regulate Ras/MAPK signaling. *Mol. Cell* **7,** 451–460.

Papadimou, E., Menard, C., Grey, C., and Puceat, M. (2005). Interplay between the retinoblastoma protein and LEK1 specifies stem cells toward the cardiac lineage. *EMBO J.* **24,** 1750–1761.

Park, M., and Krause, M. W. (1999). Regulation of postembryonic G(1) cell cycle progression in *Caenorhabditis elegans* by a cyclin D/CDK-like complex. *Development* **126,** 4849–4860.

Pearson, A., and Greenblatt, J. (1997). Modular organization of the E2F1 activation domain and its interaction with general transcription factors TBP and TFIIH. *Oncogene* **15,** 2643–2658.

Pennaneach, V., Salles-Passador, I., Munshi, A., Brickner, H., Regazzoni, K., Dick, F., Dyson, N., Chen, T. T., Wang, J. Y., Fotedar, R., and Fotedar, A. (2001). The large subunit of replication factor C promotes cell survival after DNA damage in an LxCxE motif- and Rb-dependent manner. *Mol. Cell* **7,** 715–727.

Polager, S., Ofir, M., and Ginsberg, D. (2008). E2F1 regulates autophagy and the transcription of autophagy genes. *Oncogene* **27,** 4860–4864.

Prieur, A., and Peeper, D. S. (2008). Cellular senescence in vivo: A barrier to tumorigenesis. *Curr. Opin. Cell Biol.* **20,** 150–155.

Qian, Y. W., and Lee, E. Y. (1995). Dual retinoblastoma-binding proteins with properties related to a negative regulator of Ras in yeast. *J. Biol. Chem.* **270,** 25507–25513.

Qian, Y. W., Wang, Y. C., Hollingsworth, R. E., Jr., Jones, D., Ling, N., and Lee, E. Y. (1993). A retinoblastoma-binding protein related to a negative regulator of Ras in yeast. *Nature* **364,** 648–652.

Ren, B., Cam, H., Takahashi, Y., Volkert, T., Terragni, J., Young, R. A., and Dynlacht, B. D. (2002). E2F integrates cell cycle progression with DNA repair, replication, and G(2)/M checkpoints. *Genes Dev.* **16,** 245–256.

Robertson, K. D., Ait-Si-Ali, S., Yokochi, T., Wade, P. A., Jones, P. L., and Wolffe, A. P. (2000). DNMT1 forms a complex with Rb, E2F1 and HDAC1 and represses transcription from E2F-responsive promoters. *Nat. Genet.* **25,** 338–342.

Royzman, I., Austin, R. J., Bosco, G., Bell, S. P., and Orr-Weaver, T. L. (1999). ORC localization in *Drosophila* follicle cells and the effects of mutations in dE2F and dDP. *Genes Dev.* **13,** 827–840.

Rubin, S. M., Gall, A.-L., Zheng, N., and Pavletich, N. P. (2005). Structure of the Rb C-terminal domain bound to E2F1-DP1: A mechanism for phosphorylation-induced E2F release. *Cell* **123,** 1093–1106.

Ruiz, S., Santos, M., Segrelles, C., Leis, H., Jorcano, J. L., Berns, A., Paramio, J. M., and Vooijs, M. (2004). Unique and overlapping functions of pRb and p107 in the control of proliferation and differentiation in epidermis. *Development* **131,** 2737–2748.

Sage, J., Mulligan, G. J., Attardi, L. D., Miller, A., Chen, S., Williams, B., Theodorou, E., and Jacks, T. (2000). Targeted disruption of the three Rb-related genes leads to loss of G(1) control and immortalization. *Genes Dev.* **14,** 3037–3050.

Sage, J., Miller, A. L., Perez-Mancera, P. A., Wysocki, J. M., and Jacks, T. (2003). Acute mutation of retinoblastoma gene function is sufficient for cell cycle re-entry. *Nature* **424,** 223–228.

Sage, C., Huang, M., Karimi, K., Gutierrez, G., Vollrath, M. A., Zhang, D. S., Garcia-Anoveros, J., Hinds, P. W., Corwin, J. T., Corey, D. P., and Chen, Z. Y. (2005). Proliferation of functional hair cells in vivo in the absence of the retinoblastoma protein. *Science* **307,** 1114–1118.

Saijo, M., Sakai, Y., Kishino, T., Niikawa, N., Matsuura, Y., Morino, K., Tamai, K., and Taya, Y. (1995). Molecular cloning of a human protein that binds to the retinoblastoma protein and chromosomal mapping. *Genomics* **27,** 511–519.

Sanchez Mde, L., Caro, E., Desvoyes, B., Ramirez-Parra, E., and Gutierrez, C. (2008). Chromatin dynamics during the plant cell cycle. *Semin. Cell Dev. Biol.* **19,** 537–546.

Sang, Y., Wu, M. F., and Wagner, D. (2009). The stem cell–chromatin connection. *Semin. Cell Dev. Biol.* **20,** 1143–1148.

Schneider, J. W., Gu, W., Zhu, L., Mahdavi, V., and Nadal-Ginard, B. (1994). Reversal of terminal differentiation mediated by p107 in Rb−/− muscle cells. *Science* **264,** 1467–1471.

Shan, B., Zhu, X., Chen, P. L., Durfee, T., Yang, Y., Sharp, D., and Lee, W. H. (1992). Molecular cloning of cellular genes encoding retinoblastoma- associated proteins: Identification of a gene with properties of the transcription factor E2F. *Mol. Cell. Biol.* **12,** 5620–5631.

Sharma, S., Kelly, T. K., and Jones, P. A. (2010). Epigenetics in cancer. *Carcinogenesis* **31,** 27–36.

Shen, W. H. (2002). The plant E2F-Rb pathway and epigenetic control. *Trends Plant Sci.* **7,** 505–511.

Sherr, C. J., and McCormick, F. (2002). The RB and p53 pathways in cancer. *Cancer Cell* **2,** 103–112.

Siddiqui, H., Fox, S. R., Gunawardena, R. W., and Knudsen, E. S. (2007). Loss of RB compromises specific heterochromatin modifications and modulates HP1alpha dynamics. *J. Cell. Physiol.* **211,** 131–137.

Sijen, T., and Plasterk, R. H. (2003). Transposon silencing in the *Caenorhabditis elegans* germ line by natural RNAi. *Nature* **426,** 310–314.

Singh, P., Coe, J., and Hong, W. (1995). A role for retinoblastoma protein in potentiating transcriptional activation by the glucocorticoid receptor. *Nature* **374,** 562–565.

Sotillo, R., Hernando, E., Diaz-Rodriguez, E., Teruya-Feldstein, J., Cordon-Cardo, C., Lowe, S. W., and Benezra, R. (2007). Mad2 overexpression promotes aneuploidy and tumorigenesis in mice. *Cancer Cell* **11,** 9–23.

Spike, B. T., Dirlam, A., Dibling, B. C., Marvin, J., Williams, B. O., Jacks, T., and Macleod, K. F. (2004). The Rb tumor suppressor is required for stress erythropoiesis. *EMBO J.* **23,** 4319–4329.

Sterner, J. M., Dew-Knight, S., Musahl, C., Kornbluth, S., and Horowitz, J. M. (1998). Negative regulation of DNA replication by the retinoblastoma protein is mediated by its association with MCM7. *Mol. Cell. Biol.* **18,** 2748–2757.

Strober, B. E., Dunaief, J. L., Guha, J. L., and Goff, S. P. (1996). Functional interactions between the hBRM/hBRG1 transcriptional activators and the pRB family of proteins. *Mol. Cell. Biol.* **16,** 1576–1583.

Suzuma, K., Takahara, N., Suzuma, I., Isshiki, K., Ueki, K., Leitges, M., Aiello, L. P., and King, G. L. (2002). Characterization of protein kinase C beta isoform's action on retinoblastoma protein phosphorylation, vascular endothelial growth factor-induced endothelial cell proliferation, and retinal neovascularization. *Proc. Natl. Acad. Sci. USA* **99**(2), 721–726.

Takaki, T., Fukasawa, K., Suzuki-Takahashi, I., Semba, K., Kitagawa, M., Taya, Y., and Hirai, H. (2005). Preferences for phosphorylation sites in the retinoblastoma protein of D-type cyclin-dependent kinases, Cdk4 and Cdk6, in vitro. *J. Biochem.* **137,** 381–386.

Takemoto, A., Maeshima, K., Ikehara, T., Yamaguchi, K., Murayama, A., Imamura, S., Imamoto, N., Yokoyama, S., Hirano, T., Watanabe, Y., Hanaoka, F., Yanagisawa, J., *et al.* (2009). The chromosomal association of condensin II is regulated by a noncatalytic function of PP2A. *Nat. Struct. Mol. Biol.* **16,** 1302–1308.

Templeton, D. J., Park, S. H., Lanier, L., and Weinberg, R. A. (1991). Nonfunctional mutants of the retinoblastoma protein are characterized by defects in phosphorylation, viral oncoprotein association, and nuclear tethering. *Proc. Natl. Acad. Sci. USA* **88,** 3033–3037.

Thomas, D. M., Carty, S. A., Piscopo, D. M., Lee, J. S., Wang, W. F., Forrester, W. C., and Hinds, P. W. (2001). The retinoblastoma protein acts as a transcriptional coactivator required for osteogenic differentiation. *Mol. Cell* **8,** 303–316.

Tracy, K., Dibling, B. C., Spike, B. T., Knabb, J. R., Schumacker, P., and Macleod, K. F. (2007). BNIP3 is an RB/E2F target gene required for hypoxia-induced autophagy. *Mol. Cell. Biol.* **27,** 6229–6242.

Trojer, P., Li, G., Sims, R. J., 3rd, Vaquero, A., Kalakonda, N., Boccuni, P., Lee, D., Erdjument-Bromage, H., Tempst, P., Nimer, S. D., Wang, Y. H., and Reinberg, D. (2007). L3MBTL1, a histone-methylation-dependent chromatin lock. *Cell* **129,** 915–928.

Trouche, D., Le Chalony, C., Muchardt, C., Yaniv, M., and Kouzarides, T. (1997). RB and hbrm cooperate to repress the activation functions of E2F1. *Proc. Natl. Acad. Sci. USA* **94,** 11268–11273.

Tsai, K. Y., Hu, Y., Macleod, K. F., Crowley, D., Yamasaki, L., and Jacks, T. (1998). Mutation of E2f-1 suppresses apoptosis and inappropriate S phase entry and extends survival of Rb-deficient mouse embryos. *Mol. Cell* **2,** 293–304.

van den Heuvel, S., and Dyson, N. J. (2008). Conserved functions of the pRB and E2F families. *Nat. Rev. Mol. Cell Biol.* **9,** 713–724.

van Harn, T., Foijer, F., van Vugt, M., Banerjee, R., Yang, F., Oostra, A., Joenje, H., and te Riele, H. (2010). Loss of Rb proteins causes genomic instability in the absence of mitogenic signaling. *Genes Dev.* **24,** 1377–1388.

Vance, K. W., Shaw, H. M., Rodriguez, M., Ott, S., and Goding, C. R. (2010). The retinoblastoma protein modulates Tbx2 functional specificity. *Mol. Biol. Cell,* **21,** 2770–2779.

Vandel, L., Nicolas, E., Vaute, O., Ferreira, R., Ait-Si-Ali, S., and Trouche, D. (2001). Transcriptional repression by the retinoblastoma protein through the recruitment of a histone methyltransferase. *Mol. Cell. Biol.* **21,** 6484–6494.

Vanderluit, J. L., Ferguson, K. L., Nikoletopoulou, V., Parker, M., Ruzhynsky, V., Alexson, T., McNamara, S. M., Park, D. S., Rudnicki, M., and Slack, R. S. (2004). p107 regulates neural precursor cells in the mammalian brain. *J. Cell Biol.* **166,** 853–863.

Vaute, O., Nicolas, E., Vandel, L., and Trouche, D. (2002). Functional and physical interaction between the histone methyl transferase Suv39H1 and histone deacetylases. *Nucleic Acids Res.* **30,** 475–481.

Verschure, P. J., van der Kraan, I., de Leeuw, W., van der Vlag, J., Carpenter, A. E., Belmont, A. S., and van Driel, R. (2005). In vivo HP1 targeting causes large-scale chromatin condensation and enhanced histone lysine methylation. *Mol. Cell. Biol.* **25,** 4552–4564.

Vooijs, M., van, d.V., te, R. H., and Berns, A. (1998). Flp-mediated tissue-specific inactivation of the retinoblastoma tumor suppressor gene in the mouse. *Oncogene* **17,** 1–12.

Wang, D., Kennedy, S., Conte, D., Jr., Kim, J. K., Gabel, H. W., Kamath, R. S., Mello, C. C., and Ruvkun, G. (2005). Somatic misexpression of germline P granules and enhanced RNA interference in retinoblastoma pathway mutants. *Nature* **436,** 593–597.

Wang, H., Bauzon, F., Ji, P., Xu, X., Sun, D., Locker, J., Sellers, R. S., Nakayama, K., Nakayama, K. I., Cobrinik, D., and Zhu, L. (2010). Skp2 is required for survival of aberrantly proliferating Rb1-deficient cells and for tumorigenesis in Rb1+/− mice. *Nat. Genet.* **42,** 83–88.

Weinberg, R. A. (1995). The retinoblastoma protein and cell cycle control. *Cell* **81,** 323–330.

Welch, P. J., and Wang, J. Y. J. (1993). A c-terminal protein-binding domain in the retinoblastoma protein regulates nuclear c-Abl tyrosine kinase in the cell cycle. *Cell* **75,** 779–790.

Wenzel, P. L., Wu, L., de Bruin, A., Chong, J. L., Chen, W. Y., Dureska, G., Sites, E., Pan, T., Sharma, A., Huang, K., Ridgway, R., Mosaliganti, K., *et al.* (2007). Rb is critical in a mammalian tissue stem cell population. *Genes Dev.* **21,** 85–97.

Wikenheiser-Brokamp, K. A. (2004). Rb family proteins differentially regulate distinct cell lineages during epithelial development. *Development* **131,** 4299–4310.

Wildwater, M., Campilho, A., Perez-Perez, J. M., Heidstra, R., Blilou, I., Korthout, H., Chatterjee, J., Mariconti, L., Gruissem, W., and Scheres, B. (2005). The retinoblastoma-related gene regulates stem cell maintenance in arabidopsis roots. *Cell* **123,** 1337–1349.

Williams, J. P., Stewart, T., Li, B., Mulloy, R., Dimova, D., Classon, M., Williams, J. P., Stewart, T., Li, B., Mulloy, R., Dimova, D., and Classon, M. (2006). The retinoblastoma protein is required for Ras-induced oncogenic transformation. *Mol. Cell. Biol.* **26,** 1170–1182.

Wintersteiner, H. (1897). Das Neuroepithelioma Retinae. Franz Deuticke, Vienna.

Wistuba, I. I., Gazdar, A. F., and Minna, J. D. (2001). Molecular genetics of small cell lung carcinoma. *Semin. Oncol.* **28,** 3–13.

Wong, S., and Weber, J. D. (2007). Deacetylation of the retinoblastoma tumour suppressor protein by SIRT1. *Biochem. J.* **407,** 451–460.

Wu, L., Timmers, C., Maiti, B., Saavedra, H. I., Sang, L., Chong, G. T., Nuckolls, F., Giangrande, P., Wright, F. A., Field, S. J., Greenberg, M. E., Orkin, S., *et al.* (2001). The E2F1-3 transcription factors are essential for cellular proliferation. *Nature* **414,** 457–462.

Wu, L., de Bruin, A., Saavedra, H. I., Starovic, M., Trimboli, A., Yang, Y., Opavska, J., Wilson, P., Thompson, J. C., Ostrowski, M. C., Rosol, T. J., Woollett, L. A., *et al.* (2003). Extra-embryonic function of Rb is essential for embryonic development and viability. *Nature* **421,** 942–947.

Xia, X., Cheng, A., Lessor, T., Zhang, Y., and Hamburger, A. W. (2001). Ebp1, an ErbB-3 binding protein, interacts with Rb and affects Rb transcriptional regulation. *J. Cell. Physiol.* **187,** 209–217.

Xiao, B., Spencer, J., Clements, A., Ali-Khan, N., Mittnacht, S., Broceno, C., Burghammer, M., Perrakis, A., Marmorstein, R., and Gamblin, S. J. (2003). Crystal structure of the retinoblastoma tumor suppressor protein bound to E2F and the molecular basis of its regulation. *Proc. Natl. Acad. Sci. USA* **100,** 2363–2368.

Xu, X. L., Fang, Y., Lee, T. C., Forrest, D., Gregory-Evans, C., Almeida, D., Liu, A., Jhanwar, S. C., Abramson, D. H., and Cobrinik, D. (2009). Retinoblastoma has properties of a cone precursor tumor and depends upon cone-specific MDM2 signaling. *Cell* **137,** 1018–1031.

Yao, G., Lee, T. J., Mori, S., Nevins, J. R., You, L., Yao, G., Lee, T. J., Mori, S., Nevins, J. R., and You, L. (2008). A bistable Rb-E2F switch underlies the restriction point. *Nat. Cell Biol.* **10,** 476–482.

Yu, B. D., Becker-Hapak, M., Snyder, E. L., Vooijs, M., Denicourt, C., and Dowdy, S. F. (2003). Distinct and nonoverlapping roles for pRB and cyclin D: Cyclin-dependent kinases 4/6 activity in melanocyte survival. *Proc. Natl. Acad. Sci. USA* **100,** 14881–14886.

Zacksenhaus, E., Jiang, Z., Chung, D., Marth, J. D., Phillips, R. A., and Gallie, B. L. (1996). pRb controls proliferation, differentiation, and death of skeletal muscle cells and other lineages during embryogenesis. *Genes Dev.* **10,** 3051–3064.

Zarkowska, T., and Mittnacht, S. (1997). Differential phosphorylation of the retinoblastoma protein by G1/S cyclin-dependent kinases. *J. Biol. Chem.* **272,** 12738–12746.

Zhang, J., Gray, J., Wu, L., Leone, G., Rowan, S., Cepko, C. L., Zhu, X., Craft, C. M., and Dyer, M. A. (2004). Rb regulates proliferation and rod photoreceptor development in the mouse retina. *Nat. Genet.* **36,** 351–360.

Zhang, R., Poustovoitov, M. V., Ye, X., Santos, H. A., Chen, W., Daganzo, S. M., Erzberger, J. P., Serebriiskii, I. G., Canutescu, A. A., Dunbrack, R. L., Pehrson, J. R., Berger, J. M., *et al.* (2005). Formation of MacroH2A-containing senescence-associated heterochromatin foci and senescence driven by ASF1a and HIRA. *Dev. Cell* **8,** 19–30.

Zheng, L., Flesken-Nikitin, A., Chen, P. L., and Lee, W. H. (2002). Deficiency of Retinoblastoma gene in mouse embryonic stem cells leads to genetic instability. *Cancer Res.* **62,** 2498–2502.

Zhou, Z., Flesken-Nikitin, A., Corney, D. C., Wang, W., Goodrich, D. W., Roy-Burman, P., and Nikitin, A. Y. (2006). Synergy of p53 and Rb deficiency in a conditional mouse model for metastatic prostate cancer. *Cancer Res.* **66,** 7889–7898.

Zhou, Z., Flesken-Nikitin, A., and Nikitin, A. Y. (2007). Prostate cancer associated with p53 and Rb deficiency arises from the stem/progenitor cell-enriched proximal region of prostatic ducts. *Cancer Res.* **67,** 5683–5690.

CHAPTER SIX

Genetic Alterations Targeting Lymphoid Development in Acute Lymphoblastic Leukemia

J. Racquel Collins-Underwood *and* Charles G. Mullighan

Contents

Abstract

While more than 80% of children with acute lymphoblastic leukemia (ALL) are cured with current chemotherapeutic regimens, a significant proportion of this patient population is at high risk of relapse. Recent advances in genomic profiling have identified novel genetic alterations that target key growth and development pathways in ALL and that influence the risk of treatment outcome. Notably, deletions and sequence mutations of lymphoid transcription factors are central events in the pathogenesis of B-lymphoid leukemia. Here, we describe these modern molecular techniques and their application to leukemia research. We also discuss the genetic lesions identified from these studies and how novel therapeutics directed at these targets may improve survival of pediatric ALL patients at high risk for relapse.

Department of Pathology, St. Jude Children's Research Hospital, Memphis, Tennessee, USA

Current Topics in Developmental Biology, Volume 94
ISSN 0070-2153, DOI: 10.1016/B978-0-12-380916-2.00006-1

1. Introduction

Acute lymphoblastic leukemia (ALL) is a neoplasm of lymphocyte progenitors in which immature lymphoid cells proliferate in the bone marrow, blood, and other organs. If untreated, ALL results in death, most commonly from bone marrow failure. ALL is the most common childhood cancer (Stiller, 2004) with peak prevalence occurring between the ages of 2 and 5 years (Pui *et al.*, 2008). Over the last five decades, cure rates for pediatric ALL have increased from less than 5% to greater than 80% (Pui *et al.*, 2004, 2008). However, up to one-quarter of children fail therapy and relapse, which carries a dismal prognosis (Einsiedel *et al.*, 2005; Rivera *et al.*, 2005). The survival rate for relapsed ALL is only 30%, making relapsed ALL the fourth most common childhood cancer and the leading cause of cancer-related death in young people (Einsiedel *et al.*, 2005).

ALL is characterized by a number of recurring chromosomal alterations, including aneuploidy and rearrangements that are important determinants of leukemogenesis, and while several genetic alterations are associated with the risk of relapse, the biologic factors determining treatment failure are poorly understood. Furthermore, while many of these gross chromosomal rearrangements are important in leukemia initiation, alone they are usually not sufficient to generate a full leukemic phenotype, suggesting that cooperating oncogenic lesions are required. Consequently, significant improvements in treatment outcomes will not be achieved by escalation of existing therapies, which are largely nontargeted and accompanied by substantial side effects due to dose-limiting toxicities. Rather, novel therapeutic approaches directed against rational targets are essential to improve survival of pediatric ALL patients at high risk for relapse, and this, in turn, requires a detailed understanding of the genetic lesions contributing to leukemogenesis and treatment failure.

2. Genetics of ALL

Because of the ease of access to tumor material and the successful implementation of cytogenetic techniques to identify gross chromosomal alterations, acute leukemia is one of the best genetically characterized diseases (Harrison, 2009; Rowley, 2008). These cytogenetic advances have enabled the detection of multiple chromosomal abnormalities in pediatric ALL, including aneuploidy and chromosomal translocations. Aneuploidy is an important prognostic factor in pediatric ALL as high hyperdiploidy (greater than 50 chromosomes) is present in approximately 30% of pediatric ALL cases and is associated with favorable outcome (Sutcliffe *et al.*, 2005).

Conversely, hypodiploidy (less than 45 chromosomes) is present in ~6% of pediatric ALL cases, and is associated with poor outcome (Harrison *et al.*, 2004). Structural alterations include translocations, deletions, insertions, and inversions, and common alterations in ALL; are t(12;21) [*ETV6-RUNX1* or *TEL-AML1*], t(1;19) [*TCF3-PBX1*], t(9;22) [*BCR–ABL1*], and *MLL* rearrangement in B-progenitor ALL; and rearrangement of the *TLX1* (*HOX11*), *TLX3* (*HOX11L2*), *LYL1*, *TAL1*, and *MLL* genes in T-lineage ALL (T-ALL; Graux *et al.*, 2006; Harrison, 2009; Harrison and Foroni, 2002; Pui *et al.*, 2004; Raimondi, 2006). These translocations frequently involve transcriptional regulators of hematopoiesis or kinases. For example, the t(12;21) chromosomal translocation leads to the fusion of *ETV6* and *RUNX1* genes, both of which are essential regulators of hematopoiesis and results in repressed transcription and disordered B-lineage lymphocyte development (Hiebert *et al.*, 1996). Likewise, the t(9;22) translocation, or the Philadelphia (Ph) chromosome, is the result of the reciprocal translocation between chromosomes 9 and 22. The resulting fusion of the BCR signaling protein to the ABL1 nonreceptor tyrosine kinase leads to constitutive tyrosine kinase activity and interaction with other signaling pathways involved in cell differentiation, proliferation, and survival (e.g., Ras pathway; McLaughlin *et al.*, 1987; Ren, 2005). These genetic alterations are important initiating events in leukemogenesis, but do not completely explain the genetic basis of leukemia. The observation that several of these alterations may be detected *in utero* and for years prior to the onset of leukemia and commonly do not alone induce leukemia in experimental models suggest that additional genetic or epigenetic alterations must contribute to leukemogenesis (Andreasson *et al.*, 2001; Wiemels *et al.*, 1999, 2006).

3. Lymphoid Development

B-progenitor ALL is the most common type of childhood leukemia and accounts for 70% of pediatric ALL cases. B-ALL is characterized by a block in lymphoid development at the pro- to pre-B cell stage, and multiple recurring genetic alterations targeting the early stages of B-lymphoid development have been identified in this disease, suggesting that these alterations are important in the development of B-lymphoid leukemia.

Hematopoietic stem cells (HSCs) in the fetal liver and postnatal bone marrow develop into B lymphocytes through a series of intermediate B cell progenitors with progressively decreased lineage potential (Adolfsson *et al.*, 2005). The process of differentiation involves chromatin remodeling that allows genes to be appropriately positioned within chromatin such that there is increased expression of lineage-associated genes (specification) and repression of lineage-inappropriate genes (commitment; Busslinger, 2004).

Networks of transcription factors coordinate these transcriptional events as well as the associated chromatin-remodeling factors, and together, these actions result in commitment to differentiation through a specific lineage (Dias *et al.*, 2008).

In the earliest stage of B cell development, multipotent, self-renewing HSCs give rise to multipotent progenitors (MPPs), characterized by the loss of self-renewal capacity but preservation of multilineage differentiation potential (Adolfsson *et al.*, 2001). Expression of the tyrosine kinase receptor Flt3 in a subset of MPPs further restricts cells to the lymphoid lineage. Flt3-expressing MPPs are known as lymphoid multipotent progenitors (LMPPs), which demarcates the initial phase of B-lineage priming as common myeloid progenitors (CMPs), precursor progenitors for erythroid and myeloid lineages, are also progeny of MPPs. LMPPs retain lymphoid and some myeloid potential but lose erythromegakaryocytic potential. B-lineage priming in LMPPs is mediated by *Ikzf1* (encoding Ikaros), a zinc-finger transcription factor responsible for restricting the self-renewal process in early stages of hematopoiesis and advancing B cell development at later stages (Ng *et al.*, 2009). Ikaros lineage priming functions overlap with those of the Ets family transcription factor, *Sfpi1* (encoding PU.1, Purine box factor 1), with low PU.1 concentration favoring B cell fate (DeKoter and Singh, 2000). Together, Ikaros and PU.1 also control the development of lymphoid progenitors by regulating the expression of essential signaling receptors: Flt3, cKit, and interleukin-7 receptor alpha (Il7ra; DeKoter *et al.*, 2002; Nichogiannopoulou *et al.*, 1999).

The LMPP population contains early lymphoid progenitors (ELPs), lymphoid-restricted cells characterized by expression of terminal deoxynucleotidyl transferase (TdT) and recombination-activating genes *Rag1/2*, which initiate rearrangement at the immunoglobulin heavy-chain (Igh) locus (Igarashi *et al.*, 2002). ELPs further differentiate into lymphoid-restricted common lymphoid progenitors (CLPs). CLPs are characterized by increased expression of Il7ra, lack all myeloid potential, and give rise to dendritic cells, natural killer cells, T cells, and B cells. This stage of B-lineage priming is mediated by the basic helix-loop-helix transcription factor E2a (E box-binding protein 2a; encoded by *Tcf3*), which is required for the proper formation of CLPs (Zhuang *et al.*, 1994). E2a initiates and maintains expression of B cell-specific genes, including *Ebf1* (early B cell factor 1) and *Pax5* (paired box gene 5). Ebf1 is an atypical basic helix-loop-helix transcription factor that is essential for B cell specification, initiating a precommitment process in which B cell-specific genes are primed for later expression (Lin and Grosschedl, 1995). Together, E2a and Ebf1 coordinately activate the B cell gene expression program and Igh gene rearrangements at the onset of B lymphopoiesis, while Ikaros mediates chromatin accessibility required for V(D)J recombination. At this stage of B cell development, CLPs are enriched for B-lineage-specified cells that are not yet firmly committed to the B cell fate (Northrup and Allman, 2008). Pax5, directly regulated by Ebf1, is the key transcription factor involved in commitment to the B lineage (Nutt *et al.*, 1999). Pax5 initiates and maintains B cell identity by repressing the transcription of lineage-inappropriate genes and simultaneously activating the expression of B cell lineage-associated genes, including *Ebf1*, *Cd19*, *Cd79a*, *Blnk*, *Igll1*, and *VpreB1* (Cobaleda *et al.*, 2007).

The first distinct B-biased cells arising from CLPs are characterized by expression of the B cell-associated marker Ptprc (also known as Cd45 or B220) and activation of B cell lineage-associated genes (Rumfelt *et al.*, 2006). These cells are termed CLP-2 or pre- to pro-B cells and represent the first stage of B cell differentiation. Cells committed to the B cell lineage, identified by expression of the Pax5 target Cd19 and initiation of V(D)J recombination, are termed pro-B cells that differentiate to pre-B cells upon completion of V(D)J recombination and assembly of the pre-BCR (B cell receptor). Once surface BCR and Igm are expressed, the pre-B cell differentiates to an immature B cell, which emigrates from the bone marrow into the periphery and becomes a mature B cell, the central mediator of humoral immunity. Notably, the majority of B-progenitor ALL cases is characterized by a block in differentiation at the pro- to pre-B cell stage. Until the advent of high-resolution genomic studies, the genetic alterations leading to this block were largely unknown.

4. Genome-Wide Profiling of Genetic Alterations in Cancer

Since the completion of the human genome project, the identification of genetic alterations in cancer in a genome-wide fashion has become possible. Microarray approaches to profile gene expression, DNA copy number alterations (CNAs), and loss-of-heterozygosity (LOH) are the most widely used and have been extremely informative, particularly in ALL.

Microarray-based gene expression profiling has been a valuable approach to characterize abnormalities in key signaling pathways in the leukemic cells, and may be used clinically to diagnose and classify leukemia and predict outcome (Haferlach *et al.*, 2009; Ross *et al.*, 2003, 2004; Wouters *et al.*, 2009; Yeoh *et al.*, 2002). However, it can be difficult to identify leukemogenic, or "driver" genetic lesions from the extensive lists of differentially expressed genes characteristic of each leukemia subtype. Therefore, genome-wide profiling of genetic alterations is of great interest in this disease. Several microarray platforms have been widely used in the analysis of cancer genomes that differ in resolution, technical performance, and the ability to detect DNA deletions, gains, and copy neutral LOH (CN-LOH; Davies *et al.*, 2005; Maciejewski and Mufti, 2008; Mullighan and Downing, 2009;

Strefford *et al.*, 2007). The earliest platforms were bacterial artificial chromosome (BAC) arrays, in which large DNA probes were spotted onto arrays that were hybridized with both test and reference DNA. Indeed, BAC arrays have successfully identified key submicroscopic lesions, including *ETV6* (TEL) and *CDKN2A/CDKN2B* (INK4/ARF) deletions, 9q34 duplications in T-ALL, and intrachromosomal amplification of chromosome 21 (Davidsson *et al.*, 2007; Huhta *et al.*, 1999; Jalali *et al.*, 2008; Kuchinskaya *et al.*, 2007, 2008; Larramendy *et al.*, 1998a,b; Lilljebjorn *et al.*, 2007; Lundin *et al.*, 2007; Pinkel and Albertson, 2005; Rabin *et al.*, 2008; Schoumans *et al.*, 2006; Steinemann *et al.*, 2008; Strefford *et al.*, 2006, 2007; Van Vlierberghe *et al.*, 2008). While BAC arrays have been highly informative, they have limited ability to detect focal CNAs—the hallmarks of ALL—because of the large (up to 200 kb) probe size (Hehir-Kwa *et al.*, 2007; Lo *et al.*, 2008; Mullighan *et al.*, 2007). Consequently, the most detailed insights into genetic alterations in acute leukemia have been obtained from higher resolution microarray studies of diagnostic leukemic samples.

Oligonucleotide arrays that comprise up to millions of short (20–80 mer) nucleotide probes are now more commonly used. These arrays may be used for comparative genomic hybridization (array-CGH), in which test and reference DNA are hybridized to an array (e.g., those manufactured by Agilent Technologies or Roche NimbleGen), or single channel arrays in which either a single test or reference sample is hybridized alone (e.g., Affymetrix or Illumina). These single channel arrays include widely used nucleotide polymorphism arrays that genotype up to a million single-nucleotide polymorphisms (SNPs), enabling both genome-wide association studies (GWAS), detection of CN-LOH, and identification of both inherited and tumor-acquired copy number alterations. Current SNP array platforms examine over 1.87 million SNP and copy number loci (Affymetrix) or over 1 million SNPs (Illumina; Matsuzaki *et al.*, 2004; Peiffer *et al.*, 2006). While most studies of genetic variants in leukemia have examined acquired genetic changes, oligonucleotide arrays are also being used to examine associations of inherited genetic variants, including copy number polymorphisms (CNP) and SNP genotypes, with tumor susceptibility, treatment responsiveness, and outcome (French *et al.*, 2009; Yang *et al.*, 2009).

The identification of all genetic alterations in leukemia is critically dependent upon array resolution. Lower resolution SNP arrays, initially designed as genotyping tools, may provide suboptimal coverage of many genes as the distribution of markers is not even across the genome (Xavier and Rioux, 2008). Careful bioinformatic analysis is also required in order to appropriately normalize raw array data, correct for aneuploidy, and generate sensitive and accurate calls of DNA CNAs, many of which are focal and may not be evident upon visual inspection of data. Commercial and freely available tools are available for these analyses (Mullighan *et al.*, 2007; Olshen *et al.*, 2004; Pounds *et al.*, 2009).

5. Genome-Wide Profiling of Genetic Alterations in ALL

The first SNP array study to demonstrate the feasibility of genome-wide analysis of genetic alterations in ALL utilized an 11,000-feature array, which detected focal deletions in *CDKN2A/B* resulting in LOH (Irving *et al.*, 2005). Since 2007, studies from St Jude and other groups have utilized higher resolution SNP and oligonucleotide array-CGH platforms to examine large cohorts of ALL samples, the results of which have provided critical new insights into the biology of ALL. Most of those studies utilized 100,000–500,000 feature Affymetrix SNP or Agilent oligonucleotide arrays with an average intermarker resolution of 5–25 kb. However, contemporary platforms now contain up to 2 million probes. For example, the SNP 6.0 platform (Affymetrix) incorporates 900,000 SNP probes and 900,000 CNP probes, which may be analyzed together to yield an intermarker distance of approximately 1 kb. Similarly, non-SNP oligonucleotide array-CGH platforms employ over 2 million probes (e.g., Roche NimbleGen).

Genome-wide analysis of pediatric B-ALL utilizing SNP 6.0 arrays suggest that disruption of genes involved in normal B cell development and differentiation play a critical role in leukemogenesis in pediatric B-ALL (Mullighan *et al.*, 2007). The mutations most commonly involve only a single copy of the affected gene; however, multiple mutations involving this pathway are common in high-risk B-ALL. Further, a higher number of lesions in the pathway is associated with poor outcome (Mullighan *et al.*, 2009a), suggesting the degree of "block" in B cell differentiation induced by mutations in this pathway not only contributes to leukemogenesis, but also treatment responsiveness.

6. Genetic Alterations Targeting Lymphoid Development in B-ALL

Several studies of B-ALL have now identified a high frequency of alterations in genes with key roles in normal lymphoid development. In the initial study from St Jude, over 40% of B-ALL cases harbored at least one mutation in this pathway (Mullighan *et al.*, 2007). More recent studies utilizing higher resolution arrays in different cohorts have found that over two-thirds of cases were affected, which suggests that a lesion in this pathway may be a common feature in the pathogenesis of B-ALL. The *PAX5* gene, encoding a paired box transcription factor required for B-lymphoid lineage commitment, maturation, and the suppression of differentiation into other lineages (Busslinger, 2004; Nutt and Kee, 2007;

Nutt *et al.*, 1999, 2001), was most frequently involved. In over one-third of cases, *PAX5* was targeted by deletion, focal internal amplification, sequence mutation, and translocation (An *et al.*, 2008; Bousquet *et al.*, 2007; Cazzaniga *et al.*, 2001; Kawamata *et al.*, 2008; Nebral *et al.*, 2007, 2009; Strehl *et al.*, 2003). Other genes involved include the IKAROS family of transcription factors [*IKZF1* (IKAROS), *IKZF2* (HELIOS), *IKZF3* (AIOLOS)], *EBF1* (early B cell factor), *TCF3* (transcription factor 3), *LEF1* (lymphoid enhancer-binding factor-1), *RAG1/2* (recombinase-activating genes 1/2), *BLNK* (B cell linker), and *VPREB1* (pre-B cell receptor). Deletions were most common in each case, but sequence mutation and translocation were also observed for *PAX5* as well as infrequent sequence mutations in *IKZF1*.

The frequency and pattern of B cell developmental alterations is also associated with ALL subtype. Hypodiploid ALL cases commonly have *PAX5* deletions, concomitant point mutations, and additional lesions in this pathway. Notably, hypodiploidy is associated with poor outcome, and studies in other cohorts have observed that an increasing number of lesions targeting the B-lymphoid development pathway is associated with poor outcome (Mullighan *et al.*, 2009a). Of interest, the lymphoid transcription factors targeted by genetic alteration are involved in relatively early stages of lymphoid development (Busslinger, 2004; Nutt and Kee, 2007) up to the stage of arrested maturation characteristic of ALL, suggesting that these alterations contribute to leukemogenesis. Accordingly, mouse models examining the role of *Ikzf1* or *Pax5* haploinsufficiency in the pathogenesis of ALL support a key role of these lesions in the leukemia development (Collins-Underwood *et al.*, 2009; Dang *et al.*, 2008; Miller *et al.*, 2008). In a retroviral bone marrow transplant model of *BCR–ABL1* ALL, *Ikzf1*, or *Pax5* haploinsufficiency reduces the latency of ALL (Collins-Underwood *et al.*, 2009; Miller *et al.*, 2008). Similarly, *Pax5* haploinsufficiency following chemical or retroviral mutagenesis significantly increases the penetrance of B-ALL (Dang *et al.*, 2008). In each model, the development of leukemia was accompanied by the acquisition of secondary genetic changes that are also observed in the human disease, including deletion of *Cdkn2a/b* and deletion or sequence mutation of the second *Pax5* allele. These studies suggest that these lesions cooperate in the development of the full leukemic phenotype.

7. *IKZF1* Alterations in *BCR–ABL1* Positive ALL

Because relapse occurs across the spectrum of ALL subtypes, identification of novel genetic markers that can predict outcome and treatment response is a major goal in leukemia genomics. Relapse is most often observed in high-risk ALL subtypes such as *BCR–ABL1* positive,

MLL-rearranged, and hypodiploid ALL. *BCR–ABL1* leukemia is particularly interesting as expression of the chimeric *BCR–ABL1* oncoprotein, encoded by the t(9;22) Philadelphia chromosome, is a hallmark of two diseases with distinct phenotypes and therapeutic responsiveness: chronic myeloid leukemia (CML), an expansion of relatively mature granulocytes that typically responds well to tyrosine kinase inhibitors (TKIs; Goldman and Melo, 2003) and *BCR–ABL1 de novo* ALL which usually does not show a durable response to TKI monotherapy, and has a dismal prognosis (Ribeiro *et al.*, 1987). While the basis of this dichotomy is incompletely understood, it has been attributed to the *BCR–ABL1* isoform (p185/p190 vs. p210) and the lineage and maturational stage of the leukemia initiating cell (Huntly *et al.*, 2004; Melo, 1996; Savona and Talpaz, 2008).

To examine the role of cooperating genetic alterations in leukemia development, we used SNP arrays to study large cohorts of ALL and CML patients, including 43 *BCR–ABL1 de novo* ALL cases and 128 samples obtained from 90 CML patients, including chronic phase ($n = 64$), accelerated phase ($n = 15$), and blast crisis ($n = 22$ myeloid; $n = 9$ lymphoid; Mullighan *et al.*, 2008). This study identified less than one CNA per chronic phase CML case with the only recurring lesions observed at the breakpoints of *BCR* and *ABL1*. In contrast, *IKZF1* (IKAROS) deletion is especially common in *de novo BCR–ABL1* ALL (36 of 43 cases) and at the progression of CML to lymphoid blast crisis, but not chronic phase CML. Furthermore, progression to lymphoid blast crisis was accompanied by the acquisition of lesions also seen at *de novo BCR–ABL1* ALL such as *CDKN2A/B*, *PAX5*, *C20orf94*, *RB1*, *MEF2C*, and *EBF1* deletions. Conversely, *TP53* deletion or sequence mutation was only observed in myeloid, but not lymphoid blast crisis CML. These findings suggest that the phenotype of *BCR–ABL1* leukemia is critically dependent on the presence and nature of concomitant genetic lesions and further suggest that IKAROS alteration is a key event in the pathogenesis of *BCR–ABL1* lymphoid leukemia.

IKAROS is a member of a family of zinc-finger transcription factors required for the development of all lymphoid lineages with complex, context-dependent functions including transcriptional regulation and chromatin remodeling (Georgopoulos, 2002; Georgopoulos *et al.*, 1992, 1994). *IKZF1* deletions involved either loss of the entire locus or loss of a subset of exons, most commonly coding exons 3–6, which results in expression of a dominant-negative isoform (IK6) that lacks the N-terminal DNA-binding zinc fingers but retains the C-terminal dimerization zinc fingers (Molnar and Georgopoulos, 1994). Interestingly, previous studies have also observed frequent expression of the dominant-negative IK6 transcript in ALL (Nakase *et al.*, 2000; Nishii *et al.*, 2000; Olivero *et al.*, 2000; Sun *et al.*, 1999a,b,c; Takanashi *et al.*, 2002), which may be attributed to aberrant posttranscriptional splicing (Klein *et al.*, 2006). Moreover, expression of a dominant-negative *Ikzf1* allele predisposed mice to the development of

T-lineage leukemias (Winandy *et al.*, 1995). SNP array profiling studies of both ALL and CML have further shown that expression of these aberrant transcripts is determined by the presence of *IKZF1* deletions that involve the exons corresponding to those deleted in the aberrant *IKZF1* transcripts and IKZF1 protein (Iacobucci *et al.*, 2008; Mullighan *et al.*, 2008). Thus, either loss of IKAROS or the expression of aberrant IKAROS transcripts may be a key event in the development of *BCR–ABL1* lymphoid leukemia. Indeed, our experimental modeling of *Ikzf1* alteration demonstrates that *Ikzf1* haploinsufficiency cooperates with *BCR–ABL1* to induce an aggressive pre-B leukemia that recapitulates human *BCR–ABL1* ALL, suggesting that IKAROS loss is important in the pathogenesis of this disease (Collins-Underwood *et al.*, 2009, unpublished data).

8. *IKZF1* Alteration in High-Risk "*BCR–ABL1*-Like" ALL

Further studies have demonstrated associations between *IKZF1* genetic alterations and poor outcome in B-ALL. In a recent genome-wide study of over 200 high-risk B-ALL BCR-ABL1 negative cases (the Children's Oncology Group P9906 cohort), SNP array analysis, gene expression profiling and candidate and gene resequencing was performed to ascertain whether an association existed between genetic alterations and outcome in this high-risk group. The study excluded known high-risk subtypes (e.g., *BCR–ABL1*, hypodiploid, and infant ALL) and identified over 50 recurring CNAs, most commonly involving genes of the B cell development pathway. This study also demonstrated a striking association between *IKZF1* mutation or deletion and poor outcome in B-ALL (Mullighan *et al.*, 2009a). Children with *IKZF1* alteration in leukemic cells at diagnosis had a 5-year cumulative incidence of relapse of 75% compared to 25% without *IKZF1* alteration. *IKZF1* alteration was also associated with high levels of minimal residual disease (MRD) and remained significantly associated with poor outcome in multivariable analyses incorporating age, leukocyte count, subtype, and MRD levels. Moreover, *IKZF1* alteration was associated with short disease-free survival and high incidence of relapse in BCR-ABL1 positive ALL (Martinelli *et al.*, 2009). These results were recently confirmed in the Dutch DCOG-ALL9 cohort (Kuiper *et al.*, 2010). These data suggest that detection of *IKZF1* alteration at diagnosis may serve as a prognostic indicator by identifying patients at high risk for treatment failure and relapse. These findings support those from *BCR–ABL1* positive leukemia that alteration of *IKZF1* is a key determinant of leukemogenesis and response to therapy. Notably, the gene expression profile of the poor outcome, *IKZF1*-altered cases in the

P9906 study was strikingly similar to that of *BCR–ABL1* positive ALL (Mullighan *et al.*, 2009a). A similar subtype of "BCR–ABL1-like" ALL enriched for genetic alterations targeting B-lymphoid development has also been described (Den Boer *et al.*, 2009).

The similarity of the gene expression profiles of *IKZF1*-deleted *BCR–ABL1* positive and *BCR–ABL1*-negative ALL suggests that perturbation of IKZF1 activity may directly influence the leukemic transcriptome and subsequently the degree of differentiation of the leukemic cell. Accordingly, the gene expression profile of *IKZF1*-mutated *BCR–ABL1*-negative ALL exhibits enrichment for HSC genes and reduced expression of B cell signaling genes (Mullighan *et al.*, 2009a). Another potential explanation is that *IKZF1* mutated, BCR–ABL1-like cases may harbor mutations that result in activation of downstream signaling pathways similar to those activated by BCR–ABL1, raising the possibility that aberrant tyrosine kinase signaling may also be involved in this high-risk leukemia subtype. Recent data have shown that this indeed is the case. A significant proportion of BCR–ABL1-like ALL cases have genetic alterations that result in aberrant cytokine receptor signaling, particularly activating Janus kinase (JAK) mutations and rearrangement of *CRLF2* (encoding the lymphoid cytokine receptor gene *c*ytokine *r*eceptor-*l*ike *f*actor 2) (Mullighan *et al.*, 2009b,c).

Genomic resequencing of targets of DNA CNAs, dysregulated genes, and a subset of receptor and nonreceptor tyrosine kinases was performed in 187 high-risk B-ALL cases from the P9906 cohort described above. This study identified somatic *JAK1*, *JAK2*, and *JAK3* mutations in 20 (10.7%) cases with mutations found most often at or near R683 in the pseudokinase domain of JAK2, but also in the kinase domain of JAK2 and the pseudokinase domain of JAK1 (Mullighan *et al.*, 2009c). The JAK pseudokinase domain is thought to negatively regulate the activity of the kinase domain (Saharinen and Silvennoinen, 2002) as well as mediate protein–protein interactions (Russo *et al.*, 1996). Thus, the pseudokinase domain mutations are predicted to disrupt the structure and dynamics of the loops at the interlobe interface, which may result in loss of its inhibitory activity. Accordingly, the kinase domain mutations may result in constitutive activation of JAK tyrosine kinase activity. Notably, the JAK2 V617F mutation, which is commonly observed in the myeloproliferative disorders (Baxter *et al.*, 2005; James *et al.*, 2005; Kralovics *et al.*, 2005; Levine *et al.*, 2005) has not been identified in B-ALL although its homolog, JAK1 V658F, has been identified (Mullighan *et al.*, 2009b). The presence of JAK mutations was associated with *IKZF1* mutations, a BCR–ABL1-like gene expression profile, and poor outcome. Notably, JAK2 mutations (again, most commonly at R683 in the pseudokinase domain) had also recently been reported in up to one-quarter of cases of B-progenitor ALL associated with Down syndrome (Bercovich *et al.*, 2008; Kearney *et al.*, 2008; Malinge *et al.*, 2007); however, most of the P9906 high-risk ALL cases with JAK mutations were not associated with Down syndrome. JAK1 pseudokinase mutations have also been described in T-lineage ALL, although more commonly in adults than in children (Flex *et al.*, 2008; Mullighan *et al.*, 2009c). Similar to the JAK2 V617F mutation, the JAK1 and JAK2 mutations observed in ALL are transforming *in vitro*, conferring cytokine-independent growth and constitutive Jak–Stat activation when introduced into the Ba/F3, a murine cytokine-dependent cell line commonly used to examine the transforming effects of kinase mutations, expressing the erythropoietin or thrombopoietin receptors (Bercovich *et al.*, 2008; Gaikwad *et al.*, 2009; Mullighan *et al.*, 2009c).

The JAKs are key mediators of hematopoietic cytokine receptor signal transduction (Baker *et al.*, 2007; Ihle and Gilliland, 2007; Vainchenker *et al.*, 2008). The identification of distinct JAK mutations in myeloproliferative diseases and ALL suggested that different mutated JAK alleles may interact with different downstream signaling pathways and influence the disease lineage. Recent studies have shown that the presence of JAK mutations in ALL are associated with chromosomal alterations resulting in overexpression of the cytokine receptor CRLF2 (or TSLPR, *t*hymic *s*tromal *l*ympho*p*oietin *r*eceptor), highlighting a new pathway of perturbed lymphoid signaling in ALL.

SNP array profiling of the high-risk ALL cohort demonstrated that many of the JAK-mutated cases harbored focal DNA CNAs, most commonly interstitial deletions, involving a cluster of hematopoietic cytokine receptor genes including *IL3RA* (interleukin 3 receptor alpha) and *CSF2RA* (GM-CSF receptor) at the pseudoautosomal region 1 (PAR1) at Xp/Yp. These alterations were adjacent to the *CRLF2* locus at PAR1 and were associated with markedly elevated expression of CRLF2 (Harvey *et al.*, 2010; Mullighan *et al.*, 2009b). Notably, Russell, Harrison and colleagues had also identified dysregulated expression of CRLF2 arising from rearrangement of *CRLF2* into the immunoglobulin heavy chain locus (*IGH@-CRLF2*), or associated with the PAR1 deletion, in a subset of B-progenitor ALL (Russell *et al.*, 2009). Defining the precise limits of the PAR1 deletion was difficult due to poor microarray probe coverage of the PAR1 region, but mapping using high-resolution microarrays and long range genomic PCR showed that the deletion extended from intron 1 of *P2RY8* (encoding the purinergic receptor gene P2Y, G-protein coupled, 8) to immediately upstream of the first coding exon of *CRLF2*. The deletion breakpoints were tightly clustered and resulted in a novel fusion transcript, *P2RY8-CRLF2*, in which the first, noncoding exon of *P2RY8* is fused to the entire coding region of CRLF2 (Mullighan *et al.*, 2009b). This rearrangement appears to represent a form of "promoter swapping". *P2RY8* is a member of a family of purinergic receptor genes that is expressed in hematopoietic cells, including leukemic blasts, and has previously been identified as a rare target of translocation to *SOX5* in lymphoma (Storlazzi *et al.*, 2007).

CRLF2 forms a heterodimeric receptor with interleukin receptor alpha (IL7RA) for the cytokine TSLP (*t*hymic *s*tromal *l*ympho*p*oietin; Hiroyama *et al.*, 2000; Pandey *et al.*, 2000; Park *et al.*, 2000). TSLP/CRLF2 signaling has a role in dendritic cell development (Zhang *et al.*, 2001), T cell responses (Mazzucchelli *et al.*, 2008; Ziegler and Liu, 2006), allergic inflammation (Al-Shami *et al.*, 2005; Rochman and Leonard, 2008; Zhou *et al.*, 2005), and promotes the proliferation of normal and leukemic B cells (Astrakhan *et al.*, 2007; Brown *et al.*, 2007; Ray *et al.*, 1996; Vosshenrich *et al.*, 2004), but may be dispensable for normal B-lymphoid development (Carpino *et al.*, 2004; Vosshenrich *et al.*, 2004). The downstream mediators of TSLP/CRLF2 signaling are poorly defined and may differ between human and mouse, and activation of Jak–Stat signaling has been described for the human, but not CRLF2. These differences may in part be due to limited homology of both the receptor and ligand across species (Carpino *et al.*, 2004).

CRLF2 alterations in B-ALL have now been identified by multiple independent groups (Harvey *et al.*, 2010; Hertzberg *et al.*, 2010; Mullighan *et al.*, 2009b; Russell *et al.*, 2009; Yoda *et al.*, 2010). *CRLF2* is rearranged in 5–7% of pediatric B-ALL cases, most commonly by *IGH@-CRLF2* rearrangement or the PAR1 deletion resulting in expression of *P2RY8-CRLF2*. Both alterations lead to increased cell surface expression of CRLF2 by leukemic cells, and flow cytometric analysis of CRLF2 expression may be used to detect CRLF2-rearranged cases. Less commonly, *CRLF2* is rearranged to other, as yet unknown partner genes, or harbors presumed activating mutations, most commonly F232C (Chapiro *et al.*, 2010; Yoda *et al.*, 2010). A striking observation is that *CRLF2* alteration, most commonly the PAR1 deletion, is present in over 50% of ALL associated with Down syndrome (DS-ALL; Hertzberg *et al.*, 2010; Mullighan *et al.*, 2009c), in which other chromosomal rearrangements characteristic of childhood ALL are uncommon (Forestier *et al.*, 2008). The basis for this increased frequency in DS-ALL is currently unknown.

In both DS- and non-DS-ALL, *CRLF2* rearrangement is significantly associated with the presence of activating JAK mutations (Hertzberg *et al.*, 2010; Mullighan *et al.*, 2009b; Russell *et al.*, 2009; Yoda *et al.*, 2010). Over half of *CRLF2*-rearranged cases harbor activating *JAK1* or *JAK2* mutations, and conversely, nearly all JAK-mutated cases have *CRLF2* rearrangements, suggesting that these lesions together contribute to leukemogenesis. Importantly, in non-DS-ALL, *CRLF2* alteration and JAK mutations are associated with the presence of *IKZF1* alterations, and very poor outcome (Harvey *et al.*, 2010), suggesting that JAK inhibition may be a useful therapeutic approach in these high-risk cases that at present frequently fail maximal therapy.

9. Role of IKAROS Perturbation in Leukemogenesis

The use of high-resolution SNP arrays to understand the molecular pathogenesis of leukemia led us to the unexpected finding that genetic alterations in key regulators of B cell development and differentiation were present in a substantial proportion of pediatric B-ALL cases, most notably *IKZF1*. The overall consequence of these mutations is a reduction in the level of the specific transcription factor either as a result of monoallelic deletions or the generation of altered forms of the specific protein. The identification of these abnormalities has provided important insights into normal and leukemic hematopoiesis.

The role of IKAROS in the pathogenesis of *BCR–ABL1*-positive and -negative ALL remains to be fully defined. To address our findings above that *IKZF1* deletion is a hallmark of both, many of which have activating JAK mutations and *CRLF2* rearrangement, we have performed preliminary studies demonstrating that *Ikzf1* deletion contributes to leukemogenesis in a mouse model of *BCR–ABL1*-positive ALL and that *JAK* mutations and CRLF2 overexpression are transforming *in vitro* in the context of *BCR–ABL1*-negative disease.

To test whether perturbation of IKAROS activity is a key event in the pathogenesis of high-risk leukemia, we examined the effect of Ikaros haploinsufficiency in a retroviral bone marrow transplant model of murine *BCR–ABL1* B-ALL using unmanipulated marrow from either *Ikzf1* wild-type mice or mice heterozygous for an *Ikzf1* null allele (Collins-Underwood *et al.*, 2009). Our studies demonstrate that *Ikzf1* haploinsufficiency accelerates the onset of leukemia and promotes B-lineage leukemia. Furthermore, genomic analysis of $Ikzf1^{+/-}$ mouse tumors identifies additional aberrations seen in human *BCR–ABL1 de novo* ALL, namely *Ebf1* and *Cdnk2a* deletions. Notably, the murine *BCR–ABL1* ALL gene expression signature is similar to human *BCR–ABL1 de novo* ALL and consistent with block in differentiation as we observed negative enrichment of B cell signaling genes and positive enrichment for HSC genes. Together, these data demonstrate that *Ikzf1* haploinsufficiency cooperates with *BCR–ABL1* to induce an aggressive pre-B leukemia with clinical, pathologic, genomic, and transcriptomic features recapitulating human *BCR–ABL1* ALL, suggesting that IKAROS loss is important in the pathogenesis of this disease. Experiments evaluating the interaction of *Ikzf1* loss with other targets of genomic alteration in *BCR–ABL1* ALL (e.g., *CDKN2A/B*, *PAX5*) and studies examining the role of *Ikzf1* haploinsufficiency in responsiveness to kinase inhibitors are under investigation at present. The role of the dominant negative isoform, IK6, in leukemogenesis is also currently being examined as existing data have shown that expression of IK6 impairs B-lymphoid maturation (Tonnelle *et al.*, 2001, 2009) and pre-B cell receptor signaling in *BCR–ABL1* positive ALL cells (Trageser *et al.*, 2009).

The role of IKAROS in the pathogenesis of BCR–ABL1 negative ALL also warrants further investigation. Although the role of *CRLF2* in lymphopoiesis is incompletely understood, existing data suggests that aberrant CRLF2/JAK signaling contributes to leukemogenesis. Expression of either CRLF2 or mutant JAK alleles alone into Ba/F3 cells not expressing EpoR/TpoR usually does not result in transformation (Mullighan *et al.*, 2009c). A notable exception is JAK1 V658F, the homolog of JAK2 V617F. Prior to the identification of *CRLF2* alterations in ALL, JAK mutations in ALL were shown (like the JAK2 V617F mutation observed in MPD), to transform Ba/F3 cells expressing the erythropoietin receptor (Ba/F3-EpoR cells) to cytokine-independent growth and result in constitutive Jak–Stat activation (Bercovich *et al.*, 2008; Gaikwad *et al.*, 2009; Mullighan *et al.*, 2009c), suggesting that interaction of Jak mutants with a cytokine receptor scaffold is required for transformation. Subsequent studies have shown that coexpression of JAK mutations and CRLF2 in Ba/F3 cells is transforming, and that this transformation is inhibited by either pharmacologic JAK inhibition or short hairpin RNA mediated knockdown of CRLF2 expression (Hertzberg *et al.*, 2010; Mullighan *et al.*, 2009b; Yoda *et al.*, 2010). Similarly, studies using primary murine hematopoietic progenitors have shown that enforced expression of CRLF2 alone promotes lymphoid expansion, but this is insufficient to result in the development of leukemia (Collins-Underwood *et al.*, 2009, unpublished data; Russell *et al.*, 2009). Ongoing studies modeling CRLF2 dysregulation and JAK mutations will be important not only to determine the role of these alterations in leukemogenesis, but also to provide preclinical models of ALL that faithfully recapitulate human leukemia in which to test the efficacy of pharmacologic JAK inhibitors. Importantly, these studies must also model the effects of additional genetic lesions commonly observed in CRLF2/JAK-mutated ALL, including deletion or mutation of B-lymphoid transcriptional regulators such as *IKZF1* and *PAX5* and deletion of *CDKN2A/B*.

10. Future Directions

Despite significant advances in genomic profiling of leukemia, much remains to be learned about the genetic basis for high-risk leukemia. Firstly, the high frequency of alterations of B cell development genes identified in pediatric B-ALL represents a lower limit of the true frequency and will require direct copy number analysis of those genes with low-density SNP coverage and full sequence analysis of all genes involved in B cell development and differentiation to better define the overall frequency. Secondly, the finding that *IKZF1* alteration is strongly associated with poor outcome regardless of *BCR–ABL1*, *CRLF2*, or *JAK* status suggests that additional high-risk cases with *IKZF1* alterations harbor other genetic alterations that promote therapy resistance and treatment failure and warrants further investigation of the tumor kinome. Thirdly, studies to directly assess the effect of coexpressing Ikaros, Pax5, and Ink4/Arf loss-of-function mutants, expression of dominant-negative Ikaros isoforms (IK6), and coexpression of CRLF2 rearrangement and activating JAK mutations in murine models should provide valuable insights into the ability of these lesions to cooperate in leukemogenesis. Finally, attempts to determine whether small-molecule inducers of differentiation can bypass the block resulting from the identified lesions, and whether these molecules, in turn, would trigger a leukemia cell-specific apoptotic response, would lead to new therapeutic approaches for pediatric B-ALL.

There remains a substantial proportion of ALL cases that lack known cytogenetic alterations and fail therapy, and the frequency of these cases rises with increasing age. Compared to childhood leukemia, there is a lack of detailed, high-resolution genomic profiling data from adolescent and adult ALL (Paulsson *et al.*, 2008; Usvasalo *et al.*, 2010), which has a markedly inferior outcome to that of childhood ALL. The frequency of *BCR–ABL1* positive ALL rises progressively with increasing age, but this alone does not alone account for the poor outcome of ALL in adults. Furthermore, several high-risk subtypes of leukemia have either not been studied in detail (e.g., ALL with low hypodiploidy; Harrison *et al.*, 2004; Heerema *et al.*, 1999; Pui *et al.*, 1987; Raimondi *et al.*, 2003), or have few structural genetic alterations on microarray analysis (e.g., *MLL*-rearranged leukemia; Bardini *et al.*, 2010; Mullighan *et al.*, 2007). Also, while microarray platforms have provide important insights into DNA CNAs in ALL, they do not directly detect structural rearrangements or DNA sequence alterations.

Thus, future genomic profiling studies of ALL require detailed analysis of less well-studied cohorts, and the application of novel genomic profiling technologies that interrogate both genetic and epigenetic changes. Detailed candidate gene sequencing studies in ALL have identified new mutations in B-progenitor ALL (Zhang *et al.*, 2009), suggesting that genome-wide sequencing is required to identify the full complement of genetic alterations in this disease. This is now feasible with next-generation, massively parallel sequencing of tumor nucleic acids (Mardis and Wilson, 2009). Next-generation sequencing of either tumor DNA or RNA has identified new targets of mutation in AML (Ley *et al.*, 2008; Mardis *et al.*, 2009), T-lineage ALL (Van Vlierberghe *et al.*, 2010), and lymphoma (Morin *et al.*, 2010), and has identified new targets of rearrangement in cancer (Maher *et al.*, 2009a,b), including B-lineage ALL (Mullighan *et al.*, 2009d). It is likely that as the time and cost requirements of these methods declines, sequencing-based approaches will assume greater importance in interrogating cancer genomes, and may supplant array-based methodologies.

ACKNOWLEDGMENTS

We thank colleagues at St Jude Children's Research Hospital and the Children's Oncology Group who provided invaluable contributions to these studies. Work described in this review was supported by the American Lebanese Syrian Associated Charities of St Jude Children's Research Hospital, the National Health and Medical Research Council of Australia, the American Association for Cancer Research, the American Society of Hematology, and the National Cancer Institute's TARGET (Therapeutically Applicable Research to Generate Effective Treatments) Initiative. C.G.M. is a Pew Scholar in the Biomedical Sciences.

REFERENCES

Adolfsson, J., Borge, O. J., Bryder, D., Theilgaard-Monch, K., Astrand-Grundstrom, I., Sitnicka, E., *et al.* (2001). Upregulation of Flt3 expression within the bone marrow Lin(-) Sca1(+)c-kit(+) stem cell compartment is accompanied by loss of self-renewal capacity. *Immunity* **15**(4), 659–669.

Adolfsson, J., Mansson, R., Buza-Vidas, N., Hultquist, A., Liuba, K., Jensen, C. T., *et al.* (2005). Identification of Flt3+ lympho-myeloid stem cells lacking erythro-megakaryocytic potential a revised road map for adult blood lineage commitment. *Cell* **121**(2), 295–306.

Al-Shami, A., Spolski, R., Kelly, J., Keane-Myers, A., and Leonard, W. J. (2005). A role for TSLP in the development of inflammation in an asthma model. *J. Exp. Med.* **202**(6), 829–839.

An, Q., Wright, S. L., Konn, Z. J., Matheson, E., Minto, L., Moorman, A. V., *et al.* (2008). Variable breakpoints target PAX5 in patients with dicentric chromosomes: A model for the basis of unbalanced translocations in cancer. *Proc. Natl. Acad. Sci. USA* **105**(44), 17050–17054.

Andreasson, P., Schwaller, J., Anastasiadou, E., Aster, J., and Gilliland, D. G. (2001). The expression of ETV6/CBFA2 (TEL/AML1) is not sufficient for the transformation of hematopoietic cell lines in vitro or the induction of hematologic disease in vivo. *Cancer Genet. Cytogenet.* **130**(2), 93–104.

Astrakhan, A., Omori, M., Nguyen, T., Becker-Herman, S., Iseki, M., Aye, T., *et al.* (2007). Local increase in thymic stromal lymphopoietin induces systemic alterations in B cell development. *Nat. Immunol.* **8**(5), 522–531.

Baker, S. J., Rane, S. G., and Reddy, E. P. (2007). Hematopoietic cytokine receptor signaling. *Oncogene* **26**(47), 6724–6737.

Bardini, M., Spinelli, R., Bungaro, S., Mangano, E., Corral, L., Cifola, I., *et al.* (2010). DNA copy-number abnormalities do not occur in infant ALL with t(4;11)/MLL-AF4. *Leukemia* **24**(1), 169–176.

Baxter, E. J., Scott, L. M., Campbell, P. J., East, C., Fourouclas, N., Swanton, S., *et al.* (2005). Acquired mutation of the tyrosine kinase JAK2 in human myeloproliferative disorders. *Lancet* **365**(9464), 1054–1061.

Bercovich, D., Ganmore, I., Scott, L. M., Wainreb, G., Birger, Y., Elimelech, A., *et al.* (2008). Mutations of *JAK2* in acute lymphoblastic leukaemias associated with Down's syndrome. *Lancet* **372,** 1484–1492.

Bousquet, M., Broccardo, C., Quelen, C., Meggetto, F., Kuhlein, E., Delsol, G., *et al.* (2007). A novel PAX5-ELN fusion protein identified in B-cell acute lymphoblastic leukemia acts as a dominant negative on wild-type PAX5. *Blood* **109**(8), 3417–3423.

Brown, V. I., Hulitt, J., Fish, J., Sheen, C., Bruno, M., Xu, Q., *et al.* (2007). Thymic stromal-derived lymphopoietin induces proliferation of pre-B leukemia and antagonizes mTOR inhibitors, suggesting a role for interleukin-7Ralpha signaling. *Cancer Res.* **67**(20), 9963–9970.

Busslinger, M. (2004). Transcriptional control of early B cell development. *Annu. Rev. Immunol.* **22,** 55–79.

Carpino, N., Thierfelder, W. E., Chang, M. S., Saris, C., Turner, S. J., Ziegler, S. F., *et al.* (2004). Absence of an essential role for thymic stromal lymphopoietin receptor in murine B-cell development. *Mol. Cell. Biol.* **24**(6), 2584–2592.

Cazzaniga, G., Daniotti, M., Tosi, S., Giudici, G., Aloisi, A., Pogliani, E., *et al.* (2001). The paired box domain gene PAX5 is fused to ETV6/TEL in an acute lymphoblastic leukemia case. *Cancer Res.* **61**(12), 4666–4670.

Chapiro, E., Russell, L., Lainey, E., Kaltenbach, S., Ragu, C., Della-Valle, V., *et al.* (2010). Activating mutation in the TSLPR gene in B-cell precursor lymphoblastic leukemia. *Leukemia* **24**(3), 642–645.

Cobaleda, C., Schebesta, A., Delogu, A., and Busslinger, M. (2007). Pax5: The guardian of B cell identity and function. *Nat. Immunol.* **8**(5), 463–470.

Collins-Underwood, J. R., Miller, C. B., Phillips, L. A., Downing, J. R., and Mullighan, C. G. (2009). Ikzf1 haploinsufficiency contributes to the pathogenesis of BCR–ABL1 positive acute lymphoblastic leukemia. *Blood* **114**:Abstract 678.

Dang, J., Mullighan, C. G., Phillips, L. A., Mehta, P., and Downing, J. R. (2008). Retroviral and chemical mutagenesis identifies Pax5 as a tumor suppressor in B-progenitor acute lymphoblastic leukemia. *Blood* **112**:Abstract 1798.

Davidsson, J., Andersson, A., Paulsson, K., Heidenblad, M., Isaksson, M., Borg, A., *et al.* (2007). Tiling resolution array comparative genomic hybridization, expression and methylation analyses of dup(1q) in Burkitt lymphomas and pediatric high hyperdiploid acute lymphoblastic leukemias reveal clustered near-centromeric breakpoints and overexpression of genes in 1q22–32.3. *Hum. Mol. Genet.* **16**(18), 2215–2225.

Davies, J. J., Wilson, I. M., and Lam, W. L. (2005). Array CGH technologies and their applications to cancer genomes. *Chromosome Res.* **13**(3), 237–248.

DeKoter, R. P., and Singh, H. (2000). Regulation of B lymphocyte and macrophage development by graded expression of PU.1. *Science* **288**(5470), 1439–1441.

DeKoter, R. P., Lee, H. J., and Singh, H. (2002). PU.1 regulates expression of the interleukin-7 receptor in lymphoid progenitors. *Immunity* **16**(2), 297–309.

Den Boer, M. L., van Slegtenhorst, M., De Menezes, R. X., Cheok, M. H., Buijs-Gladdines, J. G., Peters, S. T., *et al.* (2009). A subtype of childhood acute lymphoblastic leukaemia with poor treatment outcome: A genome-wide classification study. *Lancet Oncol.* **10**(2), 125–134.

Dias, S., Xu, W., McGregor, S., and Kee, B. (2008). Transcriptional regulation of lymphocyte development. *Curr. Opin. Genet. Dev.* **18**(5), 441–448.

Einsiedel, H. G., von Stackelberg, A., Hartmann, R., Fengler, R., Schrappe, M., Janka-Schaub, G., *et al.* (2005). Long-term outcome in children with relapsed ALL by risk-stratified salvage therapy: Results of trial acute lymphoblastic leukemia-relapse study of the Berlin-Frankfurt-Munster Group 87. *J. Clin. Oncol.* **23**(31), 7942–7950.

Flex, E., Petrangeli, V., Stella, L., Chiaretti, S., Hornakova, T., Knoops, L., *et al.* (2008). Somatically acquired JAK1 mutations in adult acute lymphoblastic leukemia. *J. Exp. Med.* **205**(4), 751–758.

Forestier, E., Izraeli, S., Beverloo, B., Haas, O., Pession, A., Michalova, K., *et al.* (2008). Cytogenetic features of acute lymphoblastic and myeloid leukemias in pediatric patients with Down syndrome: An iBFM-SG study. *Blood* **111**(3), 1575–1583.

French, D., Yang, W., Cheng, C., Raimondi, S. C., Mullighan, C. G., Downing, J. R., *et al.* (2009). Acquired variation outweighs inherited variation in whole genome analysis of methotrexate polyglutamate accumulation in leukemia. *Blood* **113**(19), 4512–4520.

Gaikwad, A., Rye, C. L., Devidas, M., Heerema, N. A., Carroll, A. J., Izraeli, S., *et al.* (2009). Prevalence and clinical correlates of JAK2 mutations in Down syndrome acute lymphoblastic leukaemia. *Br. J. Haematol.* **144**(6), 930–932.

Georgopoulos, K. (2002). Haematopoietic cell-fate decisions, chromatin regulation and ikaros. *Nat. Rev. Immunol.* **2**(3), 162–174.

Georgopoulos, K., Moore, D. D., and Derfler, B. (1992). Ikaros, an early lymphoid-specific transcription factor and a putative mediator for T cell commitment. *Science* **258**(5083), 808–812.

Georgopoulos, K., Bigby, M., Wang, J. H., Molnar, A., Wu, P., Winandy, S., *et al.* (1994). The Ikaros gene is required for the development of all lymphoid lineages. *Cell* **79**(1), 143–156.

Goldman, J. M., and Melo, J. V. (2003). Chronic myeloid leukemia—Advances in biology and new approaches to treatment. *N. Engl. J. Med.* **349**(15), 1451–1464.

Graux, C., Cools, J., Michaux, L., Vandenberghe, P., and Hagemeijer, A. (2006). Cytogenetics and molecular genetics of T-cell acute lymphoblastic leukemia: From thymocyte to lymphoblast. *Leukemia* **20**(9), 1496–1510.

Haferlach, T., Bacher, U., Kohlmann, A., and Haferlach, C. (2009). Discussion of the applicability of microarrays: Profiling of leukemias. *Methods Mol. Biol.* **509,** 15–33.

Harrison, C. J. (2009). Cytogenetics of paediatric and adolescent acute lymphoblastic leukaemia. *Br. J. Haematol.* **144**(2), 147–156.

Harrison, C. J., and Foroni, L. (2002). Cytogenetics and molecular genetics of acute lymphoblastic leukemia. *Rev. Clin. Exp. Hematol.* **6**(2), 91–113, [Discussion: 200–112].

Harrison, C. J., Moorman, A. V., Broadfield, Z. J., Cheung, K. L., Harris, R. L., Reza Jalali, G., *et al.* (2004). Three distinct subgroups of hypodiploidy in acute lymphoblastic leukaemia. *Br. J. Haematol.* **125**(5), 552–559.

Harvey, R. C., Mullighan, C. G., Chen, I. M., Wharton, W., Mikhail, F. M., Carroll, A. J., *et al.* (2010). Rearrangement of *CRLF2* is associated with mutation of *JAK* kinases, alteration of *IKZF1*, Hispanic/Latino ethnicity and a poor outcome in pediatric B-progenitor acute lymphoblastic leukemia. *Blood* **115**, 5312-5321.

Heerema, N. A., Nachman, J. B., Sather, H. N., Sensel, M. G., Lee, M. K., Hutchinson, R., *et al.* (1999). Hypodiploidy with less than 45 chromosomes confers adverse risk in childhood acute lymphoblastic leukemia: A report from the children's cancer group. *Blood* **94**(12), 4036–4045.

Hehir-Kwa, J. Y., Egmont-Petersen, M., Janssen, I. M., Smeets, D., van Kessel, A. G., and Veltman, J. A. (2007). Genome-wide copy number profiling on high-density bacterial artificial chromosomes, single-nucleotide polymorphisms, and oligonucleotide microarrays: A platform comparison based on statistical power analysis. *DNA Res.* **14**(1), 1–11.

Hertzberg, L., Vendramini, E., Ganmore, I., Cazzaniga, G., Schmitz, M., Chalker, J., *et al.* (2010). Down syndrome acute lymphoblastic leukemia: A highly heterogeneous disease in which aberrant expression of CRLF2 is associated with mutated JAK2: A report from the iBFM Study Group. *Blood* [Nov 24].

Hiebert, S. W., Sun, W., Davis, J. N., Golub, T., Shurtleff, S., Buijs, A., *et al.* (1996). The t(12;21) translocation converts AML-1B from an activator to a repressor of transcription. *Mol. Cell. Biol.* **16**(4), 1349–1355.

Hiroyama, T., Iwama, A., Morita, Y., Nakamura, Y., Shibuya, A., and Nakauchi, H. (2000). Molecular cloning and characterization of CRLM-2, a novel type I cytokine receptor preferentially expressed in hematopoietic cells. *Biochem. Biophys. Res. Commun.* **272**(1), 224–229.

Huhta, T., Vettenranta, K., Heinonen, K., Kanerva, J., Larramendy, M. L., Mahlamaki, E., *et al.* (1999). Comparative genomic hybridization and conventional cytogenetic analyses in childhood acute myeloid leukemia. *Leuk. Lymphoma* **35**(3–4), 311–315.

Huntly, B. J., Shigematsu, H., Deguchi, K., Lee, B. H., Mizuno, S., Duclos, N., *et al.* (2004). MOZ-TIF2, but not BCR-ABL, confers properties of leukemic stem cells to committed murine hematopoietic progenitors. *Cancer Cell* **6**(6), 587–596.

Iacobucci, I., Lonetti, A., Messa, F., Cilloni, D., Arruga, F., Ottaviani, E., *et al.* (2008). Expression of spliced oncogenic Ikaros isoforms in Philadelphia-positive acute lymphoblastic leukemia patients treated with tyrosine kinase inhibitors: Implications for a new mechanism of resistance. *Blood* **112**(9), 3847–3855.

Igarashi, H., Gregory, S. C., Yokota, T., Sakaguchi, N., and Kincade, P. W. (2002). Transcription from the RAG1 locus marks the earliest lymphocyte progenitors in bone marrow. *Immunity* **17**(2), 117–130.

Ihle, J. N., and Gilliland, D. G. (2007). Jak2: Normal function and role in hematopoietic disorders. *Curr. Opin. Genet. Dev.* **17**(1), 8–14.

Irving, J. A., Bloodworth, L., Bown, N. P., Case, M. C., Hogarth, L. A., and Hall, A. G. (2005). Loss of heterozygosity in childhood acute lymphoblastic leukemia detected by genome-wide microarray single nucleotide polymorphism analysis. *Cancer Res.* **65**(8), 3053–3058.

Jalali, G. R., An, Q., Konn, Z. J., Worley, H., Wright, S. L., Harrison, C. J., *et al.* (2008). Disruption of ETV6 in intron 2 results in upregulatory and insertional events in childhood acute lymphoblastic leukaemia. *Leukemia* **22**(1), 114–123.

James, C., Ugo, V., Le Couedic, J. P., Staerk, J., Delhommeau, F., Lacout, C., *et al.* (2005). A unique clonal JAK2 mutation leading to constitutive signalling causes polycythaemia vera. *Nature* **434**(7037), 1144–1148.

Kawamata, N., Ogawa, S., Zimmermann, M., Niebuhr, B., Stocking, C., Sanada, M., *et al.* (2008). Cloning of genes involved in chromosomal translocations by high-resolution single nucleotide polymorphism genomic microarray. *Proc. Natl. Acad. Sci. USA* **105**(33), 11921–11926.

Kearney, L., Gonzalez De Castro, D., Yeung, J., Procter, J., Horsley, S. W., Eguchi-Ishimae, M., *et al.* (2008). A specific JAK2 mutation (JAK2R683) and multiple gene deletions in Down syndrome acute lymphoblastic leukaemia. *Blood* **113,** 646–648.

Klein, F., Feldhahn, N., Herzog, S., Sprangers, M., Mooster, J. L., Jumaa, H., *et al.* (2006). BCR–ABL1 induces aberrant splicing of IKAROS and lineage infidelity in pre-B lymphoblastic leukemia cells. *Oncogene* **25**(7), 1118–1124.

Kralovics, R., Passamonti, F., Buser, A. S., Teo, S. S., Tiedt, R., Passweg, J. R., *et al.* (2005). A gain-of-function mutation of JAK2 in myeloproliferative disorders. *N. Engl. J. Med.* **352**(17), 1779–1790.

Kuchinskaya, E., Nordgren, A., Heyman, M., Schoumans, J., Corcoran, M., Staaf, J., *et al.* (2007). Tiling-resolution array-CGH reveals the pattern of DNA copy number alterations in acute lymphoblastic leukemia with 21q amplification: The result of telomere dysfunction and breakage/fusion/breakage cycles? *Leukemia* **21**(6), 1327–1330.

Kuchinskaya, E., Heyman, M., Nordgren, A., Schoumans, J., Staaf, J., Borg, A., *et al.* (2008). Array-CGH reveals hidden gene dose changes in children with acute lymphoblastic leukaemia and a normal or failed karyotype by G-banding. *Br. J. Haematol.* **140**(5), 572–577.

Kuiper, R. P., Waanders, E., van der Velden, V. H., van Reijmersdal, S. V., Venkatachalam, R., Scheijen, B., van Dongen, J. J., Veerman, A. J., van Leeuwen, F. N., A, Geurts van Kessel, and Hoogerbrugge, P. M. (2010). *IKZF1* deletions predict relapse in uniformly treated pediatric precursor B-ALL. *Leukemia* **24**, 1258-1264.

Larramendy, M. L., Huhta, T., Heinonen, K., Vettenranta, K., Mahlamaki, E., Riikonen, P., *et al.* (1998a). DNA copy number changes in childhood acute lymphoblastic leukemia. *Haematologica* **83**(10), 890–895.

Larramendy, M. L., Huhta, T., Vettenranta, K., El-Rifai, W., Lundin, J., Pakkala, S., *et al.* (1998b). Comparative genomic hybridization in childhood acute lymphoblastic leukemia. *Leukemia* **12**(10), 1638–1644.

Levine, R. L., Wadleigh, M., Cools, J., Ebert, B. L., Wernig, G., Huntly, B. J., *et al.* (2005). Activating mutation in the tyrosine kinase JAK2 in polycythemia vera, essential thrombocythemia, and myeloid metaplasia with myelofibrosis. *Cancer Cell* **7**(4), 387–397.

Ley, T. J., Mardis, E. R., Ding, L., Fulton, B., McLellan, M. D., Chen, K., *et al.* (2008). DNA sequencing of a cytogenetically normal acute myeloid leukaemia genome. *Nature* **456**(7218), 66–72.

Lilljebjorn, H., Heidenblad, M., Nilsson, B., Lassen, C., Horvat, A., Heldrup, J., *et al.* (2007). Combined high-resolution array-based comparative genomic hybridization and expression profiling of ETV6/RUNX1-positive acute lymphoblastic leukemias reveal a high incidence of cryptic Xq duplications and identify several putative target genes within the commonly gained region. *Leukemia* **21**(10), 2137–2144.

Lin, H., and Grosschedl, R. (1995). Failure of B-cell differentiation in mice lacking the transcription factor EBF. *Nature* **376**(6537), 263–267.

Lo, K. C., Bailey, D., Burkhardt, T., Gardina, P., Turpaz, Y., and Cowell, J. K. (2008). Comprehensive analysis of loss of heterozygosity events in glioblastoma using the 100K SNP mapping arrays and comparison with copy number abnormalities defined by BAC array comparative genomic hybridization. *Genes Chromosom. Cancer* **47**(3), 221–237.

Lundin, C., Heidenblad, M., Strombeck, B., Borg, A., Hovland, R., Heim, S., *et al.* (2007). Tiling resolution array CGH of dic(7;9)(p11 approximately 13;p11 approximately 13) in B-cell precursor acute lymphoblastic leukemia reveals clustered breakpoints at 7p11.2 approximately 12.1 and 9p13.1. *Cytogenet. Genome Res.* **118**(1), 13–18.

Maciejewski, J. P., and Mufti, G. J. (2008). Whole genome scanning as a cytogenetic tool in hematologic malignancies. *Blood* **112**(4), 965–974.

Maher, C. A., Kumar-Sinha, C., Cao, X., Kalyana-Sundaram, S., Han, B., Jing, X., *et al.* (2009a). Transcriptome sequencing to detect gene fusions in cancer. *Nature* **458**(7234), 97–101.

Maher, C. A., Palanisamy, N., Brenner, J. C., Cao, X., Kalyana-Sundaram, S., Luo, S., *et al.* (2009b). Chimeric transcript discovery by paired-end transcriptome sequencing. *Proc. Natl. Acad. Sci. USA* **106**(30), 12353–12358.

Malinge, S., Ben-Abdelali, R., Settegrana, C., Radford-Weiss, I., Debre, M., Beldjord, K., *et al.* (2007). Novel activating JAK2 mutation in a patient with Down syndrome and B-cell precursor acute lymphoblastic leukemia. *Blood* **109**(5), 2202–2204.

Mardis, E. R., and Wilson, R. K. (2009). Cancer genome sequencing: A review. *Hum. Mol. Genet.* **18**(R2), R163–R168.

Mardis, E. R., Ding, L., Dooling, D. J., Larson, D. E., McLellan, M. D., Chen, K., *et al.* (2009). Recurring mutations found by sequencing an acute myeloid leukemia genome. *N. Engl. J. Med.* **361**(11), 1058–1066.

Martinelli, G., Iacobucci, I., Storlazzi, C. T., Vignetti, M., Paoloni, F., Cilloni, D., *et al.* (2009). IKZF1 (Ikaros) deletions in BCR–ABL1-positive acute lymphoblastic leukemia are associated with short disease-free survival and high rate of cumulative incidence of relapse: A GIMEMA AL WP report. *J. Clin. Oncol.* **27**(31), 5202–5207.

Matsuzaki, H., Dong, S., Loi, H., Di, X., Liu, G., Hubbell, E., *et al.* (2004). Genotyping over 100, 000 SNPs on a pair of oligonucleotide arrays. *Nat. Methods* **1**(2), 109–111.

Mazzucchelli, R., Hixon, J. A., Spolski, R., Chen, X., Li, W. Q., Hall, V. L., *et al.* (2008). Development of regulatory T cells requires IL-7Ralpha stimulation by IL-7 or TSLP. *Blood* **112**(8), 3283–3292.

McLaughlin, J., Chianese, E., and Witte, O. N. (1987). In vitro transformation of immature hematopoietic cells by the P210 BCR/ABL oncogene product of the Philadelphia chromosome. *Proc. Natl. Acad. Sci. USA* **84**(18), 6558–6562.

Melo, J. V. (1996). The diversity of BCR-ABL fusion proteins and their relationship to leukemia phenotype. *Blood* **88**(7), 2375–2384.

Miller, C. B., Mullighan, C. G., Su, X., Ma, J., Wang, M., Zhang, J., *et al.* (2008). Pax5 haploinsufficiency cooperates with BCR–ABL1 to induce acute lymphoblastic leukemia. *Blood* **112**:Abstract 293.

Molnar, A., and Georgopoulos, K. (1994). The Ikaros gene encodes a family of functionally diverse zinc finger DNA-binding proteins. *Mol. Cell. Biol.* **14**(12), 8292–8303.

Morin, R. D., Johnson, N. A., Severson, T. M., Mungall, A. J., An, J., Goya, R., *et al.* (2010). Somatic mutations altering EZH2 (Tyr641) in follicular and diffuse large B-cell lymphomas of germinal-center origin. *Nat. Genet.* **42**(2), 181–185.

Mullighan, C. G., and Downing, J. R. (2009). Genome-wide profiling of genetic alterations in acute lymphoblastic leukemia: Recent insights and future directions. *Leukemia* **23**(7), 1209–1218.

Mullighan, C. G., Goorha, S., Radtke, I., Miller, C. B., Coustan-Smith, E., Dalton, J. D., *et al.* (2007). Genome-wide analysis of genetic alterations in acute lymphoblastic leukaemia. *Nature* **446**(7137), 758–764.

Mullighan, C. G., Miller, C. B., Radtke, I., Phillips, L. A., Dalton, J., Ma, J., *et al.* (2008). BCR–ABL1 lymphoblastic leukaemia is characterized by the deletion of Ikaros. *Nature* **453**(7191), 110–114.

Mullighan, C. G., Su, X., Zhang, J., Radtke, I., Phillips, L. A., Miller, C. B., *et al.* (2009a). Deletion of IKZF1 and Prognosis in Acute Lymphoblastic Leukemia. *N. Engl. J. Med.* **360,** 470–480.

Mullighan, C. G., Collins-Underwood, J. R., Phillips, L. A., Loudin, M. G., Liu, W., Zhang, J., *et al.* (2009b). Rearrangement of CRLF2 in B-progenitor- and Down syndrome-associated acute lymphoblastic leukemia. *Nat. Genet.* **41**(11), 1243–1246.

Mullighan, C. G., Zhang, J., Harvey, R. C., Collins-Underwood, J. R., Schulman, B. A., Phillips, L. A., *et al.* (2009c). JAK mutations in high-risk childhood acute lymphoblastic leukemia. *Proc. Natl. Acad. Sci. USA* **106,** 9414–9418.

Mullighan, C. G., Morin, R. D., Zhang, J., Hirst, M., Zhao, Y., Yan, C., *et al.* (2009d). Next generation transcriptomic resequencing identifies novel genetic alterations in high-risk (HR) childhood acute lymphoblastic leukemia (ALL): A report from the Children's Oncology Group (COG) HR ALL TARGET Project [ASH Annual Meeting Abstracts]. *Blood* **114**(22), 704.

Nakase, K., Ishimaru, F., Avitahl, N., Dansako, H., Matsuo, K., Fujii, K., *et al.* (2000). Dominant negative isoform of the Ikaros gene in patients with adult B-cell acute lymphoblastic leukemia. *Cancer Res.* **60**(15), 4062–4065.

Nebral, K., Konig, M., Harder, L., Siebert, R., Haas, O. A., and Strehl, S. (2007). Identification of PML as novel PAX5 fusion partner in childhood acute lymphoblastic leukaemia. *Br. J. Haematol.* **139**(2), 269–274.

Nebral, K., Denk, D., Attarbaschi, A., Konig, M., Mann, G., Haas, O. A., *et al.* (2009). Incidence and diversity of PAX5 fusion genes in childhood acute lymphoblastic leukemia. *Leukemia* **23**(1), 134–143.

Ng, S. Y., Yoshida, T., Zhang, J., and Georgopoulos, K. (2009). Genome-wide lineage-specific transcriptional networks underscore Ikaros-dependent lymphoid priming in hematopoietic stem cells. *Immunity* **30**(4), 493–507.

Nichogiannopoulou, A., Trevisan, M., Neben, S., Friedrich, C., and Georgopoulos, K. (1999). Defects in hemopoietic stem cell activity in Ikaros mutant mice. *J. Exp. Med.* **190**(9), 1201–1214.

Nishii, K., Kita, K., Miwa, H., Shikami, M., Taniguchi, M., Usui, E., *et al.* (2000). Expression of B cell-associated transcription factors in B-cell precursor acute lymphoblastic leukemia cells: Association with PU.1 expression, phenotype, and immunogenotype. *Int. J. Hematol.* **71**(4), 372–378.

Northrup, D. L., and Allman, D. (2008). Transcriptional regulation of early B cell development. *Immunol. Res.* **42**(1–3), 106–117.

Nutt, S. L., and Kee, B. L. (2007). The transcriptional regulation of B cell lineage commitment. *Immunity* **26**(6), 715–725.

Nutt, S. L., Heavey, B., Rolink, A. G., and Busslinger, M. (1999). Commitment to the B-lymphoid lineage depends on the transcription factor Pax5. *Nature* **401**(6753), 556–562.

Nutt, S. L., Eberhard, D., Horcher, M., Rolink, A. G., and Busslinger, M. (2001). Pax5 determines the identity of B cells from the beginning to the end of B-lymphopoiesis. *Int. Rev. Immunol.* **20**(1), 65–82.

Olivero, S., Maroc, C., Beillard, E., Gabert, J., Nietfeld, W., Chabannon, C., *et al.* (2000). Detection of different Ikaros isoforms in human leukaemias using real-time quantitative polymerase chain reaction. *Br. J. Haematol.* **110**(4), 826–830.

Olshen, A. B., Venkatraman, E. S., Lucito, R., and Wigler, M. (2004). Circular binary segmentation for the analysis of array-based DNA copy number data. *Biostatistics* **5**(4), 557–572.

Pandey, A., Ozaki, K., Baumann, H., Levin, S. D., Puel, A., Farr, A. G., *et al.* (2000). Cloning of a receptor subunit required for signaling by thymic stromal lymphopoietin. *Nat. Immunol.* **1**(1), 59–64.

Park, L. S., Martin, U., Garka, K., Gliniak, B., Di Santo, J. P., Muller, W., *et al.* (2000). Cloning of the murine thymic stromal lymphopoietin (TSLP) receptor: Formation of a functional heteromeric complex requires interleukin 7 receptor. *J. Exp. Med.* **192**(5), 659–670.

Paulsson, K., Cazier, J. B., Macdougall, F., Stevens, J., Stasevich, I., Vrcelj, N., *et al.* (2008). Microdeletions are a general feature of adult and adolescent acute lymphoblastic leukemia: Unexpected similarities with pediatric disease. *Proc. Natl. Acad. Sci. USA* **105**(18), 6708–6713.

Peiffer, D. A., Le, J. M., Steemers, F. J., Chang, W., Jenniges, T., Garcia, F., *et al.* (2006). High-resolution genomic profiling of chromosomal aberrations using Infinium whole-genome genotyping. *Genome Res.* **16**(9), 1136–1148.

Pinkel, D., and Albertson, D. G. (2005). Comparative genomic hybridization. *Annu. Rev. Genomics Hum. Genet.* **6,** 331–354.

Pounds, S., Cheng, C., Mullighan, C., Raimondi, S. C., Shurtleff, S., and Downing, J. R. (2009). Reference alignment of SNP microarray signals for copy number analysis of tumors. *Bioinformatics* **25**(3), 315–321.

Pui, C. H., Williams, D. L., Raimondi, S. C., Rivera, G. K., Look, A. T., Dodge, R. K., *et al.* (1987). Hypodiploidy is associated with a poor prognosis in childhood acute lymphoblastic leukemia. *Blood* **70**(1), 247–253.

Pui, C. H., Relling, M. V., and Downing, J. R. (2004). Acute lymphoblastic leukemia. *N. Engl. J. Med.* **350**(15), 1535–1548.

Pui, C.-H., Robison, L. L., and Look, A. T. (2008). Acute lymphoblastic leukaemia. *Lancet* **371**(9617), 1030–1043, [2008/3/28].

Rabin, K. R., Man, T. K., Yu, A., Folsom, M. R., Zhao, Y. J., Rao, P. H., *et al.* (2008). Clinical utility of array comparative genomic hybridization for detection of chromosomal abnormalities in pediatric acute lymphoblastic leukemia. *Pediatr. Blood Cancer* **51**(2), 171–177.

Raimondi, S. C. (ed.) (2006). Cytogenetics of Acute Leukemias, pp. 235–271. Cambridge University Press, Cambridge.

Raimondi, S. C., Zhou, Y., Mathew, S., Shurtleff, S. A., Sandlund, J. T., Rivera, G. K., *et al.* (2003). Reassessment of the prognostic significance of hypodiploidy in pediatric patients with acute lymphoblastic leukemia. *Cancer* **98**(12), 2715–2722.

Ray, R. J., Furlonger, C., Williams, D. E., and Paige, C. J. (1996). Characterization of thymic stromal-derived lymphopoietin (TSLP) in murine B cell development in vitro. *Eur. J. Immunol.* **26**(1), 10–16.

Ren, R. (2005). Mechanisms of BCR-ABL in the pathogenesis of chronic myelogenous leukaemia. *Nat. Rev. Cancer* **5**(3), 172–183.

Ribeiro, R. C., Abromowitch, M., Raimondi, S. C., Murphy, S. B., Behm, F., and Williams, D. L. (1987). Clinical and biologic hallmarks of the Philadelphia chromosome in childhood acute lymphoblastic leukemia. *Blood* **70**(4), 948–953.

Rivera, G. K., Zhou, Y., Hancock, M. L., Gajjar, A., Rubnitz, J., Ribeiro, R. C., *et al.* (2005). Bone marrow recurrence after initial intensive treatment for childhood acute lymphoblastic leukemia. *Cancer* **103**(2), 368–376.

Rochman, Y., and Leonard, W. J. (2008). Thymic stromal lymphopoietin: A new cytokine in asthma. *Curr. Opin. Pharmacol.* **8**(3), 249–254.

Ross, M. E., Zhou, X., Song, G., Shurtleff, S. A., Girtman, K., Williams, W. K., *et al.* (2003). Classification of pediatric acute lymphoblastic leukemia by gene expression profiling. *Blood* **102**(8), 2951–2959.

Ross, M. E., Mahfouz, R., Onciu, M., Liu, H. C., Zhou, X., Song, G., *et al.* (2004). Gene expression profiling of pediatric acute myelogenous leukemia. *Blood* **104**(12), 3679–3687.

Rowley, J. D. (2008). Chromosomal translocations: Revisited yet again. *Blood* **112**(6), 2183–2189.

Rumfelt, L. L., Zhou, Y., Rowley, B. M., Shinton, S. A., and Hardy, R. R. (2006). Lineage specification and plasticity in CD19- early B cell precursors. *J. Exp. Med.* **203**(3), 675–687.

Russell, L. J., Capasso, M., Vater, I., Akasaka, T., Bernard, O. A., Calasanz, M. J., *et al.* (2009). Deregulated expression of cytokine receptor gene, CRLF2, is involved in lymphoid transformation in B-cell precursor acute lymphoblastic leukemia. *Blood* **114** (13), 2688–2698.

Russo, A. A., Jeffrey, P. D., Patten, A. K., Massague, J., and Pavletich, N. P. (1996). Crystal structure of the p27Kip1 cyclin-dependent-kinase inhibitor bound to the cyclin A-Cdk2 complex. *Nature* **382**(6589), 325–331.

Saharinen, P., and Silvennoinen, O. (2002). The pseudokinase domain is required for suppression of basal activity of Jak2 and Jak3 tyrosine kinases and for cytokine-inducible activation of signal transduction. *J. Biol. Chem.* **277**(49), 47954–47963.

Savona, M., and Talpaz, M. (2008). Getting to the stem of chronic myeloid leukaemia. *Nat. Rev. Cancer* **8**(5), 341–350.

Schoumans, J., Johansson, B., Corcoran, M., Kuchinskaya, E., Golovleva, I., Grander, D., *et al.* (2006). Characterisation of dic(9;20)(p11-13;q11) in childhood B-cell precursor acute lymphoblastic leukaemia by tiling resolution array-based comparative genomic hybridisation reveals clustered breakpoints at 9p13.2 and 20q11.2. *Br. J. Haematol.* **135** (4), 492–499.

Steinemann, D., Cario, G., Stanulla, M., Karawajew, L., Tauscher, M., Weigmann, A., *et al.* (2008). Copy number alterations in childhood acute lymphoblastic leukemia and their association with minimal residual disease. *Genes Chromosom. Cancer* **47**(6), 471–480.

Stiller, C. A. (2004). Epidemiology and genetics of childhood cancer. *Oncogene* **23**(38), 6429–6444.

Storlazzi, C. T., Albano, F., Lo Cunsolo, C., Doglioni, C., Guastadisegni, M. C., Impera, L., *et al.* (2007). Upregulation of the SOX5 by promoter swapping with the P2RY8 gene in primary splenic follicular lymphoma. *Leukemia* **21**(10), 2221–2225.

Strefford, J. C., van Delft, F. W., Robinson, H. M., Worley, H., Yiannikouris, O., Selzer, R., *et al.* (2006). Complex genomic alterations and gene expression in acute lymphoblastic leukemia with intrachromosomal amplification of chromosome 21. *Proc. Natl. Acad. Sci. USA* **103**(21), 8167–8172.

Strefford, J. C., Worley, H., Barber, K., Wright, S., Stewart, A. R., Robinson, H. M., *et al.* (2007). Genome complexity in acute lymphoblastic leukemia is revealed by array-based comparative genomic hybridization. *Oncogene* **26**(29), 4306–4318.

Strehl, S., Konig, M., Dworzak, M. N., Kalwak, K., and Haas, O. A. (2003). PAX5/ETV6 fusion defines cytogenetic entity dic(9;12)(p13;p13). *Leukemia* **17**(6), 1121–1123.

Sun, L., Crotty, M. L., Sensel, M., Sather, H., Navara, C., Nachman, J., *et al.* (1999a). Expression of dominant-negative Ikaros isoforms in T-cell acute lymphoblastic leukemia. *Clin. Cancer Res.* **5**(8), 2112–2120.

Sun, L., Goodman, P. A., Wood, C. M., Crotty, M. L., Sensel, M., Sather, H., *et al.* (1999b). Expression of aberrantly spliced oncogenic ikaros isoforms in childhood acute lymphoblastic leukemia. *J. Clin. Oncol.* **17**(12), 3753–3766.

Sun, L., Heerema, N., Crotty, L., Wu, X., Navara, C., Vassilev, A., *et al.* (1999c). Expression of dominant-negative and mutant isoforms of the antileukemic transcription factor Ikaros in infant acute lymphoblastic leukemia. *Proc. Natl. Acad. Sci. USA* **96**(2), 680–685.

Sutcliffe, M. J., Shuster, J. J., Sather, H. N., Camitta, B. M., Pullen, J., Schultz, K. R., *et al.* (2005). High concordance from independent studies by the Children's Cancer Group (CCG) and Pediatric Oncology Group (POG) associating favorable prognosis with combined trisomies 4, 10, and 17 in children with NCI Standard-Risk B-precursor Acute Lymphoblastic Leukemia: A Children's Oncology Group (COG) initiative. *Leukemia* **19**(5), 734–740.

Takanashi, M., Yagi, T., Imamura, T., Tabata, Y., Morimoto, A., Hibi, S., *et al.* (2002). Expression of the Ikaros gene family in childhood acute lymphoblastic leukaemia. *Br. J. Haematol.* **117**(3), 525–530.

Tonnelle, C., Bardin, F., Maroc, C., Imbert, A. M., Campa, F., Dalloul, A., *et al.* (2001). Forced expression of the Ikaros 6 isoform in human placental blood CD34(+) cells impairs their ability to differentiate toward the B-lymphoid lineage. *Blood* **98**(9), 2673–2680.

Tonnelle, C., Dijon, M., Moreau, T., Garulli, C., Bardin, F., and Chabannon, C. (2009). Stage specific over-expression of the dominant negative Ikaros 6 reveals distinct role of Ikaros throughout human B-cell differentiation. *Mol. Immunol.* **46**(8–9), 1736–1743.

Trageser, D., Iacobucci, I., Nahar, R., Duy, C., von Levetzow, G., Klemm, L., *et al.* (2009). Pre-B cell receptor-mediated cell cycle arrest in Philadelphia chromosome-positive acute lymphoblastic leukemia requires IKAROS function. *J. Exp. Med.* **206**(8), 1739–1753.

Usvasalo, A., Elonen, E., Saarinen-Pihkala, U. M., Raty, R., Harila-Saari, A., Koistinen, P., *et al.* (2010). Prognostic classification of patients with acute lymphoblastic leukemia by using gene copy number profiles identified from array-based comparative genomic hybridization data. *Leuk. Res.* **34**(11), 1476–1482.

Vainchenker, W., Dusa, A., and Constantinescu, S. N. (2008). JAKs in pathology: Role of Janus kinases in hematopoietic malignancies and immunodeficiencies. *Semin. Cell Dev. Biol.* **19,** 385–393.

Van Vlierberghe, P., van Grotel, M., Tchinda, J., Lee, C., Beverloo, H. B., van der Spek, P. J., *et al.* (2008). The recurrent SET-NUP214 fusion as a new HOXA activation mechanism in pediatric T-cell acute lymphoblastic leukemia. *Blood* **111**(9), 4668–4680.

Van Vlierberghe, P., Palomero, T., Khiabanian, H., Van der Meulen, J., Castillo, M., Van Roy, N., *et al.* (2010). PHF6 mutations in T-cell acute lymphoblastic leukemia. *Nat. Genet.* **42**(4), 338–342.

Vosshenrich, C. A., Cumano, A., Muller, W., Di Santo, J. P., and Vieira, P. (2004). Pre-B cell receptor expression is necessary for thymic stromal lymphopoietin responsiveness in the bone marrow but not in the liver environment. *Proc. Natl. Acad. Sci. USA* **101**(30), 11070–11075.

Wiemels, J. L., Ford, A. M., Van Wering, E. R., Postma, A., and Greaves, M. (1999). Protracted and variable latency of acute lymphoblastic leukemia after TEL-AML1 gene fusion in utero. *Blood* **94**(3), 1057–1062.

Williams, R. T., Roussel, M. F., and Sherr, C. J. (2006). Arf gene loss enhances oncogenicity and limits imatinib response in mouse models of Bcr-Abl-induced acute lymphoblastic leukemia. *Proc. Natl. Acad. Sci. USA* **103**(17), 6688–6693.

Winandy, S., Wu, P., and Georgopoulos, K. (1995). A dominant mutation in the Ikaros gene leads to rapid development of leukemia and lymphoma. *Cell* **83**(2), 289–299.

Wouters, B. J., Lowenberg, B., and Delwel, R. (2009). A decade of genome-wide gene expression profiling in acute myeloid leukemia: Flashback and prospects. *Blood* **113**(2), 291–298.

Xavier, R. J., and Rioux, J. D. (2008). Genome-wide association studies: A new window into immune-mediated diseases. *Nat. Rev. Immunol.* **8**(8), 631–643.

Yang, J. J., Cheng, C., Yang, W., Pei, D., Cao, X., Fan, Y., *et al.* (2009). Genome-wide interrogation of germline genetic variation associated with treatment response in childhood acute lymphoblastic leukemia. *JAMA* **301**(4), 393–403.

Yeoh, E. J., Ross, M. E., Shurtleff, S. A., Williams, W. K., Patel, D., Mahfouz, R., *et al.* (2002). Classification, subtype discovery, and prediction of outcome in pediatric acute lymphoblastic leukemia by gene expression profiling. *Cancer Cell* **1**(2), 133–143.

Yoda, A., Yoda, Y., Chiaretti, S., Bar-Natan, M., Mani, K., Rodig, S. J., *et al.* (2010). Functional screening identifies CRLF2 in precursor B-cell acute lymphoblastic leukemia. *Proc. Natl. Acad. Sci. USA* **107**(1), 252–257.

Zhang, W., Wang, J., Wang, Q., Chen, G., Zhang, J., Chen, T., *et al.* (2001). Identification of a novel type I cytokine receptor CRL2 preferentially expressed by human dendritic cells and activated monocytes. *Biochem. Biophys. Res. Commun.* **281**(4), 878–883.

Zhang, J., Mullighan, C. G., Harvey, R. C., Buetow, K. E., Carroll, W. L., Chen, I.-M., *et al.* (2009). Mutations in the RAS signaling, B-cell development, TP53/RB1, and JAK signaling pathways are common in high risk B-precursor childhood acute lymphoblastic leukemia (ALL): A report from the Children's Oncology Group (COG) high-risk (HR) ALL TARGET Project [ASH Annual Meeting Abstract]. *Blood* **114**(22), 85.

Zhou, B., Comeau, M. R., De Smedt, T., Liggitt, H. D., Dahl, M. E., Lewis, D. B., *et al.* (2005). Thymic stromal lymphopoietin as a key initiator of allergic airway inflammation in mice. *Nat. Immunol.* **6**(10), 1047–1053.

Zhuang, Y., Soriano, P., and Weintraub, H. (1994). The helix-loop-helix gene E2A is required for B cell formation. *Cell* **79**(5), 875–884.

Ziegler, S. F., and Liu, Y. J. (2006). Thymic stromal lymphopoietin in normal and pathogenic T cell development and function. *Nat. Immunol.* **7**(7), 709–714.

CHAPTER SEVEN

Myogenesis and Rhabdomyosarcoma: The Jekyll and Hyde of Skeletal Muscle

Raya Saab,* Sheri L. Spunt,† *and* Stephen X. Skapek‡

Contents

* Children's Cancer Center of Lebanon, Department of Pediatrics, American University of Beirut, Beirut, Lebanon
† Department of Oncology, St. Jude Children's Research Hospital, Memphis, Tennessee, USA
‡ Department of Pediatrics, Section of Hematology/Oncology, University of Chicago and Comer Children's Hospital, Chicago, Illinois, USA

Current Topics in Developmental Biology, Volume 94
ISSN 0070-2153, DOI: 10.1016/B978-0-12-380916-2.00007-3

Abstract

Rhabdomyosarcoma, a neoplasm composed of skeletal myoblast-like cells, represents the most common soft tissue sarcoma in children. The application of intensive chemotherapeutics and refined surgical and radiation therapy approaches have improved survival for children with localized disease over the past 3 decades; however, these approaches have not improved the dismal outcome for children with metastatic and recurrent rhabdomyosarcoma. Elegant studies have defined the molecular mechanisms driving skeletal muscle lineage commitment and differentiation, and the machinery that couples differentiation with irreversible cell proliferation arrest. Further, detailed molecular analyses indicate that rhabdomyosarcoma cells have lost the capacity to fully differentiate when challenged to do so in experimental models. We review the intersection of normal skeletal muscle developmental biology and the molecular genetic defects in rhabdomyosarcoma with the underlying premise that understanding how the differentiation process has gone awry will lead to new treatment strategies aimed at promoting myogenic differentiation and concomitant cell cycle arrest.

1. Introduction

Soft tissue sarcomas (STS) account for about 7% of pediatric cancers and 1% of cancers in adults (Spunt and Pappo, 2006). Approximately half the pediatric STS cases are rhabdomyosarcoma, as compared to fewer than 2% of adult STS. A slight male predominance has been noted in epidemiologic studies, but the disease appears to be distributed equally across racial groups (Horner *et al.*, 2010). The primary tumor can arise virtually anywhere in the body, but head and neck (30%), extremity (20%), and genitourinary (15%) sites predominate (Horner *et al.*, 2010; Fig. 7.1). Approximately 25% of patients present with distant metastatic disease, typically involving the lungs, bone marrow, and bones; an additional 30% have involvement of regional lymph nodes (Horner *et al.*, 2010).

Rhabdomyosarcoma can be divided into two major histological subtypes: so-called embryonal and alveolar rhabdomyosarcoma (see more below). In childhood, the embryonal subtype is most common and predominates at favorable anatomic sites such as the orbit, other head and neck sites, and the genitourinary tract. The alveolar subtype occurs in both

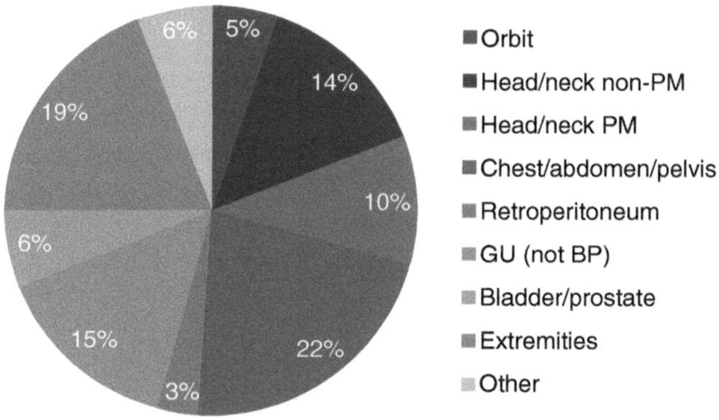

Figure 7.1 Chart depicts relative frequency of various primary tumor sites in children and young adults with rhabdomyosarcoma. Data are from SEER as published by Sultan *et al.* (2009).

children and adults, and it is more common at extremity sites and carries an overall inferior prognosis.

The outcome of patients with rhabdomyosarcoma depends on a number of factors including histological subtype; favorable versus unfavorable primary site; tumor size ≤ 5 or > 5 cm; presence or absence metastatic disease; and the extent of surgical resection performed at the time of initial presentation (Meza *et al.*, 2006). In the United States and Europe, these factors form the basis for the risk stratification schemes used to assign children into high, intermediate, and low risk groups, which are used to guide risk-based treatments. Standard therapy includes 6–12 months of systemic chemotherapy with combinations of different drugs depending on the risk classification (Crist *et al.*, 2001). In addition, treatment typically includes surgery, ionizing radiation, or a combination of both to achieve local control of the primary tumor. Multimodality therapy like this has resulted in 5-year survival rates of less than 30%, $\sim$70%, and greater than 90% for the high, intermediate, and low risk groups, respectively (Arndt *et al.*, 2009; Crist *et al.*, 2001; Oberlin *et al.*, 2008). Children with recurrent rhabdomyosarcoma are rarely cured, especially if disease recurs following intensive, initial therapy (Pappo *et al.*, 1999).

The long-term consequences of therapy can be significant (Meyer and Spunt, 2004). Depending on the chemotherapy regimen used, late effects may include infertility, second malignant neoplasia, renal insufficiency, and cardiomyopathy. Surgery may result in organ or tissue loss, impaired function, and cosmetic deficits. Radiotherapy can disrupt normal growth and development and organ function, and it also carries a risk of secondary neoplasia. Given the less-than-optimal survival rates—especially for those with advanced-stage or recurrent disease—and the chance for significant acute and long-term side effects, it's safe to say that the management of rhabdomyosarcoma remains a significant challenge. We contend that novel therapeutics based on better understanding of the disease biology will be needed to improve the clinical outcome for these patients.

In this review, we discuss aspects of rhabdomyosarcoma biology, and we provide an overview of how the disease could be viewed in the context of normal stages of skeletal myogenesis. We focus on the initial stages in the transition from a proliferating myoblast to a postmitotic myocyte because this developmental process is defunct in rhabdomyosarcoma. Understanding why the normal process has gone awry may point toward more effective and less toxic treatment options.

2. Pathology and Oncogenic Mechanisms of Rhabdomyosarcoma

2.1. Alveolar and embryonal rhabdomyosarcoma subtypes

Rhabdomyosarcoma is defined histologically as a small round blue cell tumor which expresses markers of myogenic differentiation, such as MyoD, myogenin, desmin, and actin. These myogenic markers discriminate it from other soft tissue or bone sarcomas but late markers of myogenic differentiation are absent, and rhabdomyosarcoma cells do not form myotubes or functional muscle units (Sebire and Malone, 2003; Tsokos, 1994).

Most rhabdomyosarcomas can be grouped to one of two major histologic subtypes based on microscopic appearance. Embryonal rhabdomyosarcoma is composed of spindle cells within a collagenous stroma, while alveolar rhabdomyosarcoma is characterized by small round blue cells arranged around spaces morphologically reminiscent of lung alveoli, hence the name (Fig. 7.2). In addition to clinical differences noted above,

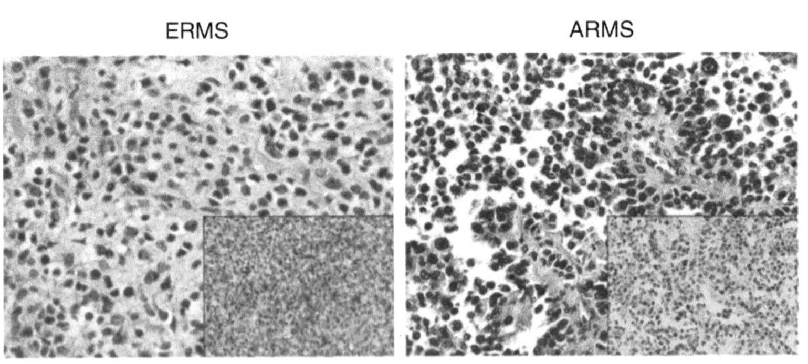

Figure 7.2 Representative hematoxylin- and eosin-stained sections of embryonal (ERMS) and alveolar (ARMS) rhabdomyosarcoma. The insets show typical immunohistochemical staining for Myogenin in each subtype.

the two major subtypes differ in cytogenetic and molecular findings. Alveolar rhabdomyosarcoma is associated in the vast majority of cases with a specific balanced translocation involving chromosomes 2 and 13 [*t*(2;13)] or, less commonly, 1 and 13 [*t*(1;13)](Barr, 1997), each of which encodes a novel fusion protein (see more below). Their specificity for alveolar rhabdomyosarcoma has led to fusion gene status becoming a widely used diagnosis test for the disease. Gene expression profiling identifies alveolar rhabdomyosarcoma tumors with positive fusion transcript to be biologically similar, and distinct from embryonal tumors and the rare alveolar histology tumors that do not express a fusion protein (Davicioni *et al.*, 2006, 2009).

2.2. Oncogenic mechanisms in alveolar rhabdomyosarcoma

A wide range of oncogenic aberrations are implicated in rhabdomyosarcoma (Table 7.1). The loci involved on chromosomes 1 and 2 in the translocations *t* (1;13) and *t*(2;13) in alveolar rhabdomyosarcoma encode two paired box transcription factors PAX7 and PAX3, respectively; their translocation with chromosome 13 juxtaposes them to the FKHR (now called FOXO1) gene, a member of the Forkhead transcription factor family (Barr *et al.*, 1993; Davis *et al.*, 1994; Galili *et al.*, 1993; Saito *et al.*, 1993; Fig. 7.3). The chimeric genes encode a fusion protein containing the PAX3 or PAX7 DNA binding domain and the C-terminal FOXO1 transcriptional activation domain (Barr *et al.*, 1998; Davis *et al.*, 1995; Fitzgerald *et al.*, 2000). The fusion proteins have more potent transactivating functions than either PAX3 or PAX7 alone (Bennicelli *et al.*, 1996; Fredericks *et al.*, 1995); this difference maps to a functional domain in the amino terminus of PAX3 that limits PAX3 activity, but not the activity of PAX3–FOXO1 (Bennicelli *et al.*, 1996).

The primary role these fusion genes play in rhabdomyosarcoma genesis is evidenced by findings in a growing number of experimental models. For instance, PAX3–FOXO1 expression in mouse mesenchymal stem cells promotes rhabdomyosarcoma-like tumors in cooperation with other oncogenic stimuli like SV40 T antigen expression (Ren *et al.*, 2008). Members of the Capecchi laboratory generated a conditional knock-in mouse model in which a mouse *Pax3–Foxo1* allele is activated solely in skeletal muscle cells (by virtue of tissue-specific expression of Cre recombinase). This mouse develops a tumor resembling alveolar rhabdomyosarcoma at morphological and molecular levels (Keller *et al.*, 2004a). The low tumor incidence in the animal greatly increases when bred into *p53* or *Ink4a/Arf* deficient backgrounds, thus showing the need for cooperating mutations subverting both p53- and RB-dependent tumor suppressor mechanisms.

Exactly how PAX3- and PAX7-containing fusion proteins promote rhabdomyosarcoma is still emerging. Transcriptional activity is critical, though, because ectopic expression of PAX3 fused to a KRAB transcriptional repressor reverses PAX–FOXO1-driven gene expression and impairs

Table 7.1 Summary of common genetic defects in rhabdomyosarcoma

	Gene	ARMS	ERMS	References
Fusion transcript	PAX3–FKHR or PAX7–FKHR	+	–	Barr *et al.* (1993), Galili *et al.* (1993), Davis (1994)
Locus	11p15	–	+	Besnard-Guerin *et al.* (1996), Visser *et al.* (1997)
Oncogenes	RAS	–	+	Chardin *et al.* (1985), Stratton *et al.* (1989), Wilke (1993), Yoo (1999)
	N-MYC	+	+	Garson *et al.* (1986), Dias et al. (1990a,b), Driman *et al.* (1994), Bayani *et al.* (1995), Hachitanda (1998)
	IGF2	+/–	+	El-Badry *et al.* (1990), Zhan *et al.* (1994), Minniti (1994), Makawita (2009)
	C-MET	+	+	Ferracini *et al.* (1996)
	EGFR	+	+	Ganti *et al.* (2006), Armistead *et al.* (2007)
	FGFR4	+	+	Taylor *et al.* (2009)
P53 pathway	MDM2	+	+	Keleti *et al.* (1996), Taylor *et al.* (2009)
	ARF	+	+	Iolascon *et al.* (1996)
	P53	+	+	Felix *et al.* (1992), Diller *et al.* (1995), Taylor *et al.* (2009)
RB pathway	CDK4	+	+	Khatib *et al.* (1993), Knudsen *et al.* (1998), Berner *et al.* (1996)
	CDKN2A	+	+	Iolascon *et al.* (1996)
	RB	+	+	Kohashi *et al.* (2008)

soft agar and tumorigenic growth of Rh30 rhabdomyosarcoma in SCID mice (Fredericks *et al.*, 2000). This and the fact that RNAi-mediated PAX3–FOXO1 inhibition decreases migration and C-MET expression in Rh30 and Rh41 cells shows that rhabdomyosarcoma cells may be "addicted" to PAX–FOXO1.

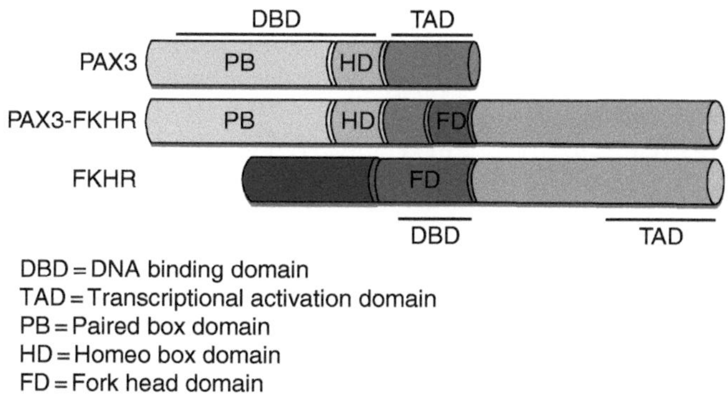

Figure 7.3 Schematic diagram depicting the wild-type proteins and fusion product of the *t*(2;13) in alveolar rhabdomyosarcoma, in which the amino-terminal DNA-binding domain of PAX3 is fused in-frame to the carboxyl terminal transcription activation domain of FOXO1. Other functional domains are noted.

2.3. Oncogenic mechanisms in embryonal rhabdomyosarcoma

Although not associated with a recurrent chromosomal rearrangement, a variety of genetic abnormalities occur in embryonal rhabdomyosarcoma. The most consistent one is loss of heterozygosity on chromosome 11 (Besnard-Guerin *et al.*, 1996; Koufos *et al.*, 1985; Visser *et al.*, 1997), with the smallest region localized to 11p15.5 (Scrable *et al.*, 1987). Inherited alterations of the 11p15.5 region also occur in Beckwith-Wiedemann syndrome (BWS), an overgrowth syndrome predisposing children to embryonal rhabdomyosarcoma and other cancers (Wiedemann *et al.*, 1983). Several lines of evidence now suggest that embryonal rhabdomyosarcoma is associated with loss of imprinting at this genomic locus. This could either (a) inactivate a tumor suppressor by allelic loss of the active maternal allele and retention of the inactive paternal allele or (2) double an oncogene dosage by expression from two alleles (Koi *et al.*, 1993; Loh *et al.*, 1992; Scrable *et al.*, 1989). Candidate genes at this genomic locus include the H19 gene product, $p57^{Kip2}$ (CDKN1C), SLC22AIL (BWR1A), and IGF2 (Hao *et al.*, 1993; Matsuoka *et al.*, 1996; Schwienbacher *et al.*, 1998; Tycko, 1994), which is highly expressed in rhabdomyosarcoma cells (El-Badry *et al.*, 1990 ;Zhan *et al.*, 1994). Of the 11p15.5 genes, $p57^{Kip2}$ and IGF2 are particularly interesting because of their relationship with normal developmental programs arresting cell proliferation and enhancing muscle gene expression, respectively (see more below).

Embryonal rhabdomyosarcoma cell lines and tumor samples also harbor activating mutations in the RAS oncogene (Chardin *et al.*, 1985; Stratton *et al.*, 1989). Activating mutations of N-RAS may occur in up to 20% of embryonal rhabdomyosarcoma, although mutations in H-RAS and K-RAS seem to be rather rare (Martinelli *et al.*, 2009; Takahashi *et al.*, 2004). Oncogenic RAS expression in precursor myogenic cells, in concert with hTERT and SV40-T/t antigens, results in tumors that resemble tumors with embryonal histology (Linardic *et al.*, 2005). Ectopic expression of RAS also leads to an embryonal rhabdomyosarcoma-like tumor in zebrafish (Langenau *et al.*, 2007). Tumors in this model display a RAS-activation gene expression signature that is enriched in human samples of embryonal rhabdomyosarcoma.

Two other oncogenic pathways validated in mouse embryonal rhabdomyosarcoma models involve cellular signaling. The importance of the SHH/PTCH/GLI1 pathway was implied by the fact that patients with Gorlin Syndrome, caused by *PTCH* mutation (Hahn *et al.*, 1996), frequently develop rhabdomyosarcoma. Mice that are haploinsufficient for the orthologous *Ptch* gene also develop rhabdomyosarcoma which expresses both mouse Gli1 and Igf2, the latter of which is essential for sarcoma formation (Hahn *et al.*, 1998, 2000). Another direct example comes from c-MET, a growth factor receptor that is expressed in and enhances migration of embryonal (and alveolar) cell lines (Ferracini *et al.*, 1996). Constitutive c-MET activation by transgenic expression of hepatocyte growth factor (aka: scatter factor; HGF/SF) in the mouse causes rhabdomyosarcoma (Sharp *et al.*, 2002); as with PAX–FOXO1 above, tumor incidence increases in p53 or *Ink4a/Arf* deficient backgrounds in the mouse.

2.4. Additional genetic abnormalities common to both subtypes

EGFR is detectably expressed in both rhabdomyosarcoma subtypes, although somewhat more frequently in the embryonal subtype (Armistead *et al.*, 2007; Ganti *et al.*, 2006). The mTOR pathway is activated in primary RMS samples (Dobashi *et al.*, 2009). Activating mutations of FGFR4 are found in 7% of rhabdomyosarcoma samples, including both subtypes (Taylor *et al.*, 2009). Comparative genomic hybridization studies have found a number of other genetic abnormalities in both rhabdomyosarcoma subtypes (Bridge *et al.*, 2000; Gordon *et al.*, 2000, 2001;Weber-Hall *et al.*, 1996). In embryonal disease, areas of chromosomal gain include Chr 2, 8, 12, 13 (Anderson *et al.*, 1999). Amplification sites identified in the alveolar subtype include regions encoding GLI1, CDK4, HDM2, and MYCN (12q13 for GLI1/HDM2/CDK4, and 2p24 for MYCN; Berner *et al.*, 1996; Dias et al., 1990a; Driman *et al.*, 1994; Keleti *et al.*, 1996; Meddeb *et al.*, 1996). More focal amplification of MYCN (Bayani *et al.*, 1995; Garson *et al.*, 1986) occurs in both types. Finally, inactivation of critical tumor suppressor pathways involving p53 and RB seem important, as the mouse genetics studies indicate. p53 is either directly mutated or indirectly inactivated by ARF silencing or HDM2 expression (Diller *et al.*, 1995; Felix *et al.*, 1992; Mulligan *et al.*, 1990; Stratton *et al.*, 1990); the RB pathway is blocked by amplification of CDK4 or loss of CDKN2A/B (Berner *et al.*, 1996; Iolascon *et al.*, 1996; Khatib *et al.*, 1993; Knudsen *et al.*, 1998), or by RB gene mutation (Kohashi *et al.*, 2008).

3. Cell-Intrinsic Regulation of Skeletal Myogenesis

In order to understand the relationships between skeletal myoblasts and rhabdomyosarcoma cells, one must understand skeletal muscle lineage specification, which largely occurs in the dermomyotome of the paraxial somites (reviewed in Buckingham and Vincent, 2009), and the processes that actually drive muscle differentiation. Leading from the identification of one critical regulator, MyoD, many aspects of this evolutionarily conserved process have been defined in great detail using relatively simple cell culture models—in which myogenic differentiation can be induced over a 48–72-h period in cultured myoblasts (Fig. 7.4)—that are complemented by genetic and histology studies using a variety of model organisms. Because cancer biology is largely driven by cell autonomous properties, our review focuses on the cell-intrinsic events that occur during mouse skeletal myogenesis, a particularly well-characterized system.

3.1. Myogenic bHLH transcription factors

A major breakthrough came from the use of a subtractive hybridization strategy to clone a single cDNA that promotes muscle differentiation when expressed in nonmyogenic 10T1/2 fibroblasts (Davis *et al.*, 1987). This

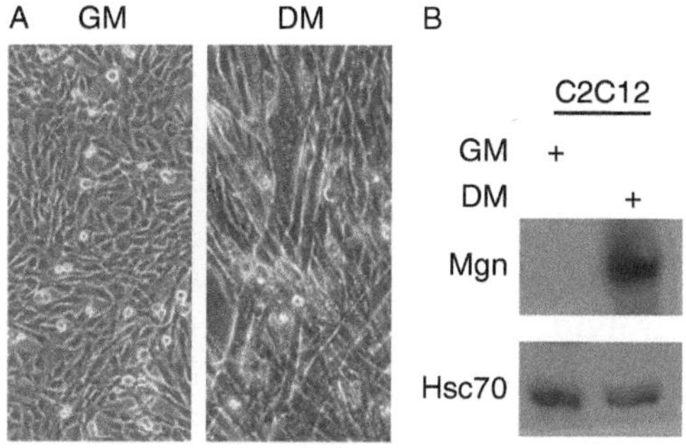

Figure 7.4 Representative photomicrographs (A) and Western blotting (B) shows morphological changes (elongation and multinucleated myotubes) and increased expression of Myogenin (Mgn) in C2C12 myoblasts cultured in 20% FBS/DMEM (GM) versus 2% HS/DMEM supplemented with insulin (DM) for 72 h. Heat shock 70 (Hsc70) serves as protein loading control.

cDNA, called MyoD, was the first of a family of four basic-helix-loop-helix (bHLH) transcription factors that include Myf5 (Braun *et al.*, 1989), Myogenin (Wright *et al.*, 1989), and Mrf4 (Rhodes and Konieczny, 1989). These myogenic bHLH proteins form heterodimers with broadly expressed E-proteins; MyoD/E heterodimers bind to distinct "E-boxes" (CANNTG) found in the regulatory elements of a wide variety of muscle-specific genes (Lassar *et al.*, 1991; Murre *et al.*, 1989).

Epistasis experiments have defined the hierarchical relationships among the myogenic bHLH proteins. MyoD-null and Myf5-null mice have seemingly normal muscle at birth (Braun *et al.*, 1992; Rudnicki *et al.*, 1992), however, Myf5$^{-/-}$ animals die at birth with severe rib anomalies. In contrast to the single knockouts, MyoD, Myf5-double knockout animals have an embryonic lethal phenotype with no myoblast development, showing the necessity of at least one of the two genes for muscle lineage commitment (Rudnicki *et al.*, 1993). The embryonic lethality was felt to represent functional compensation in the absence of either gene, but recent lineage tracing studies and selective ablation indicates that the two genes control specific lineages that cooperatively form the musculature (Haldar *et al.*, 2008). In contrast, Myogenin-deficient embryos die at birth with normal myoblast numbers but severely compromised or absent mature muscle (Hasty *et al.*, 1993; Nabeshima *et al.*, 1993). Replacing the Myf5 gene with Myogenin cDNA rescues the perinatal lethal phenotype in Myf5$^{-/-}$ pups, but it does not rescue the phenotype of MyoD, Myf5-double knockout embryos (Wang and Jaenisch, 1997; Wang *et al.*, 1996). Hence, although Myogenin displays some functional redundancy with Myf5, it *initiates* myogenic differentiation less efficiently than MyoD. The key functional difference maps to an amino-terminal histidine/cysteine rich domain and a carboxy-terminal alpha-helix conserved in MyoD and Myf5 but not Myogenin (Bergstrom and Tapscott, 2001). Mrf4 is most highly expressed in adult mouse skeletal muscle (Rhodes and Konieczny, 1989) and its knockout prevents the developmentally timed repression of Myogenin in mature muscle (Zhang *et al.*, 1995). Both of these findings suggest a role for Mrf4 in myocyte maturation. However, closer analysis of the phenotype of several different Myf5-null alleles illuminated a previously unrecognized role for Mrf4 (which coincidentally is adjacent to Myf5 on mouse chromosome 10). When the different Myf5$^{-/-}$ lines were bred into a MyoD$^{-/-}$ background, the aforementioned absence of skeletal muscle in the double knockout embryos depended on the presence or absence of Mrf4 expression in the E9.5 somite. Myf5, MyoD-double mutants retaining Mrf4 expression displayed normal skeletal myoblasts (Kassar-Duchossoy *et al.*, 2004). Although it is not a muscle-specific gene, the transcription factor PAX3 also lies upstream of MyoD, as evidenced, in part, by (a) the ability of ectopically expressed PAX3 to promote myogenesis in certain tissues in the developing chick embryo (Maroto *et al.*, 1997) and (b) the

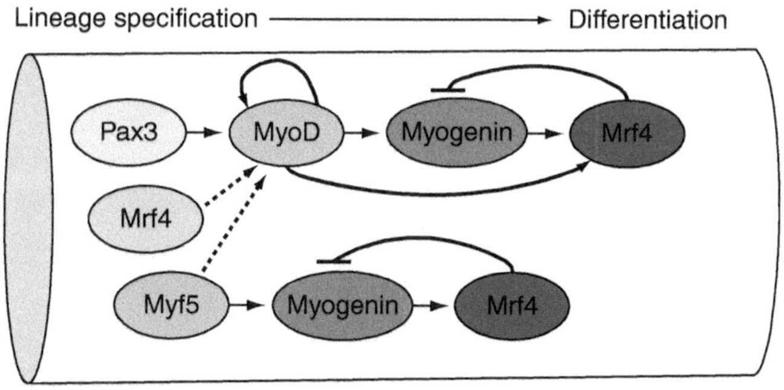

Figure 7.5 Schematic diagram depicting current understanding of epistatic relationships between PAX3 and various skeletal muscle-specific bHLH proteins. Note that Mrf4 acts as both early and late in skeletal muscle lineage specification and differentiation.

absence of myoblasts in PAX3$^{-/-}$, Myf5$^{-/-}$ mouse embryos (Tajbakhsh *et al.*, 1997). The general model emerging from this work (Fig. 7.5) is strongly conserved in evolution, but it is important to acknowledge the existence of certain species-specific differences, as well as differences in the relative importance of individual regulators in epaxial and hypaxial myotome and within certain muscle fiber types (reviewed in Buckingham and Vincent, 2009).

3.2. Controlling myogenic regulatory factors

Mere expression of myogenic bHLH proteins is not sufficient to initiate the differentiation program. Indeed, *MyoD* and *Myf5* are expressed in the dermomyotome long before muscle differentiation begins. Numerous regulatory systems have been uncovered to both negatively and positively activate the differentiation program. In some cases, the regulation is clearly linked to extracellular signals, but we again focus on cell-intrinsic events controlling the process (Fig. 7.6).

3.3. Negative regulation of MyoD/E heterodimers

Focusing on mouse MyoD as a model, the simplest regulation is at the level of its capacity to form heterodimers with E-proteins. The prototypical bHLH inhibitory mechanism involves the Id proteins which contain HLH motifs but lack a basic region required for DNA binding (Benezra *et al.*, 1990; Ruzinova and Benezra, 2003); as such, when expressed in myoblasts, Id proteins prevent muscle gene expression by titrating proteins like E2A from MyoD/E heterodimers (Atherton *et al.*, 1996; Jen *et al.*, 1992). The four mouse Id genes are generally expressed highest in midgestation

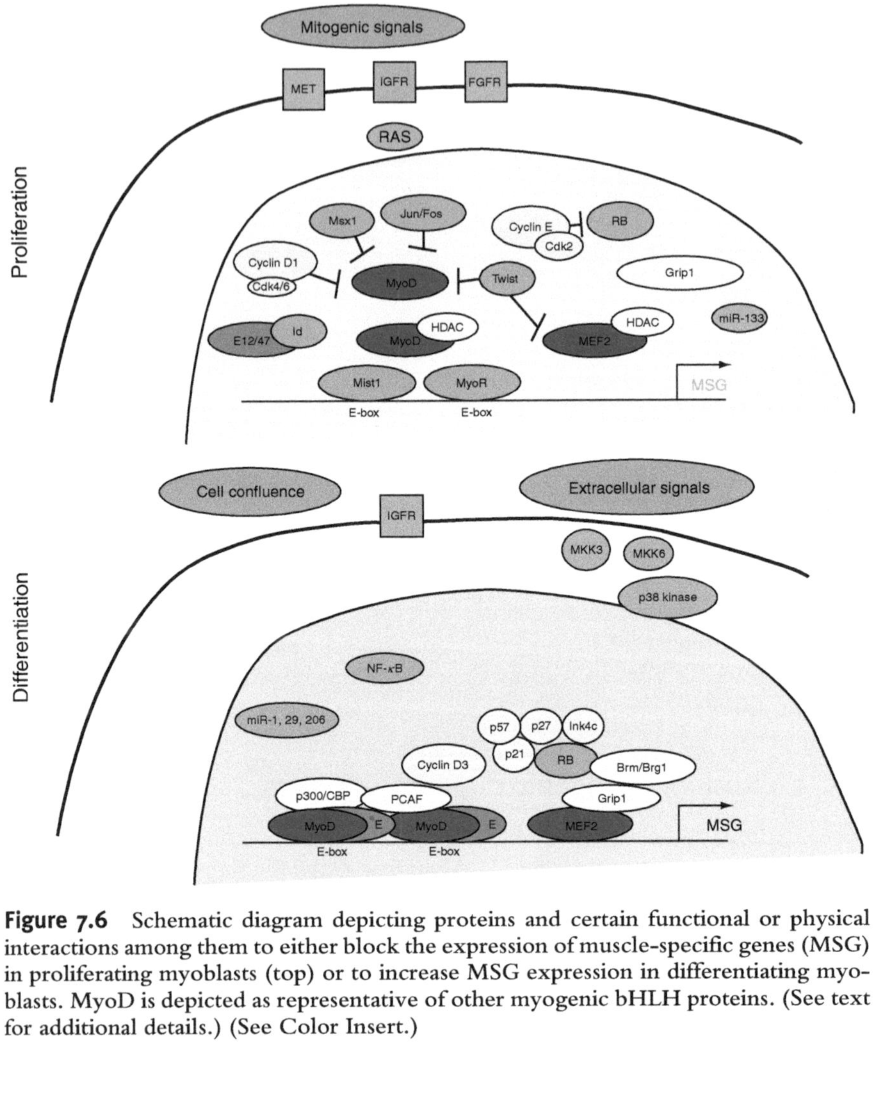

Figure 7.6 Schematic diagram depicting proteins and certain functional or physical interactions among them to either block the expression of muscle-specific genes (MSG) in proliferating myoblasts (top) or to increase MSG expression in differentiating myoblasts. MyoD is depicted as representative of other myogenic bHLH proteins. (See text for additional details.) (See Color Insert.)

and are downregulated during myogenic differentiation (Ruzinova and Benezra, 2003; Wang *et al.*, 1992). The different proteins interact to varying degrees with myogenic bHLH proteins in yeast two-hybrid models, implying that they may play distinct roles during different phases of differentiation (Langlands *et al.*, 1997). But overlapping expression patterns and functional redundancy has made it difficult to glean the exact roles of individual Id proteins (as was accomplished for myogenic bHLH proteins). While Id proteins have now been shown to have a variety of biochemical properties, the capacity of forced dimers between MyoD and E47 to foster differentiation despite Id expression supports their primary role as regulators of MyoD/E heterodimer formation (Neuhold and Wold, 1993).

3.4. Negative regulation by other mechanisms

Other proteins appear to directly interfere with myogenic bHLH proteins but do more than act as mere "dominant negative" proteins like the Ids. For example, mammalian Twist is a bHLH protein that interferes with MyoD/E binding, and it also blocks functional interactions between MyoD and Mef2 transcription factors (see more below; Spicer *et al.*, 1996). (In contrast, *Drosophila* Twist positively regulates myogenesis (Cripps *et al.*, 1998). This finding highlights the fact that some details regarding myogenesis are not conserved in all metazoans.) The bHLH proteins MyoR (Lu *et al.*, 1999) and Mist1 (Lemercier *et al.*, 1998) bind to E-boxes to repress muscle gene expression in cultured myoblasts. Their biology is incompletely understood, though: Mouse genetic studies indicate that MyoR is actually required for facial muscle development (Lu *et al.*, 1999) whereas Mist1 seems primarily required for normal pancreas development and physiology (Pin *et al.*, 2001). The homeobox-containing protein Msx1 can also repress muscle gene expression, likely by extinguishing MyoD (Odelberg *et al.*, 2000; Woloshin *et al.*, 1995). In experimental models, Msx1 fosters "dedifferentiation" of mature myotubes. I-mf was isolated from an E9.5 to E10.5 mouse embryo cDNA library by virtue of its interaction with MyoD in a yeast two-hybrid screen (Chen *et al.*, 1996); its expression in the sclerotome and functional capacity to prevent nuclear localization of myogenic bHLH proteins provide a novel mechanism to inhibit myogenic differentiation. Lastly, c-JUN, JUN-b, and v-FOS—which are generally expressed in proliferating cells, including myoblasts—can impede muscle-specific transcription by myogenic bHLH proteins (Bengal *et al.*, 1992; Li *et al.*, 1992a). A direct interaction between these growth-promoting transcription factors and myogenic bHLH proteins has been demonstrated, thereby providing an additional mechanism to prevent differentiation in proliferating cells.

3.5. Positive regulators

MyoD/E protein heterodimers act cooperatively with each other, robustly activating the expression of promoters with more than one E-box (Weintraub *et al.*, 1991). The capacity for MyoD/E heterodimers to act at muscle-specific promoters and not at many other nonmuscle-specific promoters lies in the presence of three amino acids in the junction between the basic domain and the first helix; when these amino acids are incorporated into the nonmyogenic E12 bHLH protein, it becomes myogenic (Davis and Weintraub, 1992). MyoD/E heterodimers also act with other sequence-specific

DNA binding proteins such as homeodomain proteins Pbx/Meis and the Mef2 proteins. In the former case, Pbx/Meis binding to a subset of muscle promoters, including that driving Myogenin, recruit MyoD to the relevant E-box (Berkes *et al.*, 2004). Pbx/Meis localization of MyoD depends on the cysteine/histidine rich domain and helix 3 in MyoD which are required for efficient myogenesis initiation (see above). Several myogenic bHLH proteins have been shown to synergistically act with Mef2 proteins to activate a variety of muscle-specific promoters (Li and Capetanaki, 1994; Molkentin *et al.*, 1996; Naidu *et al.*, 1995); in this capacity, Mef2 proteins help orchestrate the activation of later enhancers, "fine-tuning" muscle gene expression (Guerrero *et al.*, 2010). Transcriptional activation by MyoD/E-proteins coincides with recruitment of histone acetyltransferase activity p300 (Eckner *et al.*, 1996; Roth *et al.*, 2003) and PCAF (Sartorelli *et al.*, 1999), which target histones and other proteins like Rb and Mef2 proteins (Ma *et al.*, 2005; Nguyen *et al.*, 2004). Not unexpectedly, MyoD associated with histone deacetylases provides a mechanism to silence these same promoters in proliferating myoblasts until differentiation begins (Puri *et al.*, 2001).

Chromatin remodeling SWI/SNF proteins Brg1 and Brm are also essential for MyoD to open the Myogenin promoter in cultured cells (de *et al.*, 2001). It is interesting that the retinoblastoma gene product, Rb, forms a complex with Brg1 to promote cell cycle arrest (Dunaief *et al.*, 1994), because Rb is required for both cell cycle arrest accompanying muscle differentiation (Novitch *et al.*, 1996; Schneider *et al.*, 1994) and the robust expression of muscle-specific genes, especially those driven by Mef2 (Novitch *et al.*, 1999; Skapek *et al.*, 1996; Zacksenhaus *et al.*, 1996). Rb was found in DNA binding complexes with myogenic bHLH proteins (Gu *et al.*, 1993), but exactly how it promotes differentiation is still not totally clear (De *et al.*, 2006).

3.6. Regulatory kinases

Less direct regulation of muscle gene expression stems from a wide variety of posttranslational modifications in myogenic bHLH and Mef2 proteins, influencing biochemical properties like protein stability, subcellular localization, and capacity to interact with other proteins (reviewed in Puri and Sartorelli, 2000). Two particularly well-characterized regulators include p38 MAPK and Cyclin/Cdk activity. A variety of cell culture-based models have shown that p38 MAPK activation is required for normal differentiation (Cabane *et al.*, 2003; Cuenda and Cohen, 1999; Wu *et al.*, 2000; Zetser *et al.*, 1999). Its prodifferentiation effects include: (a) promoting cell cycle exit (Perdiguero *et al.*, 2007), (b) enhancing SWI/SNF chromatin remodeling (Simone *et al.*, 2004), (c) increasing MyoD/E protein heterodimerization (Lluis *et al.*, 2005), and (d) enhancing Mef2 function by direct phosphorylation (Wu *et al.*, 2000). The identity of extracellular signals that activate p38 MAPK are less clear, though.

In contrast to these activating effects, certain G_1 Cyclins and their catalytic Cdk partners are largely known for their capacity to impede differentiation in proliferating myoblasts. This was first evidenced by the observation that Cyclin D1 and its associated kinase activity decreases when myoblasts differentiate *in vitro* (Rao and Kohtz, 1995; Skapek *et al.*, 1995), whereas ectopic expression of Cyclin D1 or E hinders muscle gene expression (Rao *et al.*, 1994; Skapek *et al.*, 1995, 1996). The inhibition is not strictly dependent on phosphorylation of Rb, the best characterized Cyclin D/Cdk target (Skapek *et al.*, 1996). Instead, some evidence supports the concept that Cyclin-dependent phosphorylation of MyoD alters its stability (Kitzmann *et al.*, 1999; Song *et al.*, 1998). Further, Cyclin D/Cdk4 complexes can alter the interaction of Mef2 with Grip-1, a steroid receptor family coactivator that interacts with Mef2 (Lazaro *et al.*, 2002). Once the differentiation process is initiated, induction of Cyclin D3 (Cenciarelli *et al.*, 1999; De *et al.*, 2007), and certain Cdk inhibitors like $p21^{Cip1}$ (Halevy *et al.*, 1995; Parker *et al.*, 1995; Wang and Walsh, 1996), $p57^{Kip2}$ (De *et al.*, 2007; Figliola and Maione, 2004; Reynaud *et al.*, 1999; Zhang *et al.*, 1999), and $p18^{Ink4c}$ (Franklin and Xiong, 1996) help preserve Rb protein activation and cell cycle arrest, stabilize myogenic bHLH proteins, and prevent apoptosis in differentiating myocytes.

3.7. MicroRNA regulation of myogenesis

Our knowledge of how noncoding microRNAs (miRNAs) influence aspects of muscle development has rapidly grown over the last several years. These ~22 bp RNA species have the capacity to regulate the expression of a wide range of genes, coordinating complex expression routines needed for organogenesis and normal physiology. The seminal findings in muscle included the observation that miR-1, -133, and -206 were induced in differentiating mouse myoblasts in culture and in maturing cardiac and skeletal muscle in the mouse (Chen *et al.*, 2006; Kim *et al.*, 2006). Experimental manipulations of miR-1 and -133 show their capacity to promote differentiation and drive cell proliferation, respectively, in myoblasts. This is accomplished by miR-1 repressing HDAC4, a negative regulator of Mef2C, and miR-133 blocking SRFs, negative regulators of cell proliferation. DNA polymerase α is among the targets of miR-206, which also promotes muscle differentiation (Kim *et al.*, 2006). Given that miR-133 seems to block myogenic differentiation in cultured C2C12 cells, it is hard to understand why it is robustly induced during the process, and seems to be a direct target of muscle regulatory factors (see below).

Additional miRNAs have subsequently been implicated in a range of developmental and physiological processes in skeletal muscle, layering further control on the expression of certain muscle genes. They include miR-221, and -222 (targeting $p27^{Kip1}$; Cardinali *et al.*, 2009); miR-208a/b and -499 (targeting fast and slow myosins; van *et al.*, 2009); miR-214 (targeting Ezh2; Juan *et al.*, 2009); miR-206 (targeting c-Met as well as DNApolα; Yan *et al.*, 2009). As might be expected, certain miRNAs that promote skeletal myogenesis (miR-1, -206, -133, and -29) are direct targets of myogenic bHLH and Mef2 proteins (Liu *et al.*, 2007; Sweetman *et al.*, 2008) as well as NK-kappaB, which helps control the transition from myoblast to differentiating myocyte (Wang *et al.*, 2008).

3.8. Positive, feed-forward loops guide myogenesis

Shortly after MyoD was identified, it was found to directly induce Myogenin as well as its own promoter (Hollenberg *et al.*, 1993; Thayer *et al.*, 1989). This led to a concept that myogenic differentiation could be initiated and it would proceed as a cascade without additional input. A series of elegant studies from the Tapscott laboratory have shown that it is not quite that simple (Bergstrom *et al.*, 2002; Penn *et al.*, 2004; Tapscott, 2005). MyoD can directly activate over 600 genes with no intervening protein synthesis. However, it does so in a manner that is orchestrated by (a) timing of MyoD binding to specific promoters, (b) altered acetylation of underlying histones correlating with gene expression, (c) regulated association of Mef2 proteins, which are among the targets of MyoD, and (d) regulation of p38 MAPK activity, which promotes Mef2 protein localization and also that of RNA polymerase. Without positive "feed-forward" loops, like p38 MAPK activation, differentiation goes awry. It is postulated that a system of regulated, transcriptional subroutines that are initiated (but not carried forward completely) by a single gene like MyoD, allows for both an evolutionary quantum leap introducing myogenesis as well as many additional modifications in the system to foster complex physiological and anatomic differences across the range of skeletal muscle subtypes.

4. How are Myoblasts Related to Rhabdomyosarcoma Cells?

Morphology, the variable expression of certain skeletal muscle proteins like desmin and myosin, and the detection of ultrastructural features like actin/myosin complexes are long-standing observations supporting the skeletal myoblast-like nature of rhabdomyosarcoma cells. A major molecular finding solidifying this concept was the observation that MyoD was expressed in 16 of 16 human rhabdomyosarcoma tumors and only 3 of 33 other childhood cancer specimens, which include positive staining in only 3 of 8 Wilms' tumor specimens (Dias et al., 1990b). Numerous follow-up studies have shown that virtually all rhabdomyosarcoma display nuclear staining for MyoD or Myogenin, whereas the only staining with either antibody in other childhood cancers likely represents entrapped skeletal muscle (Sebire and Malone, 2003).

Further molecular evidence linking myoblasts to rhabdomyosarcoma cells comes from the above-mentioned *t*(2;13) and *t*(1;13) translocations found in alveolar tumors. As indicated above, PAX3 is required for MyoD expression during mouse embryogenesis. Similarly PAX7 is expressed in the developing dermomyotome from which myogenic cells arise (Jostes *et al.*, 1990). Interestingly, ectopic expression of PAX3–FOXO1 (but not PAX3) is capable of reprogramming NIH 3T3 cells into myoblast-like cells, as evidenced by the induction of MyoD, Myogenin, and other muscle functional or structural genes (Khan *et al.*, 1999). But constitutive expression of PAX3 or PAX3–FOXO1 usually blocks terminal differentiation (Epstein *et al.*, 1995). Taken together, these findings are consistent with alveolar rhabdomyosarcoma cells arising either from cells with innate myogenic potential or those artificially committed to become myoblast-like by virtue of PAX–FOXO1 expression.

Microarray technology has been used to profile the genes expressed in human rhabdomyosarcoma (Baer *et al.*, 2004; Davicioni *et al.*, 2006, 2009). Somewhat surprisingly, the expression signatures correspond with the presence or absence of a PAX–FOXO1 fusion transcript status, but not with histological subtype. Despite the microscopic appearance, "alveolar" tumors that do not expression a fusion transcript are more closely related to the embryonal subtype—at least from the perspective of gene expression. The fusion gene-negative rhabdomyosarcomas can also be clustered into groups with more or less differentiated signatures, but this seems to have less clinical relevance (Davicioni *et al.*, 2009). How these rhabdomyosarcoma gene expression signatures relate to the expression signatures of myoblasts induced to differentiate *in vitro* is not yet established.

5. Is There a Rhabdomyosarcoma "Stem Cell"? What Type of Cell Gives Rise to Rhabdomyosarcoma?

Identifying a rhabdomyosarcoma stem cell, or a rhabdomyosarcoma-initiating cell, could provide novel targeted therapeutic approaches for more effective disease control. Although there has been no proof to date regarding the existence of such a tumor-initiating cell, a few studies have demonstrated the presence of a stem-like population of cells within sarcoma cell lines, including rhabdomyosarcoma. These cells are termed the side-population (or SP) fraction, due to their identification on FACS by Hoechst 33342 dye exclusion; ABC transporters, which contribute to the multidrug resistant phenotype, mediate this dye exclusion (Das *et al.*, 2008; Komuro *et al.*, 2007; Tsuchida *et al.*, 2008). However, more work is necessary to prove whether the SP cells are indeed rhabdomyosarcoma stem cells, and whether they exist in primary human rhabdomyosarcoma tumors.

The rhabdomyosarcoma cell of origin also remains controversial. On one hand, finding rhabdomyosarcoma presenting with diffuse bone marrow involvement and no clear primary tumor, and at paratesticular, gall bladder, or parameningeal sites where skeletal muscle does not normally exist, supports the possibility that the tumor might arise from a mesenchymal progenitor or stem cell with the capacity to be pushed down the skeletal muscle lineage (such as by PAX–FOXO1 fusion gene expression). However, rhabdomyosarcoma also frequently arises directly within skeletal muscle in the limb or trunk, implying a direct origin from skeletal muscle. Experimental evidence provides some support for each of these possibilities, which do not need to be mutually exclusive.

5.1. MSC origin of rhabdomyosarcoma?

As mentioned above, developmental studies showing that PAX3 lies genetically upstream of MyoD and that PAX3–FOXO1 induces myogenic genes in nonmyogenic fibroblasts has led some to propose that mesenchymal stem cells may give rise to alveolar tumors (reviewed in Charytonowicz *et al.*, 2009). However, several attempts to generate mouse rhabdomyosarcoma in which the PAX3–FOXO1 transgene is driven by PAX3 or PAX7 promoters have resulted in developmental defects, but not tumor formation (Anderson *et al.*, 2001; Keller *et al.*, 2004b; Lagutina *et al.*, 2002; Relaix *et al.*, 2003).

Given the importance of cooperating mutations in rhabdomyosarcoma, failure in these mouse models does not formally exclude premyogenic PAX3- or PAX7-expressing cells giving rise to the disease. Indeed, directly forcing expression of PAX3–FOXO1 in cultured mouse MSCs only fosters tumorigenic growth in the presence of coexpressed SV40 T antigen (Ren *et al.*, 2008).

5.2. Rhabdomyosarcoma arising from myoblasts or satellite cells?

Other evidence supports the potential myogenic origin for rhabdomyosarcoma. In this regard, myogenic suspects include embryonic myoblasts; muscle satellite cells, a myogenic precursor cell residing in mature myofibers; or a myocyte that is transformed at some stage of myogenesis. There is no reason to exclude the possibility that different rhabdomyosarcoma subtypes could have distinct origins. The idea that satellite cells might be the source is intriguing because these normally quiescent cells are primed to proliferate in response to pathological states (Wagers and Conboy, 2005), and they express c-MET, which further links them to oncogenic pathways that are activated in some tumors (Anastasi *et al.*, 1997; Cornelison and Wold, 1997; Tatsumi *et al.*, 1998). PAX7 expression in embryonal rhabdomyosarcoma is also consistent with their arising from satellite cells (Tiffin *et al.*, 2003). That maturing myoblasts could lead to rhabdomyosarcoma comes, in part, from work showing that genetically manipulated myoblasts in which (a) antitumor effects of RB and p53 are corrupted, and (b) telomeres are maintained by hTERT, and (c) oncogenic RAS and MYC are expressed, leads to "embryonal" rhabdomyosarcoma when implanted into immunodeficient mice (Linardic *et al.*, 2005). It is particularly interesting that engineering the same genetic hits in human fetal skeletal myoblasts gives rise to undifferentiated sarcomas with no myogenic features (Linardic *et al.*, 2005). Hence, a different source cell could alter the ultimate rhabdomyosarcoma phenotype.

5.3. Mature myocytes dedifferentiating in rhabdomyosarcoma?

A commonly held belief is that rhabdomyosarcoma arises from a progenitor cell, in part because many classical studies demonstrated that mammalian myogenic differentiation results in irreversible cell cycle arrest (reviewed in Lassar *et al.*, 1994). Indeed, when differentiated myocytes are able to overcome the proliferation arrest, such as in serum-stimulated, $RB^{-/-}$ myocytes, progression through mitosis is blocked (Novitch *et al.*, 1999). It should be noted that RB is required for cell cycle exit and robust muscle gene expression, but preserving the arrested state is independent of RB and other related "pocket" proteins (p107 and p130; Camarda *et al.*, 2004). Nonetheless, the molecular machinery to "dedifferentiate" mature mammalian myocytes does exist. First, ectopic expression of the transcriptional repressor Msx1 in differentiated C2C12 myotubes leads to dramatic morphological changes and the emergence of proliferating, mononuclear cells (Odelberg *et al.*, 2000). Even more remarkably, applying a microtubule-binding chemical compound Myoseverin (or any of several other microtubule poisons) dissolves the myotube cytoskeleton, leading to cleavage of single, proliferating cells from the myotube (Rosania *et al.*, 2000). Viewed in this light, the finding that deregulated PAX3–FOXO1 in Mrf4-expressing cells causes rhabdomyosarcoma could be consistent with tumor arising from a more differentiated cell, as suggested by Keller *et al.* (2004a). In *Drosophila*, transgenic expression of PAX7–FOXO1 using the myosin heavy chain promoter allows expansion of myoblast-like "tumor" cells that seemingly separate from multinucleated myotubes and migrate to distant sites (Galindo *et al.*, 2006). However, caution is warranted before concluding that rhabdomyosarcoma really can arise from mature muscle. First, genetic evidence indicates that mouse Mrf4 is, in fact, expressed in a subset of very early skeletal muscle progenitors (Fig. 7.5). Second, fundamental differences in myogenesis in *Drosophila* versus mammalian myocytes may foster the apparent budding of individual PAX7–FOXO1 expressing myocytes from mature myotubes; whether this can also occur in mammalian systems is not proven.

6. How Do Oncogenic Pathways in Rhabdomyosarcoma Impede Myogenic Differentiation?

Regardless of the cellular origins—whether arising from a nonmyogenic cell or dedifferentiating from a mature myocyte—numerous studies illustrate that rhabdomyosarcoma cells largely fail to (a) exit the cell cycle, (b) undergo morphological changes, or (c) induce muscle-specific functional and structural proteins when challenged to do so when cultivated in mitogen-depleted media *in vitro*. The seminal paper showing that MyoD and E protein heterodimers can form in these cells but fail to robustly activate muscle-specific promoters was published over 15 years ago (Tapscott *et al.*, 1993). Rhabdomyosarcoma-derived cell lines seemed to be missing a critical cofactor that can be provided in heterokaryons from fusing them with mouse embryonic 10T1/2 fibroblasts. Over the years, a large amount of work has been devoted to understanding the principal differentiation defect(s) in rhabdomyosarcoma. Deregulation of nearly every step in normal myoblast development—from deregulated cell proliferation to abnormalities in muscle regulatory proteins—could contribute to this phenotype. Here we focus on four areas that have been explored in experimental models and, to some degree, demonstrated in human tumor samples.

6.1. Deregulated Cyclins/Cdk/RB

Long before MyoD was identified, it was clear that manipulations disrupting cell cycle exist, like expression of Rous sarcoma virus, disrupted myogenic differentiation (Holtzer *et al.*, 1975). At a molecular level, this seems to translate into inability to repress Cyclin/Cdk activity [especially Cyclin D1/Cdk4(6)], and failure to activate RB and the related p107 and p130. Rhabdomyosarcoma cell lines and tumor specimens reflect this by expression of D-type Cyclins (Zhang *et al.*, 2004) and Cdks, including Cdk4 which is detectable by immunohistochemical staining in ~82% of alveolar and 63% of embryonal cases (Saab *et al.*, 2006). It is also notable that the Cdk4 gene is a component of the aforementioned chromosome 12p13 amplicon that is common to both rhabdomyosarcoma subtypes. The RB gene itself undergoes homozygous deletion relatively rarely in rhabdomyosarcoma (6 of 27 embryonal and 2 of 20 alveolar cases; Kohashi *et al.*, 2008). The importance of deregulating the Cyclin/Cdk/RB pathway is also evident from the models showing that rhabdomyosarcoma development significantly increases in the absence of the Ink4a/ARF locus, encoding two proteins that normally arrest Cyclin/Cdk activity (Keller *et al.*, 2004a; Linardic *et al.*, 2007; Sharp *et al.*, 2002). When Cdk4(6) activity is blocked using PD 0332991, cell proliferation arrests in a panel of rhabdomyosarcoma cell lines, and the Rh30 alveolar rhabdomyosarcoma line undergoes morphological changes and increased expression of Myogenin (Saab *et al.*, 2006), demonstrating the capacity for reactivation of this normal developmental program (downregulation of Cyclin D-associated Cdk activity) to promote muscle gene expression, if only to a small degree.

6.2. Uncontrolled mitogenic signaling

There are many examples in which deregulated growth factor receptors are implicated in rhabdomyosarcoma genesis (see above), and additional cases when their deregulation impedes muscle differentiation. We focus on three—FGFs, HGF/SF, and IGFs—because they play key roles *in vivo*. FGFs and HGF/SF likely act dually to block myogenic differentiation and promote myoblast migration at several sites in the developing mouse embryo (Brand-Saberi *et al.*, 1996; Marics *et al.*, 2002). IGFs have more complex roles in that they foster myoblast proliferation prior to differentiation (Coolican *et al.*, 1997), yet they also provide autocrine activation of MyoD (Wilson and Rotwein, 2006) and they enhance muscle gene expression which ultimately leads to myocyte hypertrophy *in vitro* and in mouse models (Musaro and Rosenthal, 1999; Musaro *et al.*, 1999). Their links to rhabdomyosarcoma are clear in that FGFR4 has activating mutations in ~7% of rhabdomyosarcoma cases (Taylor *et al.*, 2009); the HGF/SF receptor, c-MET, is induced by PAX–FOXO1 (Ginsberg *et al.*, 1998; Relaix *et al.*, 2003); and loss of imprinting of IGF2 provides an autocrine growth signal in human rhabdomyosarcoma cell lines (Zhan *et al.*, 1994). Controlling these pathways is critical for the transition from proliferating myoblast to a postmitotic, differentiating myocyte. For example, FGFs and their receptors decrease during skeletal muscle maturation *in vitro* (Moore *et al.*, 1991); FGFR4 expression in particular is high in the embryo and it decreases in adult muscle (Sogos *et al.*, 1998). It seems likely that failed dampening of proliferation signals could bolster Cyclin/Cdk expression and activity while crippling RB protein function; for example, Cyclin D1 is a target of deregulated HGF/SF in the mouse model (Merlino and Helman, 1999). However, the signals can also more directly hinder myogenic regulatory factor activity, such as by FGF-driven phosphorylation of Myogenin, which blocks its DNA binding activity (Li *et al.*, 1992b).

6.3. Failed activation of p38 MAPK

In contrast to these inhibitory programs, activation of p38 MAPK plays a positive role fostering cell cycle exit and muscle gene expression. Strikingly, p38 MAPK is not activated in most rhabdomyosarcoma cell lines when they are cultured in differentiation-promoting conditions (Puri *et al.*, 2000; S. X. Skapek, unpublished). As expected, the ectopic expression of an activated form of MKK6 (MKK6EE) in a subset of rhabdomyosarcoma cell lines enhances muscle gene expression and arrests cell proliferation (Puri *et al.*, 2000). How commonly this pathway is deregulated in human rhabdomyosarcoma and whether impaired p38 MAPK activation in myoblasts is sufficient to promote rhabdomyosarcoma is not established.

6.4. Defects in myogenic regulatory factors

Given that the engine driving muscle differentiation lies in transcriptional regulators, their inactivity might contribute to failed terminal differentiation in rhabdomyosarcoma. Two examples of this type of mechanistic defect have been uncovered in rhabdomyosarcoma. One centers on the bHLH protein Twist which, as noted above, is expressed in the developing mouse somite to diminish both bHLH- and MEF2-dependent gene expression (Spicer *et al.*, 1996). Immunohistochemical staining showed Twist to be expressed in 8 of 15 in a panel of human rhabdomyosarcoma samples (Maestro *et al.*, 1999). Importantly, Twist was also identified as a putative oncogene whose expression can bypass MYC-induced programmed cell death (Maestro *et al.*, 1999), and it fosters the epithelial-to-mesenchymal transition and metastasis (Yang *et al.*, 2004). Thus, deregulated Twist might contribute to several aspects of the rhabdomyosarcoma phenotype.

In the second example, MyoD/E-proteins were recently shown to form poorly in the RD rhabdomyosarcoma cell line; in this case, Musculin (aka MyoR) titrates away a limiting amount of E-proteins (Yang *et al.*, 2009). The expression of an E2A gene splice variant which lacks part of the activation domain also contributes to the weak MyoD/E protein activity in a panel of rhabdomyosarcoma cell lines and tumor samples. This work highlights an area of rhabdomyosarcoma biology that seems incompletely explored. The rhabdomyosarcoma gene expression studies to date have not found key consistent deficiencies or deregulated expression of recognized activators or inhibitors of transcription factors that promote differentiation (Baer *et al.*, 2004; Davicioni *et al.*, 2006, 2009). However, a more subtle (but potentially common) problem like expression of a splice variant of E2A might have easily gone undetected.

7. Can the Normal Differentiation Program Be Reactivated as a Novel Therapy for Rhabdomyosarcoma?

As noted at the outset, the dismal survival for children with metastatic and recurrent rhabdomyosarcoma underscores the need for new therapeutic approaches. Given that a fundamental feature of cancer is the cell's ability to evade DNA damage response pathways resulting in senescence and apoptosis, it seems unlikely that further treatment intensification by using cytotoxic agents that act via ATM/p53-dependent pathways will dramatically change the outcome. As such, developing new agents by identifying and targeting critical survival signals might be a fruitful endeavor. But because terminal skeletal muscle differentiation leads to irreversible cell cycle arrest, manipulations that primarily reactivate myogenic differentiation may also be valuable (Fig. 7.7). In principle, they should not rely on proapoptotic and DNA-damage sensing pathways that are typically dysfunctional in human cancer. Even if such prodifferentiation agents do not control rhabdomyosarcoma when used alone, they may favorably alter the cancer biology and lead to better outcomes when used with current therapies. For example, gene expression profiling of human rhabdomyosarcoma shows that higher expression of differentiation markers correlates with better prognosis in children with nonalveolar disease, at least in retrospective series (Davicioni *et al.*, 2006, 2009).

Pharmacological, genetic, or epigenetic strategies to activate the differentiation program are all tenable as long as there are no fundamental cell-intrinsic problems, like deletion or inactivating mutations in the transcriptional machinery that drives differentiation and the accompanying cell cycle arrest. This does not seem to be the case because more well-differentiated "rhabdomyoblasts" are frequently observed following standard cytotoxic agents and radiation (d'Amore *et al.*, 1994; Lowichik *et al.*, 2000). Conceivably, the prodifferentiation effects here come about because genotoxic stress causes p53-dependent induction of $p21^{Cip1}$, which would potentially shift the balance of Cyclin/Cdk activity and favor differentiation.

A major question relates to how to develop these strategies. On one hand, gaining a better understanding of the fundamental developmental biology has already illuminated attractive candidates, like drugs targeting FGFR4, C-MET, or Cdk4(6). Our own work using a Cdk4(6) inhibitor (Saab *et al.*, 2006) and others using retinoids (Barlow *et al.*, 2005) establish the principle that a pharmacological approach might be feasible. However,

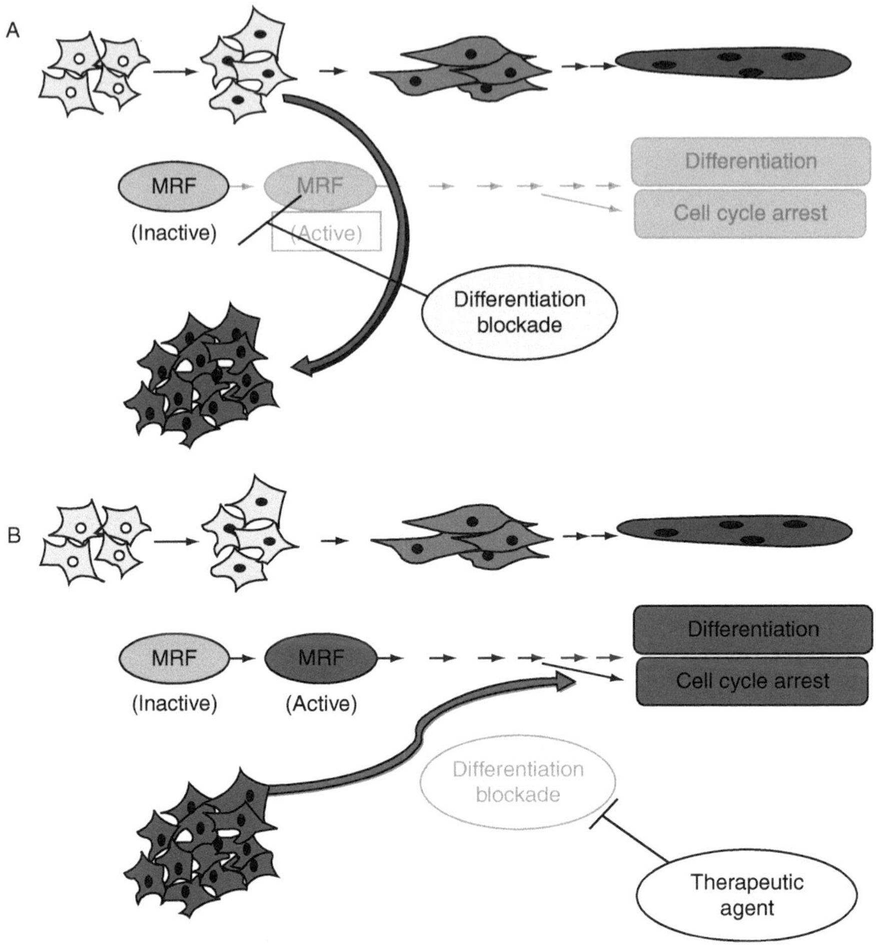

Figure 7.7 Schematic diagram depicting the general concept that a component of the phenotype of rhabdomyosarcoma is cellular and molecular evidence of arrested (or dedifferentiation) to a myoblast-like state (A). This likely correlates with a cell-intrinsic molecular change that blocks the continued, feed-forward differentiation program rendering muscle regulatory factors (MRF) incapable of promoting terminal differentiation. In principle, therapeutic interventions primarily aimed at reactivating differentiation could culminate in terminal differentiation and irreversible cell proliferation arrest. (See text for additional details.) (See Color Insert.)

these studies also illustrate the difficulty of such a "candidate gene" approach: The *in vitro* effects are not very robust and the best prodifferentiation agents are not clear.

Instead of focusing on candidates, an unbiased approach could center on developing high-throughput screens measuring muscle differentiation. This could be accomplished using cell-based assays of reporter gene expression, as has been accomplished for N-MYC and p53 (Komarov *et al.*, 1999; Lu *et al.*, 2003); morphology changes, which could be enhanced by fluorescent dyes that are specific for differentiated myocytes (Wagner *et al.*, 2008); or by quantitative measurements of a "signature" of differentiation induced genes, as was accomplished for neuroblastoma and leukemia (Hahn *et al.*, 2008; Stegmaier *et al.*, 2004). Coupling this assay with genetic manipulations, such as by siRNA libraries, could point out the key regulatory genes that will trigger the differentiation process. Given the number of enzyme-driven steps that negatively regulate myogenic differentiation, it seems reasonable that critical enzymes may be identified and be developed as novel targets for drug therapy.

Lastly, by gaining an even better understanding of where rhabdomyosarcoma tumor samples lie with respect for normal phases of the myoblast to myotube transition, the potential targets could become obvious. Such an understanding seems to be within our grasp with the emergence of even better gene expression tools like human exon arrays and next-generation sequencing technology.

ACKNOWLEDGMENTS

The authors wish to acknowledge the many researchers whose substantial contributions to the understanding of skeletal muscle development and rhabdomyosarcoma biology were not cited here. We gratefully acknowledge support to SXS from the Ted Mullin Sarcoma Research Fund at The University of Chicago; and support provided to RS and SLS from the American Lebanese and Syrian Associated Charities.

REFERENCES

Anastasi, S., Giordano, S., Sthandier, O., Gambarotta, G., Maione, R., Comoglio, P., and Amati, P. (1997). A natural hepatocyte growth factor/scatter factor autocrine loop in myoblast cells and the effect of the constitutive Met kinase activation on myogenic differentiation. *J. Cell Biol.* **137,** 1057–1068.

Anderson, J., Gordon, A., Pritchard-Jones, K., and Shipley, J. (1999). Genes, chromosomes, and rhabdomyosarcoma. *Genes Chromosomes Cancer* **26,** 275–285.

Anderson, M. J., Shelton, G. D., Cavenee, W. K., and Arden, K. C. (2001). Embryonic expression of the tumor-associated PAX3–FKHR fusion protein interferes with the developmental functions of Pax3. *Proc. Natl. Acad. Sci. USA* **98,** 1589–1594.

Armistead, P. M., Salganick, J., Roh, J. S., Steinert, D. M., Patel, S., Munsell, M., El-Naggar, A. K., Benjamin, R. S., Zhang, W., and Trent, J. C. (2007). Expression of receptor tyrosine kinases and apoptotic molecules in rhabdomyosarcoma: Correlation with overall survival in 105 patients. *Cancer* **110,** 2293–2303.

Arndt, C. A., Stoner, J. A., Hawkins, D. S., Rodeberg, D. A., Hayes-Jordan, A. A., Paidas, C. N., Parham, D. M., Teot, L. A., Wharam, M. D., Breneman, J. C., Donaldson, S. S., Anderson, J. R., *et al.* (2009). Vincristine, actinomycin, and cyclophosphamide compared with vincristine, actinomycin, and cyclophosphamide alternating

with vincristine, topotecan, and cyclophosphamide for intermediate-risk rhabdomyosarcoma: Children's oncology group study D9803. *J. Clin. Oncol.* **27,** 5182–5188.
Atherton, G. T., Travers, H., Deed, R., and Norton, J. D. (1996). Regulation of cell differentiation in C2C12 myoblasts by the Id3 helix-loop-helix protein. *Cell Growth Differ.* **7,** 1059–1066.
Baer, C., Nees, M., Breit, S., Selle, B., Kulozik, A. E., Schaefer, K. L., Braun, Y., Wai, D., and Poremba, C. (2004). Profiling and functional annotation of mRNA gene expression in pediatric rhabdomyosarcoma and Ewing's sarcoma. *Int. J. Cancer* **110,** 687–694.
Barlow, J. W., Wiley, J. C., Mous, M., Narendran, A., Gee, M. F., Goldberg, M., Sexsmith, E., and Malkin, D. (2005). Differentiation of rhabdomyosarcoma cell lines using retinoic acid. *Pediatr. Blood Cancer.*
Barr, F. G. (1997). Fusions involving paired box and fork head family transcription factors in the pediatric cancer alveolar rhabdomyosarcoma. *Curr. Top. Microbiol. Immunol.* **220,** 113–129.
Barr, F. G., Galili, N., Holick, J., Biegel, J. A., Rovera, G., and Emanuel, B. S. (1993). Rearrangement of the PAX3 paired box gene in the paediatric solid tumour alveolar rhabdomyosarcoma. *Nat. Genet.* **3,** 113–117.
Barr, F. G., Nauta, L. E., and Hollows, J. C. (1998). Structural analysis of PAX3 genomic rearrangements in alveolar rhabdomyosarcoma. *Cancer Genet. Cytogenet.* **102,** 32–39.
Bayani, J., Thorner, P., Zielenska, M., Pandita, A., Beatty, B., and Squire, J. A. (1995). Application of a simplified comparative genomic hybridization technique to screen for gene amplification in pediatric solid tumors. *Pediatr. Pathol. Lab. Med.* **15,** 831–844.
Benezra, R., Davis, R. L., Lockshon, D., Turner, D. L., and Weintraub, H. (1990). The protein Id: A negative regulator of helix-loop-helix DNA binding proteins. *Cell* **61,** 49–59.
Bengal, E., Ransone, L., Scharfmann, R., Dwarki, V. J., Tapscott, S. J., Weintraub, H., and Verma, I. M. (1992). Functional antagonism between c-Jun and MyoD proteins: A direct physical association. *Cell* **68,** 507–519.
Bennicelli, J. L., Edwards, R. H., and Barr, F. G. (1996). Mechanism for transcriptional gain of function resulting from chromosomal translocation in alveolar rhabdomyosarcoma. *Proc. Natl. Acad. Sci. USA* **93,** 5455–5459.
Bergstrom, D. A., and Tapscott, S. J. (2001). Molecular distinction between specification and differentiation in the myogenic basic helix-loop-helix transcription factor family. *Mol. Cell. Biol.* **21,** 2404–2412.
Bergstrom, D. A., Penn, B. H., Strand, A., Perry, R. L. S., Rudnicki, M. A., and Tapscott, S. J. (2002). Promoter-specific regulation of MyoD binding and signal transduction cooperate to pattern gene expression. *Mol. Cell* **9,** 587–600.
Berkes, C. A., Bergstrom, D. A., Penn, B. H., Seaver, K. J., Knoepfler, P. S., and Tapscott, S. J. (2004). Pbx marks genes for activation by MyoD indicating a role for a homeodomain protein in establishing myogenic potential. *Mol. Cell* **14,** 465–477.
Berner, J. M., Forus, A., Elkahloun, A., Meltzer, P. S., Fodstad, O., and Myklebost, O. (1996). Separate amplified regions encompassing CDK4 and MDM2 in human sarcomas. *Genes Chromosom. Cancer* **17,** 254–259.
Besnard-Guerin, C., Newsham, I., Winqvist, R., and Cavenee, W. K. (1996). A common region of loss of heterozygosity in Wilms' tumor and embryonal rhabdomyosarcoma distal to the D11S988 locus on chromosome 11p15.5. *Hum. Genet.* **97,** 163–170.
Brand-Saberi, B., Muller, T. S., Wilting, J., Christ, B., and Birchmeier, C. (1996). Scatter factor/hepatocyte growth factor (SF/HGF) induces emigration of myogenic cells at interlimb level in vivo. *Dev. Biol.* **179,** 303–308.
Braun, T., Buschhausen-Denker, G., Bober, E., Tannich, E., and Arnold, H. (1989). A novel human muscle factor related to but distinct from MyoD1 induces myogenic conversion in 10T1/2 fibroblasts. *EMBO J.* **8,** 701–709.

Braun, T., Rudnicki, M. A., Arnold, H., and Jaenisch, R. (1992). Targeted inactivation of the muscle regulatory gene myf-5 results in abnormal rib development and perinatal death. *Cell* **71,** 369–382.

Bridge, J. A., Liu, J., Weibolt, V., Baker, K. S., Perry, D., Kruger, R., Qualman, S., Barr, F., Sorensen, P., Triche, T., and Suijkerbuijk, R. (2000). Novel genomic imbalances in embryonal rhabdomyosarcoma revealed by comparative genomic hybridization and fluorescence in situ hybridization: An intergroup rhabdomyosarcoma study. *Genes Chromosomes Cancer* **27,** 337–344.

Buckingham, M., and Vincent, S. D. (2009). Distinct and dynamic myogenic populations in the vertebrate embryo. *Curr. Opin. Genet. Dev.* **19,** 444–453.

Cabane, C., Englaro, W., Yeow, K., Ragno, M., and Derijard, B. (2003). Regulation of C2C12 myogenic terminal differentiation by MKK3/p38alpha pathway. *Am. J. Physiol. Cell Physiol.* **284,** C658–C666.

Camarda, G., Siepi, F., Pajalunga, D., Bernardini, C., Rossi, R., Montecucco, A., Meccia, E., and Crescenzi, M. (2004). A pRb-independent mechanism preserves the postmitotic state in terminally differentiated skeletal muscle cells. *J. Cell Biol.* **167,** 417–423.

Cardinali, B., Castellani, L., Fasanaro, P., Basso, A., Alema, S., Martelli, F., and Falcone, G. (2009). Microrna-221 and microrna-222 modulate differentiation and maturation of skeletal muscle cells. *PLoS ONE* **4,** e7607.

Cenciarelli, C., De, S. F., Puri, P. L., Mattei, E., Ricci, L., Bucci, F., Felsani, A., and Caruso, M. (1999). Critical role played by cyclin D3 in the MyoD-mediated arrest of cell cycle during myoblast differentiation. *Mol. Cell. Biol.* **19,** 5203–5217.

Chardin, P., Yeramian, P., Madaule, P., and Tavitian, A. (1985). N-ras gene activation in the RD human rhabdomyosarcoma cell line. *Int. J. Cancer* **35,** 647–652.

Charytonowicz, E., Cordon-Cardo, C., Matushansky, I., and Ziman, M. (2009). Alveolar rhabdomyosarcoma: Is the cell of origin a mesenchymal stem cell? *Cancer Lett.* **279,** 126–136.

Chen, C. M., Kraut, N., Groudine, M., and Weintraub, H. (1996). I-mf, a novel myogenic repressor, interacts with members of the MyoD family. *Cell* **86,** 731–741.

Chen, J. F., Mandel, E. M., Thomson, J. M., Wu, Q., Callis, T. E., Hammond, S. M., Conlon, F. L., and Wang, D. Z. (2006). The role of microRNA-1 and microRNA-133 in skeletal muscle proliferation and differentiation. *Nat. Genet.* **38,** 228–233.

Coolican, S. A., Samuel, D. S., Ewton, D. Z., McWade, F. J., and Florini, J. R. (1997). The mitogenic and myogenic actions of insulin-like growth factors utilize distinct signaling pathways. *J. Biol. Chem.* **272,** 6653–6662.

Cornelison, D. D., and Wold, B. J. (1997). Single-cell analysis of regulatory gene expression in quiescent and activated mouse skeletal muscle satellite cells. *Dev. Biol.* **191,** 270–283.

Cripps, R. M., Black, B. L., Zhao, B., Lien, C. L., Schulz, R. A., and Olson, E. N. (1998). The myogenic regulatory gene Mef2 is a direct target for transcriptional activation by Twist during *Drosophila myogenesis*. *Genes Dev.* **12,** 422–434.

Crist, W. M., Anderson, J. R., Meza, J. L., Fryer, C., Raney, R. B., Ruymann, F. B., Breneman, J., Qualman, S. J., Wiener, E., Wharam, M., Lobe, T., Webber, B., *et al.* (2001). Intergroup rhabdomyosarcoma study-IV: Results for patients with nonmetastatic disease. *J. Clin. Oncol.* **19,** 3091–3102.

Cuenda, A., and Cohen, P. (1999). Stress-activated protein kinase-2/p38 and a rapamycin-sensitive pathway are required for C2C12 myogenesis. *J. Biol. Chem.* **274,** 4341–4346.

d'Amore, E. S., Tollot, M., Stracca-Pansa, V., Menegon, A., Meli, S., Carli, M., and Ninfo, V. (1994). Therapy associated differentiation in rhabdomyosarcomas. *Mod. Pathol.* **7,** 69–75.

Das, B., Tsuchida, R., Malkin, D., Koren, G., Baruchel, S., and Yeger, H. (2008). Hypoxia enhances tumor stemness by increasing the invasive and tumorigenic side population fraction. *Stem Cells* **26,** 1818–1830.

Davicioni, E., Finckenstein, F. G., Shahbazian, V., Buckley, J. D., Triche, T. J., and Anderson, M. J. (2006). Identification of a PAX-FKHR gene expression signature that defines molecular classes and determines the prognosis of alveolar rhabdomyosarcomas. *Cancer Res.* **66,** 6936–6946.

Davicioni, E., Anderson, M. J., Finckenstein, F. G., Lynch, J. C., Qualman, S. J., Shimada, H., Schofield, D. E., Buckley, J. D., Meyer, W. H., Sorensen, P. H. B., and Triche, T. J. (2009). Molecular classification of rhabdomyosarcoma—Genotypic and phenotypic determinants of diagnosis. *Am. J. Pathol.* **174,** 550–564.

Davis, R. L., and Weintraub, H. (1992). Acquisition of myogenic specificity by replacement of three amino acid residues from MyoD into E12. *Science* **256,** 1027–1030.

Davis, R. L., Weintraub, H., and Lassar, A. B. (1987). Expression of a single transfected cDNA converts fibroblasts to myoblasts. *Cell* **51,** 987–1000.

Davis, R. J., D'Cruz, C. M., Lovell, M. A., Biegel, J. A., and Barr, F. G. (1994). Fusion of PAX7 to FKHR by the variant t(1;13)(p36;q14) translocation in alveolar rhabdomyosarcoma. *Cancer Res.* **54,** 2869–2872.

Davis, R. J., Bennicelli, J. L., Macina, R. A., Nycum, L. M., Biegel, J. A., and Barr, F. G. (1995). Structural characterization of the FKHR gene and its rearrangement in alveolar rhabdomyosarcoma. *Hum. Mol. Genet.* **4,** 2355–2362.

de, l.S., I, Carlson, K. A., and Imbalzano, A. N. (2001). Mammalian SWI/SNF complexes promote MyoD-mediated muscle differentiation. *Nat. Genet.* **27,** 187–190.

De, F. G., Comes, F., and Simone, C. (2006). pRb: Master of differentiation. Coupling irreversible cell cycle withdrawal with induction of muscle-specific transcription. *Oncogene* **25,** 5244–5249.

De, S. F., Albini, S., Mezzaroma, E., Baron, L., Felsani, A., and Caruso, M. (2007). pRb-dependent cyclin D3 protein stabilization is required for myogenic differentiation. *Mol. Cell. Biol.* **27,** 7248–7265.

Dias, P., Kumar, P., Marsden, H. B., Gattamaneni, H. R., Heighway, J., and Kumar, S. (1990a). N-myc gene is amplified in alveolar rhabdomyosarcomas (RMS) but not in embryonal RMS. *Int. J. Cancer* **45,** 593–596.

Dias, P., Parham, D. M., Shapiro, D. N., Webber, B. L., and Houghton, P. J. (1990b). Myogenic regulatory protein (MyoD1) expression in childhood solid tumors: Diagnostic utility in rhabdomyosarcoma. *Am. J. Pathol.* **137,** 1283–1291, Reference type: Generic.

Diller, L., Sexsmith, E., Gottlieb, A., Li, F. P., and Malkin, D. (1995). Germline p53 mutations are frequently detected in young children with rhabdomyosarcoma. *J. Clin. Invest.* **95,** 1606–1611.

Dobashi, Y., Suzuki, S., Sato, E., Hamada, Y., Yanagawa, T., and Ooi, A. (2009). EGFR-dependent and independent activation of Akt/mTOR cascade in bone and soft tissue tumors. *Mod. Pathol.* **22,** 1328–1340.

Driman, D., Thorner, P. S., Greenberg, M. L., Chilton-MacNeill, S., and Squire, J. (1994). MYCN gene amplification in rhabdomyosarcoma. *Cancer* **73,** 2231–2237.

Dunaief, J. L., Strober, B. E., Guha, S., Khavari, P. A., Alin, K., Luban, J., Begemann, M., Crabtree, G. R., and Goff, S. P. (1994). The retinoblastoma protein and BRG1 form a complex and cooperate to induce cell cycle arrest. *Cell* **79,** 119–130.

Eckner, R., Yao, T.-P., Oldread, E., and Livingston, D. M. (1996). Interaction and functional collaboration of p300/CBP and bHLH proteins in muscle and B-cell differentiation. *Genes Dev.* **10,** 2478–2490.

El-Badry, O. M., Minniti, C., Kohn, E. C., Houghton, P. J., Daughaday, W. H., and Helman, L. J. (1990). Insulin-like growth factor II acts as an autocrine growth and motility factor in human rhabdomyosarcoma tumors. *Cell Growth Differ.* **1,** 325–331.

Epstein, J. A., Lam, P., Jepeal, L., Maas, R. L., and Shapiro, D. N. (1995). Pax3 inhibits myogenic differentiation of cultured myoblast cells. *J. Biol. Chem.* **270,** 11719–11722.

Felix, C. A., Kappel, C. C., Mitsudomi, T., Nau, M. M., Tsokos, M., Crouch, G. D., Nisen, P. D., Winick, N. J., and Helman, L. J. (1992). Frequency and diversity of p53 mutations in childhood rhabdomyosarcoma. *Cancer Res.* **52,** 2243–2247.

Ferracini, R., Olivero, M., Di Renzo, M. F., Martano, M., De, G. C., Nanni, P., Basso, G., Scotlandi, K., Lollini, P. L., and Comoglio, P. M. (1996). Retrogenic expression of the MET proto-oncogene correlates with the invasive phenotype of human rhabdomyosarcomas. *Oncogene* **12,** 1697–1705.

Figliola, R., and Maione, R. (2004). MyoD induces the expression of p57Kip2 in cells lacking p21Cip1/Waf1: Overlapping and distinct functions of the two cdk inhibitors. *J. Cell. Physiol.* **200,** 468–475.

Fitzgerald, J. C., Scherr, A. M., and Barr, F. G. (2000). Structural analysis of PAX7 rearrangements in alveolar rhabdomyosarcoma. *Cancer Genet. Cytogenet.* **117,** 37–40.

Franklin, D. S., and Xiong, Y. (1996). Induction of the CDK inhibitor p18INK4C and its predominant association with CDK4 and CDK6 during myogenic differentiation. *Mol. Biol. Cell* **7,** 1587–1599.

Fredericks, W. J., Galili, N., Mukhopadhyay, S., Rovera, G., Bennicelli, J., Barr, F. G., and Rauscher, F. J., 3rd (1995). The Pax3–FKHR fusion protein created by the t(2;13) translocation in alveolar rhabdomyosarcoma is a more potent transcriptional activator than Pax3. *Mol. Cell. Biol.* **15,** 1522–1535.

Fredericks, W. J., Ayyanathan, K., Herlyn, M., Friedman, J. R., and Rauscher, F. J., III (2000). An engineered PAX3–KRAB transcriptional repressor inhibits the malignant phenotype of alveolar rhabdomyosarcoma cells harboring the endogenous PAX3–FKHR oncogene. *Mol. Cell. Biol.* **20,** 5019–5031.

Galili, N., Davis, R. J., Fredericks, W. J., Mukhopadhyay, S., Rauscher, F. J., III, Emanuel, B. S., Rovera, G., and Barr, F. G. (1993). Fusion of a fork head domain gene to PAX3 in the solid tumour alveolar rhabdomyosarcoma. *Nat. Genet.* **5,** 230–235.

Galindo, R. L., Allport, J. A., and Olson, E. N. (2006). A *Drosophila* model of the rhabdomyosarcoma initiator PAX7–FKHR. *Proc. Natl. Acad. Sci. USA* **103,** 13439–13444.

Ganti, R., Skapek, S. X., Zhang, J., Fuller, C. E., Wu, J., Billups, C. A., Breitfeld, P. P., Dalton, J. D., Meyer, W. H., and Khoury, J. D. (2006). Expression and genomic status of EGFR and ErbB-2 in alveolar and embryonal rhabdomyosarcoma. *Mod. Pathol.* **19,** 1213–1220.

Garson, J. A., Clayton, J., McIntyre, P., and Kemshead, J. T. (1986). N-myc oncogene amplification in rhabdomyosarcoma at release. *Lancet* **1,** 1496.

Ginsberg, J. P., Davis, R. J., Bennicelli, J. L., Nauta, L. E., and Barr, F. G. (1998). Up-regulation of MET but not neural cell adhesion molecule expression by the PAX3–FKHR fusion protein in alveolar rhabdomyosarcoma. *Cancer Res.* **58,** 3542–3546.

Gordon, A. T., Brinkschmidt, C., Anderson, J., Coleman, N., Dockhorn-Dworniczak, B., Pritchard-Jones, K., and Shipley, J. (2000). A novel and consistent amplicon at 13q31 associated with alveolar rhabdomyosarcoma. *Genes Chromosomes Cancer* **28,** 220–226.

Gordon, T., McManus, A., Anderson, J., Min, T., Swansbury, J., Pritchard-Jones, K., and Shipley, J. (2001). Cytogenetic abnormalities in 42 rhabdomyosarcoma: A United Kingdom Cancer Cytogenetics Group Study. *Med. Pediatr. Oncol.* **36,** 259–267.

Gu, W., Schneider, J. W., Condorelli, G., Kaushal, S., Mahdavi, V., and Nadal-Ginard, B. (1993). Interaction of myogenic factors and the retinoblastoma protein mediates muscle cell commitment and differentiation. *Cell* **72,** 309–324.

Guerrero, L., Marco-Ferreres, R., Serrano, A. L., Arredondo, J. J., and Cervera, M. (2010). Secondary enhancers synergise with primary enhancers to guarantee fine-tuned muscle gene expression. *Dev. Biol.* **337,** 16–28.

Hachitanda, Y., Toyoshima, S., Akazawa, K., and Tsuneyoshi, M. (1998). N-myc gene amplification in rhabdomyosarcoma detected by fluorescence in situ hybridization: its correlation with histologic features. *Mod. Pathol.* **11**(12), 1222–1227.

Hahn, H., Wicking, C., Zaphiropoulous, P. G., Gailani, M. R., Shanley, S., Chidambaram, A., Vorechovsky, I., Holmberg, E., Unden, A. B., Gillies, S., Negus, K., Smyth, I., *et al.* (1996). Mutations of the human homolog of Drosophila patched in the nevoid basal cell carcinoma syndrome. *Cell* **85,** 841–851.

Hahn, H., Wonjnowski, L., Zimmer, A. M., Hall, J., Miller, G., and Zimmer, A. (1998). Rhabdomyosarcomas and radiation hypersensitivity in a mouse model of Gorlin syndrome. *Nat. Med.* **4,** 619–622.

Hahn, H., Wojnowski, L., Specht, K., Kappler, R., Calzada-Wack, J., Potter, D., Zimmer, A., Muller, U., Samson, E., Quintanilla-Martinez, L., and Zimmer, A. (2000). Patched target Igf2 is indispensable for the formation of medulloblastoma and rhabdomyosarcoma. *J. Biol. Chem.* **275,** 28341–28344.

Hahn, C. K., Ross, K. N., Warrington, I. M., Mazitschek, R., Kanegai, C. M., Wright, R. D., Kung, A. L., Golub, T. R., and Stegmaier, K. (2008). Expression-based screening identifies the combination of histone deacetylase inhibitors and retinoids for neuroblastoma differentiation. *Proc. Natl. Acad. Sci. USA* **105,** 9751–9756.

Haldar, M., Karan, G., Tvrdik, P., and Capecchi, M. R. (2008). Two cell lineages, myf5 and myf5-independent, participate in mouse skeletal myogenesis. *Dev. Cell* **14,** 437–445.

Halevy, O., Novitch, B. G., Spicer, D. B., Skapek, S. X., Rhee, J., Hannon, G. J., Beach, D., and Lassar, A. B. (1995). Correlation of terminal cell cycle arrest of skeletal muscle with induction of p21 by MyoD. *Science* **267,** 1018–1021.

Hao, Y., Crenshaw, T., Moulton, T., Newcomb, E., and Tycko, B. (1993). Tumour-suppressor activity of H19 RNA. *Nature* **365,** 764–767.

Hasty, P., Bradley, A., Morris, J. H., Edmondson, D. G., Venuti, J. M., Olson, E. N., and Klein, W. H. (1993). Muscle deficiency and neonatal death in mice with a targeted mutation in the myogenin gene. *Nature* **364,** 501–506.

Hollenberg, S. M., Cheng, P. F., and Weintraub, H. (1993). Use of a conditional MyoD transcription factor in studies of MyoD trans-activation and muscle differentation. *Proc. Natl. Acad. Sci. USA* **90,** 8028–8032.

Holtzer, H., Biehl, J., Yeoh, G., Meganathan, R., and Kaji, A. (1975). Effect of oncogenic virus on muscle differentiation. *Proc. Natl. Acad. Sci. USA* **72,** 4051–4055.

Horner, M. J., Ries, L. A. G., Krapcho, M., Neyman, N., Aminou, R., Howlader, N., Altekruse, S. F., Feuer, E. J., Huang, L., Mariotto, A., Miller, B. A., and Lewis, D. R., *et al.* (eds.), (2010). SEER Cancer Statistics Review, 1975–2006, National Cancer Institute, Bethesda, MD. http://seer.cancer.gov/csr/1975_2006/based on November 2008 SEER data submission, posted to the SEER web site, 2009.

Iolascon, A., Faienza, M. F., Coppola, B., Rosolen, A., Basso, G., Della, R. F., and Schettini, F. (1996). Analysis of cyclin-dependent kinase inhibitor genes (CDKN2A, CDKN2B, and CDKN2C) in childhood rhabdomyosarcoma. *Genes Chromosomes Cancer* **15,** 217–222.

Jen, Y., Weintraub, H., and Benezra, R. (1992). Overexpression of Id protein inhibits the muscle differentiation program: In vivo association of Id with E2A proteins. *Genes Dev.* **6,** 1466–1479.

Jostes, B., Walther, C., and Gruss, P. (1990). The murine paired box gene, Pax7, is expressed specifically during the development of the nervous and muscular system. *Mech. Dev.* **33,** 27–37.

Juan, A. H., Kumar, R. M., Marx, J. G., Young, R. A., and Sartorelli, V. (2009). Mir-214-dependent regulation of the polycomb protein Ezh2 in skeletal muscle and embryonic stem cells. *Mol. Cell* **36,** 61–74.

Kassar-Duchossoy, L., Gayraud-Morel, B., Gomes, D., Rocancourt, D., Buckingham, M., Shinin, V., and Tajbakhsh, S. (2004). Mrf4 determines skeletal muscle identity in Myf5: Myod double-mutant mice. *Nature* **431,** 466–471.

Keleti, J., Quezado, M. M., Abaza, M. M., Raffeld, M., and Tsokos, M. (1996). The MDM2 oncoprotein is overexpressed in rhabdomyosarcoma cell lines and stabilizes wild-type p53 protein. *Am. J. Pathol.* **149,** 143–151.

Keller, C., Arenkiel, B. R., Coffin, C. M., El-Bardessy, N., DePinho, R. A., and Capecchi, M. R. (2004a). Alveolar rhabdomyosarcomas in conditional Pax3:Fkhr mice: Cooperativity of Ink4a/ARF and Trp53 loss of function. *Genes Dev.* **18,** 2614–2626.

Keller, C., Hansen, M. S., Coffin, C. M., and Capecchi, M. R. (2004b). Pax3:Fkhr interferes with embryonic Pax3 and Pax7 function: Implications for alveolar rhabdomyosarcoma cell of origin. *Genes Dev.* **18,** 2608–2613.

Khan, J., Bittner, M. L., Saal, L. H., Teichmann, U., Azorsa, D. O., Gooden, G. C., Pavan, W. J., Trent, J. M., and Meltzer, P. S. (1999). cDNA microarrays detect activation of a myogenic transcription program by the PAX3–FKHR fusion oncogene. *Proc. Natl. Acad. Sci. USA* **96,** 13264–13269.

Khatib, Z. A., Matsushime, H., Valentine, M., Shapiro, D. N., Sherr, C. J., and Look, A. T. (1993). Coamplification of the CDK4 gene with MDM2 and GLI in human sarcomas. *Cancer Res.* **53,** 5535–5541.

Kim, H. K., Lee, Y. S., Sivaprasad, U., Malhotra, A., and Dutta, A. (2006). Muscle-specific microRNA miR-206 promotes muscle differentiation. *J. Cell Biol.* **174,** 677–687.

Kitzmann, M., Vandromme, M., Schaeffer, V., Carnac, G., Labbe, J. C., Lamb, N., and Fernandez, A. (1999). cdk1- and cdk2-mediated phosphorylation of MyoD Ser200 in growing C2 myoblasts: Role in modulating MyoD half-life and myogenic activity. *Mol. Cell. Biol.* **19,** 3167–3176.

Knudsen, E. S., Pazzagli, C., Born, T. L., Bertolaet, B. L., Knudsen, K. E., Arden, K. C., Henry, D. R., and Feramiso, J. R. (1998). Elevated cyclins and cyclin-dependent kinase activity in the rhabdomyosarcoma cell line RD. *Cancer Res.* **58,** 2042–2049. Reference type: Generic.

Kohashi, K., Oda, Y., Yamamoto, H., Tamiya, S., Takahira, T., Takahashi, Y., Tajiri, T., Taguchi, T., Suita, S., and Tsuneyoshi, M. (2008). Alterations of RB1 gene in embryonal and alveolar rhabdomyosarcoma: Special reference to utility of pRB immunoreactivity in differential diagnosis of rhabdomyosarcoma subtype. *J. Cancer Res. Clin. Oncol.* **134,** 1097–1103.

Koi, M., Johnson, L. A., Kalikin, L. M., Little, P. F., Nakamura, Y., and Feinberg, A. P. (1993). Tumor cell growth arrest caused by subchromosomal transferable DNA fragments from chromosome 11. *Science* **260,** 361–364.

Komarov, P. G., Komarova, E. A., Kondratov, R. V., Christov-Tselkov, K., Coon, J. S., Chernov, M. V., and Gudkov, A. V. (1999). A chemical inhibitor of p53 that protects mice from the side effects of cancer therapy. *Science* **285,** 1733–1737.

Komuro, H., Saihara, R., Shinya, M., Takita, J., Kaneko, S., Kaneko, M., and Hayashi, Y. (2007). Identification of side population cells (stem-like cell population) in pediatric solid tumor cell lines. *J. Pediatr. Surg.* **42,** 2040–2045.

Koufos, A., Hansen, M. F., Copeland, N. G., Jenkins, N. A., Lampkin, B. C., and Cavenee, W. K. (1985). Loss of heterozygosity in three embryonal tumours suggests a common pathogenetic mechanism. *Nature* **316,** 330–334.

Lagutina, I., Conway, S. J., Sublett, J., and Grosveld, G. C. (2002). Pax3–FKHR knock-in mice show developmental aberrations but do not develop tumors. *Mol. Cell. Biol.* **22,** 7204–7216.

Langenau, D. M., Keefe, M. D., Storer, N. Y., Guyon, J. R., Kutok, J. L., Le, X., Goessling, W., Neuberg, D. S., Kunkel, L. M., and Zon, L. I. (2007). Effects of RAS on the genesis of embryonal rhabdomyosarcoma. *Genes Dev.* **21,** 1382–1395.

Langlands, K., Yin, X., Anand, G., and Prochownik, E. V. (1997). Differential interactions of Id proteins with basic-helix-loop-helix transcription factors. *J. Biol. Chem.* **272,** 19785–19793.

Lassar, A. B., Davis, R. L., Wright, W. E., Kadesch, T., Murre, C., Voronova, A., Baltimore, D., and Weintraub, H. (1991). Functional activity of myogenic HLH proteins requires hetero-oligomerization with E12/E47-like proteins in vivo. *Cell* **66,** 305–315.

Lassar, A. B., Skapek, S. X., and Novitch, B. (1994). Regulatory mechanisms that coordinate skeletal muscle differentiation and cell cycle withdrawal. *Curr. Opin. Cell Biol.* **6,** 788–794.

Lazaro, J. B., Bailey, P. J., and Lassar, A. B. (2002). Cyclin D-cdk4 activity modulates the subnuclear localization and interaction of MEF2 with SRC-family coactivators during skeletal muscle differentiation. *Genes Dev.* **16,** 1792–1805. Reference type: Generic.

Lemercier, C., To, R. Q., Carrasco, R. A., and Konieczny, S. F. (1998). The basic helix-loop-helix transcription factor Mist1 functions as a transcriptional repressor of myoD. *EMBO J.* **17,** 1412–1422.

Li, H., and Capetanaki, Y. (1994). An E box in the desmin promoter cooperates with the E box and MEF-2 sites of a distal enhancer to direct muscle-specific transcription. *EMBO J.* **13,** 3580–3589.

Li, L., Chambard, J. C., Karin, M., and Olson, E. N. (1992a). Fos and Jun repress transcriptional activation by myogenin and MyoD: The amino terminus of Jun can mediate repression. *Genes Dev.* **6,** 676–689.

Li, L., Zhou, J., James, G., Heller-Harrison, R., Czech, M. P., and Olson, E. N. (1992b). FGF inactivates myogenic helix-loop-helix proteins through phosphorylation of a conserved protein kinase C site in their DNA-binding domains. *Cell* **71,** 1181–1194.

Linardic, C. M., Downie, D. L., Qualman, S., Bentley, R. C., and Counter, C. M. (2005). Genetic modeling of human rhabdomyosarcoma. *Cancer Res.* **65,** 4490–4495.

Linardic, C. M., Naini, S., Herndon, J. E., Kesserwan, C., Qualman, S. J., and Counter, C. M. (2007). The PAX3–FKHR fusion gene of rhabdomyosarcoma cooperates with loss of p16INK4A to promote bypass of cellular senescence. *Cancer Res.* **67,** 6691–6699.

Liu, N., Williams, A. H., Kim, Y., McAnally, J., Bezprozvannaya, S., Sutherland, L. B., Richardson, J. A., Bassel-Duby, R., and Olson, E. N. (2007). An intragenic MEF2-dependent enhancer directs muscle-specific expression of microRNAs 1 and 133. *Proc. Natl. Acad. Sci. USA* **104,** 20844–20849.

Lluis, F., Ballestar, E., Suelves, M., Esteller, M., and Munoz-Canoves, P. (2005). E47 phosphorylation by p38 MAPK promotes MyoD/E47 association and muscle-specific gene transcription. *EMBO J.* **24,** 974–984.

Loh, W. E., Jr., Scrable, H. J., Livanos, E., Arboleda, M. J., Cavenee, W. K., Oshimura, M., and Weissman, B. E. (1992). Human chromosome 11 contains two different growth suppressor genes for embryonal rhabdomyosarcoma. *Proc. Natl. Acad. Sci. USA* **89,** 1755–1759.

Lowichik, A., Zhou, H., Pysher, T. J., Smith, L., Lemons, R., and Coffin, C. M. (2000). Therapy associated changes in childhood tumors. *Adv. Anat. Pathol.* **7,** 341–359.

Lu, J., Webb, R., Richardson, J. A., and Olson, E. N. (1999). MyoR: A muscle-restricted basic helix-loop-helix transcription factor that antagonizes the actions of MyoD. *Proc. Natl. Acad. Sci. USA* **96,** 552–557.

Lu, X., Pearson, A., and Lunec, J. (2003). The MYCN oncoprotein as a drug development target. *Cancer Lett.* **197,** 125–130.

Ma, K., Chan, J. K., Zhu, G., and Wu, Z. (2005). Myocyte enhancer factor 2 acetylation by p300 enhances its DNA binding activity, transcriptional activity, and myogenic differentiation. *Mol. Cell. Biol.* **25,** 3575–3582.

Maestro, R., Dei Tos, A. P., Hamamori, Y., Krasnokutsky, S., Sartorelli, V., Kedes, L., Doglioni, C., Beach, D. H., and Hannon, G. J. (1999). *twist* is a potential oncogene that inhibits apoptosis. *Genes Dev.* **13,** 2207–2217.

Makawita, S., Ho, M., Durbin, A. D., Thorner, P. S., Malkin, D., and Somers, G. R. (2009). Expression of insulin-like growth factor pathway proteins in rhabdomyosarcoma: IGF-2 expression is associated with translocation-negative tumors. *Pediatr. Dev. Pathol.* **12**(2), 127–135.

Marics, I., Padilla, F., Guillemot, J. F., Scaal, M., and Marcelle, C. (2002). FGFR4 signaling is a necessary step in limb muscle differentiation. *Development* **129,** 4559–4569.

Maroto, M., Reshef, R., Munsterberg, A. E., Koester, S., Goulding, M., and Lassar, A. B. (1997). Ectopic Pax-3 activates MyoD and Myf-5 expression in embryonic mesoderm and neural tissue. *Cell* **89,** 139–148.

Martinelli, S., McDowell, H. P., Vigne, S. D., Kokai, G., Uccini, S., Tartaglia, M., and Dominici, C. (2009). RAS signaling dysregulation in human embryonal rhabdomyosarcoma. *Genes Chromosomes Cancer* **48,** 975–982.

Matsuoka, S., Thompson, J. S., Edwards, M. C., Bartletta, J. M., Grundy, P., Kalikin, L. M., Harper, J. W., Elledge, S. J., and Feinberg, A. P. (1996). Imprinting of the gene encoding a human cyclin-dependent kinase inhibitor, p57KIP2, on chromosome 11p15. *Proc. Natl. Acad. Sci. USA* **93,** 3026–3030.

Meddeb, M., Valent, A., Danglot, G., Nguyen, V. C., Duverger, A., Fouquet, F., Terrier-Lacombe, M. J., Oberlin, O., and Bernheim, A. (1996). MDM2 amplification in a primary alveolar rhabdomyosarcoma displaying a t(2;13)(q35;q14). *Cytogenet. Cell Genet.* **73,** 325–330.

Merlino, G., and Helman, L. J. (1999). Rhabdomyosarcoma—Working out the pathways. *Oncogene* **18,** 5340–5348.

Meyer, W. H., and Spunt, S. L. (2004). Soft tissue sarcomas of childhood. *Cancer Treat. Rev.* **30,** 269–280.

Meza, J. L., Anderson, J., Pappo, A. S., and Meyer, W. H. (2006). Analysis of prognostic factors in patients with nonmetastatic rhabdomyosarcoma treated on intergroup rhabdomyosarcoma studies III and IV: The Children's Oncology Group. *J. Clin. Oncol.* **24,** 3844–3851.

Minniti, C. P., Tsokos, M., Newton, W. A., Jr, and Helman, L. J. (1994). Specific expression of insulin-like growth factor-II in rhabdomyosarcoma tumor cells. *Am J Clin Pathol.* **101**(2), 198–203.

Molkentin, J. D., Firulli, A. B., Black, B. L., Martin, J. F., Hustad, C. M., Copeland, N., Jenkins, N., Lyons, G., and Olson, E. N. (1996). MEF2B is a potent transactivator expressed in early myogenic lineages. *Mol. Cell. Biol.* **16,** 3814–3824.

Moore, J. W., Dionne, C., Jaye, M., and Swain, J. L. (1991). The mRNAs encoding acidic FGF, basic FGF and FGF receptor are coordinately downregulated during myogenic differentiation. *Development* **111,** 741–748.

Mulligan, L. M., Matlashewski, G. J., Scrable, H. J., and Cavenee, W. K. (1990). Mechanisms of p53 loss in human sarcomas. *Proc. Natl. Acad. Sci. USA* **87,** 5863–5867.

Murre, C., McCaw, P. S., Vaessin, H., Caudy, M., Jan, L. Y., Jan, Y. N., Cabrera, C. V., Buskin, J. N., Hauschka, S. D., and Lassar, A. B. (1989). Interactions between heterologous helix-loop-helix proteins generate complexes that bind specifically to a common DNA sequence. *Cell* **58,** 537–544.

Musaro, A., and Rosenthal, N. (1999). Maturation of the myogenic program is induced by postmitotic expression of insulin-like growth factor I. *Mol. Cell. Biol.* **19,** 3115–3124.

Musaro, A., McCullagh, K. J. A., Naya, F. J., Olson, E. N., and Rosenthal, N. (1999). IGF-1 induces skeletal myocyte hypertrophy through calcineurin in association with GATA-2 and NF-ATc1. *Nature* **400,** 581–585.

Nabeshima, Y., Hanaoka, K., Hayasaka, M., Esumi, E., Li, S., Nonaka, I., and Nabeshima, Y. (1993). Myogenin gene disruption results in perinatal lethality because of severe muscle defect. *Nature* **364,** 532–535.

Naidu, P. S., Ludolph, D. C., To, R. Q., Hinterberger, T. J., and Konieczny, S. F. (1995). Myogenin and MEF2 function synergistically to activate the MRF4 promoter during myogenesis. *Mol. Cell. Biol.* **15,** 2707–2718.

Neuhold, L. A., and Wold, B. (1993). Tethering MyoD to E47 generates a dominant positive myogenic factor insulated from negative regulation by Id. *Cell* **74,** 1033–1042.

Nguyen, D. X., Baglia, L. A., Huang, S. M., Baker, C. M., and McCance, D. J. (2004). Acetylation regulates the differentiation-specific functions of the retinoblastoma protein. *EMBO J.* **23,** 1609–1618.

Novitch, B. G., Mulligan, G. J., Jacks, T., and Lassar, A. B. (1996). Skeletal muscle cells lacking the retinoblastoma protein display defects in muscle gene expression and accumulate in S and G2 phases of the cell cycle. *J. Cell Biol.* **135,** 441–456.

Novitch, B. G., Spicer, D. B., Kim, P. S., Cheung, W. L., and Lassar, A. B. (1999). pRb is required for MEF2-dependent gene expression as well as cell-cycle arrest during skeletal muscle differentiation. *Curr. Biol.* **9,** 449–459.

Oberlin, O., Rey, A., Lyden, E., Bisogno, G., Stevens, M. C., Meyer, W. H., Carli, M., and Anderson, J. R. (2008). Prognostic factors in metastatic rhabdomyosarcomas: Results of a pooled analysis from United States and European cooperative groups. *J. Clin. Oncol.* **26,** 2384–2389.

Odelberg, S. J., Kollhoff, A., and Keating, M. T. (2000). Dedifferentiation of mammalian myotubes induced by msx1. *Cell* **103,** 1099–1109.

Pappo, A. S., Anderson, J. R., Crist, W. M., Wharam, M. D., Breitfeld, P. P., Hawkins, D., Raney, R. B., Womer, R. B., Parham, D. M., Qualman, S. J., and Grier, H. E. (1999). Survival after relapse in children and adolescents with rhabdomyosarcoma: A report from the Intergroup Rhabdomyosarcoma Study Group. *J. Clin. Oncol.* **17,** 3487–3493.

Parker, S. B., Eichele, G., Zhang, P., Rawls, A., Sands, A. T., Bradley, A., Olson, E. N., Harper, J. W., and Elledge, S. J. (1995). p53-Independent expression of $p21^{Cip1}$ in muscle and other terminally differentiating cells. *Science* **267,** 1024–1027.

Penn, B. H., Bergstrom, D. A., Dilworth, F. J., Bengal, E., and Tapscott, S. J. (2004). A MyoD-generated feed-forward circuit temporally patterns gene expression during skeletal muscle differentiation. *Genes Dev.* **18,** 2348–2353.

Perdiguero, E., Ruiz-Bonilla, V., Gresh, L., Hui, L., Ballestar, E., Sousa-Victor, P., Baeza-Raja, B., Jardi, M., Bosch-Comas, A., Esteller, M., Caelles, C., Serrano, A. L., *et al.* (2007). Genetic analysis of p38 MAP kinases in myogenesis: Fundamental role of p38alpha in abrogating myoblast proliferation. *EMBO J.* **26,** 1245–1256.

Pin, C. L., Rukstalis, J. M., Johnson, C., and Konieczny, S. F. (2001). The bHLH transcription factor Mist1 is required to maintain exocrine pancreas cell organization and acinar cell identity. *J. Cell Biol.* **155,** 519–530.

Puri, P. L., and Sartorelli, V. (2000). Regulation of muscle regulatory factors by DNA-binding, interacting proteins, and post-transcriptional modifications. *J. Cell. Physiol.* **185,** 155–173. Reference type: Generic.

Puri, P. L., Wu, Z., Zhang, P., Wood, L. D., Bhakta, K. S., Han, J., Feramisco, J. R., Karin, M., and Wang, J. Y. (2000). Induction of terminal differentiation by constitutive activation of p38 MAP kinase in human rhabdomyosarcoma cells. *Genes Dev.* **14,** 574–584.

Puri, P. L., Iezzi, S., Stiegler, P., Chen, T. T., Schiltz, R. L., Muscat, G. E., Giordano, A., Kedes, L., Wang, J. Y., and Sartorelli, V. (2001). Class I histone deacetylases sequentially interact with MyoD and pRb during skeletal myogenesis. *Mol. Cell* **8,** 885–897.

Rao, S. S., and Kohtz, D. S. (1995). Positive and negative regulation of D-type cyclin expression in skeletal myoblasts by basic fibroblast cactor beta. A role for cyclin D1 in control of myoblast differentiation. *J. Biol. Chem.* **270,** 4093–4100.

Rao, S. S., Chu, C., and Kohtz, D. S. (1994). Ectopic expression of cyclin D1 prevents activation of gene transcription by myogenic basic helix-loop-helix regulators. *Mol. Cell. Biol.* **14,** 5259–5267.

Relaix, F., Polimeni, M., Rocancourt, D., Ponzetto, C., Schafer, B. W., and Buckingham, M. (2003). The transcriptional activator PAX3–FKHR rescues the defects of Pax3 mutant mice but induces a myogenic gain-of-function phenotype with ligand-independent activation of Met signaling in vivo. *Genes Dev.* **17,** 2950–2965.

Ren, Y. X., Finckenstein, F. G., Abdueva, D. A., Shahbazian, V., Chung, B., Weinberg, K. I., Triche, T. J., Shimada, H., and Anderson, M. J. (2008). Mouse mesenchymal stem cells expressing PAX-FKHR form alveolar rhabdomyosarcomas by cooperating with secondary mutations. *Cancer Res.* **68,** 6587–6597.

Reynaud, E. G., Pelpel, K., Guillier, M., Leibovitch, M. P., and Leibovitch, S. A. (1999). p57(Kip2) stabilizes the MyoD protein by inhibiting cyclin E-Cdk2 kinase activity in growing myoblasts. *Mol. Cell. Biol.* **19,** 7621–7629.

Rhodes, S. J., and Konieczny, S. F. (1989). Identification of MRF4: A new member of the muscle regulatory factor gene family. *Genes Dev.* **3,** 2050–2061.

Rosania, G. R., Chang, Y. T., Perez, O., Sutherlin, D., Dong, H., Lockhart, D. J., and Schultz, P. G. (2000). Myoseverin, a microtubule-binding molecule with novel cellular effects. *Nat. Biotechnol.* **18,** 304–308.

Roth, J. F., Shikama, N., Henzen, C., Desbaillets, I., Lutz, W., Marino, S., Wittwer, J., Schorle, H., Gassmann, M., and Eckner, R. (2003). Differential role of p300 and CBP acetyltransferase during myogenesis: p300 acts upstream of MyoD and Myf5. *EMBO J.* **22,** 5186–5196.

Rudnicki, M. A., Braun, T., Hinuma, S., and Jaenisch, R. (1992). Inactivation of MyoD in mice leads to up-regulation of the myogenic HLH gene Myf-5 and results in apparently normal muscle development. *Cell* **71,** 383–390.

Rudnicki, M. A., Schnegelsberg, P. N., Stead, R. H., Braun, T., Arnold, H. H., and Jaenisch, R. (1993). MyoD or Myf-5 is required for the formation of skeletal muscle. *Cell* **75,** 1351–1359.

Ruzinova, M. B., and Benezra, R. (2003). Id proteins in development, cell cycle and cancer. *Trends Cell Biol.* **13,** 410–418.

Saab, R., Bills, J. L., Miceli, A. P., Anderson, C. M., Khoury, J. D., Fry, D. W., Navid, F., Houghton, P. J., and Skapek, S. X. (2006). Pharmacological inhibition of cyclin-dependent kinase 4/6 activity arrests proliferation in myoblasts and rhabdomyosarcoma-derived cells. *Mol. Cancer Ther.* **5,** 1299–1308.

Saito, H., Inazawa, J., Saito, S., Kasumi, F., Koi, S., Sagae, S., Kudo, R., Saito, J., Noda, K., and Nakamura, Y. (1993). Detailed deletion mapping of chromosome 17q in ovarian and breast cancer: 2 cM region on 17q21.3 often and commonly deleted in tumors. *Cancer Res.* **53,** 3382–3385.

Sartorelli, V., Puri, P. L., Hamamori, Y., Ogryzko, V., Chung, G., Nakatani, Y., Wang, J. Y., and Kedes, L. (1999). Acetylation of MyoD directed by PCAF is necessary for the execution of the muscle program. *Mol. Cell* **4,** 725–734.

Schneider, J. W., Gu, W., Zhu, L., Mahdavi, V., and Nadal-Ginard, B. (1994). Reversal of terminal differentiation mediated by p107 in RB−/− muscle cells. *Science* **264,** 1467–1471. Reference type: Journal (Full).

Schwienbacher, C., Sabbioni, S., Campi, M., Veronese, A., Bernardi, G., Menegatti, A., Hatada, I., Mukai, T., Ohashi, H., Barbanti-Brodano, G., Croce, C. M., and Negrini, M. (1998). Transcriptional map of 170-kb region at chromosome 11p15.5: Identification and mutational analysis of the BWR1A gene reveals the presence of mutations in tumor samples. *Proc. Natl. Acad. Sci. USA* **95,** 3873–3878.

Scrable, H. J., Witte, D. P., Lampkin, B. C., and Cavenee, W. K. (1987). Chromosomal localization of the human rhabdomyosarcoma locus by mitotic recombination mapping. *Nature* **329,** 645–647.

Scrable, H., Cavenee, W., Ghavimi, F., Lovell, M., Morgan, K., and Sapienza, C. (1989). A model for embryonal rhabdomyosarcoma tumorigenesis that involves genome imprinting. *Proc. Natl. Acad. Sci. USA* **86,** 7480–7484.

Sebire, N. J., and Malone, M. (2003). Myogenin and MyoD1 expression in paediatric rhabdomyosarcomas. *J. Clin. Pathol.* **56,** 412–416.

Sharp, R., Recio, J. A., Jhappan, C., Otsuka, T., Liu, S., Yu, Y., Liu, W., Anver, M., Navid, F., Helman, L. J., DePinho, R. A., and Merlino, G. (2002). Synergism between Ink4a/Arf inactivation and aberrrant HGF/SF signaling in rhabdomyosarcomagenesis. *Nat. Med.* **8,** 1276–1280.

Simone, C., Forcales, S. V., Hill, D. A., Imbalzano, A. N., Latella, L., and Puri, P. L. (2004). p38 pathway targets SWI-SNF chromatin-remodeling complex to muscle-specific loci. *Nat. Genet.* **36,** 738–743.

Skapek, S. X., Rhee, J., Spicer, D. B., and Lassar, A. B. (1995). Inhibition of myogenic differentiation in proliferating myoblasts by cyclin D1-dependent kinase. *Science* **267,** 1022–1024.

Skapek, S. X., Rhee, J., Kim, P. S., Novitch, B. G., and Lassar, A. B. (1996). Cyclin-mediated inhibition of muscle gene expression via a mechanism that is independent of pRB hyperphosphorylation. *Mol. Cell. Biol.* **16,** 7043–7053.

Sogos, V., Balaci, L., Ennas, M. G., Dell'era, P., Presta, M., and Gremo, F. (1998). Developmentally regulated expression and localization of fibroblast growth factor receptors in the human muscle. *Dev. Dyn.* **211,** 362–373.

Song, A., Wang, Q., Goebl, M. G., and Harrington, M. A. (1998). Phosphorylation of nuclear MyoD is required for its rapid degradation. *Mol. Cell. Biol.* **18,** 4994–4999.

Spicer, D. B., Rhee, J., Cheung, W. L., and Lassar, A. B. (1996). Inhibition of myogenic bHLH and MEF2 transcription factors by the bHLH protein twist. *Science* **272,** 1476–1480.

Spunt, S. L., and Pappo, A. S. (2006). Childhood nonrhabdomyosarcoma soft tissue sarcomas are not adult-type tumors. *J. Clin. Oncol.* **24,** 1958–1959.

Stegmaier, K., Ross, K. N., Colavito, S. A., O'Malley, S., Stockwell, B. R., and Golub, T. R. (2004). Gene expression-based high-throughput screening(GE-HTS) and application to leukemia differentiation. *Nat. Genet.* **36,** 257–263.

Stratton, M. R., Fisher, C., Gusterson, B. A., and Cooper, C. S. (1989). Detection of point mutations in N-ras and K-ras genes of human embryonal rhabdomyosarcomas using oligonucleotide probes and the polymerase chain reaction. *Cancer Res.* **49,** 6324–6327.

Stratton, M. R., Moss, S., Warren, W., Patterson, H., Clark, J., Fisher, C., Fletcher, C. D., Ball, A., Thomas, M., and Gusterson, B. A. (1990). Mutation of the p53 gene in human soft tissue sarcomas: Association with abnormalities of the RB1 gene. *Oncogene* **5,** 1297–1301.

Sultan, I., Qaddoumi, I., Yaser, S., Rodriguez-Galindo, C., and Ferrari, A. (2009). Comparing adult and pediatric rhabdomyosarcoma in the surveillance, epidemiology and end results program, 1973 to 2005: An analysis of 2, 600 patients. *J. Clin. Oncol.* **27,** 3391–3397.

Sweetman, D., Goljanek, K., Rathjen, T., Oustanina, S., Braun, T., Dalmay, T., and Munsterberg, A. (2008). Specific requirements of MRFs for the expression of muscle specific microRNAs, miR-1, miR-206 and miR-133. *Dev. Biol.* **321,** 491–499.

Tajbakhsh, S., Rocancourt, D., Cossu, G., and Buckingham, M. (1997). Redefining the genetic hierarchies controlling skeletal myogenesis: Pax-3 and Myf-5 act upstream of MyoD. *Cell* **89,** 127–138.

Takahashi, Y., Oda, Y., Kawaguchi, K., Tamiya, S., Yamamoto, H., Suita, S., and Tsuneyoshi, M. (2004). Altered expression and molecular abnormalities of cell-cycle-regulatory proteins in rhabdomyosarcoma. *Mod. Pathol.* **17,** 660–669.

Tapscott, S. J. (2005). The circuitry of a master switch: Myod and the regulation of skeletal muscle gene transcription. *Development* **132,** 2685–2695.

Tapscott, S. J., Thayer, M. J., and Weintraub, H. (1993). Deficiency in rhabdomyosarcomas of a factor required for MyoD activity and myogenesis. *Science* **259,** 1450–1453. Reference type: Journal (Full).

Tatsumi, R., Anderson, J. E., Nevoret, C. J., Halevy, O., and Allen, R. E. (1998). HGF/SF is present in normal adult skeletal muscle and is capable of activating satellite cells. *Dev. Biol.* **194,** 114–128.

Taylor, J. G., Cheuk, A. T., Tsang, P. S., Chung, J. Y., Song, Y. K., Desai, K., Yu, Y., Chen, Q. R., Shah, K., Youngblood, V., Fang, J., Kim, S. Y., *et al.* (2009). Identification of FGFR4-activating mutations in human rhabdomyosarcomas that promote metastasis in xenotransplanted models. *J. Clin. Invest.* **119,** 3395–3407.

Thayer, M. J., Tapscott, S. J., Davis, R. L., Wright, W. E., Lassar, A. B., and Weintraub, H. (1989). Positive autoregulation of the myogenic determination gene MyoD1. *Cell* **58,** 241–248.

Tiffin, N., Williams, R. D., Shipley, J., and Pritchard-Jones, K. (2003). Pax7 expression in embryonal rhabdomyosarcoma suggests an origin in muscle satellite cells. *Br. J. Cancer* **89,** 327–332.

Tsokos, M. (1994). The diagnosis and classification of childhood rhabdomyosarcoma. *Semin. Diagn. Pathol.* **11,** 26–38.

Tsuchida, R., Das, B., Yeger, H., Koren, G., Shibuya, M., Thorner, P. S., Baruchel, S., and Malkin, D. (2008). Cisplatin treatment increases survival and expansion of a highly tumorigenic side-population fraction by upregulating VEGF/Flt1 autocrine signaling. *Oncogene* **27,** 3923–3934.

Tycko, B. (1994). Genomic imprinting: Mechanism and role in human pathology. *Am. J. Pathol.* **144,** 431–443.

van,, R. E., Quiat, D., Johnson, B. A., Sutherland, L. B., Qi, X., Richardson, J. A., Kelm, R. J., Jr., and Olson, E. N. (2009). A family of microRNAs encoded by myosin genes governs myosin expression and muscle performance. *Dev. Cell* **17,** 662–673.

Visser, M., Sijmons, C., Bras, J., Arceci, R. J., Godfried, M., Valentijn, L. J., Voute, P. A., and Baas, F. (1997). Allelotype of pediatric rhabdomyosarcoma. *Oncogene* **15,** 1309–1314.

Wagers, A. J., and Conboy, I. M. (2005). Cellular and molecular signatures of muscle regeneration: Current concepts and controversies in adult myogenesis. *Cell* **122,** 659–667.

Wagner, B. K., Carrinski, H. A., Ahn, Y. H., Kim, Y. K., Gilbert, T. J., Fomina, D. A., Schreiber, S. L., Chang, Y. T., and Clemons, P. A. (2008). Small-molecule fluorophores to detect cell-state switching in the context of high-throughput screening. *J. Am. Chem. Soc.* **130,** 4208–4209.

Wang, Y., and Jaenisch, R. (1997). Myogenin can substitute for Myf5 in promoting myogenesis but less efficiently. *Development* **124,** 2507–2513.

Wang, J., and Walsh, K. (1996). Resistance to apoptosis conferred by Cdk inhibitors during myocyte differentiation. *Science* **273,** 359–361.

Wang, Y., Benezra, R., and Sassoon, D. A. (1992). Id expression during mouse development: A role in morphogenesis. *Dev. Dyn.* **194,** 222–230.

Wang, Y., Schnegelsberg, P. N. J., Dausman, J., and Jaenisch, R. (1996). Functional redundancy of the muscle-specific transcription factors Myf5 and myogenin. *Nature* **1996,** 823–825.

Wang, H., Garzon, R., Sun, H., Ladner, K. J., Singh, R., Dahlman, J., Cheng, A., Hall, B. M., Qualman, S. J., Chandler, D. S., Croce, C. M., and Guttridge, D. C. (2008). NF-kappaB-YY1-miR-29 regulatory circuitry in skeletal myogenesis and rhabdomyosarcoma. *Cancer Cell* **14,** 369–381.

Weber-Hall, S., Anderson, J., McManus, A., Abe, S., Nojima, T., Pinkerton, R., Pritchard-Jones, K., and Shipley, J. (1996). Gains, losses, and amplification of genomic material in rhabdomyosarcoma analyzed by comparative genomic hybridization. *Cancer Res.* **56,** 3220–3224.

Weintraub, H., Dwarki, V., Davis, R., Hollenberg, S., Snider, L., Lassar, A., and Tapscott, S. (1991). Muscle-specific transcriptional activation by MyoD. *Genes Dev.* **5,** 1377–1386.

Wiedemann, H. R., Burgio, G. R., Aldenhoff, P., Kunze, J., Kaufmann, H. J., and Schirg, E. (1983). The proteus syndrome. Partial gigantism of the hands and/or feet, nevi, hemihypertrophy, subcutaneous tumors, macrocephaly or other skull anomalies and possible accelerated growth and visceral affections. *Eur. J. Pediatr.* **140,** 5–12.

Wilson, E. M., and Rotwein, P. (2006). Control of MyoD function during initiation of muscle differentiation by an autocrine signaling pathway activated by insulin-like growth factor-II. *J. Biol. Chem.* **281,** 29962–29971.

Wilke, W., Maillet, M., and Robinson, R. (1993). H-ras-1 point mutations in soft tissue sarcomas. *Mod. Pathol.* **6**(2), 129–132.

Woloshin, P., Song, K., Degnin, C., Killary, A. M., Goldhamer, D. J., Sassoon, D., and Thayer, M. J. (1995). MSX1 inhibits myoD expression in fibroblast x 10T1/2 cell hybrids. *Cell* **82,** 611–620.

Wright, W. E., Sassoon, D. A., and Lin, V. K. (1989). Myogenin, a factor regulating myogenesis, has a domain homologous to MyoD. *Cell* **56,** 607–617.

Wu, Z., Woodring, P. J., Bhakta, K. S., Tamura, K., Wen, F., Feramisco, J. R., Karin, M., Wang, J. Y., and Puri, P. L. (2000). p38 and extracellular signal-regulated kinases regulate the myogenic program at multiple steps. *Mol. Cell. Biol.* **20,** 3951–3964.

Yan, D., Dong, X. E., Chen, X., Wang, L., Lu, C., Wang, J., Qu, J., and Tu, L. (2009). MicroRNA-1/206 targets c-Met and inhibits rhabdomyosarcoma development. *J. Biol. Chem.* **284,** 29596–29604.

Yang, J., Mani, S. A., Donaher, J. L., Ramaswamy, S., Itzykson, R. A., Come, C., Savangner, P., Gitelman, I., Richardson, A., and Weinberg, R. A. (2004). Twist, a master regulator of morphogenesis, plays an essential role in tumor metastasis. *Cell* **117,** 927–939.

Yang, Z., MacQuarrie, K. L., Analau, E., Tyler, A. E., Dilworth, F. J., Cao, Y., Diede, S. J., and Tapscott, S. J. (2009). MyoD and E-protein heterodimers switch rhabdomyosarcoma cells from an arrested myoblast phase to a differentiated state. *Genes Dev.* **23,** 694–707.

Yoo, J., and Robinson, R. A. (1999). H-ras and K-ras mutations in soft tissue sarcoma: comparative studies of sarcomas from Korean and American patients. *Cancer* **86**(1), 58–63.

Zacksenhaus, E., Jiang, Z., Chung, D., Marth, J. D., Phillips, R. A., and Gallie, B. L. (1996). pRb controls proliferation, differentiation, and death of skeletal muscle cells and other lineages during embryogenesis. *Genes Dev.* **10,** 3051–3064.

Zetser, A., Gredinger, E., and Bengal, E. (1999). p38 mitogen-activated protein kinase pathway promotes skeletal muscle differentiation. Participation of the Mef2c transcription factor. *J. Biol. Chem.* **274,** 5193–5200.

Zhan, S., Shapiro, D. N., and Helman, L. J. (1994). Activation of an imprinted allele of the insulin-like growth factor II gene implicated in rhabdomyosarcoma. *J. Clin. Invest.* **94,** 445–448.

Zhang, W., Behringer, R. R., and Olson, E. N. (1995). Inactivation of the myogenic bHLH gene MRF4 results in the up-regulation of myogenin and rib anomalies. *Genes Dev.* **9,** 1388–1399.

Zhang, P., wong, C., Liu, D., Finegold, M., Harper, J. W., and Elledge, S. J. (1999). p21 (CIP1) and p57(KIP2) control muscle differentiation at the myogenin step. *Genes Dev.* **13,** 213–224.

Zhang, J., Hu, S., Schofield, D. E., Sorensen, P. H. B., and Triche, T. J. (2004). Selective usage of D-type cyclins by Ewing's tumors and rhabdomyosarcomas. *Cancer Res.* **64,** 6026–6034.

CHAPTER EIGHT

Cerebellum: Development and Medulloblastoma

Martine F. Roussel* *and* Mary E. Hatten†

Contents

* Department of Tumor Cell Biology and Genetics, St. Jude Children's Research Hospital, Memphis, Tennessee, USA
† Laboratory of Developmental Neurobiology, The Rockefeller University, New York, USA

Current Topics in Developmental Biology, Volume 94
ISSN 0070-2153, DOI: 10.1016/B978-0-12-380916-2.00008-5

Abstract

In the last 20 years, it has become clear that developmental genes and their regulators, noncoding RNAs including microRNAs and long-noncoding RNAs, within signaling pathways play a critical role in the pathogenesis of cancer. Many of these pathways were first identified in genetic screens in *Drosophila* and other lower organisms. Mammalian orthologs were subsequently identified and genes within the pathways cloned and found to regulate cell growth. Genes and pathways expressed during embryonic development, including the Notch, Wnt/β-Catenin, TGF-β/BMP, Shh/Patched, and Hippo pathways are mutated, lost, or aberrantly regulated in a wide variety of human cancers, including skin, breast, blood, and brain cancers, including medulloblastoma. These biochemical pathways affect cell fate determination, axis formation, and patterning during development and regulate tissue homeostasis and regeneration in adults.

Medulloblastoma, the most common malignant nervous system tumor in childhood, are thought to arise from disruptions in cerebellar development [reviewed by Marino, S. (2005)]. Defining the extracellular cues and intracellular signaling pathways that control cerebellar neurogenesis, especially granule cell progenitor (GCP) proliferation and differentiation has been useful for developing models to unravel the mechanisms underlying medulloblastoma formation and growth. In this chapter, we will review the development of the cerebellar cortex, highlighting signaling pathways of potential relevance to tumorigenesis.

1. Introduction

In their classical treatise on brain tumors, Bailey and Cushing wrote, "the histogenesis of the brain furnishes the indispensable background for an understanding of its tumors" (Bailey and Cushing, 1926). The idea that tumors form from specific populations of immature neurons suggests that common mechanisms underlie development and tumor formation. In the developing cerebellum, precursors of the granule cell are thought to give rise to medulloblastomas, (Bailey and Cushing, 1925) the most common childhood primary CNS tumor (Packer *et al.*, 1999). Medulloblastomas arise in the cerebellar vermis and spread rapidly through the cerebrospinal pathways, where they form tumors of variable size along the ventricles (Packer *et al.*, 1999). Medulloblastomas are a diverse set of tumors as evidenced by several criteria including differing histopathologies (Louis *et al.*, 2007). The most aggressive forms of the disease occur in infants and young children. Although current therapy cure a large proportion of patients, long-term survivors are at significant risk of cognitive and psychological deficits (Levisohn *et al.*, 2000) due to the effects of current treatment protocols that include resection, irradiation, and chemotherapy.

2. Cerebellar Development

The cerebellar cortex is a remarkably simple laminar structure, with two principal neurons, the granule cell and the Purkinje cell, and a diverse set of interneurons, which modulate the output of the Purkinje cell to the cerebellar nuclei (Palay and Chan-Palay, 1974). The most prevalent neuronal subclass within the cerebellum, indeed within the entire mammalian central nervous system, is the cerebellar granule neuron. Granule neurons serve an essential role in coordinating afferent input to and motor output from the cerebellum through their excitatory connections with Purkinje neurons (Ito, 2006; Fig. 8.1). While the role of the cerebellum in sensorimotor functions, balance control, and the vestibular ocular reflex have long been appreciated (Ito, 2006), recent studies have revealed a role for the cerebellum in a wide range of cognitive functions, including feed-forward sensory-motor learning, speech, and spatial memory (Boyden *et al.*, 2004; De Zeeuw and Yeo, 2005; du Lac *et al.*, 1995; Fiez, 1996; Fiez and Petersen, 1998; Schmahmann and Caplan, 2006; Timmann *et al.*, 1999). Notably, a loss of spatial memory and other cognitive functions have been reported in children after successful tumor resection (Levisohn *et al.*, 2000). Although the function of the cerebellum in learning and memory is complex, the remarkably simple architectonics of the cerebellum make it an attractive model system for studying CNS tumors, especially developmental tumors such as medulloblastoma.

2.1. Embryonic cerebellar development: Neurogenesis and patterning in the cerebellar anlagen

Fate mapping and transplantation studies indicate that the cerebellar territory arises from rhombomere 1, delineated by the expression of *Hoxa2* (posterior) and *Otx2* (anterior) (Wingate and Hatten, 1999). Secreted proteins of the Wnt (McMahon and Bradley, 1990) and fibroblast growth factor (FGF) families (Chi *et al.*, 2003; Crossley *et al.*, 1996; Martinez *et al.*, 1999), including Wnt1, Fgf8, and Fgf15, control the expression of transcription factors that delineate rhombomere 1 (*Otx2* and *Gbx1*; Joyner *et al.*, 2000) and of genes required to establish the cerebellar territory, including the homeobox genes *En1* and *Pax2/5/8* (Joyner, 1996; Joyner *et al.*, 1989).

The cerebellum, like the cerebrum, consists of an outer cortical structure and set of subcortical nuclei, the cerebellar nuclei, which project to cerebellar targets (Fig. 8.2). During cerebellar histogenesis, a complex pattern of neurogenesis and cell movements generates the cerebellar cortex and cerebellar nuclei (Hatten and Heintz, 1995; Morales and Hatten, 2006). Classical studies indicate that the dorsomedial ventricular zone (VZ) along the

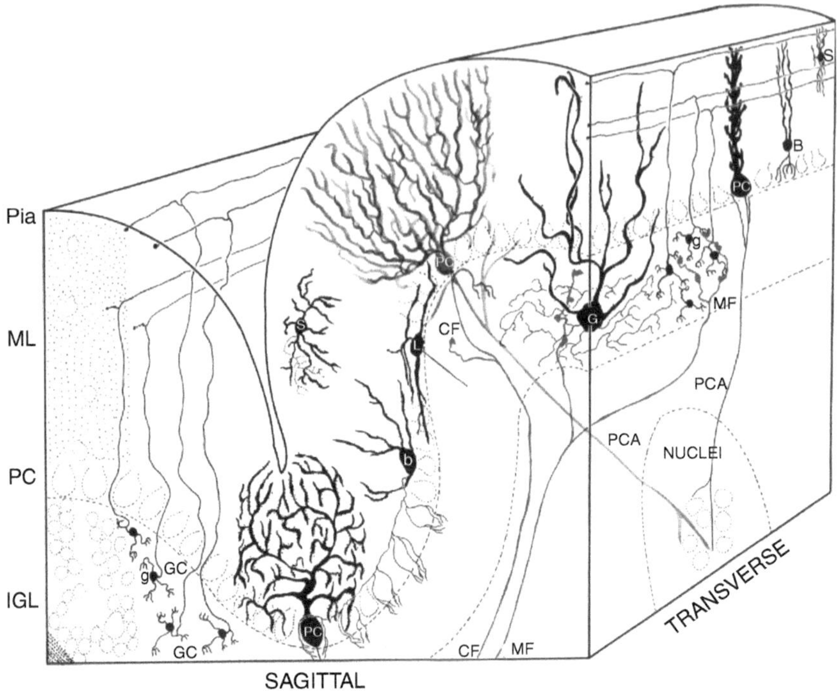

Figure 8.1 Neurons of the cerebellar cortex. Four granule cells (GC) are shown at the lower left of the sagittal view of the cerebellar cortex. Short, claw-like dendrites extend from the granule cell soma, which projects an ascending axon that bifurcates in a T. The resultant branches extend through the molecular layer (ML) parallel to the long axis of the folium in the transverse plane and form synaptic connections with the Purkinje cell (PC). The dendritic arbors of the PC extend in the sagittal plane. Mossy fiber (MF) afferents from the pontine nucleus and other sources synapse with the dendrites of the granule cells (in the transverse plane). The axons of the Golgi (G) interneuron, the mossy fiber axons, and granule cell dendrites form the synaptic glomeruli of the internal granule cell layer (IGL). PCs align with the cell soma of Bergmann glia (B) in the Purkinje cell layer (PC), as well as several classes of cerebellar interneurons, including basket (b), Lugaro (L), and candelabra (not shown) neurons. Stellate interneurons (S) are located among the parallel fiber axons and Purkinje cell dendrites in the molecular layer (ML). Afferent axons from the olivary nucleus form single climbing fiber projections (CF) onto the Purkinje neurons. Purkinje neurons, the sole output neuron of the cerebellar cortex, project axons (PCA) to the neurons of the cerebellar nuclei (Nuclei). Modified from Palay and Chan-Palay (1974).

fourth ventricle gives rise to the principal output neuron of the cerebellar cortex, the Purkinje cell, neurons of the cerebellar nuclei, and more than half a dozen types of cerebellar interneurons, including Golgi, basket, and stellate cells (Dino *et al.*, 2000; Laine and Axelrad, 2002; Palay and Chan-Palay, 1974). A secondary germinal zone forms along the anterior aspect of the rhombic lip, which generates the cerebellar granule neuron as well as a

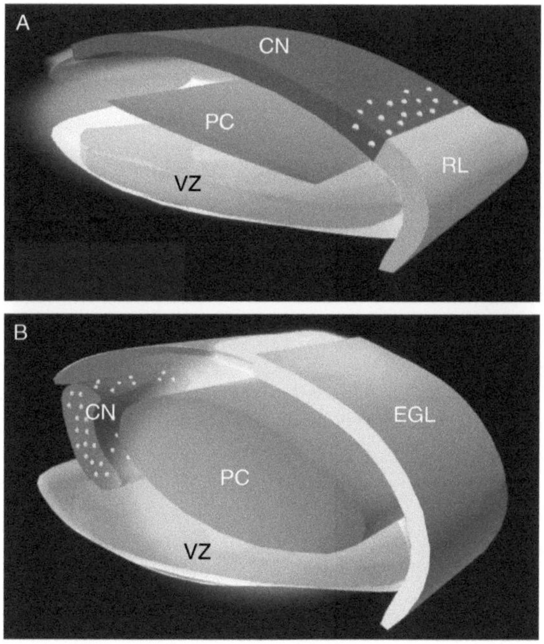

Figure 8.2 A model for the generation and migratory pathways of progenitors of the cerebellar cortex and the cerebellar nuclei. In the mouse, progenitors of the cerebellar nuclei (CNP) are generated in the VZ of the emerging cerebellar anlagen. Between E10 and E12.5, this progenitor population migrates radially along the glial fiber system (data not shown) to establish a superficial layer (dark gray) Subsequently, progenitors of the Purkinje cell (PCP) are generated in the VZ (E11–E14.5) and migrate radially to form a zone (PC, medium gray) beneath the CNPs. A second germinal zone, the anterior rhombic lip (RL, light gray), generates a subpopulation of precursors of neurons of the cerebellar nuclei, precerebellar nuclei (data not shown), and the granule neurons of the cerebellar cortex. At E12.5, a layer of CNPs (dark gray) generated in the VZ occupy the surface of the embryonic cerebellum. Between E12.5 and E14.5, several populations of RL progenitors move up onto the surface of the anlagen (RL, light gray). These include the bulk of the progenitor population, the precursors of the granule neuron (light gray), and a subpopulation of cells that intercalate among the VZ-derived CNPs (gray zone with white dots). By E14.5, the CNPs (containing both VZ and RL derived progenitors) have migrated off of the surface and formed the nascent cerebellar nuclei. By E16, the cerebellar nuclei (gray zone with white dots) settle beneath the emerging cerebellar cortex (containing the precursors of the granule neurons (EGL, light gray)), the Purkinje neuron (PC, medium gray zone), and the interneurons. Thus, a complex pattern of histogenesis and migration set forth the architecture of the murine cerebellum. From Morales and Hatten (2006).

subpopulation of neurons of the cerebellar nuclei (Fink *et al.*, 2006; Wang *et al.*, 2005) and neurons of several precerebellar nuclei of the "cerebellar system" (Dymecki and Tomasiewicz, 1998; Machold and Fishell, 2005; Wang *et al.*, 2005; Wingate, 2001; Wingate and Hatten, 1999).

In the mouse, the earliest cerebellar progenitors exit the cell cycle in the VZ at approximately embryonic (E) day E10.25 and generate neurons of the cerebellar nuclei (Fig. 8.3). Studies on transcription factor expression during cerebellar histogenesis show that postmitotic precursors of neurons of the cerebellar nuclei express the transcription factors *Lhx2/9*, *Meis 1/2*, and *Irx3* (Morales and Hatten, 2006). By E11.2, this pool of progenitors migrates radially along the emerging glial fiber system to form a superficial zone across the dorsal surface of the cerebellar anlagen. Between E11 and E14, postmitotic precursors of the Purkinje neuron, identified by expression of the LIM transcription factors Lhx1 and Lhx5, migrate away from the VZ along radial glial fibers and assemble into a broad zone in the core of the anlagen (Morales and Hatten, 2006). Recent genetic loss of function studies and fate mapping analyses demonstrate that *Ptf1a* expression is required to generate the progenitors of cerebellar GABAergic neurons (Purkinje cells and interneurons) in the cerebellar ventricular zone (Hoshino *et al.*, 2005; Pascual *et al.*, 2007). In humans, *PTF1A* mutations are associated with cerebellar agenesis (Aldinger and Elsen, 2008).

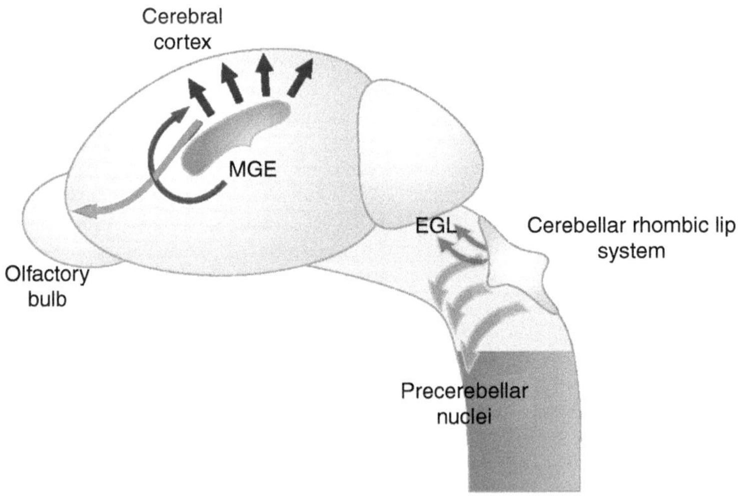

Figure 8.3 Transverse migratory pathways from the rhombic lip of the cerbellar territory establish the precerebellar nuclei in the brainstem. Progenitors of the cerebellum arise along the rhombic lip at the midbrain/hindbrain junction and migrate dorsally, (dark gray arrows). Progenitors of the precerebellar nuclei in the brainstem arise from more caudal portions of the rhombic lip and migrate ventrally along a netrin1 pathway (light gray arrows). In the forebrain, transverse neuronal migrations include dorsal migration of progenitors from the medial ganglionic eminence (MGE) into the cortex (dark gray arrow), and the posterior to anterior migration of progenitors in the rostral migratory stream (light gray arrow). Radial migrations in the cerebral cortex are also illustrated (black arrows). Adapted from Hatten (2002).

2.2. The rhombic lip, a secondary germinal zone

By E12.5, a secondary neurogenic zone appears along the anterior aspect of the rhombic lip. Cells in this zone express the basic helix-loop-helix (bHLH) transcription factor *Atoh1/Math1*, (referred as Math1 throughout) (Alder *et al.*, 1996; Machold and Fishell, 2005; Wang *et al.*, 2005) the zinc finger protein *Zic1* (Aruga, 2004), and *Meis1* as well as the markers *Pde1c*, and *Pcsk9* (Morales and Hatten, 2006). At this stage, proliferating pools of progenitors migrate out of the rhombic lip and spread across the surface of the cerebellar anlagen to form the external granule layer (EGL). As this morphogenetic movement out of the rhombic lip begins, postmitotic precursors of the cerebellar nuclei along the surface of the anlagen migrate toward the rostral aspect of the anlage (E12.5–E15.5) and inward to a position beneath the emerging zone of Purkinje cell precursors to form the cerebellar nuclei. Fate mapping experiments confirm that a subpopulation of *Math1*-positive progenitors in the rhombic lip migrate with the pool of cerebellar nuclei progenitors into the deeper aspect of the anlagen as a subpopulation of the neurons of the cerebellar nuclei (Fink *et al.*, 2006; Morales and Hatten, 2006; Wang *et al.*, 2005). The anterior rhombic lip also generates neuronal precursors that migrate ventrally where they form the lateral pontine nucleus and cochlear nucleus, hindbrain nuclei of the "cerebellar system" (Dymecki and Tomasiewicz, 1998; Machold and Fishell, 2005; Morales and Hatten, 2006; Wang *et al.*, 2005; Wingate, 2001; Wingate and Hatten, 1999).

The vast majority of rhombic lip derivatives migrate onto the dorsal surface where they form the EGL (Miale and Sidman, 1961; Palay and Chan-Palay, 1974; Ramon y Cajal, 1995), a zone of proliferating GCPs that generates the cerebellar granule cell (Alder *et al.*, 1996; Alder *et al.*, 1999), the most abundant neuron in the brain (Fig. 8.4). In man, some 45 billion granule neurons are produced during cerebellar development, out of approximately 110 billion neurons in the human brain. Molecular genetic studies demonstrate that *Math1* (Ben-Arie *et al.*, 1997) and *Mycn* (Knoepfler *et al.*, 2002) are required for granule cell specification and the expansion of the pool of granule cerebellar progenitors (GCPs) in the early postnatal period of development. Studies on the transcriptional regulation of *Math1*, by Johnson and colleagues show that Zic1 binds a conserved site within the sequence of the *Math1* enhancer region, and represses *Math1* transcription by blocking the autoregulatory activity of Math1 (Ebert *et al.*, 2003; Fig. 8.4).

The specification of granule cell identity also depends on extracellular signals that dorsalize the neural tube, including the bone morphogenic protein (BMP) family members Gdf7, Bmp6, and Bmp7 (Hogan, 1996). To analyze the role of BMPs in granule cell specification, we examined the pattern of expression of genes encoding BMP family members in the anterior rhombic lip and adjacent tissues (Alder *et al.*, 1999; Fig. 8.5). These experiments showed that Bmp7 was expressed in the anterior

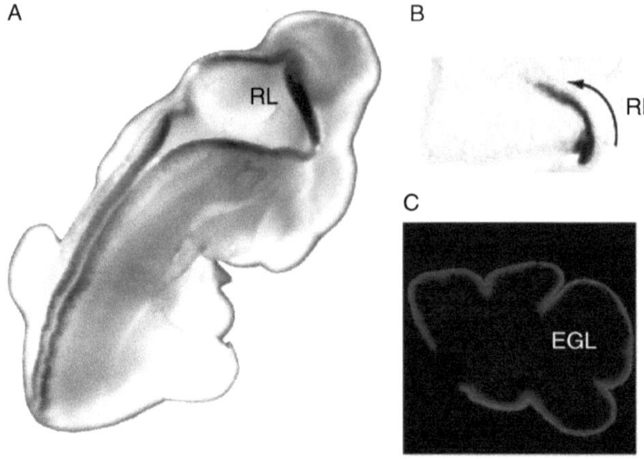

Figure 8.4 Math1 Expression in the developing cerebellum. *Math1* is expressed in the rhombic lip (RL) of the cerebellar territory and the dorsal ridge of the developing CNS in the E12.5 mouse embryo (A). At E14.5 *Math1* is expressed in RL progenitors forming the EGL of the developing anlagen (B). In the postnatal cerebellar cortex, *Math1* is expressed in proliferating GCPs located in the superficial aspect of the EGL. Photos in panels (A) and (C) were kindly provided by Dr. Jane Johnson.

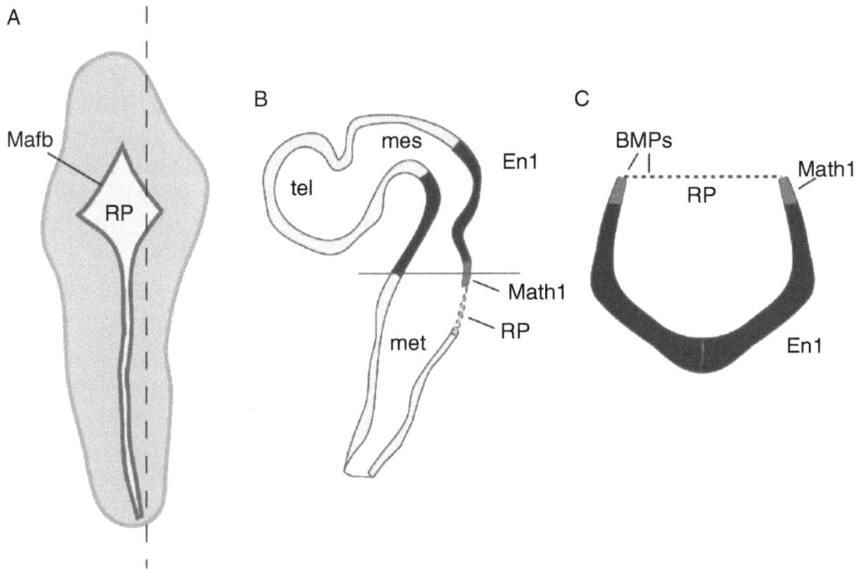

Figure 8.5 Expression of Bmps in the embryonic cerebellar anlagen. Schematic diagram of an E11 mouse embryo (A), sagittal section (B) through level indicated by dashed line in (A), and a transverse section (C) at the level indicated by the black line in (B). In (A), the roof plate (RP), shown in dark gray, is revealed by expression of the roof plate marker Mafb (Millonig *et al.*, 2000), across the rautenbreite and the dorsal ridge of the embryo. In (B) and (C) *En1/2* expression, which marks the cerebellar territory, is shown in black, and the RP is represented as a dotted line. *Gdf7*, *Bmp6*, and *Bmp7* are expressed along the dorsal edges of the neural tube and in the overlying RP. The only region of overlapping expression between *En1/2* (black) and *Math1* (gray) is in the rhombic lip, which is adjacent to the RP (for details see Alder *et al.*, 1999). mes, mesencephalon; met, metencephalon; tel, telencephalon.

rhombic lip and that Bmp6 and Gdf7 were expressed in cells of the dorsal neuroepithelium adjacent to the rhombic lip (Alder *et al.*, 1999). Treatment of progenitors from the ventral region of the mes/met region where the cerebellar territory forms with Bmp7 induced the expression of both *En1* and *Math1*. In addition, transplantation studies showed the generation of GCPs from neural tube explants treated with Bmp7. These studies support the hypothesis that in the developing cerebellum, Bmp signaling pathways play a critical role in GCP differentiation (Alder *et al.*, 1999). This view is consistent with recent genetic analyses showing that mice with a targeted deletion of the Bmp type 1 receptor genes *Bmpr1a* and *Bmpr1b* have a severe loss of granule neurons, resulting in a smaller cerebellar cortex without foliation (Qin *et al.*, 2006).

Other studies, employing both gain of function and loss of function genetic approaches in mice, provide evidence that Notch1 activation increases Math1 expression by antagonizing Bmp receptor signaling (Machold *et al.*, 2007). These studies further suggest that the level of Notch1 activity in cerebellar rhombic lip progenitors regulates their fate. These and other studies have identified changing patterns of gene expression during GCP development (Hatten and Heintz, 1995; Kuhar *et al.*, 1993) that include cell cycle regulators (Cyclins D1,D2/Ccnd1,2), genes required for glial-guided migration of GCPs (Zheng *et al.*, 1996), the GABAβ6 receptor, and the Lynx family of proteins involved in modulating the activity of a subset of parallel fiber-Purkinje cell circuits (Ibanez-Tallon *et al.*, 2002; Miwa *et al.*, 1999, 2006).

2.3. Cerebellar radial glia

Although radial glia are required for the migration of several populations of neuronal progenitors, including Purkinje cell progenitors, in embryonic cerebellar histogenesis (Morales and Hatten, 2006) and Bergmann glia guide the migration of postmitotic GCPs in postnatal cerebellar development (Edmondson and Hatten, 1987; Hatten, 2002; Rakic, 1971), relatively little attention has been given to the role of neuron–glial interactions in cerebellar neurogenesis (Fig. 8.6). Early studies on neuron–glia interactions in cultured GCPs and in cultured glioblastoma cells showed that neurons regulate the proliferation of cerebellar glial cells and of human glioblastoma cell lines (Hatten *et al.*, 1984). Molecular genetic studies on transgenic lines of mice expressing the radial glial gene *Blbp* (Anthony *et al.*, 2004; Feng *et al.*, 1994) indicate that radial glia are neurogenic in all brain regions, including the cerebellum (Anthony *et al.*, 2004). Biochemical and molecular genetic studies by Heintz and colleagues further demonstrate that Blbp is a target of Notch signaling (Anthony *et al.*, 2005). Recent genetic studies show that mice lacking *Mycn* have uncontrolled proliferation of GCPs and impaired differentiation of cerebellar glial cells (D'Arca *et al.*,

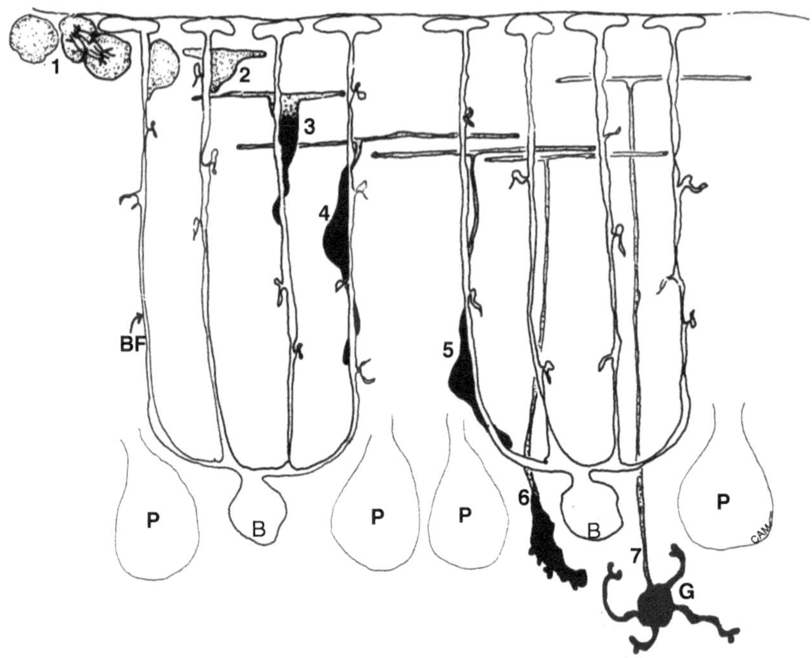

Figure 8.6 GCP migration and development. GCPs divide in the superficial layer of the neonatal cerebellar cortex (1), after which postmitotic GCPs begin to extend axons (2) and a descending, leading process (black) (3) which migrates along the radial fibers of Bergmann glia (B) by closely opposing the cell soma to the glial fiber (4,5). At the level of the Purkinje neurons (P) GCPs detach from the glial fiber (6) and move into a deeper layer where they differentiate into mature granule neurons (G) with short, claw-like dendrites that form connections with mossy fiber afferents (not shown). Drawing kindly provided by Dr. Carol A. Mason.

2010). These studies raise the question as to whether neuron–glial interactions regulate cerebellar neurogenesis, especially granule cell neurogenesis, as well as the formation and growth of medulloblastoma and glioblastoma.

2.4. Postnatal cerebellar development: GCP proliferation and migration

After birth, between the second and fourth days postnatal (P2–P4), a number of signaling pathways promote GCP proliferation (Wechsler-Reya and Scott, 1999). At peak periods of proliferation (P5–8) more than a million GCPs can be isolated from a single mouse, providing an unparalleled resource for studies of primary neural progenitor populations (Hatten, 1985). Between birth and the end of the second postnatal week, GCPs exit the cell cycle and move into the inner regions of the EGL where they extend parallel fibers and migrate along the radial fibers of Bergmann glial cells (Edmondson and Hatten, 1987; Rakic, 1971) to a position beneath the Purkinje cells, where they form the inner granule layer (IGL). The spatio-temporal dynamics of GCP proliferation in the superficial zone of the EGL, cell cycle exit, formation of parallel fibers and migration into the IGL are potentially relevant to tumorigenesis. Classical [^{3}H]-thymidine labeling studies by Fujita (Fujita, 1967; Fujita *et al.*, 1966) show temporal regulation of the step-wise progression of GCPs through phases of rapid proliferation to cell cycle exit and differentiation. In the mouse, the transit time of postmitotic GCPs across the EGL and the molecular layer (ML) is 21 and 4 h, respectively. As GCPs exit the cell cycle, they down regulate expression of the bHLH gene *Atoh1/Math1* (Ben-Arie *et al.*, 1996; Helms *et al.*, 2000) and upregulate expression of *NeuroD1* (Lee *et al.*, 2000; Pan *et al.*, 2009), *Zic1,3* (Aruga, 2004; Aruga *et al.*, 1996), and the tumor suppressor cyclin-dependent kinase inhibitory protein p27^{Kip1} (Ayrault *et al.*, 2009). During their movement from the mantle of the EGL into the deeper layers of the EGL, GCPs extend long, parallel fiber axons that express the GPI-linked axonal glycoprotein Tag1 (Furley *et al.*, 1990), among others. Recent studies by Luo *et al.* (Espinosa and Luo, 2008), discussed below, suggest that clones of GCPs stack their parallel fibers in a chronological order that relates to the timing of their penultimate cell division. The importance of correlated timing of axon outgrowth and gene expression is underscored by the anatomical studies of Rivas, who demonstrated Tag1 immunoreactive of parallel fibers peaks during the first 3 days after the GCPs become postmitotic. Tag1 levels decrease dramatically with the differentiation of Purkinje into the ML and formation of parallel fiber synapses with Purkinje cell dendrites (Stottmann and Rivas, 1998).

The timing of GCP migration away from the EGL is another important determinant of cerebellar patterning. GCPs first begin to exit the cell cycle and migrate across the ML into the IGL in the late embryonic period. Several classes of genes that modify the timing of glial-guided GCP migration out of the EGL have potential importance to medulloblastoma. First, in mice lacking the G Protein-coupled chemokine receptor CXCR4, which is broadly expressed in both the CNS and immune system, GCPs migrate along Bergmann glial fibers prematurely, with large numbers of cells pouring down the Bergmann glia in the late embryonic period. The premature migration of GCPs from the EGL into the cerebellar anlage, prevents the massive expansion of GCPs that normally occurs in the postnatal period. Several studies indicate a potential relevance of CXCR4 to medulloblastoma, as blocking CXCR4-mediated cyclin AMP suppression inhibits tumor growth (Yang *et al.*, 2007). Further support for the idea that CXCR4 is critical for the progression of brain tumors comes from studies showing that small molecule inhibitors of CXCR4 inhibit intracranial tumor growth (Rubin *et al.*, 2003).

Molecular genetic studies on genes involved in glial-guided migration, including *Astn1* and *Pex5*, raise the possibility that slowed migration or altered patterns of GCP survival and differentiation cause ectopic zones of GCPs that fail to migrate into the IGL. In the case of mice lacking *Astn1* or *Pex5*, GCP migration along Bergmann glial fibers is slowed markedly, causing the dendritic arbors of young Purkinje neurons to tilt out of the sagittal plane (Adams *et al.*, 2002; Faust, 2003; Faust and Hatten, 1997). In addition, the slowed migration results in ectopic zones of GCPs in the EGL that persist into late childhood, a time by which the EGL has normally dissipated. Since Purkinje neurons provide the mitogen sonic hedgehog (Shh) for GCP expansion in the EGL, it is possible that the growth rates of GCPs depend on the spatiotemporal dynamics of GCP and Purkinje neuron maturation, and the assembly of the cerebellar circuit. Finally, it is worth noting that Purkinje cells also provide BDNF, a neurotrophin that promotes the stepwise differentiation of postmitotic GCPs, including migration along Bergmann fibers (Borghesani *et al.*, 2002). Indeed, mice lacking BDNF also have defects in cerebellar patterning that include ectopic zones of immature GCPs in the superficial aspect of the cerebellar cortex (Schwartz *et al.*, 1997). All of these perturbations of the normal spatiotemporal program of GCP expansion and differentiation could be relevant to medulloblastoma formation in childhood.

Although a number of studies have demonstrated that the EGL gives rise to a single class of neurons, the granule cells (Alder *et al.*, 1996, 1999; Hallonet *et al.*, 1990; Zhang and Goldman, 1996), recent studies by Luo and colleagues using the mosaic analysis with double markers (MADM) show that GCPs undergo predominantly symmetric divisions during EGL development and that clonally related granule cells exit the cell cycle within a narrow time frame (Espinosa and Luo, 2008). In agreement with the earlier studies by Fujita (Fujita, 1967; Fujita *et al.*, 1966), these studies showed a progressive slowing of GCP proliferation just before birth, and a rapid expansion of clonally related GCPs just prior to cell cycle exit in the postnatal period (Espinosa and Luo, 2008). These studies show that GCP expansion produces distinct clones of granule cells (GCs). Studies on a large number of BAC transgenic lines of mice in the GENSAT Project also provide molecular genetic evidence for subpopulations of GCPs and of cerebellar granule cells (Hatten and Heintz, unpublished; www.gensat.org). These subpopulations of GCPs may be relevant to some of the subtypes of medulloblastomas (Gilbertson and Ellison, 2008).

2.5. Differentiation of ES cells toward a granule cell identity

As a proof of principle for the role of key developmental regulators in granule cell development (Hatten and Heintz, 1995), we treated E14 embryonic stem cells, as well as ES cell lines derived from *Tg(Pde1c–Egfp)* transgenic mice generated using BAC methodology (Gong *et al.*, 2003), with "cerebellar organizers" (Fgf8, RA), followed by the local signals that "dorsalize" the cerebellar anlage (*Wnt1*, *Wnt3a*, *Bmp7*, *Gdf7*, *Bmp6*), and mitogens that expand the cerebellar granule cell progenitor cell population (*Shh*, *Jag1*) (Salero and Hatten, 2007). To achieve terminal differentiation, we cultured differentiated ES cells in medium conditioned by purified cerebellar glial cells. After these treatments, which parallel known steps of GCP differentiation *in vivo*, treated ES cells expressed multiple markers of granule neurons (*En1*, *Math1*, *Zic1*, *Zic2*, *Pax6*, and $GABAa_6$). To test the behavior of these differentiated *Tg*(*Pde1c–Egfp*) ES cells into the EGL of the cerebella of living, neonatal mice, a stereotaxic injection system was used to follow, the migration and development of EGFP-positive ES cells *in vivo*. After 3 days to 2 weeks, the cells migrated into the IGL, the position where mature granule neurons function in the cerebellar cortex. Thus, ES cell differentiation confirms the role of local signals in normal cerebellar neuron specification (Salero and Hatten, 2007), and provides an initial strategy for CNS cell replacement therapy.

2.6. Mitogenic pathways that promote GCP proliferation

A number of mitogenic pathways maintain the rapid expansion of the pool of GCPs in the EGL. A role for Shh as a GCP mitogen was suggested by studies of *Patched* (*Ptch*), a gene that encodes a Shh-binding protein that functions as an antagonist of Shh signaling (Fig. 8.7). Since mutations in *PTCH1* occur in sporadic human medulloblastoma and promote medulloblastoma in mouse models, Wechsler-Reya and Scott tested the idea that Shh signaling regulates cerebellar growth. Their studies identified Shh as a key mitogen for GCP expansion (Wechsler-Reya and Scott, 1999), and numerous subsequent studies indicate that SHH signaling is defective in ~25% of human medulloblastoma (reviewed in Gilbertson and Ellison, 2008). Activation of the Shh signaling pathway leads to increased transcription of the primary downstream mediator, the zinc finger protein Gli. In the embryonic cerebellar development, Shh signaling regulates midbrain/hindbrain development through positive regulation of the Gli activators, Gli1 and Gli2 (GliA) and inhibition of the Gli3 repressor (Gli3R). The levels of Gli3R control the overall growth of the midbrain/hindbrain territory and play a critical role in sustaining the activity of the anterior posterior (AP) organizer Fgf8 in the isthmus region (Blaess *et al.*, 2006). In the early postnatal period, Shh signaling is a primary driver of the dramatic expansion of the GCP precursor pool. Analysis of the role of Gli family members in medulloblastoma, by inactivation of Gli alleles in heterozygote *Pcth1* mice, suggests that Gli1 both provides a marker of Shh pathway activation and functions in medulloblastoma formation from GCPs (Kimura *et al.*, 2005).

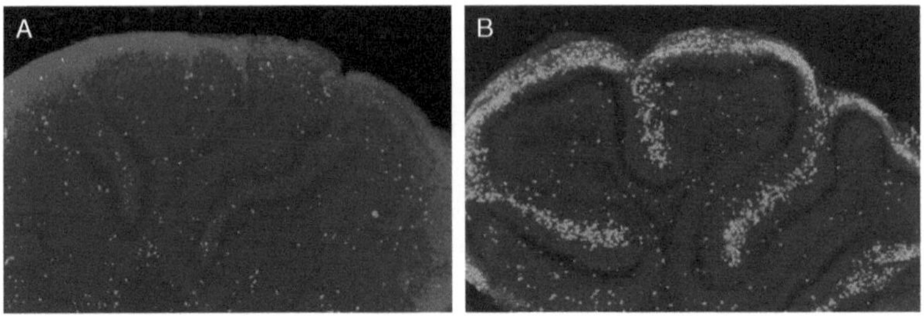

Figure 8.7 Mitogenic effect of sonic hedgehog (Shh) on the proliferation of granule cell progenitors (GCPs) of the developing cerebellum. The addition of Shh to early postnatal cerebellar cortex (B) results in a large increase in the number of dividing cells, as measured by incorporation of bromodeoxyuridine (BrdU), a deoxythymidine analog (light gray). Adapted from Wechsler-Reya and Scott, (1999).

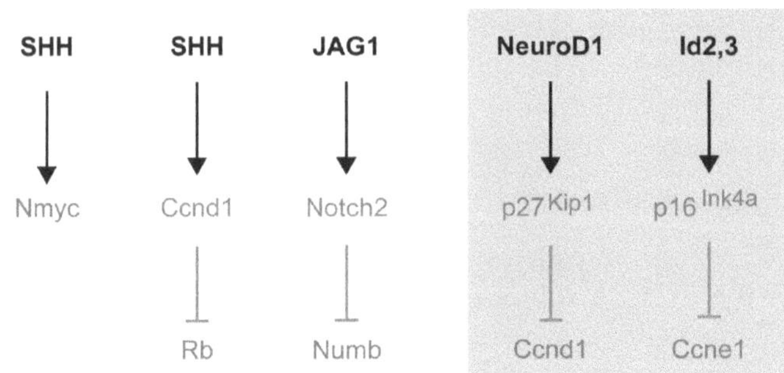

Figure 8.8 Mitogenic pathway and their inhibitors in GCPs of the developing cerebellum. Schema of the primary mitogens (Shh and JAG1) in the early postnatal cerebellum.

Shh appears to regulate GCP proliferation by several mechanisms (Fig. 8.8). First, Shh signaling regulates the expression of the cell cycle regulators cyclin D1 (Ccnd1), cyclin D2 (Ccnd2), and cyclin E (Ccne) (Kenney and Rowitch, 2000). During cerebellar development, early postnatal GCPs express Ccnd1 while GCPs generated during the peak of GCP neurogenesis express both Ccnd1 and Ccnd2. Mice lacking *Ccnd1* have slowed GCP proliferation and cerebellar development (Pogoriler *et al.*, 2006). Interestingly mice lacking both *Ccnd1* and *Ptch1* have a reduced incidence of medulloblastoma, suggesting that a loss of *Ccnd1* inhibits medulloblastoma formation (Pogoriler *et al.*, 2006). Second, activation of the Shh pathway upregulates the expression of the proto-oncogene Mycn that when overexpressed, promotes cell autonomous upregulation of *Ccnd1* mRNA and protein independently of Shh signaling (Kenney *et al.*, 2003). Thus, Shh controls cerebellar growth and GCP proliferation by multiple mechanisms.

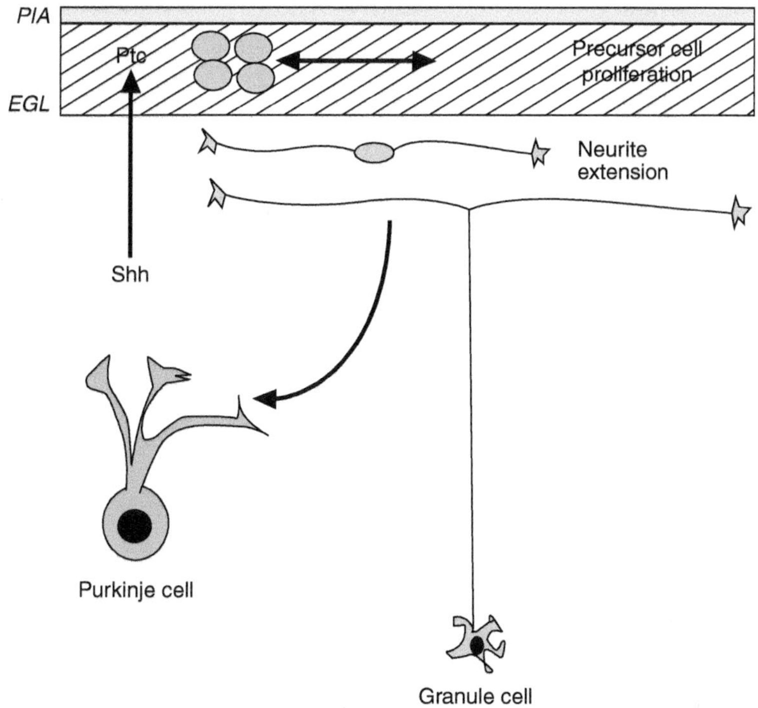

Figure 8.9 Model for the action of sonic hedgehog in cerebellar development. Sonic hedgehog (Shh) is produced by Purkinje neurons during the developmental stage when they are extending their dendritic tree and beginning to form synaptic connections with the axons of granule neurons. Granule cell precursors bind Shh via the protein Patched (Ptc), which releases Ptc's inhibition of Smoothened and induces a cascade of signals leading to proliferation. Granule cells later exit the cell cycle, extend neurites, and form synaptic connections with Purkinje cells. Thus, Purkinje cells appear to control the number of granule cells entering the cerebellar circuit. Adapted from Hatten (1999).

The importance of Shh to cerebellar histogenesis is also underscored by elegant genetic analyses of Joyner and colleagues, who demonstrated that the levels of Shh signaling control the foliation patterning of the cerebellar cortex (Corrales *et al.*, 2006). At present, the regulation of Purkinje cell expression and release of Shh is not as well understood as the mitogenic Shh signaling pathways in GCP neurogenesis. Gene expression studies show that Purkinje neurons are the primary source of Shh in the neonatal cerebellum (Wechsler-Reya and Scott, 1999). The role of the Purkinje neuron in GCP expansion is of particular interest, given the increasing ratio of granule cells to Purkinje cells during evolution (750:1 in the mouse; 3300:1 in humans; Lange, 1975; Fig. 8.9). This suggests the possibility that the PC as the primary output neuron of the cerebellar cortex, regulates the production of interneurons in the cerebellar circuitry (Hatten, 1999) by a feedback loop that involves Shh.

A number of studies have defined a role for SHH pathway activation in desmoplastic medulloblastoma that represent ~25% of all medulloblastoma in children. Major effort has been spent identifying genes that are specifically expressed in embryonal tumors and medulloblastoma (Pomeroy *et al.*, 2002; Thompson *et al.*, 2006), especially the targets of the Gli1 expressed in medulloblastoma (Yoon *et al.*, 2009). Their studies suggest that Gli1 affects medulloblastoma growth and survival via targets that include p53, SGK1, MGMT, and NTRK2.

Notch2 signaling also stimulates granule cell proliferation (Solecki *et al.*, 2001), as activated *Notch2* maintains GCP proliferation and inhibits granule neuron differentiation. Exposure of GCPs to the Notch2 ligand, Jag1, or a constitutively activated form of the Notch2 receptor results in increased (3–5×) cell proliferation. In these cells, expression of the downstream transcription factor, Hes1, is upregulated and Hes1 overexpression has a similar ability to maintain proliferation in granule cell progenitor populations. Hes1 expression is also induced by the Shh pathway, suggesting that it is a common downstream effector of these two pathways (Solecki *et al.*, 2001). This hypothesis is consistent with gene expression studies in medulloblastoma from *Ptch1* heterozygote mice with elevated Shh signaling, showing increased expression of components of the Notch and Wnt pathways. In addition, genetic studies on mice with reduced Shh signaling report a downregulation of expression of Notch2, Jag1, and Hes1 (Dakubo *et al.*, 2006). Studies on the expression of Notch family members in medulloblastomas and other tumors also report a link between Notch2 activation and tumorigenesis. Immunocytochemical studies indicate that expression of Notch2, but not Notch 1,3,4, is detected in medulloblastoma, with increasing numbers of immunopositive cells in the tumor (Xu *et al.*, 2009). Direct studies on the effect of expressing truncated, constitutively active forms of Notch1 or Notch2 support the idea that activation of Notch2, but not Notch1, has oncogenic effects on tumor formation. Examination of the levels of Notch2, and its downstream target Hes1, mRNAs in 40 embryonal tumors showed enhanced expression in 15% of the tumors. Taken together, these findings suggest that Notch1 and Notch2 have different effects on GCP proliferation, as well as on human medulloblastoma (Fan *et al.*, 2004).

Cell cycle regulators are also important regulators of normal GCP proliferation. In the postnatal EGL, $p18^{Ink4c}$ (Cdkn2c) is transiently expressed in GCPs during cell cycle exit (Uziel *et al.*, 2005). Mycn is also required for the rapid expansion of GCP proliferation in the postnatal cerebellum (Knoepfler *et al.*, 2002). Loss of Mycn increases expression of two cyclin-dependent kinase inhibitory proteins, $p27^{Kip1}$ and $p18^{Ink4c}$ in the cerebellum, and the D-type cyclins Ccnd1 and Ccnd2 dramatically expand the GCP population in the EGL. Studies by Zindy *et al.* (2006) further demonstrate that Mycn expression and downregulation of $p18^{Ink4c}$ and $p27^{Kip1}$ are both critical for GCP expansion during cerebellar development.

2.7. Negative regulators of GCP proliferation

In addition to signaling pathways that promote growth, GCP cell cycle exit and differentiation depend on signaling pathways that provide negative growth regulation. Recent studies have described several signaling molecules that antagonize the Shh-mediated proliferation of GCPs, including bFGF (Fogarty *et al.*, 2007) and members of the BMP family. Among the latter, Bmp2 and Bmp4 are expressed in granule cell progenitors, and in postmitotic, differentiating GCPs in the EGL. *In vitro* assays indicate that Bmp2 and Bmp4, but not Bmp7, inhibit Shh-induced GCP proliferation (Rios *et al.*, 2004) via the Smad signaling pathway (Zhao *et al.*, 2008). Recent studies showed that Bmp4 antigonizes Shh signaling and induces differentiation of GCPs by rapid posttranscriptional turnover of Math1/Atoh1 (Ayrault *et al.*, 2010). Recently, we have used a combination of genetic, cell, molecular, and biochemical methods to identify Wnt3 as a novel negative regulator of GCP proliferation (Kim *et al.*, 2010, submitted). Our experiments show that Wnt3, which is provided by Bergmann glia in the developing cerebellum, blocks Shh-induced stimulation of GCP growth and slows medulloblastoma growth through a noncanonical Wnt signaling pathway.

In addition to changes in the regulation of cell cycle regulator RNAs and proteins, cell-cycle transitions are driven by ubiquitin-dependent degradation of key cell-cycle regulators. SCF (Skp1/Cullin/F-box protein) complexes and anaphase-promoting complexes (APC) represent two major classes of ubiquitin ligases whose activities are thought to regulate the G1/S and metaphase/anaphase cell-cycle transitions, respectively (Ayad *et al.*, 2005; Rankin *et al.*, 2005; Wei *et al.*, 2004). The APC complex is an E3 ubiquitin ligase required for the metaphase-to-anaphase transition and mitotic exit. GCPs provide a unique model for understanding cell cycle transitions. Recent studies by Harmey *et al.* (2009) indicate that the APC is essential for GCP proliferation and differentiation. Inhibition of APC activity by either a dominant negative approach or shRNAi depletion of the APC activator Cdh1 reduced GCP neurite outgrowth and migration in organotypic slice cultures. A clearer understanding of the role of ubiquitin-dependent degradation of GCP cell cycle regulators in normal cerebellar development could have implications for medulloblastoma growth and treatment.

2.8. Addendum: Timing of human cerebellar development

In humans, between 24 and 40 weeks of gestation, the cerebellum undergoes a remarkable rate of growth, as reviewed by Volpe (2009). During this period, as assessed by 3-D volumetric ultrasound, the volume of the cerebellar territory increases fivefold (Limperopoulos *et al.*, 2005). In humans, the major embryonic histogenic events occur by 20 weeks gestation. Between 20 and 30 weeks, the EGL forms and development accelerates as evidenced by surface foliation (Volpe, 2009). At 25 weeks, the EGL reaches peak thickness, 6–8 cells deep. GCP proliferation and migration continue between 30 and 40 weeks. A number of features of GCP expansion, including a role for SHH, provided by Purkinje cells have been documented in human cerebellar development (Carletti and Rossi, 2008). During this period, although the thickness of the EGL remains constant, the surface expands horizontally via a more than 30-fold increase in the surface area of the cerebellar cortex. In human infants, the granule cell population accounts for more than 95% of the neurons of cerebellar cortex (Andersen *et al.*, 1992), with the ratio of granule cells to Purkinje neurons being 3300:1. During the first postnatal year, as the EGL dissipates, the IGL continues to increase in size as the cerebellar circuitry matures (Volpe, 2009).

3. Medulloblastoma

Medulloblastoma is the most common malignant pediatric brain tumor with about 1000 new cases every year worldwide and a mean age between 3 and 7 years (Fogarty *et al.*, 2005). Medulloblastoma, a cancer of the cerebellum, (Fig. 8.10), is a heterogenous class of embryonal tumors, that include subgroups with genetic anomalies in developmental pathways that are critical for normal cerebellar development. In the last 10 years a variety of studies, including analyses of primary human patient medulloblastoma samples, as well as cell culture and mouse models, have identified signaling pathways that promote or suppress medulloblastoma. As we refine our understanding of the disease at the molecular level, we will no doubt improve diagnostic tools and develop new and improved targeted therapies.

3.1. Current therapy and its consequences and targeted therapy

Current therapies for all subgroups of MB regardless of the subgroup include surgical resection, radiation therapy, and chemotherapy. Although advances in the radiation oncology and chemotherapy have led to dramatic increases in

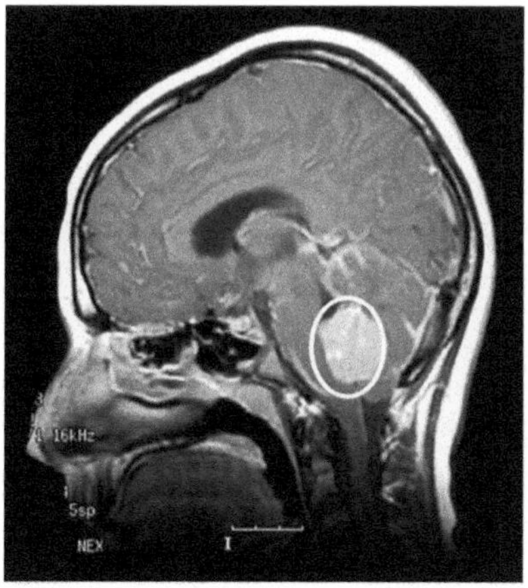

Figure 8.10 Medulloblastoma. Sagittal MRI scan following gadolinium. The tumor appears dense and is outlined by a circle. Image provided by Dr. Richard Gilbertson.

survival rates, ~70% of patients with average risk MBs, current treatments have significant morbidity and adverse secondary effects including loss of hearing, due mainly to the use of ototoxic drugs, and cognitive impairment (Gajjar *et al.*, 2006). Such toxicities have profound effects in children that even if cured, face many years of posttreatment disabilities. High risk patients fare less well and ~30% of children die of advanced disease. High risk MBs including the large cell anaplastic variants with *MYC* amplification and those associated with metastasis are associated with a poor outcome (Finlay *et al.*, 2007; Gajjar *et al.*, 2006). In contrast, a subgroup of MBs with mutations in the WNT signaling pathway (see below), found primarily in adolescents and invariably are associated with high survival rates.

Our understanding of Shh/Ptch signaling has led to the discovery of small molecule inhibitors of Smoothened (Smo) function that worked remarkably well to suppress MBs in allografts and mouse models of medulloblastoma (Berman *et al.*, 2002; Romer and Curran, 2005). These compounds are currently in clinical trials. Although these small molecule inhibitors were very effective in reducing and eliminating tumor cells in mice, a case report indicates that one patient with remarkable remission of the disease later had a recurrence of his MB associated with a mutation in *SMO*, which made the tumor cells insensitive to the drug (Rudin *et al.*, 2009). This emphasizes the need to discover novel targets and small molecule antagonists to the Shh pathway. A nasal antifungal, itracodazole, was recently found to block Smo function while arsenic trioxide was shown to antagonize Shh signaling by targeting Gli2 rather than Smoothened. Like cyclopamine and analogs, itracodazole and arsenic block proliferation of tumor cells. However, arsenic has the advantage to be active against MBs in which Smo is mutated, suggesting that it could be used therapeutically in patients in which tumors express a mutated Smo protein (Kim *et al.*, 2010 a,b). Ongoing molecular analysis of primary MBs combined with the discovery of new biomarkers for histopathology studies should provide physicians with better diagnostic and therapeutic tools. In patients with MBs with a good prognosis, including those with mutations in the WNT pathway, the treatment could be less aggressive thus lessening the negative secondary effects of the treatment. Such approaches are currently being tested in clinical trials.

3.2. Histopathology

The 2007 WHO classification of medulloblastomas defines five histopathologic subgroups or variants, including the classic subgroup with some differentiated neurons, the desmoplastic variant in which tumor cells show some differentiation and are surrounded by extracellular matrix, MBs with nodularity, and the anaplastic and large cell anaplastic forms, the most aggressive forms of the disease that invariably connote poor prognosis (Fig. 8.11; Louis *et al.*, 2007; reviewed in Gilbertson and Ellison, 2008; Northcott *et al.*, 2009). Molecular genetic analyses have identified only four subgroups that do not completely overlap with the WHO pathological classification (Kool *et al.*, 2008; Thompson *et al.*, 2006). Whereas medulloblastoma with mutations in the SHH/PTCH pathway are characterized as desmoplastic and represent ~25% of all MBs, the aggressive large cell anaplastic MBs are found within each molecularly distinct subgroup and represent ~15% of all MBs (Thompson *et al.*, 2006). MBs with a good prognosis include desmoplastic MBs and those with increased nodularity characterized with monosomy 6, mutations in the WNT pathway, and high TRKC expression. In contrast, high risk patients with medulloblastoma often have metastases and amplification or high expression of *MYC (C-MYC)* or *Mycn* which is associated with the large cell anaplastic variants, 17p loss and 1q gain (Eberhart *et al.*, 2004; von Hoff *et al.*, 2010). MBs also can spread within the cerebrospinal fluid and outside the cerebral nervous system (CNS) in bone and bone marrow in very few cases conferring an adverse prognosis (reviewed in Pizer and Clifford, 2009). The identification of new biomarkers that define subgroups of MBs and the development of antibodies to these markers will likely lead to better diagnostic tools.

3.3. Origin of medulloblastomas

Like other embryonal tumors in children, MBs are thought to arise from neuronal progenitors with defects in gene regulation or genetic anomalies in genes and proteins that regulate normal growth and development. To date, one subgroup of MBs with constitutive activation of the Shh/Ptch pathway

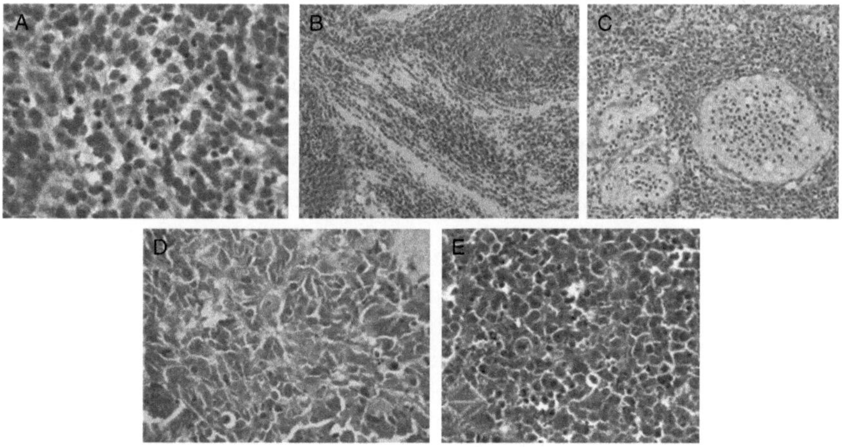

Figure 8.11 Histopathology of the five medulloblastoma variants. (A) Classic medulloblastoma marked here by a syncytial arrangement of densely packed small uniform cells with hyperchromatic nuclei. (B) Desmoplastic/nodular medulloblastoma showing nodules of differentiated neurocytic cells separated by regions containing moderately pleomorphic cells with a high growth fraction. (C) Medulloblastoma with extensive nodularity with a large irregularly shaped nodules of neurocytic cells often forming linear patterns interspersed with small collections of pleomorphic cells. (D) Anaplastic medulloblastoma characterized with a large markedly pleomorphic cells with a high mitotic count and nuclear:cytoplasmic ratio mold to each other. (E) Large cell medulloblastoma showing large cells with a single nucleolus among more pleomorphic cells with an anaplastic phenotype. (Figure kindly provided by Dr. David Ellison).

is thought to originate from granule cerebellar progenitors (GCPs) in the external granule layer (EGL) (reviewed in Behesti and Marino, 2009; Gilbertson and Ellison, 2008). The question remains as to whether the other subgroups of MBs arise from subpopulations of GCPs in the cerebellar EGL, from ectopic GCPs at developmental stages after the EGL dissipates, or from other types of cerebellar neurons. Recently, Gilbertson and colleagues developed a mouse model for MBs that recapitulate human MBs from the Wnt subgroup by targeting neuronal progenitors in the ventricular zone distinct from the GCPs using a mouse line in which the Cre recombinase is expressed in neuronal progenitors located in the ventricular zone (Blbp-Cre). By breeding this mouse line with a conditional form of a constitutively activated β-Catenin protein in a p53-null background, they were able to generate a mouse model that mirror the human subgroup with a Wnt signature (Gibson *et al.*, 2010). This is especially interesting because progenitors that express the glial gene Blbp give rise to both neurons and glial cells (Anthony and Heintz, 2008) and Blbp is a direct target of Notch1 in radial glial cells (Anthony *et al.*, 2005). Thus, different subgroups of MB may derive from different subpopulations of cerebellar progenitors, including embryonic radial glial cells, although animal models will be required to test this hypothesis.

3.4. Molecular characterization of MBs

Transcriptional profiling of messenger mRNAs in human medulloblastomas (Kool *et al.*, 2008; Thompson *et al.*, 2006) and more recently microRNAs profiling (Ferretti *et al.*, 2008; Northcott *et al.*, 2009) have identified four subgroups of medulloblastoma with distinct RNA signatures. It is likely that more tumor suppressors and possibly oncogenic mutations and amplifications will be identified as we continue to analyze an increasing number of tumors (reviewed in Lindsey *et al.*, 2005). Genetic and molecular genetic analyses of mouse and human medulloblastoma have identified changes in the regulation of signaling pathways that are important for cerebellar progenitor cell neurogenesis, including pathways that promote growth (SHH/PCTH, WNT, NOTCH, IGF/PTEN/mTOR) and negative growth regulators (BMP). Molecular changes in these pathways associated with MB include both loss of function and gain of function mutations, as well as alterations in the mRNA or protein levels of regulators of these pathways.

3.4.1. Sonic Hedgehog/Patched signaling

The Shh/Ptch signaling pathway is the most studied in medulloblastoma (Fig. 8.12). The mitogen Sonic hedgehog (Shh) drives proliferation of granule cerebellar progenitors (GCPs) by binding to the 12 transmembrane receptor Patched (Ptch) that exists in two forms, Ptch1 and Ptch2 (Hatten, 1999). In the absence of Shh, Ptch represses the function of Smoothened (Smo), a seven transmembrane G-protein-coupled receptor-like protein that activates the Gli1 and Gli2 transcription factors and inactivates the transcriptional repressor Gli3 that together regulate the transcriptional program in the cell nucleus (reviewed by Wechsler-Reya and Scott, 2001). The primary cilium, a cellular structure essential to cell proliferation and function, was recently found to concentrate components of Shh signaling in cerebellar GCPs and to be required for Shh signaling in GCP expansion (Fig. 8.13; Rohatgi *et al.*, 2007; Spassky *et al.*, 2008). While the three Gli transcription factors and their partner Suppressor of Fuse (SuFu) are present in the cilia together with Ptch1 in the absence of Shh stimulation, in the presence of the mitogen, Smo is recruited to the cilia while Ptch is degraded in the cytoplasm (Wen *et al.*, 2010).

Patients with Gorlin syndrome (also called Nevoid Basal Cell Carcinoma Syndrome) sustain germline mutations in *PTCH1* and *SUFU* that predisposes them to multiple cancers including medulloblastomas (Hahn *et al.*, 1996; Johnson *et al.*, 1996; Pastorino *et al.*, 2009). Loss of function mouse models in which *Ptch1* is deleted in the germline or conditionally

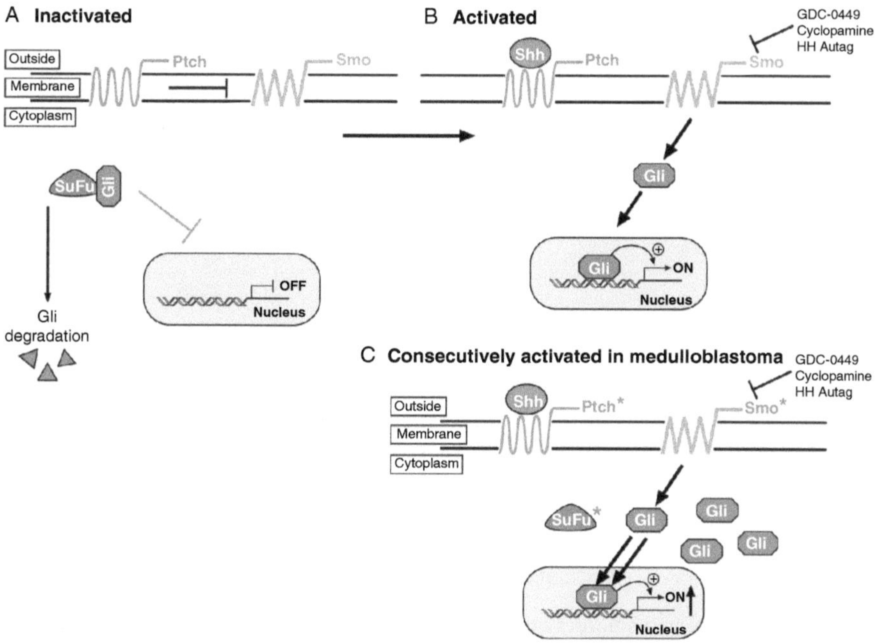

Figure 8.12 Schematic of the sonic hedgehog signaling pathway. (A) In the absence of Sonic hedgehog (Shh), the 12 transmembrane receptor Patched (Ptch) suppresses the seven transmembrane G-coupled receptor Smoothened (Smo) and the transcription factor Gli is sequestered into the cytoplasm by the Suppressor-of-Fuse (SuFu) preventing it from translocating to the nucleus and is degraded leading to a block of transcription. (B) In the presence of Shh, the suppression by Ptch is relieved and Smo activates Gli1 that is translocated to the nucleus and induces the transcription of Gli targets, including many proproliferative genes. (C) In medulloblastoma, loss of Ptch and mutations in Smo or SuFu lead to the constitutive activity of Smo, accumulation of Gli and increased transcription of proproliferative Gli-dependent targets. Asterisk (*) identifies proteins mutated in medulloblastoma.

demonstrates that loss of *Ptch1* induces medulloblastoma with histopathologic features of desmoplastic human MBs (Goodrich *et al.*, 1997; Lee *et al.*, 2003; Marino *et al.*, 2000; Oliver *et al.*, 2005). Molecular analysis of sporadic human MB primary samples revealed activation of the SHH/PTCH pathway from the loss of PTCH, and mutations in SUFU (Brugieres *et al.*, 2010), and SMO (reviewed by Rubin and Rowitch, 2002; Thompson *et al.*, 2006). Mice lacking *Sufu* develop medulloblastomas in conjunction with Trpp53 loss, demonstrating that *Sufu* functions as a tumor suppressor gene (Lee *et al.*, 2007).

The two Shh-dependent transcription factors, Gli1 and Gli2, activate the transcription of several proproliferative genes, including Mycn, cyclins D1 (Ccnd1), and D2 (Ccnd2), and cyclin E (Ccnde) (Fig. 8.5; Ciemerych

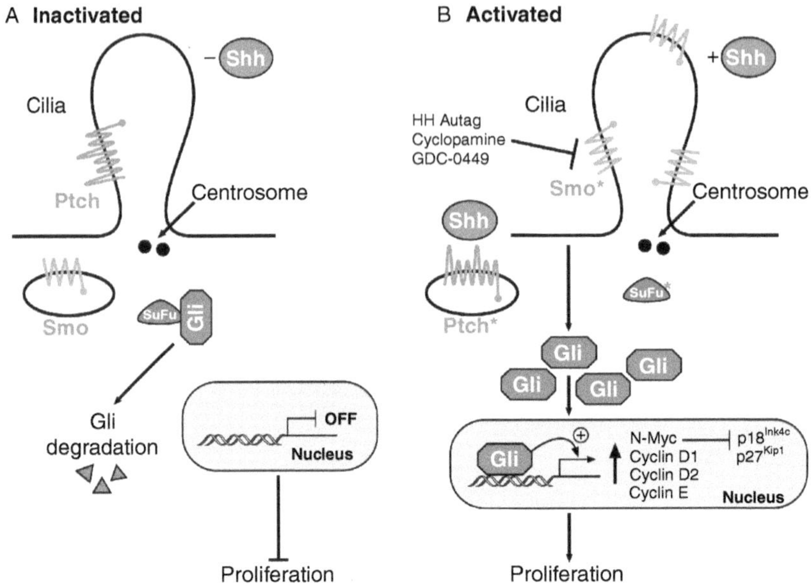

Figure 8.13 Schematic of the sonic hedgehog signaling pathway in the primary cilium. (A) In the absence of Shh, Ptch resides at the base of the cilium whereas Smo is in the cytoplasm and the pathway is off. (B) In the presence of Shh or in medulloblastoma, Smo becomes located to the cilium. In the absence of Ptch or when Smo or SuFu are mutated, Gli and the levels of targets increase. Asterisk (*) identifies proteins mutated in medulloblastoma.

et al., 2002; Hatton *et al.*, 2006; Kenney *et al.*, 2003; reviewed in Katoh and Katoh, 2009). In turn, Mycn, a basic-loop-helix (b-hlh) transcription factor induces the expression of many genes and microRNAs, among which is the miR-17~92 cluster (see below). Conversely, Mycn suppresses the expression of two cyclin-dependent kinase inhibitory proteins, $p18^{Ink4c}$ and $p27^{Kip1}$ to induce Rb and p107 phosphorylation and cell cycle progression (Fig. 8.14; Knoepfler *et al.*, 2002). While Rb and p107 are required for normal cerebellar development and GNP survival, their loss in mice in conjunction with loss of *Tpr53* induces MBs (Marino *et al.*, 2003). We found that in mice $p18^{Ink4c}$ is transiently expressed in GCPs to time their exit from the cell cycle, whereas $p27^{Kip1}$ is turned on in postmitotic GCPs in the inner EGL and in the IGL (Uziel *et al.*, 2005). Loss of *Ink4c* or *Kip1* indeed collaborates with the loss of *Ptch1* to induce medulloblastoma in mice (Uziel *et al.*, 2005). Similarly, *Ink4c* and the tumor suppressor *Trp53* co-repress the induction of tumors (Zindy *et al.*, 2003). Although *Trp53* mutations are seldom found in human MBs that represent only ~10%, Li-Fraumeni patients with familial *TP53* mutations are prone to medulloblastoma development (Malkin *et al.*, 1990; Shrivastava et al., 1990).

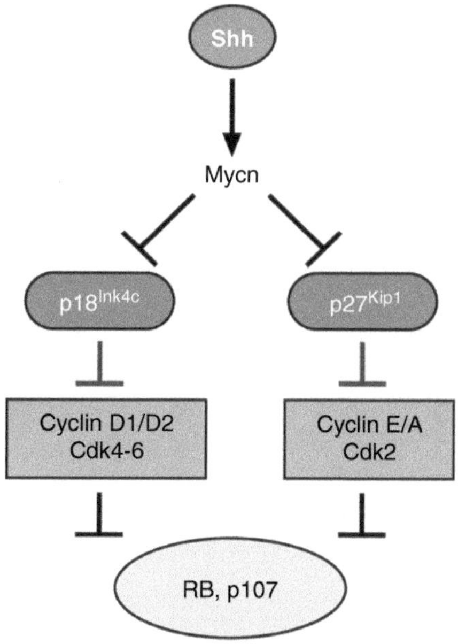

Figure 8.14 Regulation of Mycn and cell cycle proteins by sonic hedgehog signaling. Shh signaling directly induces the expression of Mycn and Cyclins D1, D2, and E. Mycn in turn suppresses the expression of two cyclin-dependent kinase (Cdk) inhibitory proteins, $p18^{Ink4c}$ and $p27^{Kip1}$. While $p18^{Ink4c}$ inhibits the activity of Cdk4 and Cdk6 in complex with cyclins D1 and D2, $p27^{Kip1}$ suppresses the activity of the Cyclin E/ and A/Cdk2 complexes. Cyclin/Cdk complexes phosphorylate Rb and p107 to allow the cells to progress through the first gap (G1) phase of the cell cycle and enter the DNA synthetic (S) phase when they replicate their DNA.

For reasons that are not entirely clear, GCPs are exquisitely sensitive to DNA damage and rely heavily on an intact *Trp53* pathway to eliminate cells that have acquired mutations during their division cycle. Several experiments illustrate this point: first, mouse models of medulloblastoma with a Shh signature were generated by mutations of genes in the DNA repair pathways, including Ligase 4, PARP1, Xrcc4 in combination with *Trp53* loss (Lee and McKinnon, 2002; Tong *et al.*, 2003; Yan *et al.*, 2006); second, 5 days old mice irradiated with nonlethal doses (~4 Gy) all develop medulloblastoma by 3 months of age (Zindy *et al.*, 2007).

Initial karyotyping and recent comparative genome hybridization (CGH) analysis revealed that the genes of the MYC family, MYC (C-MYC) on chromosome 8q24, Mycn on chromosome 2p, and MYCL1 on chromosome 1p are frequently amplified in medulloblastomas (reviewed in Northcott *et al.*, 2010). Indeed, enforced expression of Myc or Mycn in GCPs either from *Trp53*-null or *Ptch1* heterozygote mice induce medulloblastoma with full penetrance (100%) after orthotopic transplants in naïve recipient animals (Zindy *et al.*, 2007).

3.4.2. Collaboration between Shh and insulin-like growth factor

Shh and insulin-like growth factor (IGF) promote GCP growth and constitutive activation of the Shh/Ptch and IGF signaling pathways are found in medulloblastomas in mouse and man (Kenney *et al.*, 2004). IGF-I and IGF-II are potent survival factors that bind to IGF-receptors (IGF-R). IGFRs in turn activate PI3K, Akt, and mTOR, to promote mRNA translation (Fig. 8.15). Most growth factors activate mTOR by suppressing the tuberous sclerosis complex (Tsc) that include Tsc1 (hamartin) and Tsc2 (tuberin) that normally restrain mTOR activity. Tsc1 stabilizes Tsc2, and Tsc2 upregulates $p27^{Kip1}$ levels by preventing its degradation, promoting its nuclear localization and preventing cell cycle progression. Although Tsc inactivation alone has no effects, in collaboration with *Ptch1* mutation it drives MB formation in mice by increasing translation of mRNAs and possibly affecting $p27^{Kip1}$ localization (Bhatia *et al.*, 2009, 2010). Similar results were found using an *Igf-1* transgene which expression was driven in GNPs and in a *Ptch1* hererozygous background (Tanori *et al.*, 2010). Several inhibitors of the mTOR pathway (rapamycin) are currently being tested as a treatment for MBs.

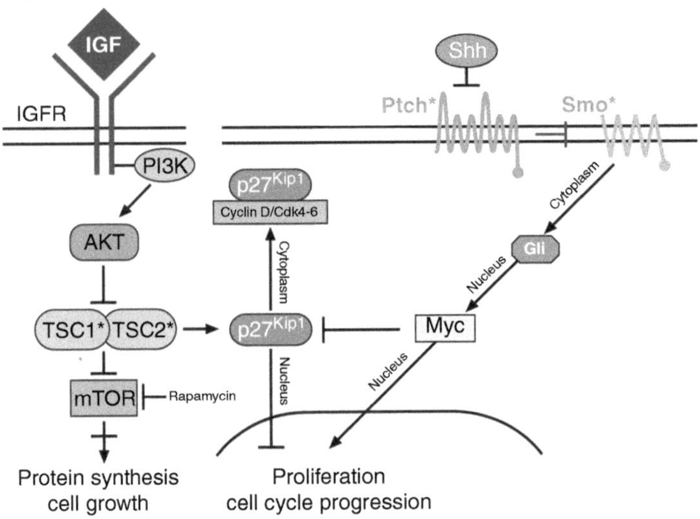

Figure 8.15 Collaboration between insulin-like growth factor and Shh signaling. IGF activates the insulin-like growth factor (IGF) receptor (IGFR) which in turns activates the phosphoinositol 3-kinase (PI3K), the AKT kinase which suppresses the tumor suppressor TSC1 and TSC2 proteins. TSC1 and TSC2 proteins induce p27Kip1 that prevent G1 progression and block mTOR which is required for protein synthesis and cell growth. Rapamycin blocks mTOR activity. Shh signaling in turn activates Mycn that suppresses $p27^{Kip1}$. Asterisk (*) identifies proteins mutated in medulloblastoma. Adapted from Bhat-ia *et al.* (2010).

3.4.3. Collaboration between Shh and Hippo pathways

The Hippo pathway is thought to limit cell proliferation and promote apoptosis in differentiating cells (reviewed in Saucedo and Edgar, 2007). Moreover, mutations in this pathway have been proposed to provide cancer cells with a competitive advantage in genetically mosaic tumors, thereby promoting proliferation. The Hippo pathway contains two serine-threonine kinases Mst1 and Mst2 that redundantly phosphorylate the serine-threonine kinases Lats1 or Lats2. Activity of these kinases is enhanced by scaffolding proteins Sav1 (Ww45) for Mst 1 and 2 and Mob1-4 for Lats proteins. Lats proteins phosphorylate two related transcriptional adaptor proteins Yes-associated protein (YAP) and TAZ. Phosphorylation of YAP and TAZ induces their interaction with the chaperone protein 14-3-3 and retention into the cytoplasm thus inhibiting their function. In the absence or the presence of reduced Hippo signaling, YAP/TAZ phosphorylation is absent or low, respectively, inducing their translocation to the nucleus where they interact with DNA binding TEA domain transcription factor TEAD (1–4), as well as several other transcription factors including RUNX2, Smad7, p73, and p53BP2 to activate the transcription of genes involved in cell proliferation and survival (reviewed in Saucedo and Edgar, 2007).

Recent studies implicate the Hippo pathway in medulloblastomas in concert with the Shh/Ptch signaling pathway (Fernandez *et al.*, 2009). Fernandez and collaborators found that YAP expression is regulated by Shh in granule neuron progenitors, that its translocation is mediated by Shh and that it regulates GCPs proliferation (Fig. 8.16). Consistent with its expression in GCPs, YAP is upregulated by overexpression and amplification

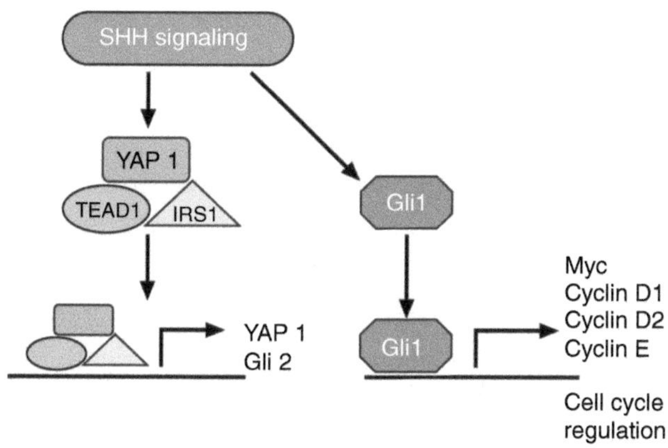

Figure 8.16 Collaboration between Shh and Hippo signaling. Shh regulates YAP1 in complex with TEAD and IRS1 that is translocated to the nucleus where it regulates YAP1 itself and Gli2. Shh signaling regulates Gli1 and its direct targets. Adapted from Fernandez *et al.* (2009).

of the locus on chromosome 11q22 in human medulloblastomas in which the Shh signaling pathway is activated. YAP is also highly overexpressed in human medulloblastomas with activation of the WNT signaling pathway. TEAD, the major partner of YAP1 is similarly overexpressed in human MBs with constitutive activation of the SHH or WNT signaling pathways. In a mouse medulloblastoma model that recapitulates the human tumors with a Shh "signature," YAP is expressed in the perivascular niche, a region of the tumor surrounding blood vessels, in which cancer stem cells reside. These YAP expressing cells were resistant to irradiation that otherwise eradicated all other cell types (Fernandez *et al.*, 2009). Since YAP may maintain "stemness" and protect tumor cells from DNA damage-induced apoptosis, it is a potential new therapeutic target in the treatment of medulloblastomas with Shh and Wnt pathway activation. However, experiments in animal models will be required to validate the function of YAP1 in MBs.

3.4.4. The Wnt signaling pathway

The canonical WNT signaling pathway is activated when Wnt binds to the seven transmembrane receptor Frizzled that activates the Disheveled protein which has two functions: one to block β-Catenin activation and nuclear localization and the other to regulate actin stress fibers (Fig. 8.17). β-Catenin protein is part of a complex of proteins containing the kinase GSK-3β, Axins 1 and 2, the adenomatous polyposis coli (APC) and Caseine kinase a, CK1a. β-Catenin levels are normally regulated by phosphorylation by GSK-3β leading to ubiquitination and proteasome-dependent degradation. Mutations in APC or in the phosphorylation site of β-Catenin that can no longer be degraded by the proteasome, induces the constitutive activation of nuclear β-Catenin that by binding to the transcription factors TCF/LEF induces unabated transcription of proproliferative target genes including cyclin D1 (Ccnd1) and c-Myc.

Deregulated WNT signaling occurs in ~10–15% of human medulloblastomas. It was first identified in patients with Turcot's syndrome in whom colon cancer and malignant neuroepithelial brain tumors including medulloblastomas were associated with mutations in *APC* (reviewed in Gilbertson and Ellison, 2008; Marino, 2005). Subsequently mutations in *BETA-CATENIN/CTNNB1* were also found in human medulloblastomas (Thompson *et al.*, 2006). Besides activated mutations of *B-CATENIN/CTNNB1*, these MBs harbor a single copy loss of chromosome 6, also called monosomy 6. Monosomy 6 represents the most significant prognostic factor for the WNT subgroup of tumors that correlates with a good outcome (Gajjar *et al.*, 2006). Finally, epigenetic silencing of the secreted frizzled-related proteins (SFRP) family of WNT inhibitors, SFRP-1,-2, and -3 is thought to elevate WNT signaling in MBs (Fig. 8.17). In addition, forced expression of these SFRP proteins reduces the proliferation and anchorage-independent growth of medulloblastoma cells, limits tumor

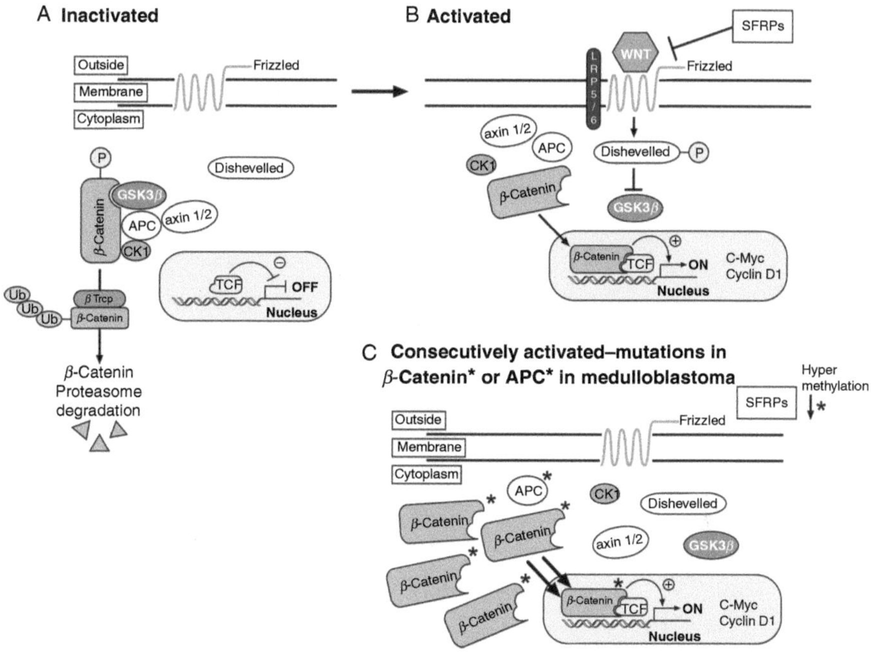

Figure 8.17 The Wnt signaling pathway. (A) In the absence of ligand, the transmembrane frizzled receptor is inactive and β-Catenin is in complex with APC, axin 1 and 2, caseine kinase 1 (CK1), and GSK-3beta. This enables GSK-3beta to phosphorylate β-Catenin that is then ubiquitinaed by the E3 ligase β-Trcp and degraded by the proteasome. (B) Upon binding of Wnt ligands, lrp5/6 and Frizzell are activated, disheveled is phosphorylated and blocks GSK-3β kinase activity. β-Catenin is no longer in a complex with APC, Axin, and CK1 and β-Catenin is translocated to the nucleus where it transcribed cyclin D1 and C-Myc. SFRPs prevent Wnt from binding to Frizzell and inhibit Wnt signaling. (C) In medulloblastomas, the Wnt pathway is constitutively activated leading to the accumulation of β-Catenin levels and increase transcription.

burden, and prolongs survival in xenografts of MB in mice (Kongkham *et al.*, 2010). The recent mouse model that recapitulates the WNT-subtype of human MB will provide a preclinical model for potential novel therapeutic intervention in patients afflicted with this subgroup of MB (Gibson *et al.*, 2010).

3.4.5. The Notch signaling pathway

Notch signaling is involved in many cellular processes, including stem cell renewal, cell fate specification, and proliferation. Intercellular Notch signaling is mediated by expression of the membrane-bound ligands Delta (1–4) and Jagged (Jag1, 2) on one cell and of the transmembrane receptors Notch (1–4) on the other cell (Fig. 8.18). Upon ligand binding, the

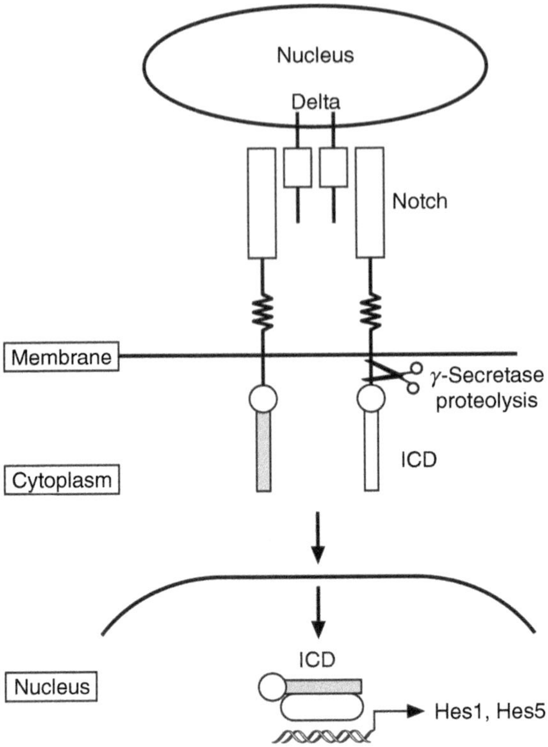

Figure 8.18 The Notch signaling pathway. Notch signaling is mediated by two cells; one providing the transmembrane ligands Delta and Jagged, the other the single transmembrane receptor, Notch, that transmits the signals. Upon ligand binding, the intracellular domain (ICD) is clipped by the γ-secretase at the membrane, is translocated to the nucleus where it enters in a complex with other proteins, and induces transcription of specific targets including the basic-helix-loop-helix transcription factors, Hes1 and Hes5.

intracellular domain (ICD) is clipped by the γ-secretase and translocated to the nucleus where it enters in complexes to activate the transcription of several targets, in particular the basic-helix-loop-helix (bHLH) transcription factors, Hes1 and Hes5. Activated Notch 2 stimulates GCP proliferation (Solecki *et al.*, 2001). Consistent with a role for NOTCH2 in GCP proliferation, high levels of NOTCH2 expression, but not of other NOTCH family members, was detected in human medulloblastoma (Fan *et al.*, 2004). In addition to Notch2, Hes1 and Jag1 were found to be highly expressed in mouse medulloblastoma models by several groups (Dakubo *et al.*, 2006; Hallahan *et al.*, 2004). However, genetic studies do not support a role for Notch1 in medulloblastoma formation in collaboration with the Shh/Ptch pathway (Julian *et al.*, 2010). Thus, components of the Notch1

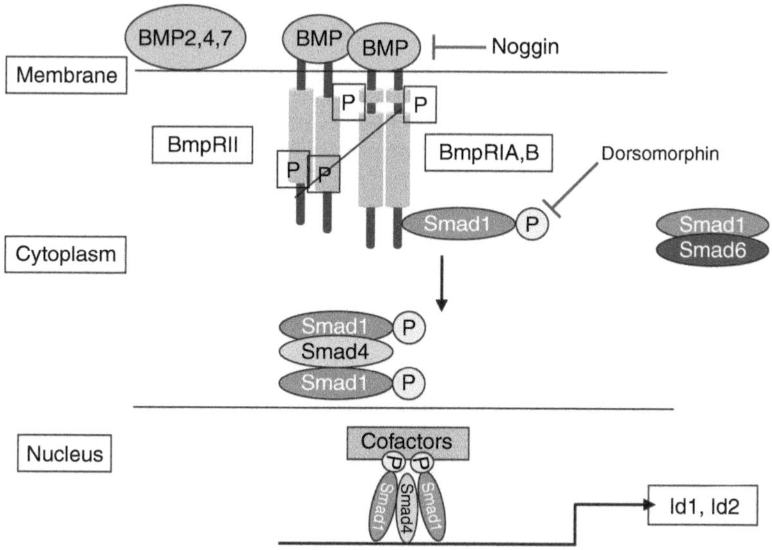

Figure 8.19 Bone morphogenic protein (BMP) signaling. BMP-2, 4, 7 bind to heterodimeric serine/threonine kinase receptors BMP-RII and BMP-RIA/B to induce their activation and *trans*-phosphorylation. This induces the recruitment of Smad1 that is sequestered by Smad6 in the cytoplasm in the absence of BMPs, and its phosphorylation. Phosphorylated Smad1 enters in complexes with Smad4 and is translocated to the nucleus where is induces the transcription of several genes including the inhibitors of DNA binding, Id1 and Id2.

and Notch2 signaling pathways may delineate important differences among subgroups of MBs.

3.4.6. The BMP2/4 signaling pathway

Unlike the growth promoting signaling pathways described above, the BMP signaling pathway inhibits proliferation and induces the differentiation of GCPs postnatally and GCP-like tumor cells. Bmps, 1–7, belong to the Transforming growth factor (TGF-β) family of cytokines (reviewed in Massague, 1996; Massague *et al.*, 2005). BMP-2,-4, and 7 bind as dimers to promote the heterodimerization and transphosphorylation of serine/threonine Bmp receptors type I (Alk3) and II (Alk6) (Fig. 8.19). Upon receptor activation, Smad1is recruited to and phosphorylated by the receptor type II (Alk6), inducing its conformational change and heterodimerization with Smad4. The heterodimer is then translocated to the nucleus where it directs the transcription of several genes including the Inhibitors of division (Ids), Id1-4, and TGF-β-early inducible gene 1 (Teig-1) (Fig. 8.19).

Several studies found that in postnatal cerebellum development and in MBs with a Shh signature Bmps oppose Shh-induced proliferation and

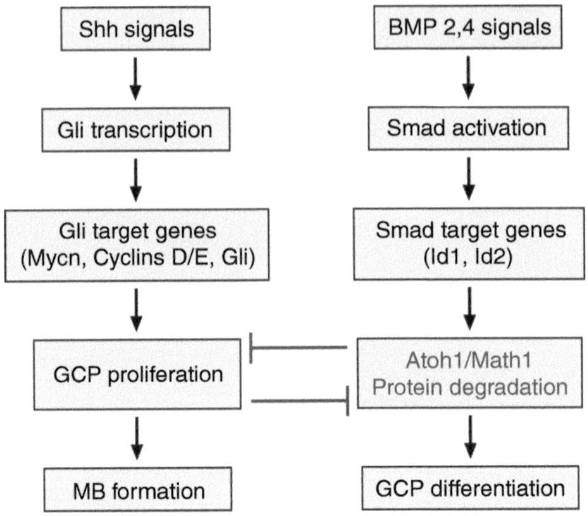

Figure 8.20 Shh and BMP signaling. Both signaling pathways antagonize each other to regulate proliferation of GCPs by regulating the levels of the bHLH transcription factor Atoh1/Math1. Whereas Shh induces proliferation, BMP inhibits proliferation and induces the irreversible differentiation of GCPs. BMP induces cell growth arrest by the rapid posttranscriptional downregulation of Atoh1/Math1 protein by the proteasome. In medulloblastomas, Atoh1/Math1 levels are high whereas the genes regulating the BMP pathway are downregulated.

induce irreversible neuronal differentiation. In 2004, Rios and collaborators showed that Bmp2 antigonizes Shh-dependent proliferation of GCPs (Rios *et al.*, 2004). In 2007, the same group found that Bmp2 antigonizes Shh-dependent proliferation by expressing Tieg-1 which directly represses *Mycn* transcription. Conversely, they found that Tieg-1 overexpression promotes cell cycle arrest and apoptosis of GCPs in the absence of other differentiation signals (Alvarez-Rodriguez *et al.*, 2007). In 2008, we found that cell cycle exit and terminal differentiation of GCPs and GCP-like tumor cells is mediated by the rapid posttranscriptional downregulation of Math1, a b-HLH- transcription factor that expression marks proliferating GNPs. Enforced Math1 expression in turn prevented differentiation (Fig. 8.20). Finally, Math1 was recently demonstrated to be essential for MB development (Flora *et al.*, 2010) and to collaborate with Gli1 to transform normal GCPs into tumor-initiating cells (Ayrault *et al.*, 2010).

Gene profiling of mouse and human MBs with a constitutively activated SHH pathway revealed that most genes within the BMP signaling pathway are downregulated in tumors compared to proliferating GCPs and neurons, whereas Math1 levels are high (Zhao *et al.*, 2008). Epigenetic studies of human medulloblastoma samples identified a tumor suppressor called

HIC1, for hypermethylated in cancer-1, the normal function of which is to suppress *MATH1*. In human medulloblastoma, the promoter of HIC1 is hypermethylated leading to inhibition of its expression and to the induction of high MATH1 protein levels (Lindsey *et al.*, 2004; reviewed by Briggs *et al.*, 2008a,b). Current CGH analysis confirmed that the most frequently affected chromosome in MBs with a SHH signature is 17 that encodes the tumor suppressor HIC1 (Ferretti *et al.*, 2005). However, it is likely that other genes are affected as the result of the frequent loss of heterozygocity on 17p.

3.5. Epigenetic silencing in MBs

Besides loss of function by gene deletions or mutations, epigenetic silencing also plays an important role in the development of MBs (reviewed in Lindsey *et al.*, 2005). Survey of the methylation status of tumor suppressors or oncogenes in human MBs has revealed several genes the disregulation of which is associated with MBs. They include SFRPs that negatively regulate the WNT pathway (Kongkham *et al.*, 2010), the S100 gene family (Lindsey *et al.*, 2007), the Kruppel-like factor 4 (Nakahara *et al.*, 2010), and HIC1 that regulates *MATH1* expression (Briggs *et al.*, 2008a,b). As these studies are expanded to an increasing number of primary tumors in the future, it is likely that more genes will be found to be regulated epigenetically and to be relevant to MB formation.

3.6. Other genes involved in medulloblastoma

A number of other genes have been associated with human medulloblastoma. However, the mechanism of action for these genes in MB or their association with specific subgroups of MBs remains to be determined. Among these, high expression of the neurotrophin TRKC receptor is associated with a good prognostic and a favorable outcome in patients with medulloblastoma. TRKC is found in desmoplastic nodular medulloblastoma and is suggested to be responsible for neuronal differentiation. Unlike TRKC, other proteins are correlated with poor prognosis and metastasis. These include ERBB2, a member of the epidermal growth factor receptor, the platelet-derived growth factor receptor, PDFGR and C-MET, the receptor tyrosine kinase for the hepatocyte growth factor (reviewed in Guessous *et al.*, 2008).

In addition, $\text{REN}^{\text{KCTD11}}$ that maps to 17p13.2 and is deleted in human medulloblastoma with a SHH "signature" is a suppressor of Hedgehog signaling and was shown by the same group to regulate proliferation of GCPs (Argenti *et al.*, 2005; Di Marcotullio *et al.*, 2006). On the other hand, expression of OTX1 and OTX2, two developmentally regulated transcription factors was found to correlate with two subgroups of medulloblastoma;

whereas overexpression of OTX1 was found in nodular/desmoplastic human tumors, OTX2 corresponded to the classic subgroup in man and mice (Adamson *et al.*, 2010; de Haas *et al.*, 2006).

3.7. MicroRNAs and medulloblastoma

In the last 10 years, microRNAs and long noncoding RNAs have emerged as novel regulators of many aspects of cellular biology and development. MicroRNAs are small noncoding 22 nucleotides long RNAs the sequence of which is complementary to those in the 3′-UTR of targeted messenger RNAs. MicroRNAs are thought to regulate hundreds of mRNA targets (reviewed in Bartel, 2004).

MicroRNAs are encoded on human and mouse chromosomes as long pri-miRNAs that are processed in the nucleus as pre-miRNAs by a processing enzyme called Drosha (Fig. 8.21). Pre-miRNAs are processed by Dicer in the cytoplasm as mature microRNAs that form complexes, called RISC, with argonaute proteins. The RISC complex recognizes the complementary sequences on the messenger RNA targets within its 3′ end, 5′ end and often coding sequences. Several microRNAs were found to be overexpressed and possibly act as oncogenes in tumors whereas others were shown to be underexpressed in tumors suggesting that their expression might suppress tumor development (reviewed in Garzon *et al.*, 2006). As such, microRNAs are being considered as targets for therapy.

Two groups identified the microRNA *miR-17∼92* cluster as an oncogene in MBs with a Shh signature (Northcott *et al.*, 2009; Uziel *et al.*, 2009). The *mir-17∼92* cluster belongs to a family of a three clusters encoded by different chromosomes in mouse and man (Ventura *et al.*, 2008) (Fig. 8.22). It encodes six unique microRNAs, *miR-17, 18, 19, 20,* and *92*, that share common seed sequences with microRNAs encoded by the two other clusters of the family, *miR-106b∼25* and *miR-106a-363*. *MiR-17∼92* was found to be a direct target of Myc and Mycn (O'Donnell *et al.*, 2005). Consistent with this finding, the *miR-17∼92* cluster is found at high levels in human medulloblastoma with high expression or amplification of MYC or Mycn. Although several important targets of the *miR-17∼92* cluster were found in B cells including E2F-1, PTEN, and Bim that all induce apoptosis (Fig. 8.23), specific bona fide targets in MBs have not yet been identified. Whereas ectopic expression of *miR-17∼92* in GCPs is not sufficient to drive tumorigenesis on its own, it does so in collaboration with *Ptch1* loss in an orthotopic transplant model in mice (Uziel *et al.*, 2009). If this cluster is required for MB development, antigomiRs to this cluster may prove to be an efficacious and novel therapeutic approach for MBs in which this cluster is overexpressed.

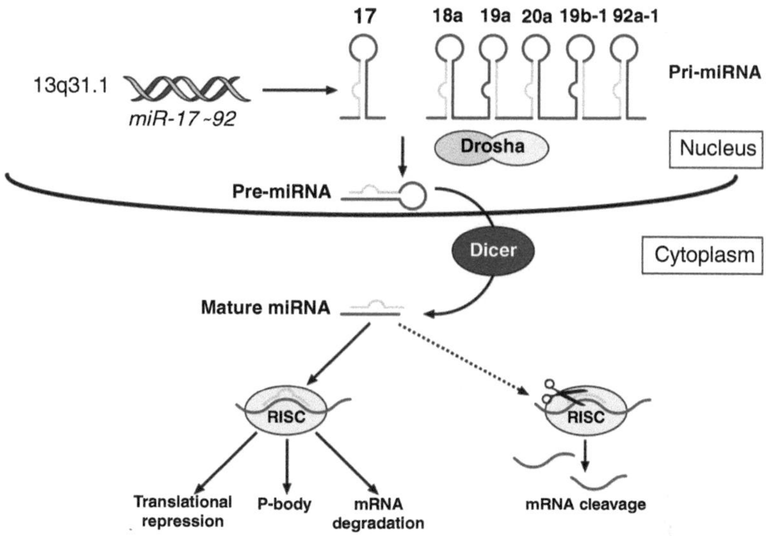

Figure 8.21 MicroRNA biogenesis and miR-17∼92 processing. MicroRNAs (miRs) are encoded by the genome on many chromosomes but do not encode proteins. The *miR-17∼92* cluster is encoded on human chromosome 13q31.1 as a long pri-miRNA and processed by the processing enzyme Drosha. Pre-miRNAs are further processed in mature miRs where they enter in a RISC complex with argonaute proteins. RISC complexes recognize the complementary sequence on the 3′ end, 5′ end or even the coding region of the messenger mRNA targets. This leads mainly to the inhibition of protein synthesis but the mRNA target could also be trapped in P bodies or degraded. Ultimately, gene expression is suppressed. Figure provided by Dr. Zindy.

3.8. Mouse models of medulloblastoma and preclinical testing of novel targeted therapies

Remarkably, all mouse models of medulloblastoma that have been developed to date mirror only one subgroup of human MBs, with a Shh signature, even though they were generated with different founding mutations (Fig. 8.24; Lee *et al.*, 2003). These include mutations in *Sufu*, *Smo*, loss of *Ptch1* alone or together with *Ink4c* or *Kip1*, *Rb*, hypermethylation of the *HIC1* promoter, loss of *Ligase 4*, *Xrcc4*, *Brca2*, and *PARP1*, often in combination with *Trp53* loss (reviewed in Behesti and Marino, 2009). In contrast, forced expression of Mycn, Myc, and *miR-17∼92*, induced MB development in combination with loss of *Ptch1* or loss of *Trp53* (Uziel *et al.*, 2009; Zindy *et al.*, 2007). Comparison of the gene profile of GCPs and tumors with that of differentiated neurons revealed that in all these mouse models, the cell of origin is a GCP (Lee *et al.*, 2003). Although TP^{53} mutations are not common in human MBs (∼10%), most of the mouse models, except those from *Ptch1*+/−, *Ink4c*−/− and *Ptch1*+/−, *Kip1*−/−, required the

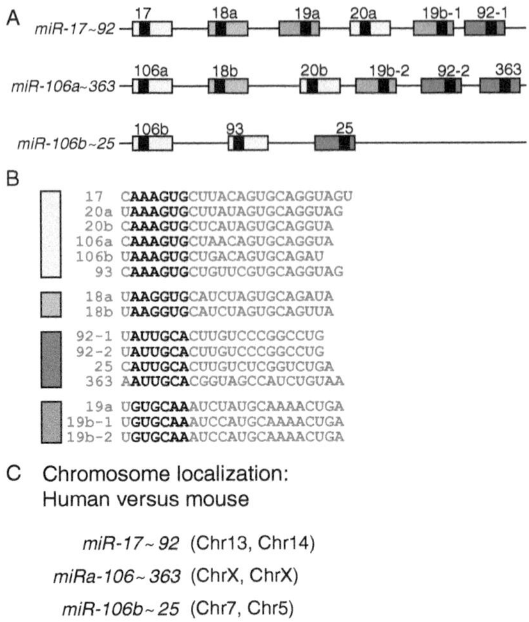

Figure 8.22 The *miR-17∼92* cluster family. The *miR-17∼92* cluster family comprises three members (A) encoding unique miRs that share sequence homology in the 6 mer "seed" sequence labeled in bold letters (B) and encoded in different chromosomes in human and mice (C). Adapted from Ventura *et al.* (2008).

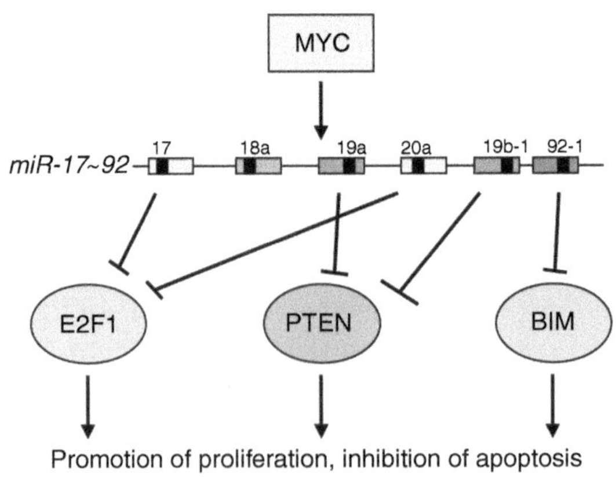

Figure 8.23 The *miR-17∼92* cluster is an MYC target. Myc directly binds to the promoter of the cluster to induce its expression. The *miR-17∼92* encodes six individual miRs that target different cell death-inducing proteins, including E2F-1, PTEN, and BIM. Figure provided by Dr. Zindy.

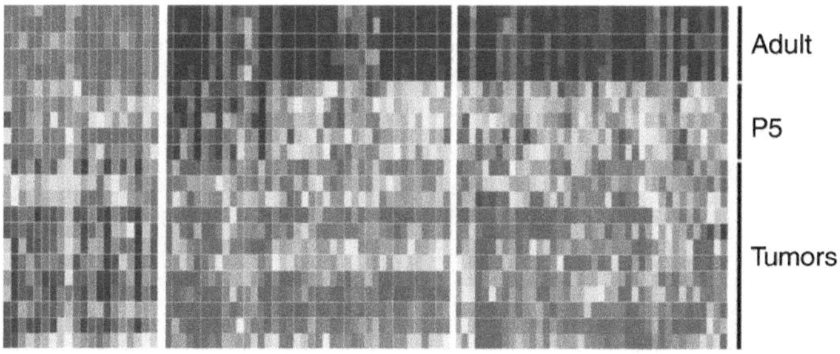

Figure 8.24 Gene profile of mouse genes in the cerebellum and medulloblastoma. Upregulated (light gray) and downregulated (dark gray) gene expressions are compared between adult mouse cerebella, GCPs purified at postnatal day P5 and in medulloblastomas spontaneously arisen from genetically engineered mice, including *Ligase4−/−*, *p53−/−*, *Ptch1+/−*, *Ptch1+/−*, *Trp53−/−*, *Trp53−/−*, *Ink4c−/−*. Note that the profile of gene expression is similar between GCPs from P5 cerebella and tumors consistent with the GCPs as the cell of origin for this MB variant. Adapted from Lee *et al.* (2003).

loss of Trp53. This may reflect the extreme sensitivity of GCPs to DNA damage that must rely on the Trp53 checkpoint to eliminate all cells that fail to properly replicate their DNA. It is plausible that mutations upstream of downstream of Trp53 affect its function without a requirement for its elimination in human tumors.

These recent findings suggest that successful development of mouse models that recapitulate the two other subgroups of human MBs, will require a better understanding of the location and identity of the GCPs as well as the alterations in gene or/and microRNAs associated with these specific tumor subgroups. This will enable the targeting of the right mutations into the right cell type at the right time.

3.9. What does the future hold?

An understanding of the Shh signaling pathway and the development of a mouse model that recapitulates the human disease has led to a therapeutic drug targeting the Shh pathway currently in clinical trial. A better understanding of the signaling pathways that are disrupted in each subgroup of medulloblastoma will enable the development of mouse models for all subgroups of human medulloblastomas and the screening of small molecules that ultimately will be used as novel therapeutic drugs. The complete sequencing of human primary medulloblastoma samples and the complete molecular analysis of these tumors will provide additional information that hopefully will lead to better targeted therapies.

ACKNOWLEDGMENTS

We thank Joshua Stokes from Biomedical Communication for the drawing of the figures; Dr. Eve-Ellen Govek for reading the chapter; and Drs. David Ellison, Richard Gilbertson, Frederique Zindy, Jane Johnson, and Carol A. Mason for providing illustrations used in this chapter. This work was funded by NIH grants CA-096832 (M.F.R), NINDS R01 NS051778-05 (M. E. H), the Children's Brain Tumor Foundation (M.F.R), the Pediatric Brain Tumor Foundation (M. F. R), the American Brain Tumor Foundation (M.F.R), the Starr Cancer Consortium (M. E. H), and the American Lebanese-Syrian Associated Charities (ALSAC) of St. Jude Children's Research Hospital (M.F.R).

REFERENCES

Adams, N. C., Tomoda, T., Cooper, M., Dietz, G., and Hatten, M. E. (2002). Mice that lack astrotactin have slowed neuronal migration. *Development* **129,** 965–972.

Adamson, D. C., Shi, Q., Wortham, M., Northcott, P. A., *et al.* (2010). OTX2 is critical for the maintenance and progression of Shh-independent medulloblastoma. *Cancer Res.* **70,** 181–191.

Alder, J., Cho, N. K., and Hatten, M. E. (1996). Embryonic precursor cells from the rhombic lip are specified to a cerebellar granule neuron identity. *Neuron* **17,** 389–399.

Alder, J., Lee, K., Jessell, T., and Hatten, M. E. (1999). Generation of cerebellar granule neurons in vivo by transplantation of BMP-treated neural progenitor cells. *Nat. Neurosci.* **2,** 535–540.

Aldinger, K. A., and Elsen, G. E. (2008). Ptf1a is a molecular determinant for both glutamatergic and GABAergic neurons in the hindbrain. *J. Neurosci.* **28,** 338–339.

Alvarez-Rodriguez, R., Barzi, M., Berenguer, J., and Pons, S. (2007). Bone morphogenetic protein 2 opposes Shh-mediated proliferation in cerebellar granule cells through a TIEG-1-based regulation of Nmyc. *J. Biol. Chem.* **282,** 37170–37180.

Andersen, B. B., Korbo, L., and Pakkenberg, B. (1992). A quantitative study of the human cerebellum with unbiased stereological techniques. *J. Comp. Neurol.* **326,** 549–560.

Anthony, T. E., and Heintz, N. (2008). Genetic lineage tracing defines distinct neurogenic. *Neural Dev.* **5,** 3–30.

Anthony, T. E., Klein, C., Fishell, G., and Heintz, N. (2004). Radial glia serve as neuronal progenitors in all regions of the central nervous system. *Neuron* **41,** 881–890.

Anthony, T. E., Mason, H. A., Gridley, T., Fishell, G., and Heintz, N. (2005). Brain lipid-binding protein is a direct target of Notch signaling in radial glial cells. *Genes Dev.* **19,** 1028–1033.

Argenti, B., Gall, R., Di Marcotulllio, L., *et al.* (2005). Hedgehog antagonist REN^{KCTD11} regulated proliferation and apoptosis of developing granule cell progenitors. *J. Neurosci.* **25,** 8338–8346.

Aruga, J. (2004). The role of Zic genes in neural development. *Mol. Cell. Neurosci.* **26,** 205–221.

Aruga, J., Nagai, T., Tokuyama, T., Hayashizaki, Y., Okazaki, Y., Chapman, V. M., and Mikoshiba, K. (1996). The mouse zic gene family. Homologues of the Drosophila pair-rule gene odd-paired. *J. Biol. Chem.* **271,** 1043–1047.

Ayad, N. G., Rankin, S., Ooi, D., Rape, M., and Kirschner, M. W. (2005). Identification of ubiquitin ligase substrates by in vitro expression cloning. *Methods Enzymol.* **399,** 404–414.

Ayrault, O., Zindy, F., Rehg, J., Sherr, C. J., and Roussel, M. F. (2009). Two tumor suppressors, p27Kip1 and patched-1, collaborate to prevent medulloblastoma. *Mol. Cancer Res.* **7,** 33–40.

Ayrault, O., Zhao, H., Zindy, F., Qu, C., Sherr, C. J., and Roussel, M. F. (2010). Atoh1 inhibits neuronal differentiation and collaborates with Gli2 to generate medulloblastoma-initiating cell. *Cancer Res.* **70,** 5618–5627.

Bailey, P., and Cushing, H. (1925). Medulloblastoma cerebelli: A common type of midcerebellar glioma of childhood. *Arch. Neurol. Psychiatry* **14,** 192–223.

Bailey, P., and Cushing, H. (1926). A Classification of the Tumors of the Glioma Group on a Histogenetic Basis with a Correlated Study of Prognosis. Lippincott, Philadelphia.

Bartel, D. P. (2004). MicroRNAs: Genomics, biogenesis, mechanisms and function. *Cell* **116,** 281–297.

Behesti, H., and Marino, S. (2009). Cerebellar granule cells: Insights into proliferation, differentiation, and role in medulloblastoma pathogenesis. *Int. J. Biochem. Cell Biol.* **41,** 435–445.

Ben-Arie, N., McCall, A., Berkman, S., Eichele, G., Bellen, H., and Zoghbi, H. (1996). Evolutionary conservation of sequence and expression of the bHLH protein atonal suggests a conserved role in neurogenesis. *Hum. Mol. Genet.* **5,** 1207–1216.

Ben-Arie, N., Bellen, H. J., Armstrong, D. L., McCall, A. E., Gordadze, P. R., Guo, Q., Matzuk, M. M., and Zoghbi, H. Y. (1997). Math1 is essential for genesis of cerebellar granule neurons. *Nature* **390,** 169–172.

Berman, D. M., Karhadkar, S. S., Hallahan, A. R., Pritchard, J. I., Eberhart, C. G., Watkins, D. N., Chen, J. K., Cooper, M. K., Taipale, J., Olson, J. M., and Beachy, P. A. (2002). Medulloblastoma growth inhibition by hedgehog pathway blockade. *Science* **297,** 1559–1561.

Bhatia, B., Northcott, P. A., Hambardzumyan, D., Govindarajan, B., Brat, D. J., Arbiser, J. L., Holland, E. C., Taylor, M. D., and Kenney, A. M. (2009). Tuberous sclerosis complex suppression in cerebellar development and medulloblastoma: Separate regulation of mammalian target of rapamycin activity and p27 Kip1 localization. *Cancer Res.* **69,** 7224–7234.

Bhatia, B., Nahle, Z., and Kenney, A. M. (2010). Double trouble: When sonic hedgehog signaling meets TSC inactivation. *Cell Cycle* **9,** 456–459.

Blaess, S., Corrales, J. D., and Joyner, A. L. (2006). Sonic hedgehog regulates Gli activator and repressor functions with spatial and temporal precision in the mid/hindbrain region. *Development* **133,** 1799–1809.

Borghesani, P. R., Peyrin, J. M., Klein, R., Rubin, J., Carter, A. R., Schwartz, P. M., Luster, A., Corfas, G., and Segal, R. A. (2002). BDNF stimulates migration of cerebellar granule cells. *Development* **129,** 1435–1442.

Boyden, E. S., Katoh, A., and Raymond, J. L. (2004). Cerebellum-dependent learning: The role of multiple plasticity mechanisms. *Ann. Rev. Neurosci.* **27,** 581–609.

Briggs, K. J., Corcoran-Schwartz, I. M., Zhang, W., Harcke, T., Devereux, W. L., Baylin, S. B., Eberhart, C. G., and Watkins, D. N. (2008a). Cooperation between the Hic1 and Ptch1 tumor suppressors in medulloblastoma. *Genes Dev.* **22,** 770–785.

Briggs, K. J., Eberhart, C. G., and Watkins, D. N. (2008b). Just say no to ATOH: How HIC1 methylation might predispose medulloblastoma to lineage addiction. *Cancer Res.* **68,** 8654–8656.

Brugieres, L., Pierron, G., Chompret, A., Paillerets, B. B., Di Rocco, F., Varlet, P., Pierre-Kahn, A., Caron, O., Grill, J., and Delattre, O. (2010). Incomplete penetrance of the predisposition to medulloblastoma associated with germ-line SUFU mutations. *J. Med. Genet.* **47,** 142–144.

Carletti, B., and Rossi, F. (2008). Neurogenesis in the cerebellum. *Neuroscientist* **14,** 91–100.

Chi, C. L., Martinez, S., Wurst, W., and Martin, G. R. (2003). The isthmic organizer signal FGF8 is required for cell survival in the prospective midbrain and cerebellum. *Development* **130,** 2633–2644.

Ciemerych, M. A., Kenney, A. M., Sicinska, E., Kalaszczynska, I., Bronson, R. T., Rowitch, D. H., Gardner, H., and Sicinski, P. (2002). Development of mice expressing a single D-type cyclin. *Genes Dev.* **16,** 3277–3289.

Corrales, J. D., Blaess, S., Mahoney, E. M., and Joyner, A. L. (2006). The level of sonic hedgehog signaling regulates the complexity of cerebellar foliation. *Development* **133,** 1811–1821.

Crossley, P. H., Martinez, S., and Martin, G. R. (1996). Midbrain development induced by FGF8 in the chick embryo. *Nature* **380,** 66–68.

Dakubo, G. D., Mazerolle, C. J., and Wallace, V. A. (2006). Expression of Notch and Wnt pathway components and activation of Notch signaling in medulloblastomas from heterozygous patched mice. *J. Neurooncol.* **79,** 221–227.

D'Arca, D., Zhao, X., Xu, W., Ramirez-Martinez, N. C., Iavarone, A., and Lasorella, A. (2010). Huwe1 ubiquitin ligase is essential to synchronize neuronal and glial differentiation in the developing cerebellum. *Proc. Natl. Acad. Sci. USA* **107,** 5875–5880.

De Haas, T., Oussoren, E., Grojkowska, W., *et al.* (2006). OTX1 and OTX2 expression correlated with the clinical pathologic classification of medulloblastoma. *J. Neuropathol. Exp. Neurol.* **65,** 176–186.

De Zeeuw, C. I., and Yeo, C. H. (2005). Time and tide in cerebellar memory formation. *Curr. Opin. Neurobiol.* **15,** 667–674.

Di Marcotullio, L., Ferretti, E., De Smaele, E., *et al.* (2006). REN^{KTCD11} is a suppressor of hedgehog signaling and is delted in human medulloblastoma. *Proc. Natl. Acad. Sci. USA* **29,** 10833–10838.

Dino, M. R., Schuerger, R. J., Liu, Y., Slater, N. T., and Mugnaini, E. (2000). Unipolar brush cell: A potential feedforward excitatory interneuron of the cerebellum. *Neuroscience* **98,** 625–636.

du Lac, S., Raymond, J. L., Sejnowski, T. J., and Lisberger, S. G. (1995). Learning and memory in the vestibulo-ocular reflex. *Annu. Rev. Neurosci.* **18,** 409–441.

Dymecki, S. M., and Tomasiewicz, H. (1998). Using Flp-recombinase to characterize expansion of Wnt1-expressing neural progenitors in the mouse. *Dev. Biol.* **201,** 57–65.

Eberhart, C. G., Kratz, J., Wang, Y., Summers, K., *et al.* (2004). Histopatholical and molecular prognostic markers in medulloblastoma: c-Myc, N-Myc, TrkC, and anaplasia. *J. Neuropathol. Exp. Neurol.* **63,** 441–449.

Ebert, P. J., Timmer, J. R., Nakada, Y., Helms, A. W., Parab, P. B., Liu, Y., Hunsaker, T. L., and Johnson, J. E. (2003). Zic1 represses Math1 expression via interactions with the Math1 enhancer and modulation of Math1 autoregulation. *Development* **130,** 1949–1959.

Edmondson, J. C., and Hatten, M. E. (1987). Glial-guided granule neuron migration in vitro: A high-resolution time-lapse video microscopic study. *J. Neurosci.* **7,** 1928–1934.

Espinosa, J. S., and Luo, L. (2008). Timing neurogenesis and differentiation: Insights from quantitative clonal analyses of cerebellar granule cells. *J. Neurosci.* **28,** 2301–2312.

Fan, X., Mikolaenko, I., Elhassan, I., Ni, X., Wang, Y., Ball, D., Brat, D. J., Perry, A., and Eberhart, C. G. (2004). Notch1 and notch2 have opposite effects on embryonal brain tumor growth. *Cancer Res.* **64,** 7787–7793.

Faust, P. L. (2003). Abnormal cerebellar histogenesis in PEX2 Zellweger mice reflects multiple neuronal defects induced by peroxisome deficiency. *J. Comp. Neurol.* **461,** 394–413.

Faust, P. L., and Hatten, M. E. (1997). Targeted deletion of the PEX2 peroxisome assembly gene in mice provides a model for Zellweger syndrome, a human neuronal migration disorder. *J. Cell Biol.* **139,** 1293–1305.

Feng, L., Hatten, M. E., and Heintz, N. (1994). Brain lipid-binding protein (BLBP): A novel signaling system in the developing mammalian CNS. *Neuron* **12,** 895–908.

Fernandez, L. A., Northcott, P. A., Dalton, J., Fraga, C., Ellison, D., Angers, S., Taylor, M. D., and Kenney, A. M. (2009). YAP1 is amplified and up-regulated in hedgehog-associated medulloblastomas and mediates Sonic hedgehog-driven neural precursor proliferation. *Genes Dev.* **23,** 2729–2741.

Ferretti, E., De Smaele, E., Di Marcotullio, L., Screpanti, I., and Gulino, A. (2005). Hedgehog checkpoints in medulloblastoma: The chromosome 17p deletion paradigm. *Trends Mol. Med.* **11,** 537–545.

Ferretti, E., De Smaele, E., Miele, E., Lavene, P., *et al.* (2008). Concerted microRNA control of Hedeghog signaling in cerebellar neuronal progenitor and tumour cell. *EMBO J.* **27,** 2616–2627.

Fiez, J. A. (1996). Cerebellar contributions to recognition. *Neuron* **16,** 13–15.

Fiez, J. A., and Petersen, S. E. (1998). Neuroimaging studies of word reading. *Proc. Natl. Acad. Sci. USA* **95,** 914–921.

Fink, A. J., Englund, C., Daza, R. A., Pham, D., Lau, C., Nivison, M., Kowalczyk, T., and Hevner, R. F. (2006). Development of the deep cerebellar nuclei: Transcription factors and cell migration from the rhombic lip. *J. Neurosci.* **26,** 3066–3076.

Finlay, J. L., Erdreich-Epstein, A., and Parker, R. J. (2007). Progress in the treatment of childhood brain tumors: No room for complacency. *Pediatr. Hematol. Oncol.* **24,** 79–84.

Flora, A., Klisch, T. J., Schuster, G., and Zoghbi, H. Y. (2010). Deletion of Atoh1 disrupts sonic hedgehog signaling in the developing cerebellum and prevents medulloblastoma. *Science* **326,** 1424–1427.

Fogarty, M. P., Kessker, J. D., and Wechsler-Reya, R. J. (2005). Morphing into cancer: The role of developmental signaling pathways in brain tumor formation. *J. Neurobiol.* **64,** 458–475.

Fogarty, M. P., Emmenegger, B. A., Grasfeder, L. L., Oliver, T. G., and Wechsler-Reya, R. J. (2007). Fibroblast growth factor blocks Sonic hedgehog signaling in neuronal precursors and tumor cells. *Proc. Natl. Acad. Sci. USA* **104,** 2973–2978.

Fujita, S. (1967). Quantitative analysis of cell proliferation and differentiation in the cortex of the postntal mouse cerebellum. *J. Cell Biol.* **32,** 277–287.

Fujita, S., Shimada, M., and Nakanuna, T. (1966). 3H-thymidine autoradiographic studies on the cell proliferation and differentiation in the external and internal granular layers of the mouse cerebellum. *J. Comp. Neurol.* **128,** 191–209.

Furley, A. J., Morton, S. B., Manalo, D., Karagogeos, D., Dodd, J., and Jessell, T. M. (1990). The axonal glycoprotein TAG-1 is an immunoglobulin superfamily member with neurite outgrowth-promoting activity. *Cell* **61,** 157–170.

Gajjar, A., Chintagumpala, M., Ashley, D., Kellie, S., Kun, L. E., Merchant, T. E., Woo, S., Wheeler, G., Ahern, V., Krasin, M. J., Fouladi, M., Broniscer, A., *et al.* (2006). Risk-adapted craniospinal radiotherapy followed by high-dose chemotherapy and stem-cell rescue in children with newly diagnosed medulloblastoma (St Jude Medulloblastoma-96): Long-term results from a prospective, multicentre trial. *Lancet Oncol.* **7,** 813–820.

Garzon, R., Fabbri, M., Cimmino, A., Calin, G. A., and Croce, C. M. (2006). MicroRNA expression and function in cancer. *Trends Mol. Med.* **12,** 580–587.

Gibson, P., Tong, Y., Robinson, G., Thompson, M., Currle, D. S., Eden, C., Hogg, T., Martin, J., Lee, Y., Zindy, F., Finkelstein, D., Pounds, S., *et al.* (2010). Subtypes of medulloblastoma have distinct developmental origins. *Nature.* [Epub ahead of print Dec 8].

Gilbertson, R. J., and Ellison, D. W. (2008). The origins of medulloblastoma subtypes. *Annu. Rev. Pathol.* **3,** 341–365.

Gong, S., Zheng, C., Doughty, M. L., Losos, K., Didkovsky, N., Schambra, U. B., Nowak, N. J., Joyner, A., Leblanc, G., Hatten, M. E., and Heintz, N. (2003). A gene expression atlas of the central nervous system based on bacterial artificial chromosomes. *Nature* **425,** 917–925.

Goodrich, L. V., Milenkovic, L., Higgins, K. M., and Scott, M. P. (1997). Altered neural cell fates and medulloblastoma in mouse patched mutants. *Science* **277,** 1109–1113.

Guessous, F., Li, Y., and Abounader, R. (2008). Signaling pathways in medulloblastoma. *J. Cell. Physiol.* **217,** 577–583.

Hahn, H., Wicking, C., Zaphiropoulous, P. G., Gailani, M. R., Shanley, S., Chidambaram, A., Vorechovsky, I., Holmberg, E., Unden, A. B., Gillies, S., Negus, K., Smyth, I., *et al.* (1996). Mutations of the human homolog of Drosophila patched in the nevoid basal cell carcinoma syndrome. *Cell* **85,** 841–851.

Hallahan, A. R., Pritchard, J. I., Hansen, S., Benson, M., Stoeck, J., Hatton, B. A., Russell, T. L., Ellenbogen, R. G., Bernstein, I. D., Beachy, P. A., and Olson, J. M. (2004). The SmoA1 mouse model reveals that notch signaling is critical for the growth and survival of sonic hedgehog-induced medulloblastomas. *Cancer Res.* **64,** 7794–7800.

Hallonet, M. E., Teillet, M. A., and Le Douarin, N. M. (1990). A new approach to the development of the cerebellum provided by the quail-chick marker system. *Development* **108,** 19–31.

Harmey, D., Smith, A., Simanski, S., Moussa, C. Z., and Ayad, N. G. (2009). The anaphase promoting complex induces substrate degradation during neuronal differentiation. *J. Biol. Chem.* **284,** 4317–4323.

Hatten, M. E. (1985). Neuronal regulation of astroglial morphology and proliferation in vitro. *J. Cell Biol.* **100,** 384–396.

Hatten, M. E. (1999). Expansion of CNS precursor pools: A new role for Sonic Hedgehog. *Neuron* **22,** 2–3.

Hatten, M. E. (2002). New directions in neuronal migration. *Science* **297,** 1660–1663.

Hatten, M. E., and Heintz, N. (1995). Mechanisms of neural patterning and specification in the developing cerebellum. *Annu. Rev. Neurosci.* **18,** 385–408.

Hatten, M. E., Mason, C. A., Liem, R. K., Edmondson, J. C., Bovolenta, P., and Shelanski, M. L. (1984). Neuron-astroglial interactions in vitro and their implications for repair of CNS injury. *Cent. Nerv. Syst. Trauma* **1,** 15–27.

Hatton, B. A., Knoepfler, P. S., Kenney, A. M., Rowitch, D. H., de Alboran, I. M., Olson, J. M., and Eisenman, R. N. (2006). N-myc is an essential downstream effector of Shh signaling during both normal and neoplastic cerebellar growth. *Cancer Res.* **66,** 8655–8661.

Helms, A. W., Abney, A. L., Ben-Arie, N., Zoghbi, H. Y., and Johnson, J. E. (2000). Autoregulation and multiple enhancers control Math1 expression in the developing nervous system. *Development* **127,** 1185–1196.

Hogan, B. L. (1996). Bone morphogenetic proteins: Multifunctional regulators of vertebrate development. *Genes Dev.* **10,** 1580–1594.

Hoshino, M., Nakamura, S., Mori, K., Kawauchi, T., Terao, M., Nishimura, Y. V., Fukuda, A., Fuse, T., Matsuo, N., Sone, M., Watanabe, M., Bito, H., *et al.* (2005). Ptf1a, a bHLH transcriptional gene, defines GABAergic neuronal fates in cerebellum. *Neuron* **47,** 201–213.

Ibanez-Tallon, I., Miwa, J. M., Wang, H. L., Adams, N. C., Crabtree, G. W., Sine, S. M., and Heintz, N. (2002). Novel modulation of neuronal nicotinic acetylcholine receptors by association with the endogenous prototoxin lynx1. *Neuron* **33,** 893–903.

Ito, M. (2006). Cerebellar circuitry as a neuronal machine. *Prog. Neurobiol.* **78,** 272–303.

Johnson, R. L., Rothman, A. L., Xie, J., Goodrich, L. V., Bare, J. W., Bonifas, J. M., Quinn, A. G., Myers, R. M., Cox, D. R., Epstein, E. H., Jr., and Scott, M. P. (1996).

Human homolog of patched, a candidate gene for the basal cell nevus syndrome. *Science* **272,** 1668–1671.

Joyner, A. L. (1996). Engrailed, Wnt and Pax genes regulate midbrain–hindbrain development. *Trends Genet.* **12,** 15–20.

Joyner, A. L., Skarnes, W. C., and Rossant, J. (1989). Production of a mutation in mouse En-2 gene by homologous recombination in embryonic stem cells. *Nature* **338,** 153–156.

Joyner, A. L., Liu, A., and Millet, S. (2000). Otx2, Gbx2 and Fgf8 interact to position and maintain a mid-hindbrain organizer. *Curr. Opin. Cell Biol.* **12,** 736–741.

Julian, E., Dave, R. K., Robson, J. P., Hallahan, A. R., and Wainwright, B. J. (2010). Canonical Notch signaling is not required for the growth of Hedgehog pathway-induced medulloblastoma. *Oncogene* **29,** 3465–3476.

Katoh, Y., and Katoh, M. (2009). Hedgehog target genes: Mechanisms of carcinogenesis induced by aberrant hedgehog signaling activation. *Curr. Mol. Med.* **9,** 873–886.

Kenney, A. M., and Rowitch, D. H. (2000). Sonic hedgehog promotes G(1) cyclin expression and sustained cell cycle progression in mammalian neuronal precursors. *Mol. Cell. Biol.* **20,** 9055–9067.

Kenney, A. M., Cole, M. D., and Rowitch, D. H. (2003). Nmyc upregulation by sonic hedgehog signaling promotes proliferation in developing cerebellar granule neuron precursors. *Development* **130,** 15–28.

Kenney, A. M., Widlund, H. R., and Rowitch, D. H. (2004). Hedgehog and PI-3 kinase signaling converge on Nmyc1 to promote cell cycle progression in cerebellar neuronal precursors. *Development* **131,** 217–228.

Kim, J., Tang, J. Y., Gong, R., Lee, J. J., Clemons, K. V., Chong, C. R., Chang, K. S., Fereshteh, M., Gardner, D., Reya, T., Liu, J. O., Epstein, E. H., *et al.* (2010a). Itraconazole, a commonly used antifungal that inhibits Hedgehog pathway activity and cancer growth. *Cancer Cell* **17,** 388–399.

Kim, J., Lee, J., Kim, J., Gardner, D., and Beachy, P. A. (2010b). Arsenic antigonizes the Hedgehog pathway by preventing ciliary accumulation and reducing stability of the Gli2 transcriptional effector. *Proc. Natl. Acad. Sci.* USA **107,** 13432–13437.

Kimura, H., Stephen, D., Joyner, A., and Curran, T. (2005). Gli1 is important for medulloblastoma formation in Ptc1+/− mice. *Oncogene* **24,** 4026–4036.

Knoepfler, P. S., Cheng, P. F., and Eisenman, R. N. (2002). N-myc is essential during neurogenesis for the rapid expansion of progenitor cell populations and the inhibition of neuronal differentiation. *Genes Dev.* **16,** 2699–2712.

Kongkham, P. N., Northcott, P. A., Croul, S. E., Smith, C. A., Taylor, M. D., and Rutka, J. T. (2010). The SFRP family of WNT inhibitors function as novel tumor suppressor genes epigenetically silenced in medulloblastoma. *Oncogene* **29,** 3017–3024.

Kool, M., Koster, J., Bunt, J., Hasselt, N. E., Lakeman, A., van Sluis, P., Troost, D., Meeteren, N. S., Caron, H. N., Cloos, J., Mrsic, A., Ylstra, B., *et al.* (2008). Integrated genomics identifies five medulloblastoma subtypes with distinct genetic profiles, pathway signatures and clinicopathological features. *PLoS ONE* **3,** e3088.

Kuhar, S. G., Feng, L., Vidan, S., Ross, M. E., Hatten, M. E., and Heintz, N. (1993). Changing patterns of gene expression define four stages of cerebellar granule neuron differentiation. *Development* **117,** 97–104.

Laine, J., and Axelrad, H. (2002). Extending the cerebellar Lugaro cell class. *Neuroscience* **115,** 363–374.

Lange, W. (1975). Cell number and cell density in the cerebellar cortex of man and some other mammals. *Cell Tissue Res.* **157,** 115–124.

Lee, Y., and McKinnon, P. J. (2002). DNA ligase IV suppresses medulloblastoma formation. *Cancer Res.* **62,** 6395–6399.

Lee, J. K., Cho, J. H., Hwang, W. S., Lee, Y. D., Reu, D. S., and Suh-Kim, H. (2000). Expression of neuroD/BETA2 in mitotic and postmitotic neuronal cells during the development of nervous system. *Dev. Dyn.* **217,** 361–367.

Lee, Y., Miller, H. L., Jensen, P., Hernan, R., Connelly, M., Wetmore, C., Zindy, F., Roussel, M. F., Curran, T., Gilbertson, R. J., and McKinnon, P. J. (2003). A molecular fingerprint for medulloblastoma. *Cancer Res.* **63,** 5428–5437.

Lee, Y., Kawagoe, R., Sasai, K., Li, Y., Russell, H. R., Curran, T., and McKinnon, P. J. (2007). Loss of suppressor-of-fused function promotes tumorigenesis. *Oncogene* **26,** 6442–6447.

Levisohn, L., Cronin-Golomb, A., and Schmahmann, J. D. (2000). Neuropsychological consequences of cerebellar tumour resection in children: Cerebellar cognitive affective syndrome in a paediatric population. *Brain* **123**(Pt 5), 1041–1050.

Limperopoulos, C., Soul, J. S., Gauvreau, K., Huppi, P. S., Warfield, S. K., Bassan, H., Robertson, R. L., Volpe, J. J., and du Plessis, A. J. (2005). Late gestation cerebellar growth is rapid and impeded by premature birth. *Pediatrics* **115,** 688–695.

Lindsey, J. C., Lusher, M. E., Anderton, J. A., Bailey, S., Gilbertson, R. J., Pearson, A. D., Ellison, D. W., and Clifford, S. C. (2004). Identification of tumour-specific epigenetic events in medulloblastoma development by hypermethylation profiling. *Carcinogenesis* **25,** 661–668.

Lindsey, J. C., Anderton, J. A., Lusher, M. E., and Clifford, S. C. (2005). Epigenetic events in medulloblastoma development. *Neurosurg. Focus* **15,** E10.

Lindsey, J. C., Lusher, M. E., Anderton, J. A., Gilbertson, R. J., Ellison, D. W., and Clifford, S. C. (2007). Epigenetic deregulation of multiple S100 gene family members by differential hypomethylation and hypermethylation events in medulloblastoma. *Br. J. Cancer* **97,** 267–274.

Louis, D. N., Ohgaki, H., Wiestler, O. D., Cavenee, W. K., Burger, P. C., Jouvet, A., Scheithauer, B. W., and Kleihues, P. (2007). The 2007 WHO classification of tumours of the central nervous system. *Acta Neuropathol.* **114,** 97–109.

Machold, R., and Fishell, G. (2005). Math1 is expressed in temporally discrete pools of cerebellar rhombic-lip neural progenitors. *Neuron* **48,** 17–24.

Machold, R. P., Kittell, D. J., and Fishell, G. J. (2007). Antagonism between Notch and bone morphogenetic protein receptor signaling regulates neurogenesis in the cerebellar rhombic lip. *Neural Dev.* **2,** 5.

Malkin, D., Li, F. P., Strong, L. C., Fraumeni, J. F., Jr., Nelson, C. E., Kim, D. H., *et al.* (1990). Germline p53 mutations in a familial syndrome of breast cancer, sarcomas, and other neoplasms. *Science* **250,** 1233–1238.

Marino, S. (2005). Medulloblastoma: Developmental mechanisms out of control. *Trends Mol. Med.* **11,** 17–22.

Marino, S., Vooijs, M., van Der Gulden, H., Jonkers, J., and Berns, A. (2000). Induction of medulloblastomas in p53-null mutant mice by somatic inactivation of Rb in the external granular layer cells of the cerebellum. *Genes Dev.* **14,** 994–1004.

Marino, S., Hoogervoorst, D., Brandner, S., and Berns, A. (2003). Rb and P107 are required for normal cerebellar development and granule cell survival but nor Purkinje cell persistence. *Development* **130,** 3359–3368.

Martinez, S., Crossley, P. H., Cobos, I., Rubenstein, J. L., and Martin, G. R. (1999). FGF8 induces formation of an ectopic isthmic organizer and isthmocerebellar development via a repressive effect on Otx2 expression. *Development* **126,** 1189–2000.

Massague, J. (1996). TGFbeta signaling: Receptors, transducers, and Mad proteins. *Cell* **85,** 947–950.

Massague, J., Seoane, J., and Wotton, D. (2005). Smad transcripton factors. *Genes Dev.* **19,** 2783–2810.

McMahon, A. P., and Bradley, A. (1990). The Wnt-1 (int-1) proto-oncogene is required for development of a large region of the mouse brain. *Cell* **62,** 1073–1085.
Miale, I., and Sidman, R. L. (1961). An autoradiographic analysis of histogenesis in the mouse cerebellum. *Exp. Neurol.* **4,** 277–296.
Millonig, J. H., Millen, K. J., and Hatten, M. E. (2000). The mouse Dreher gene Lmx1a controls formation of the roof plate in the vertebrate CNS. *Nature* **403,** 764–769.
Miwa, J. M., Ibanez-Tallon, I., Crabtree, G. W., Sancxhez, R., Sali, A., Role, L. W., and Heintz, N. (1999). lynx1, an endogenous toxin-like modulator of nicotinic acetylcholine receptors in the mammalian CNS. *Neuron* **23,** 105–114.
Miwa, J. M., Stevens, T. R., King, S. L., Caldarone, B. J., Ibanez-Tallon, I., Xiao, C., Fitzsimonds, R. M., Pavlides, C., Lester, H. A., Picciotto, M. R., and Heintz, N. (2006). The prototoxin lynx1 acts on nicotinic acetylcholine receptors to balance neuronal activity and survival in vivo. *Neuron* **51,** 587–600.
Morales, D., and Hatten, M. E. (2006). Molecular markers of neuronal progenitors in the embryonic cerebellar anlage. *J. Neurosci.* **26,** 12226–12236.
Nakahara, Y., Northcott, P. A., Li, M., Kongkham, P. N., *et al.* (2010). Genetic and epigentic inactivation of Kruppel-like factor 4 in medulloblastoma. *Neoplasia* **12,** 20–27.
Northcott, P. A., Fernandez, L. A., Hagan, J. P., Ellison, D. W., Grajkowska, W., Gillespie, Y., Grundy, R., Van Meter, T., Rutka, J. T., Croce, C. M., Kenney, A. M., and Taylor, M. D. (2009). The miR-17/92 polycistron is up-regulated in sonic hedgehog-driven medulloblastomas and induced by N-myc in sonic hedgehog-treated cerebellar neural precursors. *Cancer Res.* **69,** 3249–3255.
Northcott, P. A., Rutka, J. T., and Taylor, M. D. (2010). Genomics of medulloblastoma: From Giemsa-banding to next-generation sequencing in 20 years. *Neurosurg. Focus* **28** (E6), 1–20.
O'Donnell, K. A., Wentzel, E. A., Zeller, K. I., Dang, C. V., and Mendell, J. T. (2005). C-Myc-regulated microRNAs modulate E2F1 expression. *Nature* **435,** 839–843.
Oliver, T. G., Read, T. A., Kessler, J. D., Mehmeti, A., Wells, J. F., Huynh, T. T., Lin, S. M., and Wechsler-Reya, R. J. (2005). Loss of patched and disruption of granule cell development in a pre-neoplastic stage of medulloblastoma. *Development* **132,** 2425–2439.
Packer, R. J., Cogen, P., Vezina, G., and Rorke, L. B. (1999). Medulloblastoma: Clinical and biologic aspects. *Neuro-oncology* **1,** 232–250.
Palay, S. L., and Chan-Palay, S. (1974). The Cerebellar Cortex: Cytology and Organization. Springer-Verlag, New York.
Pan, N., Jahan, I., Lee, J. E., and Fritzsch, B. (2009). Defects in the cerebella of conditional Neurod1 null mice correlate with effective Tg(Atoh1-cre) recombination and granule cell requirements for Neurod1 for differentiation. *Cell Tissue Res.* **337,** 407–428.
Pascual, M., Abasolo, I., Mingorance-Le Meur, A., Martinez, A., Del Rio, J. A., Wright, C. V., Real, F. X., and Soriano, E. (2007). Cerebellar GABAergic progenitors adopt an external granule cell-like phenotype in the absence of Ptf1a transcription factor expression. *Proc. Natl. Acad. Sci. USA* **104,** 5193–5198.
Pastorino, L., Ghiorzo, P., Nasti, S., Battistuzzi, L., Cusano, R., Marzocchi, C., Garre, M. L., Clementi, M., and Scarra, G. B. (2009). Identification of a SUFU germline mutation in a family with Gorlin syndrome. *Am. J. Med. Genet. A* **149A,** 1539–1543.
Pizer, B. L., and Clifford, S. C. (2009). The potential impact of tumour biology on improved clinical practice for medulloblastoma: Progress towards biologically driven clinical trials. *Br. J. Neurosurg.* **23,** 364–375.
Pogoriler, J., Millen, K., Utset, M., and Du, W. (2006). Loss of cyclin D1 impairs cerebellar development and suppresses medulloblastoma formation. *Development* **133,** 3929–3937.
Pomeroy, S. L., Tamayo, P., Gaasenbeek, M., Sturla, L. M., Angelo, M., McLaughlin, M. E., Kim, J. Y., Goumnerova, L. C., Black, P. M., Lau, C.,

Allen, J. C., Zagzag, D., *et al.* (2002). Prediction of central nervous system embryonal tumour outcome based on gene expression. *Nature* **415,** 436–442.

Qin, L., Wine-Lee, L., Ahn, K. J., and Crenshaw, E. B., 3rd (2006). Genetic analyses demonstrate that bone morphogenetic protein signaling is required for embryonic cerebellar development. *J. Neurosci.* **26,** 1896–1905.

Rakic, P. (1971). Neuron-glia relationship during granule cell migration in developing cerebellar cortex. A Golgi and electronmicroscopic study in Macacus Rhesus. *J. Comp. Neurol.* **141,** 283–312.

Ramon y Cajal, S. (1995). Histology of the Nervous System of Man and Vertebrates. Oxford University Press, Oxford.

Rankin, S., Ayad, N. G., and Kirschner, M. W. (2005). Sororin, a substrate of the anaphase-promoting complex, is required for sister chromatid cohesion in vertebrates. *Mol. Cell* **18,** 185–200.

Rios, I., Alvarez-Rodriguez, R., Marti, E., and Pons, S. (2004). Bmp2 antagonizes sonic hedgehog-mediated proliferation of cerebellar granule neurones through Smad5 signalling. *Development* **131,** 3159–3168.

Rohatgi, R., Milenkovic, L., and Scott, M. P. (2007). Patched 1 regulates hedgehog signaling at the primary cilium. *Science* **317,** 372–376.

Romer, J., and Curran, T. (2005). Targeting medulloblastoma: Small-molecule inhibitors of the Sonic Hedgehog pathway as potential cancer therapeutics. *Cancer Res.* **65,** 4975–4978.

Rubin, J. B., and Rowitch, D. H. (2002). Medulloblastoma: A problem of developmental biology. *Cancer Cell* **2,** 7–8.

Rubin, J. B., Kung, A. L., Klein, R. S., Chan, J. A., Sun, Y., Schmidt, K., Kieran, M. W., Luster, A. D., and Segal, R. A. (2003). A small-molecule antagonist of CXCR4 inhibits intracranial growth of primary brain tumors. *Proc. Natl. Acad. Sci. USA* **100,** 13513–13518.

Rudin, C. M., Hann, C. L., Laterra, J., Yauch, R. L., Callahan, C. A., Fu, L., Holcomb, T., Stinson, J., Gould, S. E., Coleman, B., LoRusso, P. M., Von Hoff, D. D., *et al.* (2009). Treatment of medulloblastoma with hedgehog pathway inhibitor GDC-0449. *N. Engl. J. Med.* **361,** 1173–1178.

Salero, E., and Hatten, M. E. (2007). Differentiation of ES cells into cerebellar neurons. *Proc. Natl. Acad. Sci. USA* **104,** 2997–3002.

Saucedo, L. J., and Edgar, B. A. (2007). Filling out the Hippo pathway. *Nat. Rev. Mol. Cell Biol.* **8,** 613–621.

Schmahmann, J. D., and Caplan, D. (2006). Cognition, emotion and the cerebellum. *Brain* **129,** 290–292.

Schwartz, P. M., Borhesani, P. R., Levy, R. L., Pomeroy, S. L., and Segal, R. (1997). Abnormal cerebellar development and foliation in BDNF-/- mice reveals a role for neurotrophins in CNS patterning. *Neuron* **19,** 269–281.

Shrivastava, S., Zou, Z. Q., Pirollo, K., Blattner, W., and Chang, E. H. (1990). Germline transmission of a mutated p53 gene in a cancer-prone family with Li-Fraumeni syndrome. *Nature* **348,** 747–749.

Solecki, D. J., Liu, X. L., Tomoda, T., Fang, Y., and Hatten, M. E. (2001). Activated Notch2 signaling inhibits differentiation of cerebellar granule neuron precursors by maintaining proliferation. *Neuron* **31,** 557–568.

Spassky, N., Han, Y. G., Aguilar, A., Strehl, L., Besse, L., Laclef, C., Ros, M. R., Garcia-Verdugo, J. M., and Alvarez-Buylla, A. (2008). Primary cilia are required for cerebellar development and Shh-dependent expansion of progenitor pool. *Dev. Biol.* **317,** 246–259.

Stottmann, R. W., and Rivas, R. J. (1998). Distribution of TAG-1 and synaptophysin in the developing cerebellar cortex: Relationship to Purkinje cell dendritic development. *J. Comp. Neurol.* **395,** 121–135.

Tanori, M., Santone, M., Mancuso, M., Pasquali, E., Leonardi, S., Di Majo, V., Rebessi, S., Saran, A., and Pazzaglia, S. (2010). Developmental and oncogenic effects of insulin-like growth factor-I in Ptc1+/- mouse cerebellum. *Mol. Cancer* **9,** 53.

Thompson, M. C., Fuller, C., Hogg, T. L., Dalton, J., Finkelstein, D., Lau, C. C., Chintagumpala, M., Adesina, A., Ashley, D. M., Kellie, S. J., Taylor, M. D., Curran, T., *et al.* (2006). Genomics identifies medulloblastoma subgroups that are enriched for specific genetic alterations. *J. Clin. Oncol.* **24,** 1924–1931.

Timmann, D., Watts, S., and Hore, J. (1999). Failure of cerebellar patients to time finger opening precisely causes ball high-low inaccuracy in overarm throws. *J. Neurophysiol.* **82,** 103–114.

Tong, W. M., Ohgati, H., Hunag, H., Granier, C., Kleihues, P., and Wang, Z. Q. (2003). Null mutation of DNA strand break-binding molecule poly(ADP-ribose) polymerase causes medulloblastoms in p53-/- mice. *Am. J. Pathol.* **162,** 343–352.

Uziel, T., Zindy, F., Xie, S., Lee, Y., Forget, A., Magdaleno, S., Rehg, J. E., Calabrese, C., Solecki, D., Eberhart, C. G., Sherr, S. E., Plimmer, S., *et al.* (2005). The tumor suppressors Ink4c and p53 collaborate independently with Patched to suppress medulloblastoma formation. *Genes Dev.* **19,** 2656–2667.

Uziel, T., Karginov, F. V., Xie, S., Parker, J. S., Wang, Y.-D., Gajjar, A., He, L., Ellison, D., Gilbertson, R. G., Hannon, G., and Roussel, M. F. (2009). The miR-17 92 cluster collaborates with the sonic hedgehog pathway in medulloblastoma. *Proc. Natl. Acad. Sci. USA* **106,** 2812–2817.

Ventura, A., *et al.* (2008). Targeted deletion revelas essentail and overlapping functions af the miR-17 through 92 family of miRNA clusters. *Cell* **132,** 875–886.

Volpe, J. J. (2009). Cerebellum of the premature infant: Rapidly developing, vulnerable, clinically important. *J. Child Neurol.* **24,** 1085–1104.

von Hoff, K., Hartmann, W., and von Bueren, A. O. (2010). Large cell anaplastic medulloblastoma: Outcome according to Myc status, histopatholical and clinical risk factors. *Pediatr. Blood Cancer* **54,** 369–376.

Wang, V. Y., Rose, M. F., and Zoghbi, H. Y. (2005). Math1 expression redefines the rhombic lip derivatives and reveals novel lineages within the brainstem and cerebellum. *Neuron* **48,** 31–43.

Wechsler-Reya, R. J., and Scott, M. P. (1999). Control of neuronal precursor proliferation in the cerebellum by Sonic Hedgehog. *Neuron* **22,** 103–114.

Wechsler-Reya, R., and Scott, M. P. (2001). The developmental biology of brain tumors. *Annu. Rev. Neurosci.* **24,** 385–428.

Wei, W., Ayad, N. G., Wan, Y., Zhang, G. J., Kirschner, M. W., and Kaelin, W. G., Jr. (2004). Degradation of the SCF component Skp2 in cell-cycle phase G1 by the anaphase-promoting complex. *Nature* **428,** 194–198.

Wen, X., Lai, C. K., Evangelista, M., Hongo, J. A., de Sauvage, F. J., and Scales, S. J. (2010). Kinetics of hedgehog-dependent full-length Gli3 accumulation in primary cilia and subsequent degradation. *Mol. Cell. Biol.* **30,** 1910–1922.

Wingate, R. J. (2001). The rhombic lip and early cerebellar development. *Curr. Opin. Neurobiol.* **11,** 82–88.

Wingate, R. J., and Hatten, M. E. (1999). The role of the rhombic lip in avian cerebellum development. *Development* **126,** 4395–4404.

Xu, P., Yu, S., Jiang, R., Kang, C., Wang, G., Jiang, H., and Pu, P. (2009). Differential expression of Notch family members in astrocytomas and medulloblastomas. *Pathol. Oncol. Res.* **15,** 703–710.

Yan, C. T., Kaushal, D., Murphy, M., *et al.* (2006). XRCC4 suppresses medulloblastomas with recurrent translocations in p53-deficient mice. *Proc. Natl. Acad. Sci. USA* **103,** 7378–7383.

Yang, L., Jackson, E., Woerner, B. M., Perry, A., Piwnica-Worms, D., and Rubin, J. B. (2007). Blocking CXCR4-mediated cyclic AMP suppression inhibits brain tumor growth in vivo. *Cancer Res.* **67,** 651–658.

Yoon, J. W., Gilbertson, R., Iannaccone, S., Iannaccone, P., and Walterhouse, D. (2009). Defining a role for Sonic hedgehog pathway activation in desmoplastic medulloblastoma by identifying GLI1 target genes. *Int. J. Cancer* **124,** 109–119.

Zhang, L., and Goldman, J. E. (1996). Developmental fates and migratory pathways of dividing progenitors in the postnatal rat cerebellum. *J. Comp. Neurol.* **370,** 536–550.

Zhao, H., Ayrault, O., Zindy, F., Kim, J. H., and Roussel, M. F. (2008). Post-transcriptional down-regulation of Atoh1/Math1 by bone morphogenic proteins suppresses medulloblastoma development. *Genes Dev.* **22,** 722–727.

Zheng, C., Heintz, N., and Hatten, M. E. (1996). CNS gene encoding astrotactin, which supports neuronal migration along glial fibers. *Science* **272,** 417–419.

Zindy, F., Nilsson, L. M., Nguyen, L., Meunier, C., Smeyne, R. J., Rehg, J. E., Eberhart, C., Sherr, C. J., and Roussel, M. F. (2003). Hemangiomas, Medulloblastomas, and other tumors in Ink4c/p53-null mice. *Cancer Res.* **63,** 5420–5437.

Zindy, F., Knoepfler, P. S., Xie, S., Sherr, C. J., Eisenman, R. N., and Roussel, M. F. (2006). N-Myc and the cyclin-dependent kinase inhibitors p18Ink4c and p27Kip1 coordinately regulate cerebellar development. *Proc. Natl. Acad. Sci. USA* **103,** 11579–11583.

Zindy, F., Uziel, T., Ayrault, O., Calabrese, C., Valentine, M., Rehg, G., Gilbertson, R., Sherr, C. J., and Roussel, M. F. (2007). Genetic alterations in mouse medulloblastoma and generation of tumors de novo from primary cerebellar granule neuron precursors. *Cancer Res.* **67,** 2676–2684.

CHAPTER NINE

Rethinking Pediatric Gliomas as Developmental Brain Abnormalities

Nikkilina R. Crouse,* Sonika Dahiya,† *and* David H. Gutmann*

Contents

Abstract

The neurofibromatosis type 1 (NF1) tumor predisposition syndrome provides an illustrative example of brain tumor formation and growth in which a permissive microenvironment (stroma) is required for the expansion and maintenance of the neoplastic cells. In this chapter, we review the experimental evidence that supports the emerging concept that brain tumors are dynamic ecosystems where interactions between non-neoplastic and neoplastic cell types dictate where and when gliomas (astrocytomas) form and grow. The notion that brain tumors require a confluence of supportive stromal cell types and signals, susceptible preneoplastic/neoplastic cells, and genomic influences allows researchers and clinicians to develop strategies that effectively disrupt these critical relationships in a targeted and developmentally appropriate fashion.

1. Introduction

Brain tumors are often conceptualized as masses of neoplastic cells originating and growing in a passive brain environment. This model of tumorigenesis does not fully take into account the fact that the brain is a

* Department of Neurology, Washington University School of Medicine, St. Louis, Missouri, USA
† Department of Pathology, Washington University School of Medicine, St. Louis, Missouri, USA

Current Topics in Developmental Biology, Volume 94
ISSN 0070-2153, DOI: 10.1016/B978-0-12-380916-2.00009-7

highly specialized organ in which specific spatially and temporally restricted signals and cell types work together in a tightly orchestrated fashion to control normal brain development and maintenance. These cells and signals provide a dynamic environment in which growth/survival factors, guidance cues, and chemokines determine which cells proliferate, differentiate and migrate, and die. Alterations in the expression of these signals, the function of their receptors, or the activation of the relevant downstream growth control signaling cascades would therefore disrupt the intricate balance established in the normal brain, and would lead to inappropriate cell survival, proliferation, and migration. In this chapter, we will use neurofibromatosis type 1 (NF1) to illustrate the concept that brain tumor formation and growth in children represents a developmental abnormality in which molecular changes in non-neoplastic cells create a permissive environment for the expansion and maintenance of preneoplastic and neoplastic cells.

2. Pediatric Brain Tumors

Of all the solid cancers found in children, brain tumors are the most common, and represent the leading cause of pediatric cancer-related death (Surawicz *et al.*, 1998, 1999). Based on international estimates from the World Health Organization (WHO), low-grade glial cell malignancies (gliomas) account for >50% of all brain tumors in children from birth to 14 years old (Louis *et al.*, 2007). The most common glioma in children is the WHO grade I pilocytic astrocytoma (PA) (Sievert and Fisher, 2009), which comprise 85% of cerebellar gliomas and almost all gliomas within the optic nerve pathway (Freeman *et al.*, 1998).

PAs can arise anywhere within the neuraxis, but in the pediatric population are most commonly found in infratentorial locations, including the cerebellum and brainstem. In addition, PA tumors also frequently arise in the hypothalamus and optic pathway (optic nerve, optic chiasm, and postchiasmatic radiations). The spinal cord is infrequently involved. On neuroimaging studies, these tumors appear as well-demarcated masses (Fig. 9.1A), often with significant contrast enhancement. PA tumors can also have a bright T-1 magnetic resonance imaging (MRI) signal due to the presence of mucinous/proteinaceous fluid in associated cystic elements (Freeman *et al.*, 1998). In addition, PA tumors arising in the optic nerve (optic gliomas) are highly infiltrative and expand the nerve to produce a fusiform mass (Fig. 9.1B and C).

The histopathology of PAs can vary considerably and can pose a significant diagnostic challenge. Cerebellar PA tumors often have a biphasic histologic architecture, composed of piloid (compact) areas (Fig. 9.1D) alternating with loose mucinous (microcystic) areas (Fig. 9.1E). As mentioned above, PAs arising in the optic pathway have an exclusively compact

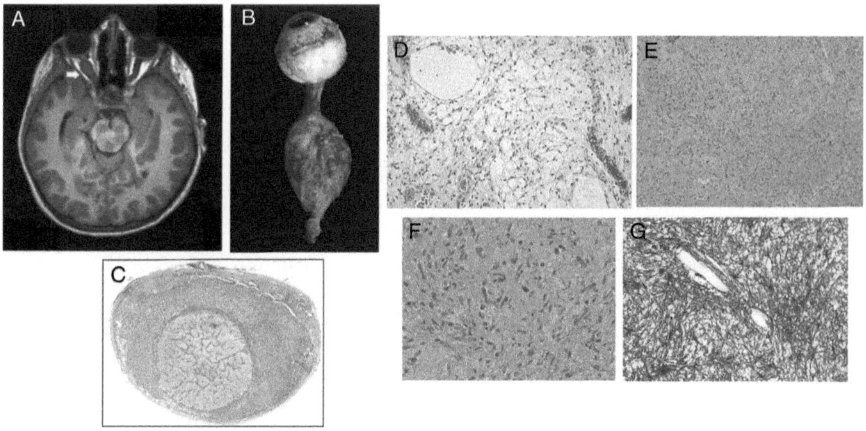

Figure 9.1 NF1-associated gliomas. (A) Magnetic resonance imaging from a young patient with a right optic nerve glioma. The arrow points to the glioma expanding the right optic nerve. (B) Enucleation specimen demonstrates a fusiform enlargement of optic nerve by a pilocytic astrocytoma. (C) Cross-section of this tumor highlights the expansile and solid nature of a pilocytic astrocytoma at this location. These tumors contain (D) loose microcystic and (E) compact areas rich in Rosenthal fibers (F). (G) The piloid neoplastic glial cells strongly express glial fibrillary acidic protein (GFAP). Images are courtesy of Dr. Robert E. Schmidt, Neuropathology Division, Washington University. (See Color Insert.)

pattern, with an inconspicuous mucinous background and abundant Rosenthal fibers (Fig. 9.1F). Furthermore, in the optic nerve, the connective tissue septa remain intact, but the fascicles are enlarged with nerve fibers being progressively replaced by tumor, such that extension into the meninges produces a fusiform enlargement. These tumors may also display marked nuclear pleiomorphism, infarct-like necrosis and microvascular proliferation; however, mitoses are rarely encountered in typical cases. Consistent with their classification as glial cell tumors, PAs express glial fibrillary acidic protein (GFAP) on routine immunostaining (Fig. 9.1G).

3. Genetics of Pediatric Gliomas

Compared to gliomas that arise in adults, considerably less is known about the signature genetic changes that drive pediatric gliomagenesis. Comprehensive genomic and genetic studies over the last several years have identified a small number of genetic alterations and mutations associated with low-grade and high-grade glioma in children (Table 9.1). Low-grade gliomas, including PAs, typically have relatively normal karyotypes and few regions of chromosomal gain or loss. The most common genetic

Table 9.1 Genetic changes found in tumors

Gene	Role in brain development	References
NF1	Neuronal neurite extension, glial proliferation, neural tube closure, hypothalamic-pituitary function, learning and memory	Andersen *et al.* (1993), Gutmann *et al.* (2000, 1995a,b), Kluwe *et al.* (2001), Marchuk *et al.* (1991), Wimmer *et al.* (2002)
BRAF	Oligodendrocyte differentiation and myelination	Forshew *et al.* (2009), Jacob *et al.* (2009), Jones *et al.* (2008), MacConaill *et al.* (2009), Robinson *et al.* (2010), Schiffman *et al.* (2010), Yu *et al.* (2009)
HIPK2	Neuronal apoptosis	Deshmukh *et al.* (2008), Jacob *et al.* (2009), Puca *et al.* (2009), Sanoudou *et al.* (2000), White *et al.* (1995), Yu *et al.* (2009), Zattara-Cannoni *et al.* (1998)
MATN2	Nerve regeneration, axonal growth	Piecha *et al.* (1999), Sanoudou *et al.* (2000), Sharma *et al.* (2006), White *et al.* (1995), Zattara-Cannoni *et al.* (1998)
TP53	Neural tube closure, neural precursor apoptosis, neural stem cell self-renewal	Gutmann *et al.* (2003), Legius *et al.* (1994), Schiffman *et al.* (2010), Verdijk *et al.* (2010)
PDGFRA	Oligodendrocyte differentiation and myelination, glial migration and proliferation	Perrone *et al.* (2009), Schiffman *et al.* (2010), Verhaak *et al.* (2009)

MET	Neuronal branching, neuronal migration, oligodendrocyte progenitor proliferation	Schiffman *et al.* (2010), Verhaak *et al.* (2009)
PTEN	Neuronal structure, synaptic plasticity, myelination, synaptogenesis, neuronal apoptosis, neurogenesis, neural stem cell self-renewal and proliferation	Gregorian *et al.* (2009), Perrone *et al.* (2009), Schiffman *et al.* (2010)
CDKN2A	Neural stem cell self-renewal and proliferation, neurogenesis	Gutmann *et al.* (2003), Schiffman *et al.* (2010)
MDM4	Neuronal apoptosis	Mancini and Moretti (2009), Schiffman *et al.* (2010), Zohrabian *et al.* (2007)
CMYC	Oligodendrocyte apoptosis and myelination, neural stem cell self-renewal	Reish *et al.* (2003), Schiffman *et al.* (2010)
WNT5B	Axonal guidance	Schiffman *et al.* (2010)
IGFR1	Neuronal dendritic growth, brain size, myelination, neuronal migration	Schiffman *et al.* (2010)
EGFR	Neural stem cell proliferation, survival and migration	Jacob *et al.* (2009), Li *et al.* (2001), Perrone *et al.* (2009), Verhaak *et al.* (2009)
PIK3CA	Neuronal structure, synaptic plasticity, myelination, synaptogenesis, neuronal apoptosis, neurogenesis, neural stem cell self-renewal and proliferation	Engelman (2009), Gibbons *et al.* (2009), Liu *et al.* (2009), Perrone *et al.* (2009), Yang *et al.* (2006)

alterations include mutation of the *BRAF* gene, gains on chromosomes 7p and 8p, and loss of *NF1* gene expression.

Of the identified genomic alterations in PA, one of the most common is rearrangement of the *BRAF* gene, a kinase with growth regulatory properties whose mutation in other cancers is oncogenic. In as many as two-thirds of PA tumors, the *BRAF* gene is fused with a gene of unknown function, *KIAA1549*, leading to the expression of abnormal *KIAA1549:BRAF* fusion transcripts (Forshew *et al.*, 2009; Jones *et al.*, 2008; Yu *et al.*, 2009). This fusion is not seen in grade II–IV gliomas, and likely represents a unique genomic change in PA. Instead, some high-grade pediatric gliomas contain a V600E point mutation within the BRAF protein sequence, rendering BRAF constitutively active (MacConaill *et al.*, 2009; Schiffman *et al.*, 2010). The significance of dysregulated BRAF signaling in the genesis of pediatric low-grade glioma is not clear, as oncogenic BRAF expression is not sufficient for gliomagenesis in mice (Robinson *et al.*, 2010).

KIAA1549:BRAF expression leads to increased ERK activation (Jacob *et al.*, 2009) similar to what is observed following oncogenic RAS expression. Consistent with the predicted consequence of deregulated BRAF function in PA, two groups have reported the presence of *KRAS* mutations in sporadic PA (Janzarik *et al.*, 2007; Sharma *et al.*, 2005). Both mutations have also been implicated in other forms of cancer (Burmer *et al.*, 1991; Gressani *et al.*, 1998).

In addition to BRAF and RAS alterations, pediatric PAs also exhibit amplification of HIPK2 (Deshmukh *et al.*, 2008) and *MATN2* (Sharma *et al.*, 2006), which correspond to previously noted cytogenetic gains on chromosomes 7p and 8p (Sanoudou *et al.*, 2000; White *et al.*, 1995; Zattara-Cannoni *et al.*, 1998). The functional significance of these molecular alterations is currently not known.

Finally, 15% of all PA tumors harbor mutations in the *NF1* tumor suppressor gene. These *NF1*-deficient PAs arise in children with the NF1 tumor predisposition syndrome. In this regard, NF1-associated PAs exhibit complete loss of *NF1* expression (Gutmann *et al.*, 2000), whereas histologically identical PAs arising sporadically in patients without NF1 retain *NF1* gene expression (Kluwe *et al.*, 2001; Wimmer *et al.*, 2002).

Taken together, PA tumors share activation of the MEK pathway resulting from *KIAA1549:BRAF* expression, mutational RAS activation, or loss of the *NF1* tumor suppressor gene (see below), suggesting that the growth control pathways deregulated by these genetic changes are important for modulating glial cell proliferation relevant to tumorigenesis. However, it should be noted that none of these PA-associated genetic alterations by themselves result in glioma formation in mice (see below), arguing that these signature PA mutations are necessary, but not sufficient, for gliomagenesis.

4. NF1 and Brain Tumors

NF1 is an autosomal dominant disorder with an estimated incidence of 1:3000 (Friedman, 1999); however, half of cases do not have a family history and represent new mutations. The disorder is characterized by the presence of neurofibromas (benign peripheral nerve sheath tumors), pigmentary changes (café-au-lait macules, skinfold freckling, and iris hamartomas), and gliomas arising within the optic pathway (Gutmann *et al.*, 1997).

Approximately 15% of patients with NF1 develop low-grade glial neoplasms largely restricted to the optic pathway, and located from just behind the eye (retrobulbar optic nerve) to the postchiasmatic optic radiations ("optic pathway" glioma, OPG). Two-thirds of NF1-associated gliomas arise in the optic pathway, 15% in the brainstem, and less than 15% in the cerebellum, cortex, and subcortical regions (Freeman *et al.*, 1998; Guillamo *et al.*, 2003). Histologically, NF1-associated PAs are indistinguishable from their sporadic counterparts; however, the clinical course of these tumors in children with NF1 is more indolent (Listernick *et al.*, 1995). Sporadic OPGs tend to progress clinically, exhibit more aggressive behavior, and nearly always require treatment, whereas only one-third of NF1-associated OPGs require treatment, and most grow slowly with some reports of spontaneous regression. Moreover, NF1-associated OPGs are nearly always tumors of early childhood (<7 years of age) and rarely continue to grow or cause symptoms after age 10 (Listernick *et al.*, 2007).

The *NF1* gene is located on chromosome 17q11.2, and spans about 350 kb of genomic DNA (Marchuk *et al.*, 1991). It contains 59 exons and encodes a 220–250 kDa protein called neurofibromin (Viskochil *et al.*, 1990; Wallace *et al.*, 1990). The *NF1* mRNA transcript is approximately 13 kb long and includes three alternatively spliced isoforms (exons 9a, 23a, and 48a), thought to reflect tissue-specific and differentiation-associated regulation (Andersen *et al.*, 1993; Costa *et al.*, 2001; Gutmann *et al.*, 1995a,b, 1999).

Similar to other inherited cancer predisposition syndromes, children with NF1 are born with one mutated (nonfunctional) and one wild-type (functional) copy of the *NF1* gene in every cell of their body. This *NF1* heterozygous state is, by itself, insufficient for tumorigenesis; inactivation of the remaining wild-type *NF1* allele in an individual cell is the rate-limiting step for subsequent tumor formation (Knudson, 1971) (Fig. 9.2). In this regard, complete loss of *NF1* gene expression in a Schwann cell would result in neurofibroma formation, while total *NF1* inactivation in a glial cell or glial progenitor leads to optic glioma development.

Initial sequence analysis of the predicted neurofibromin amino acid sequence revealed that it contains a small domain with sequence similarity

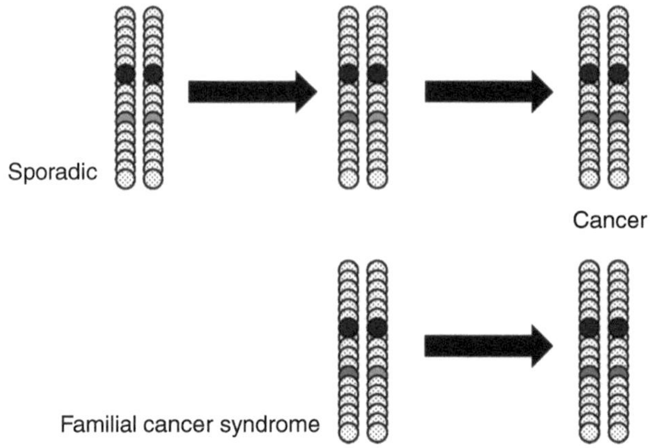

Figure 9.2 The two-hit hypothesis. Cancer in the general population (sporadic) requires the sequential inactivation of a given tumor suppressor gene in the same cell. In contrast, individuals with inherited cancer syndromes begin life with one nonfunctional copy of a given tumor suppressor gene, and develop cancer after the one remaining functional allele is inactivated. (See Color Insert.)

to a family of proteins that inhibit RAS activation (Xu *et al.*, 1990). These GTPase activating protein (GAP) molecules associate with RAS and accelerate the conversion of RAS from an active, GTP-bound form to an inactive GDP-bound form. In this fashion, GAP molecules inactivate RAS and reduce RAS-mediated growth signaling. The intrinsic GTPase activity of wild-type, but not oncogenic, RAS can be stimulated by the GAP-related domain (GRD) of neurofibromin (Ballester *et al.*, 1990; Martin *et al.*, 1990; Xu *et al.*, 1990), such that loss of neurofibromin results in RAS hyperactivation (Basu *et al.*, 1992; Bollag *et al.*, 1996; DeClue *et al.*, 1992; Kim *et al.*, 1995), and expression of the *NF1* GRD in *NF1*-deficient cells is sufficient to reverse the hyperproliferation associated with neurofibromin loss (Hiatt *et al.*, 2001). The consequences of increased RAS activation in neurofibromin-deficient cells are manifested by increased activation of various RAS effectors, including RAF-ERK and Akt-mTOR (Fig. 9.3).

Whereas neurofibromin can negatively regulate RAS in many cell types, its ability to suppress cell growth through the RAS pathway is cell type-specific. We have previously shown that loss of neurofibromin expression leads to preferential hyperactivation of only KRAS in astrocytes despite expression of all three RAS isoforms (Dasgupta *et al.*, 2005a). Moreover, only KRAS activation can substitute for neurofibromin loss in glial progenitor cells relevant to the formation of optic gliomas in genetically engineered mice. Similar results have now been reported in other *Nf1*-deficient cell types (Khalaf *et al.*, 2007; Morgan *et al.*, 2005).

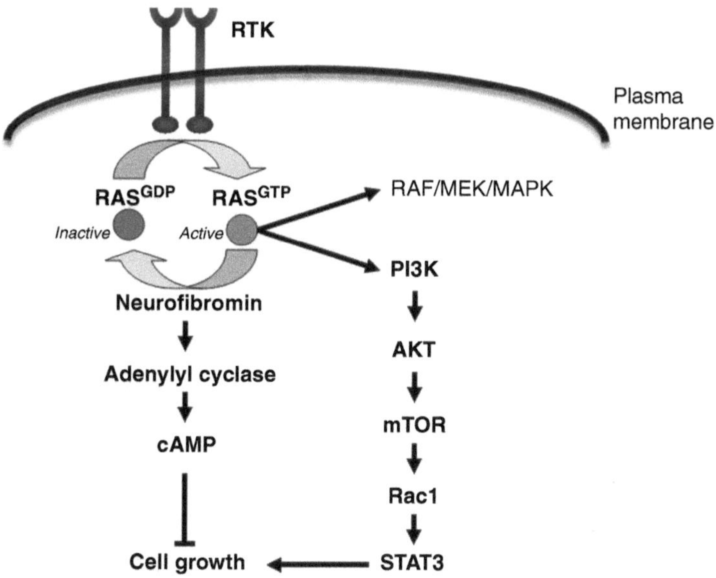

Figure 9.3 Neurofibromin growth control signaling pathways. Neurofibromin functions as a negative regulator of the RAS proto-oncogene by accelerating the conversion of active GTP-bound RAS to the inactive GDP-bound form. Active RAS initiates a signaling cascade in astrocytes involving sequential Akt/mTOR/STAT3 activation. In addition, neurofibromin positively regulates adenylyl cyclase, leading to increased intracellular cAMP levels. The absence of neurofibromin in glioma cells is associated with reduced cAMP levels and hyperactivation of the Akt/mTOR/Rac1/STAT3 signaling pathway.

Initial studies in non-nervous system cell types revealed that neurofibromin regulation of cell growth functioned predominantly through the RAS/MAPK pathway. For example, both *Nf1*-deficient and *Nf1*+/− cell growth was dependent on RAS/MAPK signaling: *Nf1*+/− mast cell (McDaniel *et al.*, 2008), *Nf1*+/− osteoclast (Yang *et al.*, 2006), and *Nf1*+/− primary vascular smooth muscle cell (Li *et al.*, 2006) function are mediated by RAS/MAPK activations. In addition, human *NF1*−/− juvenile leukemia cells (Bollag *et al.*, 1996) and *Nf1*-deficient mouse myeloid cell (Donovan *et al.*, 2002) proliferation are RAS/MAPK-dependent.

In striking contrast, neurofibromin growth regulation in astrocytes is dependent on Akt-mediated activation of the mammalian target of rapamycin (mTOR) pathway (Dasgupta *et al.*, 2005b). mTOR is a major regulator of ribosomal biogenesis and protein translation, and *Nf1*-deficient astrocytes have significantly higher levels of protein translation than their wild-type counterparts. The significance of mTOR hyperactivation to gliomagenesis was underscored by the finding of increased mTOR activity in NF1-associated human PA tumors as well as *Nf1* genetically engineered mice with optic glioma. Importantly, inhibition of mTOR activity in *Nf1* genetically engineered mice with optic glioma using the macrolide rapamycin blocks optic glioma proliferation (Hegedus *et al.*, 2008). Similar findings have been reported for other cell types, including Schwann cells (Bhola *et al.*, 2010; Johannessen *et al.*, 2005, 2008). Further studies focused on mTOR signaling have shown that neurofibromin/mTOR regulation of astrocyte growth requires Rac1 activation (Sandsmark *et al.*, 2007) and STAT3 signaling (Banerjee *et al.*, 2010), suggesting other potential therapeutic targets for NF1-associated brain tumors.

In addition to RAS/mTOR signaling, neurofibromin also regulates intracellular cyclic adenosine monophosphate (cAMP) levels in the brain (Fig. 9.3). Intracellular cAMP levels are controlled by the conversion of adenosine triphosphate (ATP) to cAMP by adenylyl cyclase (AC). We and others have shown that neurofibromin positively regulates cAMP production at the level of AC (Dasgupta *et al.*, 2003; Tong *et al.*, 2002), such that *Nf1−/−* astrocytes have lower intracellular cAMP levels. In astrocytes, elevations in intracellular cAMP levels lead to cell death through apoptosis (Warrington *et al.*, 2007). This is particularly relevant to NF1-associated optic gliomas, as the optic nerve contains high levels of expression of the CXCL12 chemokine. In addition, CXCL12 is robustly expressed in the endothelium of NF1 glioma-associated blood vessels, neuronal processes and parenchymal microglia. In contrast to wild-type astrocytes, *Nf1*-deficient astrocytes exhibit a reduced intracellular cAMP response to CXCL12, which results in inappropriate cell survival (Warrington *et al.*, 2007). This inappropriate astrocyte survival allows *Nf1−/−* glial cells to escape cell death in response to CXCL12 expression in the optic nerve environment, and in concert with other environmental (stromal) signals (see below), facilitates *Nf1−/−* glial cell transformation and culminates in glioma formation.

5. Gliomagenesis in NF1

One of the unresolved issues surrounding NF1-associated gliomagenesis is its spatial (optic pathway) and temporal (young children) pattern. Based on mouse modeling experiments by our group and others, we postulate that the unique pattern of NF1-associated gliomagenesis reflects the presence of both a permissive microenvironment and susceptible cell types.

Previous studies from our laboratory have shown that *Nf1* inactivation in glial progenitors alone in genetically engineered mice is *not* sufficient for glioma formation *in vivo* (Bajenaru *et al.*, 2002). As mentioned above, NF1 is a classic tumor predisposition syndrome in which individuals are born

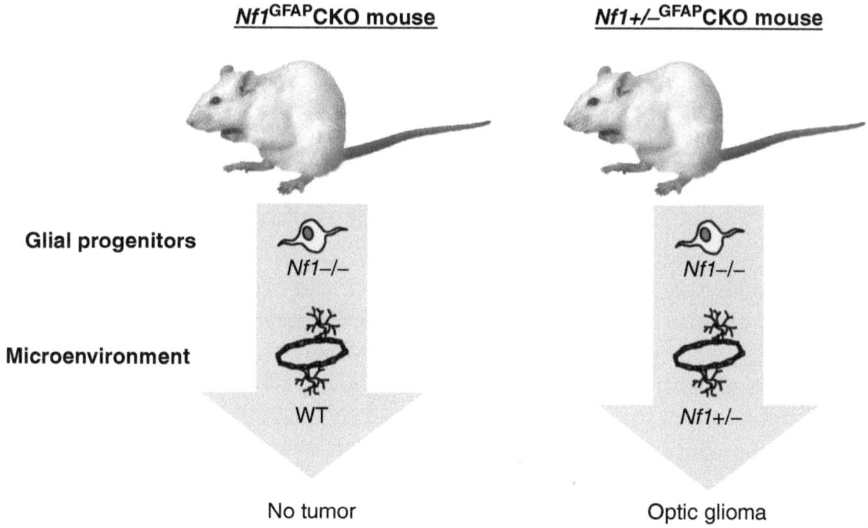

Figure 9.4 NF1 gliomagenesis in mice requires *Nf1* loss in progenitor cells coupled with a supportive *Nf1*+/– microenvironment. In *Nf1* genetically engineered mouse glioma models, *Nf1* inactivation in glial progenitors ($Nf1^{GFAP}$CKO mice) alone does not result in glioma formation. However, *Nf1*+/– mice with glial *Nf1* inactivation ($Nf1+/-^{GFAP}$CKO mice) develop glioma. These models establish the obligate role of the tumor microenvironment in NF1-associated glioma formation.

with a germline mutation in one copy of the *NF1* gene in all cells of their body. To recapitulate biallelic *Nf1* gene inactivation in glial progenitor cells in the context of germline *Nf1* heterozygosity, we and others developed *Nf1*+/– mice lacking *Nf1* gene expression in glial progenitors. Nearly 100% of these *Nf1* genetically engineered mice developed low-grade astrocytic tumors of the optic nerve (Bajenaru *et al.*, 2003; Zhu *et al.*, 2005). This finding argues that cells and signals in the *Nf1*+/– optic nerve microenvironment are necessary for gliomagenesis (Fig. 9.4).

6. The Supportive Microenvironment

The stroma in which gliomas arise is a specialized niche that provides an optimal environment for tumor initiation, growth and maintenance. This supportive microenvironment is created by a confluence of specific cell types, molecular signals, and genomic factors, which individually support the expansion of preneoplastic/neoplastic cells and promote glioma formation and continued growth.

Supportive stromal cell types. Several distinct cell types are found in the tumor microenvironment that can participate in gliomagenesis and glioma growth. Studies from our laboratory using *Nf1* genetically engineered mice have revealed increased numbers of microglia, resident brain immune system cells, in the region of the developing optic glioma prior to obvious tumor formation (Bajenaru *et al.*, 2005). Microglia play a pivotal role in maintaining brain homeostasis, and have been implicated in regulating neuronal apoptosis (Marin-Teva *et al.*, 2004), synaptogenesis (Roumier *et al.*, 2004), synaptic transmission (Coull *et al.*, 2005), and neurotrophic factor production (Elkabes *et al.*, 1996). In addition, microglia perform immunologic functions in response to brain injury and disease (Davoust *et al.*, 2008), and secrete proinflammatory cytokines, including IL-1, IL-6, and TNFα (Banati *et al.*, 1993).

The role of microglia in brain tumors, however, is less clear. Several lines of evidence support the hypothesis that microglia found in the brains of *Nf1*+/− mice are important for glioma formation and growth: First, we have shown that *Nf1*+/−, but not wild-type, microglia increase proliferation of *Nf1*−/− astrocytes *in vitro* (Daginakatte and Gutmann, 2007). Second, inactivation of microglia function using either minocycline (Daginakatte and Gutmann, 2007) or inhibition of the JNK signaling pathway in *Nf1*+/− microglia (Daginakatte *et al.*, 2008), results in reduced *Nf1* optic glioma proliferation *in vivo*. Third, genetic ablation of microglia in *Nf1* optic glioma mice results in attenuated tumor proliferation *in vivo* (Simmons *et al.*, 2011).

While microglia may be central cellular components of the glioma microenvironment, it is equally likely that other stromal cell types are necessary for creating a permissive niche for tumor formation and expansion. In this regard, endothelial cells and reactive astrocytes found in these low-grade glial neoplasms may be important contributors. For example, endothelial cells in orthotopic brain tumor xenografts expand the fraction of self-renewing cells and accelerate the initiation and growth of tumors (Calabrese *et al.*, 2007). Similarly, reactive gliosis in response to injury is partially dependent on microglia/macrophage-induced sonic hedgehog activation in astrocytes (Amankulor *et al.*, 2009), raising the intriguing possibility that the microglia in the glioma promote both reactive gliosis and endothelial cell proliferation (Lin and Pollard, 2007; Pollard, 2004). Formal demonstration of the central role of microglia in creating and maintaining the proper microenvironment for glioma formation and growth will require further study.

Supportive stromal signals. Defining how the tumor microenvironment derived glioma development and continued growth requires the identification of specific molecules present in the glioma microenvironment. Two non-mutually exclusive classes of molecules could be envisioned to play critical roles in the formation of a permissive tumor microenvironment (Fig. 9.5). Neoplastic cells release soluble factors that promote the

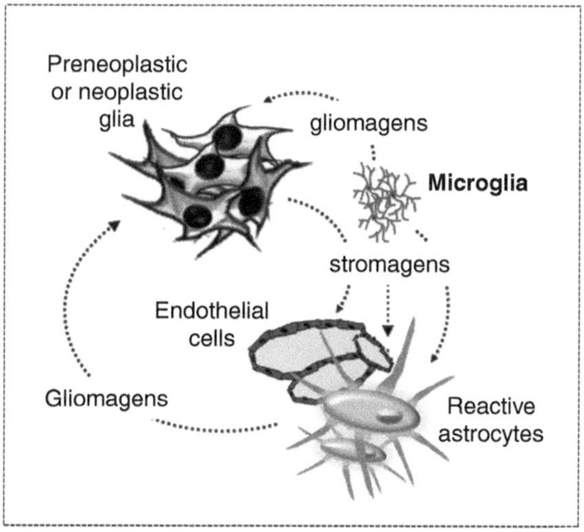

Figure 9.5 Glioma formation and growth is dictated by the presence of key stromal factors. Tumor initiation in preneoplastic/neoplastic cells accompanies loss of *Nf1* tumor suppressor gene expression. These *Nf1*-deficient glial cells are envisioned to recruit or activate microglia and other stromal cell types (endothelial calls and reactive astrocytes) through the elaboration of "stromagens." Stromagens produced by the preneoplastic/neoplastic cells as well as by *Nf1+/−* stromal cells function to create a supportive microenvironment that facilitates continued glioma growth. The expansion of the preneoplastic/neoplastic cells is further enhanced by "gliomagens" (e.g., hyaluronidase, CXCL12) provided by the supportive microenvironment that act to increase *Nf1−/−* glial cell proliferation and survival.

recruitment or activation of microglia (Kostianovsky *et al.*, 2008; Markovic *et al.*, 2009; Platten *et al.*, 2003), which in turn produce additional molecules important for endothelial cell migration and proliferation as well as astrocyte activation. These "stromagens" create the tumor microenvironment that supports the expansion of the glioma cells. In addition, microglia, reactive astrocytes, and endothelial cells release growth factors ("gliomagens") that function to increase preneoplastic/neoplastic glial cell proliferation, survival, and invasion (Markovic *et al.*, 2009; Weissenberger *et al.*, 2004; Wesolowska *et al.*, 2008). Moreover, it is highly likely that there exists a dynamic relationship between the neoplastic cellular elements and the non-neoplastic stromal cell types that further promotes the proliferation, survival, and infiltration of glioma cells.

Previous work from other laboratories has shown that tumor-associated microglia have decreased phagocytic abilities, lowered antigen presentation and a reduction in the amount of proinflammatory cytokines secreted (Flugel *et al.*, 1999; Graeber *et al.*, 2002; Parney *et al.*, 2009; Yang *et al.*, 2009). Glioma-associated microglia have also been shown to contain high levels of anti-inflammatory cytokine, IL-10, mRNA expression (Kostianovsky *et al.*, 2008). In addition, microglia secrete other factors that directly lead to tumor growth. The release of vascular endothelial growth factor (VEGF) (Lafuente *et al.*, 1999; Tsai *et al.*, 1995), hepatocyte growth factor (HGF) (Watters *et al.*, 2005), and substance P (SP) (Lai *et al.*, 2000; Rasley *et al.*, 2002) are all shown to lead to angiogenesis and tumor proliferation (Kunkel *et al.*, 2001; Lafuente *et al.*, 1999; Luber-Narod *et al.*, 1994; Martin *et al.*, 1993; Tsai *et al.*, 1995; Yamada *et al.*, 1994).

Initial studies in our laboratory using *Nf1* genetically engineered brain tumor models focused on *Nf1+/−* microglia, based on the above observations and the finding that conditioned media from *Nf1+/−*, but not wild-type, microglia increased the proliferation of *Nf1−/−* astrocytes. Using a microarray discovery approach, we found that *Nf1+/−* microglia express high levels of the meningioma-expressed antigen-5 (Mgea5) molecule, which is a member of the hyaluronidase family of enzymes that degrade extracellular matrix proteins and release bioactive growth factors (Daginakatte and Gutmann, 2007). Previous studies have shown that hyaluronidase increases astrocyte proliferation in the spinal cord (Struve *et al.*, 2005). Similarly, we showed that inhibitors of hyaluronidase block the ability of *Nf1+/−* microglia conditioned media to promote *Nf1*-deficient astrocyte growth, and purified hyaluronidase increases *Nf1−/−* astrocyte growth. In addition, we found that *Nf1+/−* microglia produce high levels of the chemokine CXCL12, which increases the survival of *Nf1*-deficient astrocytes (Warrington *et al.*, 2007). Together, these findings establish an important role for *Nf1+/−* microglia-produced molecules in the maintenance of *Nf1*-deficient astrocyte proliferation and survival. However, future studies will be required to determine whether these stromal signals are required for tumor formation. In this regard, recent studies have shown that forced CXCL12 expression in the cortex of *Nf1* optic glioma mice is not sufficient for glioma formation (Sun *et al.*, 2010), suggesting that other stromal factors likely cooperate with CXCL12 to create a permissive tumor environment.

One of these stromal determinants might be the levels of cAMP present in specific brain regions. Previous studies have demonstrated wide variations in cAMP levels in different brain regions (Warrington *et al.*, 2007). In this regard, the highest levels of cAMP were found in the cortex, whereas low levels were observed along the optic pathway. The failure of ectopic CXCL12 expression to induce glioma formation in the cortex of *Nf1* optic glioma mice might reflect these high levels of cAMP, which would counteract the effects of CXCL12 on *Nf1−/−* astrocyte growth.

Supportive genomic factors. The contribution of the genetic background to gliomagenesis is underscored by elegant studies by Reilly and colleagues which demonstrated that astrocytoma formation in *Nf1+/−; Trp53+/−* (NPCis) mice is dependent on the specific mouse genetic background: NPCis mice on the C57BL/6J genetic background are highly susceptible to glioma development, whereas 129S4/SvJae mice with the identical *Nf1+/−*; *Trp53+/−* mutation rarely form glioma (Reilly *et al.*, 2000, 2004). They further suggest that these strain differences may impact on *Nf1* gene expression, and that variations in *Nf1* mRNA expression in *Nf1+/−* cells from different genetic backgrounds may create a more permissive microenvironment for tumorigenesis (Hawes *et al.*, 2007). In addition, it is possible that some of the genetic loci responsible for these strain-dependent differences in glioma formation may represent polymorphic changes in specific genes that lead to altered expression of key stromal factors essential for gliomagenesis and glioma growth.

7. Receptive Preneoplastic Cells

In addition to the presence of a supportive microenvironment, there are likely differences in glioma susceptibility that reflect differential sensitivity of specific cell types to transformation. In this regard, the developmental age of the preneoplastic cell may be critical for transformation: Conditional inactivation of the *Nf1*, *p53*, and *Pten* tumor suppressor genes in mice results in high-grade glioma formation only if these changes occur in progenitor cells, rather than differentiated glial cells (Alcantara Llaguno *et al.*, 2009). Moreover, it is possible that progenitor cells and astrocytes from different regions of the brain respond differently to tumor suppressor gene inactivation.

In this regard, several studies have shown that astrocytes and progenitor cells from different brain regions may be distinct. For example, PA tumors from different brain regions have unique molecular signatures that reflect their brain location (Sharma *et al.*, 2007; Taylor *et al.*, 2005). These genetic "fingerprints" are also found in normal astrocytes and neural stem cells from these brain regions, raising the intriguing possibility that specific populations of site-restricted progenitor cells within the central nervous system are the cells of origin of histologically similar glial cell tumors with distinct molecular properties. With respect to NF1-associated PA, we have shown that astrocytes from different brain regions respond differently to *Nf1* gene inactivation (Yeh *et al.*, 2009). In these studies, astrocytes from the cortex, where tumors rarely form in children with NF1, have low levels of *Nf1* mRNA and protein expression, and exhibit no increased proliferation in response to *Nf1* gene inactivation *in vitro* or *in vivo*.

8. Rethinking the Two-Hit Hypothesis

One of the most time-honored and groundbreaking hypotheses that revolutionized the way we conceptualize inherited cancer syndromes was originally proposed for retinoblastoma (RB) by Dr. Alfred Knudson. Based on statistical modeling, he suggested that the increased frequency of tumors in this inherited cancer syndrome was the result of a germline mutation in the *RB* tumor suppressor gene (Knudson, 1971). With the identification of the *RB* gene, germline mutations were identified in affected families, and biallelic inactivation of the *RB* gene found in all associated tumors. Moreover, mice heterozygous for an inactivating *RB* mutation are also cancer prone, and develop tumors only upon biallelic inactivation of the *RB* gene. Similar to RB, patients with a germline inactivating *NF1* gene mutation only develop cancers following loss of the one remaining wild-type allele. Moreover, *NF1* loss results in increased neuroglial progenitor self-renewal and proliferation, and facilitates the expansion of receptive *NF1* deficient glia.

While this hypothesis has provided enormous insights into the pathogenesis of numerous hereditary cancer syndromes, it does not fully account for the roles of the local tumor and genomic environments in facilitating tumorigenesis. The findings from our group and others studying NF1-associated nervous system tumors suggest a model in which environmental influences dictate where and when tumors form in this common inherited cancer syndrome (Yang *et al.*, 2008; Zhu *et al.*, 2002). As discussed above, the obligate role of *NF1*+/− stromal cells in glioma formation illustrates the requirement for stromal cell types and signals in collaboration with a receptive *NF1*-deficient preneoplastic/neoplastic cell defined by brain region, genetic susceptibility, and developmental age (Fig. 9.6).

The requirement for both susceptible preneoplastic/neoplastic cells and a permissive environment likely explains why gliomas in children with NF1 most frequently develop in the optic pathway and grow most avidly during the early first decade of life. Specifically, optic nerve glial progenitors can respond to *NF1* gene inactivation and increase their proliferation in response to gliomagens present along the optic nerve pathway. One of these gliomagens, CXCL12, is expressed at high levels along the optic pathway in young mice, monkeys and humans, and in that fashion, provides one temporally and spatially regulated signal that allows receptive *NF1*−/− astrocytes to inappropriately survive and expand (Warrington *et al.*, 2007, 2010). Other gliomagens include MGEA5 (hyaluronidase), which is produced in abundant quantities by *NF1*+/− microglia, and can increase *NF1*-deficient astrocyte proliferation. The combination of CXCL12 and MGEA5 enhance both the proliferation and survival of susceptible *NF1*-deficient glial cells, and facilitate neoplastic transformation. It is likely that other mitogens, present in the *NF1*+/− microenvironment within the optic pathway of young children, also cooperate to restrict gliomagenesis to the location and age distribution characteristic of most NF1-associated gliomas. The identification of these gliomagens offers the opportunity to develop adjuvant treatment strategies that focus on stromal signals.

In addition, glioma susceptibility in genetically engineered *Nf1* mutant mice is determined by genomic influences encoded by polymorphic

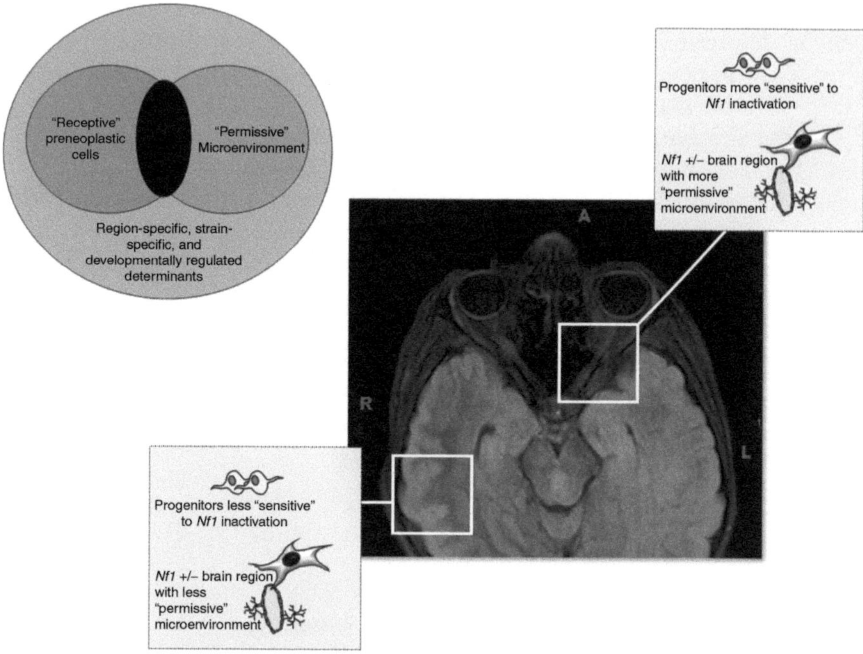

Figure 9.6 Glioma formation in NF1. Two-thirds of gliomas arise in the optic pathway of children with NF1, but rarely develop in the cortex. One model for gliomagenesis in NF1 that explains the temporal and spatial patterns of brain tumors in this inherited cancer predisposition syndrome requires the confluence of at least three conditions. First, the preneoplastic cells must be sensitive to *Nf1* gene inactivation and be capable of increasing their proliferation, survival, or migration in response to *Nf1* loss. Second, the microenvironment in which glial progenitor *Nf1* inactivation occurs must be supportive for neoplastic cell expansion. For example, *Nf1*+/− microglia have increased expression of hyaluronidase (MGEA5), and high levels of CXCL12 are found along the optic pathway in young children. Third, genomic, developmental, and tissue-specific factors also determine whether gliomas will form in the context of a receptive preneoplastic cell and a supportive tumor microenvironment. In the case of NF1, progenitor cells along the optic pathway increase their proliferation in response to *Nf1* loss. Second, *Nf1*+/− microglia present in the brain elaborate important gliomagens (e.g., CXCL12 and hyaluronidase) that facilitate the expansion of *Nf1*-deficient glial cells. Third, gliomagens, like CXCL12, are enriched along the optic pathway of young children. Finally, *Nf1* inactivation in progenitor cells, rather than in differentiated astrocytes, of susceptible mouse strains (e.g., C57BL/6J) in the presence of low levels of cAMP (e.g., optic nerve) cooperate to promote gliomagenesis specifically in the optic nerve.

differences found in different inbred strains of mice. These "modifier" loci dictate whether mice with the same genetic mutation will develop glioma, suggesting that these sequence variants are strong modifiers of glioma susceptibility. Should this be translatable to humans, it is possible that polymorphisms in our individual genetic composition provide a genomic microenvironment permissive for glioma formation. Identifying these genetic determinants in people could result in risk stratification of patients most likely to develop glioma in NF1, and lead to predictive genetic testing for specific NF1-associated features.

Lastly, the *NF1* heterozygous brain not only promotes glioma formation, but also impacts on normal neuronal function. This is particularly important, as tumors in children arise in the context of a developing and maturing nervous system composed of neurons continually establishing new connections and fortifying old ones. An intrinsic vulnerability conferred by tumor suppressor gene heterozygosity may predispose some neurons to damage and death as a direct result of tumor formation and growth as well as tumor treatment (Brown *et al.*, 2010). It is therefore vital that we more completely elucidate the signals and pathways that govern neuronal susceptibility to damage in order to minimize the impact of tumor evolution and treatment on the developing brain (Fig. 9.7).

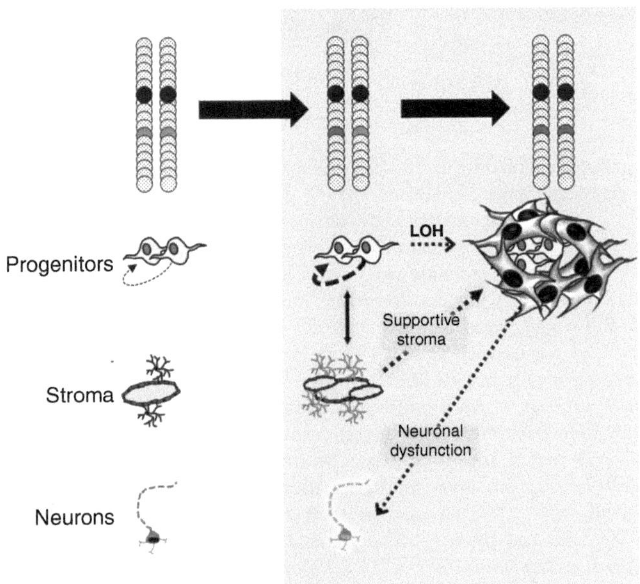

Figure 9.7 Expanded two-hit hypothesis model. In the original two-hit hypothesis, tumor suppressor gene heterozygosity resulted in a statistically increased frequency of biallelic tumor suppressor gene inactivation (second hit). In addition, tumor suppressor gene heterozygosity changes the proliferative and self-renewal potential of progenitor cells, thus facilitating this loss of heterozygosity. We now propose that tumor suppressor gene heterozygosity in non-neoplastic cells in the tumor microenvironment establishes a supportive niche for tumor formation and maintenance, which in some cases, like NF1, is required for tumorigenesis. Finally, tumor suppressor gene heterozygosity may also lead to increased neuronal dysfunction as a consequence of glioma formation and growth. (See Color Insert.)

An improved understanding of all of these contributing determinants, including the genetic background, will help to usher in an era where personalized brain tumor treatment involves identifying children in early infancy most at risk for developing glioma, and instituting therapies that target the neoplastic cells and their supportive stroma during the period of gliomagenesis in concert with agents that reduce the damage to other non-neoplastic cells of the developing brain, including neurons and oligodendrocytes.

REFERENCES

Alcantara Llaguno, S., Chen, J., Kwon, C. H., Jackson, E. L., Li, Y., Burns, D. K., Alvarez-Buylla, A., and Parada, L. F. (2009). Malignant astrocytomas originate from neural stem/progenitor cells in a somatic tumor suppressor mouse model. *Cancer Cell* **15,** 45–56.

Amankulor, N. M., Hambardzumyan, D., Pyonteck, S. M., Becher, O. J., Joyce, J. A., and Holland, E. C. (2009). Sonic hedgehog pathway activation is induced by acute brain injury and regulated by injury-related inflammation. *J. Neurosci.* **29,** 10299–10308.

Andersen, L. B., Ballester, R., Marchuk, D. A., Chang, E., Gutmann, D. H., Saulino, A. M., Camonis, J., Wigler, M., and Collins, F. S. (1993). A conserved alternative splice in the von Recklinghausen neurofibromatosis (NF1) gene produces two neurofibromin isoforms, both of which have GTPase-activating protein activity. *Mol. Cell. Biol.* **13,** 487–495.

Bajenaru, M. L., Zhu, Y., Hedrick, N. M., Donahoe, J., Parada, L. F., and Gutmann, D. H. (2002). Astrocyte-specific inactivation of the neurofibromatosis 1 gene (NF1) is insufficient for astrocytoma formation. *Mol. Cell. Biol.* **22,** 5100–5113.

Bajenaru, M. L., Hernandez, M. R., Perry, A., Zhu, Y., Parada, L. F., Garbow, J. R., and Gutmann, D. H. (2003). Optic nerve glioma in mice requires astrocyte Nf1 gene inactivation and Nf1 brain heterozygosity. *Cancer Res.* **63,** 8573–8577.

Bajenaru, M. L., Garbow, J. R., Perry, A., Hernandez, M. R., and Gutmann, D. H. (2005). Natural history of neurofibromatosis 1-associated optic nerve glioma in mice. *Ann. Neurol.* **57,** 119–127.

Ballester, R., Marchuk, D., Boguski, M., Saulino, A., Letcher, R., Wigler, M., and Collins, F. (1990). The NF1 locus encodes a protein functionally related to mammalian GAP and yeast IRA proteins. *Cell* **63,** 851–859.

Banati, R. B., Gehrmann, J., Schubert, P., and Kreutzberg, G. W. (1993). Cytotoxicity of microglia. *Glia* **7,** 111–118.

Banerjee, S., Byrd, J. N., Gianino, S. M., Harpstrite, S. E., Rodriguez, F. J., Tuskan, R. G., Reilly, K. M., Piwnica-Worms, D. R., and Gutmann, D. H. (2010). The neurofibromatosis type 1 tumor suppressor controls cell growth by regulating signal transducer and activator of transcription-3 activity in vitro and in vivo. *Cancer Res.* **70,** 1356–1366.

Basu, T. N., Gutmann, D. H., Fletcher, J. A., Glover, T. W., Collins, F. S., and Downward, J. (1992). Aberrant regulation of ras proteins in malignant tumour cells from type 1 neurofibromatosis patients. *Nature* **356,** 713–715.

Bhola, P., Banerjee, S., Mukherjee, J., Balasubramanium, A., Arun, V., Karim, Z., Burrell, K., Croul, S., Gutmann, D. H., and Guha, A. (2010). Preclinical in vivo evaluation of rapamycin in human malignant peripheral nerve sheath explant xenograft. *Int. J. Cancer* **126,** 563–571.

Bollag, G., Clapp, D. W., Shih, S., Adler, F., Zhang, Y. Y., Thompson, P., Lange, B. J., Freedman, M. H., McCormick, F., Jacks, T., and Shannon, K. (1996). Loss of NF1

results in activation of the Ras signaling pathway and leads to aberrant growth in haematopoietic cells. *Nat. Genet.* **12,** 144–148.
Brown, J. A., Gianino, S., and Gutmann, D. H. (2010). Defective cyclic AMP generation underlies the sensitivity of central nervous system neurons to neurofibromatosis-1 heterozygosity. *J. Neurosci.* (in press).
Burmer, G. C., Rabinovitch, P. S., and Loeb, L. A. (1991). Frequency and spectrum of c-Ki-ras mutations in human sporadic colon carcinoma, carcinomas arising in ulcerative colitis, and pancreatic adenocarcinoma. *Environ. Health Perspect.* **93,** 27–31.
Calabrese, C., Poppleton, H., Kocak, M., Hogg, T. L., Fuller, C., Hamner, B., Oh, E. Y., Gaber, M. W., Finklestein, D., Allen, M., Frank, A., Bayazitov, I. T., *et al.* (2007). A perivascular niche for brain tumor stem cells. *Cancer Cell* **11,** 69–82.
Costa, R. M., Yang, T., Huynh, D. P., Pulst, S. M., Viskochil, D. H., Silva, A. J., and Brannan, C. I. (2001). Learning deficits, but normal development and tumor predisposition, in mice lacking exon 23a of Nf1. *Nat. Genet.* **27,** 399–405.
Coull, J. A., Beggs, S., Boudreau, D., Boivin, D., Tsuda, M., Inoue, K., Gravel, C., Salter, M. W., and De Koninck, Y. (2005). BDNF from microglia causes the shift in neuronal anion gradient underlying neuropathic pain. *Nature* **438,** 1017–1021.
Daginakatte, G. C., and Gutmann, D. H. (2007). Neurofibromatosis-1 (Nf1) heterozygous brain microglia elaborate paracrine factors that promote Nf1-deficient astrocyte and glioma growth. *Hum. Mol. Genet.* **16,** 1098–1112.
Daginakatte, G. C., Gianino, S. M., Zhao, N. W., Parsadanian, A. S., and Gutmann, D. H. (2008). Increased c-Jun-NH2-kinase signaling in neurofibromatosis-1 heterozygous microglia drives microglia activation and promotes optic glioma proliferation. *Cancer Res.* **68,** 10358–10366.
Dasgupta, B., Dugan, L. L., and Gutmann, D. H. (2003). The neurofibromatosis 1 gene product neurofibromin regulates pituitary adenylate cyclase-activating polypeptide-mediated signaling in astrocytes. *J. Neurosci.* **23,** 8949–8954.
Dasgupta, B., Li, W., Perry, A., and Gutmann, D. H. (2005a). Glioma formation in neurofibromatosis 1 reflects preferential activation of K-RAS in astrocytes. *Cancer Res.* **65,** 236–245.
Dasgupta, B., Yi, Y., Chen, D. Y., Weber, J. D., and Gutmann, D. H. (2005b). Proteomic analysis reveals hyperactivation of the mammalian target of rapamycin pathway in neurofibromatosis 1-associated human and mouse brain tumors. *Cancer Res.* **65,** 2755–2760.
Davoust, N., Vuaillat, C., Androdias, G., and Nataf, S. (2008). From bone marrow to microglia: Barriers and avenues. *Trends Immunol.* **29,** 227–234.
DeClue, J. E., Papageorge, A. G., Fletcher, J. A., Diehl, S. R., Ratner, N., Vass, W. C., and Lowy, D. R. (1992). Abnormal regulation of mammalian p21ras contributes to malignant tumor growth in von Recklinghausen (type 1) neurofibromatosis. *Cell* **69,** 265–273.
Deshmukh, H., Yeh, T. H., Yu, J., Sharma, M. K., Perry, A., Leonard, J. R., Watson, M. A., Gutmann, D. H., and Nagarajan, R. (2008). High-resolution, dual-platform aCGH analysis reveals frequent HIPK2 amplification and increased expression in pilocytic astrocytomas. *Oncogene* **27,** 4745–4751.
Donovan, S., See, W., Bonifas, J., Stokoe, D., and Shannon, K. M. (2002). Hyperactivation of protein kinase B and ERK have discrete effects on survival, proliferation, and cytokine expression in Nf1-deficient myeloid cells. *Cancer Cell* **2,** 507–514.
Elkabes, S., DiCicco-Bloom, E. M., and Black, I. B. (1996). Brain microglia/macrophages express neurotrophins that selectively regulate microglial proliferation and function. *J. Neurosci.* **16,** 2508–2521.
Engelman, J. A. (2009). Targeting PI3K signalling in cancer: Opportunities, challenges and limitations. *Nat. Rev. Cancer* **9,** 550–562.

Flugel, A., Labeur, M. S., Grasbon-Frodl, E. M., Kreutzberg, G. W., and Graeber, M. B. (1999). Microglia only weakly present glioma antigen to cytotoxic T cells. *Int. J. Dev. Neurosci.* **17,** 547–556.

Forshew, T., Tatevossian, R. G., Lawson, A. R., Ma, J., Neale, G., Ogunkolade, B. W., Jones, T. A., Aarum, J., Dalton, J., Bailey, S., Chaplin, T., Carter, R. L., *et al.* (2009). Activation of the ERK/MAPK pathway: A signature genetic defect in posterior fossa pilocytic astrocytomas. *J. Pathol.* **218,** 172–181.

Freeman, C. R., Farmer, J. P., and Montes, J. (1998). Low-grade astrocytomas in children: Evolving management strategies. *Int. J. Radiat. Oncol. Biol. Phys.* **41,** 979–987.

Friedman, J. M. (1999). Epidemiology of neurofibromatosis type 1. *Am. J. Med. Genet.* **89,** 1–6.

Gibbons, J. J., Abraham, R. T., and Yu, K. (2009). Mammalian target of rapamycin: Discovery of rapamycin reveals a signaling pathway important for normal and cancer cell growth. *Semin. Oncol.* **36**(Suppl. 3), S3–S17.

Graeber, M. B., Scheithauer, B. W., and Kreutzberg, G. W. (2002). Microglia in brain tumors. *Glia* **40,** 252–259.

Gregorian, C., Nakashima, J., Dry, S. M., Nghiemphu, P. L., Smith, K. B., Ao, Y., Dang, J., Lawson, G., Mellinghoff, I. K., Mischel, P. S., Phelps, M., Parada, L. F., *et al.* (2009). PTEN dosage is essential for neurofibroma development and malignant transformation. *Proc. Natl. Acad. Sci. USA* **106,** 19479–19484.

Gressani, K. M., Rollins, L. A., Leone-Kabler, S., Cline, J. M., and Miller, M. S. (1998). Induction of mutations in Ki-ras and INK4a in liver tumors of mice exposed in utero to 3-methylcholanthrene. *Carcinogenesis* **19,** 1045–1052.

Guillamo, J. S., Creange, A., Kalifa, C., Grill, J., Rodriguez, D., Doz, F., Barbarot, S., Zerah, M., Sanson, M., Bastuji-Garin, S., and Wolkenstein, P. (2003). Prognostic factors of CNS tumours in Neurofibromatosis 1 (NF1): A retrospective study of 104 patients. *Brain* **126,** 152–160.

Gutmann, D. H., Geist, R. T., Rose, K., and Wright, D. E. (1995a). Expression of two new protein isoforms of the neurofibromatosis type 1 gene product, neurofibromin, in muscle tissues. *Dev. Dyn.* **202,** 302–311.

Gutmann, D. H., Geist, R. T., Wright, D. E., and Snider, W. D. (1995b). Expression of the neurofibromatosis 1 (NF1) isoforms in developing and adult rat tissues. *Cell Growth Differ.* **6,** 315–323.

Gutmann, D. H., Aylsworth, A., Carey, J. C., Korf, B., Marks, J., Pyeritz, R. E., Rubenstein, A., and Viskochil, D. (1997). The diagnostic evaluation and multidisciplinary management of neurofibromatosis 1 and neurofibromatosis 2. *JAMA* **278,** 51–57.

Gutmann, D. H., Zhang, Y., and Hirbe, A. (1999). Developmental regulation of a neuron-specific neurofibromatosis 1 isoform. *Ann. Neurol.* **46,** 777–782.

Gutmann, D. H., Donahoe, J., Brown, T., James, C. D., and Perry, A. (2000). Loss of neurofibromatosis 1 (NF1) gene expression in NF1-associated pilocytic astrocytomas. *Neuropathol. Appl. Neurobiol.* **26,** 361–367.

Gutmann, D. H., James, C. D., Poyhonen, M., Louis, D. N., Ferner, R., Guha, A., Hariharan, S., Viskochil, D., and Perry, A. (2003). Molecular analysis of astrocytomas presenting after age 10 in individuals with NF1. *Neurology* **61,** 1397–1400.

Hawes, J. J., Tuskan, R. G., and Reilly, K. M. (2007). Nf1 expression is dependent on strain background: Implications for tumor suppressor haploinsufficiency studies. *Neurogenetics* **8,** 121–130.

Hegedus, B., Banerjee, D., Yeh, T. H., Rothermich, S., Perry, A., Rubin, J. B., Garbow, J. R., and Gutmann, D. H. (2008). Preclinical cancer therapy in a mouse model of neurofibromatosis-1 optic glioma. *Cancer Res.* **68,** 1520–1528.

Hiatt, K. K., Ingram, D. A., Zhang, Y., Bollag, G., and Clapp, D. W. (2001). Neurofibromin GTPase-activating protein-related domains restore normal growth in Nf1-/- cells. *J. Biol. Chem.* **276,** 7240–7245.

Jacob, K., Albrecht, S., Sollier, C., Faury, D., Sader, E., Montpetit, A., Serre, D., Hauser, P., Garami, M., Bognar, L., Hanzely, Z., Montes, J. L., *et al.* (2009). Duplication of 7q34 is specific to juvenile pilocytic astrocytomas and a hallmark of cerebellar and optic pathway tumours. *Br. J. Cancer* **101,** 722–733.

Janzarik, W. G., Kratz, C. P., Loges, N. T., Olbrich, H., Klein, C., Schafer, T., Scheurlen, W., Roggendorf, W., Weiller, C., Niemeyer, C., Korinthenberg, R., Pfister, S., *et al.* (2007). Further evidence for a somatic KRAS mutation in a pilocytic astrocytoma. *Neuropediatrics* **38,** 61–63.

Johannessen, C. M., Reczek, E. E., James, M. F., Brems, H., Legius, E., and Cichowski, K. (2005). The NF1 tumor suppressor critically regulates TSC2 and mTOR. *Proc. Natl Acad. Sci. USA* **102,** 8573–8578.

Johannessen, C. M., Johnson, B. W., Williams, S. M., Chan, A. W., Reczek, E. E., Lynch, R. C., Rioth, M. J., McClatchey, A., Ryeom, S., and Cichowski, K. (2008). TORC1 is essential for NF1-associated malignancies. *Curr. Biol.* **18,** 56–62.

Jones, D. T., Kocialkowski, S., Liu, L., Pearson, D. M., Backlund, L. M., Ichimura, K., and Collins, V. P. (2008). Tandem duplication producing a novel oncogenic BRAF fusion gene defines the majority of pilocytic astrocytomas. *Cancer Res.* **68,** 8673–8677.

Khalaf, W. F., Yang, F. C., Chen, S., White, H., Bessler, W., Ingram, D. A., and Clapp, D. W. (2007). K-ras is critical for modulating multiple c-kit-mediated cellular functions in wild-type and Nf1+/- mast cells. *J. Immunol.* **178,** 2527–2534.

Kim, H. A., Rosenbaum, T., Marchionni, M. A., Ratner, N., and DeClue, J. E. (1995). Schwann cells from neurofibromin deficient mice exhibit activation of p21ras, inhibition of cell proliferation and morphological changes. *Oncogene* **11,** 325–335.

Kluwe, L., Hagel, C., Tatagiba, M., Thomas, S., Stavrou, D., Ostertag, H., von Deimling, A., and Mautner, V. F. (2001). Loss of NF1 alleles distinguish sporadic from NF1-associated pilocytic astrocytomas. *J. Neuropathol. Exp. Neurol.* **60,** 917–920.

Knudson, A. G., Jr. (1971). Mutation and cancer: Statistical study of retinoblastoma. *Proc. Natl. Acad. Sci. USA* **68,** 820–823.

Kostianovsky, A. M., Maier, L. M., Anderson, R. C., Bruce, J. N., and Anderson, D. E. (2008). Astrocytic regulation of human monocytic/microglial activation. *J. Immunol.* **181,** 5425–5432.

Kunkel, P., Muller, S., Schirmacher, P., Stavrou, D., Fillbrandt, R., Westphal, M., and Lamszus, K. (2001). Expression and localization of scatter factor/hepatocyte growth factor in human astrocytomas. *Neuro. Oncol.* **3,** 82–88.

Lafuente, J. V., Adan, B., Alkiza, K., Garibi, J. M., Rossi, M., and Cruz-Sanchez, F. F. (1999). Expression of vascular endothelial growth factor (VEGF) and platelet-derived growth factor receptor-beta (PDGFR-beta) in human gliomas. *J. Mol. Neurosci.* **13,** 177–185.

Lai, J. P., Zhan, G. X., Campbell, D. E., Douglas, S. D., and Ho, W. Z. (2000). Detection of substance P and its receptor in human fetal microglia. *Neuroscience* **101,** 1137–1144.

Legius, E., Dierick, H., Wu, R., Hall, B. K., Marynen, P., Cassiman, J. J., and Glover, T. W. (1994). TP53 mutations are frequent in malignant NF1 tumors. *Genes Chromosom. Cancer* **10,** 250–255.

Li, J., Perry, A., James, C. D., and Gutmann, D. H. (2001). Cancer-related gene expression profiles in NF1-associated pilocytic astrocytomas. *Neurology* **56,** 885–890.

Li, F., Munchhof, A. M., White, H. A., Mead, L. E., Krier, T. R., Fenoglio, A., Chen, S., Wu, X., Cai, S., Yang, F. C., and Ingram, D. A. (2006). Neurofibromin is a novel regulator of RAS-induced signals in primary vascular smooth muscle cells. *Hum. Mol. Genet.* **15,** 1921–1930.

Lin, E. Y., and Pollard, J. W. (2007). Tumor-associated macrophages press the angiogenic switch in breast cancer. *Cancer Res.* **67,** 5064–5066.

Listernick, R., Darling, C., Greenwald, M., Strauss, L., and Charrow, J. (1995). Optic pathway tumors in children: The effect of neurofibromatosis type 1 on clinical manifestations and natural history. *J. Pediatr.* **127,** 718–722.

Listernick, R., Ferner, R. E., Liu, G. T., and Gutmann, D. H. (2007). Optic pathway gliomas in neurofibromatosis-1: Controversies and recommendations. *Ann. Neurol.* **61,** 189–198.

Liu, P., Cheng, H., Roberts, T. M., and Zhao, J. J. (2009). Targeting the phosphoinositide 3-kinase pathway in cancer. *Nat. Rev. Drug Discov.* **8,** 627–644.

Louis, D. N., Ohgaki, H., Wiestler, O. D., Cavenee, W. K., Burger, P. C., Jouvet, A., Scheithauer, B. W., and Kleihues, P. (2007). The 2007 WHO classification of tumours of the central nervous system. *Acta Neuropathol.* **114,** 97–109.

Luber-Narod, J., Kage, R., and Leeman, S. E. (1994). Substance P enhances the secretion of tumor necrosis factor-alpha from neuroglial cells stimulated with lipopolysaccharide. *J. Immunol.* **152,** 819–824.

MacConaill, L. E., Campbell, C. D., Kehoe, S. M., Bass, A. J., Hatton, C., Niu, L., Davis, M., Yao, K., Hanna, M., Mondal, C., Luongo, L., Emery, C. M., *et al.* (2009). Profiling critical cancer gene mutations in clinical tumor samples. *PLoS ONE* **4,** e7887.

Mancini, F., and Moretti, F. (2009). Mitochondrial MDM4 (MDMX): An unpredicted role in the p53-mediated intrinsic apoptotic pathway. *Cell Cycle* **8,** 3854–3859.

Marchuk, D. A., Saulino, A. M., Tavakkol, R., Swaroop, M., Wallace, M. R., Andersen, L. B., Mitchell, A. L., Gutmann, D. H., Boguski, M., and Collins, F. S. (1991). cDNA cloning of the type 1 neurofibromatosis gene: Complete sequence of the NF1 gene product. *Genomics* **11,** 931–940.

Marin-Teva, J. L., Dusart, I., Colin, C., Gervais, A., van Rooijen, N., and Mallat, M. (2004). Microglia promote the death of developing Purkinje cells. *Neuron* **41,** 535–547.

Markovic, D. S., Vinnakota, K., Chirasani, S., Synowitz, M., Raguet, H., Stock, K., Sliwa, M., Lehmann, S., Kalin, R., van Rooijen, N., Holmbeck, K., Heppner, F. L., *et al.* (2009). Gliomas induce and exploit microglial MT1-MMP expression for tumor expansion. *Proc. Natl. Acad. Sci. USA* **106,** 12530–12535.

Martin, G. A., Viskochil, D., Bollag, G., McCabe, P. C., Crosier, W. J., Haubruck, H., Conroy, L., Clark, R., O'Connell, P., Cawthon, R. M., *et al.* (1990). The GAP-related domain of the neurofibromatosis type 1 gene product interacts with ras p21. *Cell* **63,** 843–849.

Martin, F. C., Anton, P. A., Gornbein, J. A., Shanahan, F., and Merrill, J. E. (1993). Production of interleukin-1 by microglia in response to substance P: Role for a non-classical NK-1 receptor. *J. Neuroimmunol.* **42,** 53–60.

McDaniel, A. S., Allen, J. D., Park, S. J., Jaffer, Z. M., Michels, E. G., Burgin, S. J., Chen, S., Bessler, W. K., Hofmann, C., Ingram, D. A., Chernoff, J., and Clapp, D. W. (2008). Pak1 regulates multiple c-Kit mediated Ras-MAPK gain-in-function phenotypes in Nf1+/- mast cells. *Blood* **112,** 4646–4654.

Morgan, D., Gordon, M. N., Tan, J., Wilcock, D., and Rojiani, A. M. (2005). Dynamic complexity of the microglial activation response in transgenic models of amyloid deposition: Implications for Alzheimer therapeutics. *J. Neuropathol. Exp. Neurol.* **64,** 743–753.

Parney, I. F., Waldron, J. S., and Parsa, A. T. (2009). Flow cytometry and in vitro analysis of human glioma-associated macrophages. *Lab. Invest. J. Neurosurg.* **110,** 572–582.

Perrone, F., Da Riva, L., Orsenigo, M., Losa, M., Jocolle, G., Millefanti, C., Pastore, E., Gronchi, A., Pierotti, M. A., and Pilotti, S. (2009). PDGFRA, PDGFRB, EGFR, and downstream signaling activation in malignant peripheral nerve sheath tumor. *Neuro Oncol* **11,** 725–736.

Piecha, D., Muratoglu, S., Morgelin, M., Hauser, N., Studer, D., Kiss, I., Paulsson, M., and Deak, F. (1999). Matrilin-2, a large, oligomeric matrix protein, is expressed by a great variety of cells and forms fibrillar networks. *J. Biol. Chem.* **274,** 13353–13361.

Platten, M., Kretz, A., Naumann, U., Aulwurm, S., Egashira, K., Isenmann, S., and Weller, M. (2003). Monocyte chemoattractant protein-1 increases microglial infiltration and aggressiveness of gliomas. *Ann. Neurol.* **54,** 388–392.

Pollard, J. W. (2004). Tumour-educated macrophages promote tumour progression and metastasis. *Nat. Rev. Cancer* **4,** 71–78.

Puca, R., Nardinocchi, L., Sacchi, A., Rechavi, G., Givol, D., and D'Orazi, G. (2009). HIPK2 modulates p53 activity towards pro-apoptotic transcription. *Mol. Cancer* **8,** 85.

Rasley, A., Bost, K. L., Olson, J. K., Miller, S. D., and Marriott, I. (2002). Expression of functional NK-1 receptors in murine microglia. *Glia* **37,** 258–267.

Reilly, K. M., Loisel, D. A., Bronson, R. T., McLaughlin, M. E., and Jacks, T. (2000). Nf1; Trp53 mutant mice develop glioblastoma with evidence of strain-specific effects. *Nat. Genet.* **26,** 109–113.

Reilly, K. M., Tuskan, R. G., Christy, E., Loisel, D. A., Ledger, J., Bronson, R. T., Smith, C. D., Tsang, S., Munroe, D. J., and Jacks, T. (2004). Susceptibility to astrocytoma in mice mutant for Nf1 and Trp53 is linked to chromosome 11 and subject to epigenetic effects. *Proc. Natl. Acad. Sci. USA* **101,** 13008–13013.

Reish, O., Orlovski, A., Mashevitz, M., Sher, C., Libman, V., Rosenblat, M., and Avivi, L. (2003). Modified allelic replication in lymphocytes of patients with neurofibromatosis type 1. *Cancer Genet. Cytogenet.* **143,** 133–139.

Robinson, J. P., VanBrocklin, M. W., Guilbeault, A. R., Signorelli, D. L., Brandner, S., and Holmen, S. L. (2010). Activated BRAF induces gliomas in mice when combined with Ink4a/Arf loss or Akt activation. *Oncogene* **29,** 335–344.

Roumier, A., Bechade, C., Poncer, J. C., Smalla, K. H., Tomasello, E., Vivier, E., Gundelfinger, E. D., Triller, A., and Bessis, A. (2004). Impaired synaptic function in the microglial KARAP/DAP12-deficient mouse. *J. Neurosci.* **24,** 11421–11428.

Sandsmark, D. K., Zhang, H., Hegedus, B., Pelletier, C. L., Weber, J. D., and Gutmann, D. H. (2007). Nucleophosmin mediates mammalian target of rapamycin-dependent actin cytoskeleton dynamics and proliferation in neurofibromin-deficient astrocytes. *Cancer Res.* **67,** 4790–4799.

Sanoudou, D., Tingby, O., Ferguson-Smith, M. A., Collins, V. P., and Coleman, N. (2000). Analysis of pilocytic astrocytoma by comparative genomic hybridization. *Br. J. Cancer* **82,** 1218–1222.

Schiffman, J. D., Hodgson, J. G., VandenBerg, S. R., Flaherty, P., Polley, M. Y., Yu, M., Fisher, P. G., Rowitch, D. H., Ford, J. M., Berger, M. S., Ji, H., Gutmann, D. H., *et al.* (2010). Oncogenic BRAF mutation with CDKN2A inactivation is characteristic of a subset of pediatric malignant astrocytomas. *Cancer Res.* **70,** 512–519.

Sharma, M. K., Zehnbauer, B. A., Watson, M. A., and Gutmann, D. H. (2005). RAS pathway activation and an oncogenic RAS mutation in sporadic pilocytic astrocytoma. *Neurology* **65,** 1335–1336.

Sharma, M. K., Watson, M. A., Lyman, M., Perry, A., Aldape, K. D., Deak, F., and Gutmann, D. H. (2006). Matrilin-2 expression distinguishes clinically relevant subsets of pilocytic astrocytoma. *Neurology* **66,** 127–130.

Sharma, M. K., Mansur, D. B., Reifenberger, G., Perry, A., Leonard, J. R., Aldape, K. D., Albin, M. G., Emnett, R. J., Loeser, S., Watson, M. A., Nagarajan, R., and Gutmann, D. H. (2007). Distinct genetic signatures among pilocytic astrocytomas relate to their brain region origin. *Cancer Res.* **67,** 890–900.

Sievert, A. J., and Fisher, M. J. (2009). Pediatric low-grade gliomas. *J. Child Neurol.* **24,** 1397–1408.

Simmons, G. W., Pong, W. W., Emnett, R. J., White, C. R., Gianino, S. M., Rodriguez, F. J., and Gutmann, D. H. (2011). Neurofibromatosis-1 heterozygosity increases microglia in a spatially- and temporally-restricted pattern relevant to mouse optic glioma formation and growth. *J. Neuropathol. Exp. Neurol.* (in press).

Struve, J., Maher, P. C., Li, Y. Q., Kinney, S., Fehlings, M. G., Kuntz, C. T., and Sherman, L. S. (2005). Disruption of the hyaluronan-based extracellular matrix in spinal cord promotes astrocyte proliferation. *Glia* **52,** 16–24.

Sun, T., Gianino, S., Jackson, E., Piwnica-Worms, D. R., Gutmann, D. H., and Rubin, J. B. (2010). CXCL12 alone is insufficient for gliomagenesis in *Nf1* mutant mice. *J. Neuroimmunol.* (in press).

Surawicz, T. S., Davis, F., Freels, S., Laws, E. R., Jr., and Menck, H. R. (1998). Brain tumor survival: Results from the National Cancer Data Base. *J. Neurooncol.* **40,** 151–160.

Surawicz, T. S., McCarthy, B. J., Kupelian, V., Jukich, P. J., Bruner, J. M., and Davis, F. G. (1999). Descriptive epidemiology of primary brain and CNS tumors: Results from the Central Brain Tumor Registry of the United States, 1990-1994. *Neuro Oncol* **1,** 14–25.

Taylor, M. D., Poppleton, H., Fuller, C., Su, X., Liu, Y., Jensen, P., Magdaleno, S., Dalton, J., Calabrese, C., Board, J., Macdonald, T., Rutka, J., Guha, A., Gajjar, A., Curran, T., and Gilbertson, R. J. (2005). Radial glia cells are candidate stem cells of ependymoma. *Cancer Cell* **8,** 323–335.

Tong, J., Hannan, F., Zhu, Y., Bernards, A., and Zhong, Y. (2002). Neurofibromin regulates G protein-stimulated adenylyl cyclase activity. *Nat. Neurosci.* **5,** 95–96.

Tsai, J. C., Goldman, C. K., and Gillespie, G. Y. (1995). Vascular endothelial growth factor in human glioma cell lines: Induced secretion by EGF, PDGF-BB, and bFGF. *J. Neurosurg.* **82,** 864–873.

Verdijk, R. M., den Bakker, M. A., Dubbink, H. J., Hop, W. C., Dinjens, W. N., and Kros, J. M. (2010). TP53 mutation analysis of malignant peripheral nerve sheath tumors. *J. Neuropathol. Exp. Neurol.* **69,** 16–26.

Verhaak, R. G., Hoadley, K. A., Purdom, E., Wang, V., Qi, Y., Wilkerson, M. D., Miller, C. R., Ding, L., Golub, T., Mesirov, J. P., Alexe, G., Lawrence, M., *et al.* (2009). Integrated genomic analysis identifies clinically relevant subtypes of glioblastoma characterized by abnormalities in PDGFRA, IDH1, EGFR, and NF1. *Cancer Cell* **17,** 98–110.

Viskochil, D., Buchberg, A. M., Xu, G., Cawthon, R. M., Stevens, J., Wolff, R. K., Culver, M., Carey, J. C., Copeland, N. G., Jenkins, N. A., *et al.* (1990). Deletions and a translocation interrupt a cloned gene at the neurofibromatosis type 1 locus. *Cell* **62,** 187–192.

Wallace, M. R., Marchuk, D. A., Andersen, L. B., Letcher, R., Odeh, H. M., Saulino, A. M., Fountain, J. W., Brereton, A., Nicholson, J., Mitchell, A. L., *et al.* (1990). Type 1 neurofibromatosis gene: Identification of a large transcript disrupted in three NF1 patients. *Science* **249,** 181–186.

Warrington, N. M., Woerner, B. M., Daginakatte, G. C., Dasgupta, B., Perry, A., Gutmann, D. H., and Rubin, J. B. (2007). Spatiotemporal differences in CXCL12 expression and cyclic AMP underlie the unique pattern of optic glioma growth in neurofibromatosis type 1. *Cancer Res.* **67,** 8588–8595.

Warrington, N. M., Gianino, S. M., Jackson, E., Goldhoff, P., Garbow, J. R., Piwnica-Worms, D., Gutmann, D. H., and Rubin, J. B. (2010). Cyclic AMP suppression is sufficient to induce gliomagenesis in a mouse model of neurofibromatosis-1. *Cancer Res.* **70,** 5717–5727.

Watters, J. J., Schartner, J. M., and Badie, B. (2005). Microglia function in brain tumors. *J. Neurosci. Res.* **81,** 447–455.

Weissenberger, J., Loeffler, S., Kappeler, A., Kopf, M., Lukes, A., Afanasieva, T. A., Aguzzi, A., and Weis, J. (2004). IL-6 is required for glioma development in a mouse model. *Oncogene* **23,** 3308–3316.

Wesolowska, A., Kwiatkowska, A., Slomnicki, L., Dembinski, M., Master, A., Sliwa, M., Franciszkiewicz, K., Chouaib, S., and Kaminska, B. (2008). Microglia-derived TGF-beta as an important regulator of glioblastoma invasion—An inhibition of TGF-beta-dependent effects by shRNA against human TGF-beta type II receptor. *Oncogene* **27,** 918–930.

White, F. V., Anthony, D. C., Yunis, E. J., Tarbell, N. J., Scott, R. M., and Schofield, D. E. (1995). Nonrandom chromosomal gains in pilocytic astrocytomas of childhood. *Hum. Pathol.* **26,** 979–986.

Wimmer, K., Eckart, M., Meyer-Puttlitz, B., Fonatsch, C., and Pietsch, T. (2002). Mutational and expression analysis of the NF1 gene argues against a role as tumor suppressor in sporadic pilocytic astrocytomas. *J. Neuropathol. Exp. Neurol.* **61,** 896–902.

Xu, G. F., Lin, B., Tanaka, K., Dunn, D., Wood, D., Gesteland, R., White, R., Weiss, R., and Tamanoi, F. (1990). The catalytic domain of the neurofibromatosis type 1 gene product stimulates ras GTPase and complements ira mutants of *S. cerevisiae*. *Cell* **63,** 835–841.

Yamada, T., Tsubouchi, H., Daikuhara, Y., Prat, M., Comoglio, P. M., McGeer, P. L., and McGeer, E. G. (1994). Immunohistochemistry with antibodies to hepatocyte growth factor and its receptor protein (c-MET) in human brain tissues. *Brain Res.* **637,** 308–312.

Yang, F. C., Chen, S., Robling, A. G., Yu, X., Nebesio, T. D., Yan, J., Morgan, T., Li, X., Yuan, J., Hock, J., Ingram, D. A., and Clapp, D. W. (2006). Hyperactivation of p21ras and PI3K cooperate to alter murine and human neurofibromatosis type 1-haploinsufficient osteoclast functions. *J. Clin. Invest.* **116,** 2880–2891.

Yang, F. C., Ingram, D. A., Chen, S., Zhu, Y., Yuan, J., Li, X., Yang, X., Knowles, S., Horn, W., Li, Y., Zhang, S., Yang, Y., *et al.* (2008). Nf1-dependent tumors require a microenvironment containing Nf1+/− and c-kit-dependent bone marrow. *Cell* **135,** 437–448.

Yang, I., Han, S. J., Kaur, G., Crane, C., and Parsa, A. T. (2010). The role of microglia in central nervous system immunity and glioma immunology. *J. Clin. Neurosci.* **17,** 6–10.

Yeh, T. H., Lee da, Y., Gianino, S. M., and Gutmann, D. H. (2009). Microarray analyses reveal regional astrocyte heterogeneity with implications for neurofibromatosis type 1 (NF1)-regulated glial proliferation. *Glia* **57,** 1239–1249.

Yu, J., Deshmukh, H., Gutmann, R. J., Emnett, R. J., Rodriguez, F. J., Watson, M. A., Nagarajan, R., and Gutmann, D. H. (2009). Alterations of BRAF and HIPK2 loci predominate in sporadic pilocytic astrocytoma. *Neurology* **73,** 1526–1531.

Zattara-Cannoni, H., Gambarelli, D., Lena, G., Dufour, H., Choux, M., Grisoli, F., and Vagner-Capodano, A. M. (1998). Are juvenile pilocytic astrocytomas benign tumors? A cytogenetic study in 24 cases. *Cancer Genet. Cytogenet.* **104,** 157–160.

Zhu, Y., Ghosh, P., Charnay, P., Burns, D. K., and Parada, L. F. (2002). Neurofibromas in NF1: Schwann cell origin and role of tumor environment. *Science* **296,** 920–922.

Zhu, Y., Harada, T., Liu, L., Lush, M. E., Guignard, F., Harada, C., Burns, D. K., Bajenaru, M. L., Gutmann, D. H., and Parada, L. F. (2005). Inactivation of NF1 in CNS causes increased glial progenitor proliferation and optic glioma formation. *Development* **132,** 5577–5588.

Zohrabian, V. M., Nandu, H., Gulati, N., Khitrov, G., Zhao, C., Mohan, A., Demattia, J., Braun, A., Das, K., Murali, R., and Jhanwar-Uniyal, M. (2007). Gene expression profiling of metastatic brain cancer. *Oncol. Rep.* **18,** 321–328.

CHAPTER TEN

Tumor Macrophages: Protective and Pathogenic Roles in Cancer Development

Joseph E. Qualls *and* Peter J. Murray

Contents

Abstract

Macrophage infiltration often occurs in cancer and has stimulated new efforts to define macrophage function within solid tumors. The macrophage, a myelophagocytic cell of the immune system, is at the front line of pathogen defense, wound healing, and maintaining homeostasis within the body. However, increased macrophage numbers during cancer generally correlates with poor prognosis. This chapter will focus on the function of myelophagocytic cells within solid tumors, their potential roles in tumor progression, and the possibilities of their manipulation as a component of cancer therapy.

Departments of Infectious Diseases and Immunology, St. Jude Children's Research Hospital, Memphis, TN, USA

Current Topics in Developmental Biology, Volume 94
ISSN 0070-2153, DOI: 10.1016/B978-0-12-380916-2.00010-3

1. Introduction

Virchow noted that solid tumors are associated with inflammatory reactions in the eighteenth century (Balkwill and Mantovani, 2001). Since Virchow's era, however, the link between inflammation and cancer, and especially the infiltration of solid tumors with myelophagocytic cells, was largely ignored. Contemporary triggers for renewed interest in understanding and manipulating cancer-associated inflammation came from the recognition that *Helicobacter pylori* drives inflammation in stomach cancer, that cervical cancer is linked to inflammation, that commonly used anti-inflammatory drugs have protective effects against some cancers, and that inflammation often goes hand-in-hand with poor prognosis for some cancers (Coussens and Werb, 2002). By contrast to the current view that inflammation is associated with poor outcomes in cancer, the first attempt at clinical intervention in cancer inflammation was by Coley (1893) who tried to provoke a massive inflammatory response by administering bacterial culture filtrates containing toxins and other proinflammatory products. Coley found that his "toxin" elicited potent inflammatory responses could cure some patients with terminal cancers: a finding that still drives research into harnessing the body's natural antipathogen responses to fight cancer (Coley, 1893; McCarthy, 2006).

Today, myelophagocytic cell infiltration is recognized as a "hallmark" of cancer, and has engendered new efforts to define myelophagocytic cell function within solid tumors, and to test whether the overall process of inflammation can be manipulated to improve cancer therapy outcomes (Balkwill and Mantovani, 2001). This chapter will not detail the history and recent advances in inflammation and cancer that have been reviewed extensively (Balkwill and Mantovani, 2001; Colotta *et al.*, 2009; Coussens and Werb, 2002; Mantovani *et al.*, 2008). Instead, we will focus on summarizing research on myelophagocytic cell infiltration into solid tumors and the potential roles of these cells, and how they might be manipulated.

2. Brief Summary of Clinical Observations of Cancer Inflammation, Prognostic Value of Cancer Inflammation, and Attempts to Manipulate the Inflammatory Environment

Clinical observations concerning inflammation and cancer, as well as the use of inflammatory markers as diagnostic tools are a vast area that is constantly being updated. For the purpose of this review, it is important to establish that many types of cancers are linked to inflammatory responses that vary depending on the type of cancer and the individual (Cole, 2009; Pierce *et al.*, 2009). For example, extensive inflammation occurs in breast cancers with poor prognosis. Pathologists use histological examination to estimate the extent of lymphocytic, granulocytic, and macrophage infiltration within tumors. In some cases, sophisticated approaches that combine microarray and pathological examination of cancers have been used to estimate links between inflammation and prognosis. It seems likely that measurement of molecular and cellular inflammatory parameters will have an increasingly large impact on clinical diagnosis.

Despite the fact that variance in the degree and type of inflammation between cancers and individuals will be broad, there are two obvious distinctions that can be made. The first is cancer inflammation associated with infection. Examples of this type of inflammation include stomach cancer and *H. pylori*, hepatocarcinoma and hepatitis C infection, and colon cancer and Crohn's Disease driven by the aberrant response to the gut flora. By contrast, the second type of cancer inflammation is not linked to any known infectious agent: breast and prostate cancers, head and neck cancer, and childhood solid tumors such as osteosarcoma and neuroblastoma. The latter examples might be associated with yet to be discovered infectious agents. However, it seems more likely that these solid tumors elicit an inflammatory response associated with tissue damage and repair. Here we overview the role of macrophages in both types of inflammatory response.

3. What Is a Macrophage?

3.1. The source of macrophages

Macrophages, like all blood cells, are originally derived from hematopoietic stem cells in the bone marrow. Macrophage differentiation begins with the development of the common myeloid precursor, which has the ability to develop into all types of myeloid cells, including monocytes, which are thought to eventually become macrophages that seed tissues, dendritic cells, plasmacytoid dendritic cells, osteoclasts, Langerhans cells, and microglia (Geissmann *et al.*, 2010). Our present knowledge of the precise steps of myeloid lineage development is sketchy because the relationship between the end-product cells and their originating stem cell is unclear. Furthermore, the plasticity of the different myeloid lineages to form one type of cell and then shift into another type depending on the environment is also controversial (Geissmann *et al.*, 2010). Lineage tracing experiments will be necessary to address these issues. Finally, another major unknown concerns whether macrophages can divide within tissues or, by contrast, whether all tissue macrophages are postmitotic.

Following differentiation in the bone marrow, monocytes are released into circulation. We now know that the spleen serves as a reservoir for large numbers of monocytes (Swirski *et al.*, 2009). Once monocytes migrate into the surrounding tissues, including growing cancers, they become macrophages as defined by cell-surface markers and functional criteria. Most researchers believe that monocytes seed tissues and then develop into mature macrophages. However, the relationship between the circulating monocyte pools and tissue macrophages is an evolving field that has yet to yield definitive information on which types of monocytes (or other myeloid precursors) become macrophages (Geissmann *et al.*, 2010).

It is important to point out that there are many subtypes of macrophages and dendritic cells that can be classified based on tissue location, function, and cell-surface markers such as CD11b, F4/80, CD68, CD8, and CD11c. The problem with attempting to enumerate and classify distinct subtypes of myelophagocytic cell has been recently discussed by Hume, who also noted that the number of types of myelophagocytic cell is proportional to the number of markers or assays used to make the definition (Hume, 2008). As such, this review will use the generic term "tumor-associated macrophage" (TAM) to define any macrophage that infiltrates a tumor and "macrophage" to denote tissue macrophages. We note, however, that many subtypes of TAMs likely exist, and TAM populations likely change with changing conditions inside a growing or regressing tumor.

3.2. Tissue macrophages

The main function of tissue macrophages is to patrol their tissue environment. Tissue macrophages use cell-surface receptors and phagocytosis to detect and consume dead and dying cells, pathogens, and to repair and remodel stressed tissues. Macrophages are typically the first immune cells to encounter infectious agents or damaged cells and tissues and are instrumental in modeling the subsequent immune response via the production of cytokines, chemokines, and secreted proteases. An important concept for tumor immunology is that the vast majority of tissue macrophages are immunosuppressive in that they cannot readily be provoked to secrete large amounts of proinflammatory factors. In most cases, tissue macrophages also possess potent, active immunosuppressive activities (Hume, 2008). Examples of the anti-inflammatory surveillance functions of tissue macrophages include patrol of immune privilege sites such as the eye and testis, controlled inflammation of the gut flora, homeostasis of the liver by Kupffer cells, and most remarkably, the surveillance function of alveolar macrophages in the lung. As long as we breathe, our lungs are continuously bombarded with innumerable particles, fungal spores, bacteria, and viruses that for the most part do not elicit any obvious inflammatory effects. Why are macrophages immunosuppressive when it seems that the vast majority of researchers study macrophage activation by pathogens? The primary function of tissue macrophages is to maintain homeostasis: if most macrophages were readily activated to make proinflammatory cytokines, then extensive tissue destruction would occur, even for innocuous stimuli such as engulfment of apoptotic cells. Instead, specialized subsets of macrophages are likely committed to proinflammatory activation and, once activated, are subject to overlapping anti-inflammatory check-and-balances that constrain their activity. The *in vivo* activation of macrophages is a very different scenario from *ex vivo* activation of differentiated macrophages exposed to experimental stimulation with pathogens or their products. Divorced from the tissue niche, macrophages cultured *in vitro* make abundant proinflammatory mediators. At this stage, however, there are no markers to readily distinguish a macrophage that can be activated to make proinflammatory mediators *in vivo*.

Inflammation is characterized by heat, pain, redness, and swelling. These characteristics are due to the action of the cytokines and chemokines on the surrounding tissue and vasculature making it more suitable for additional immune cell entry via vasculature dilation and permeation. Secretion of chemokines from macrophages established chemotactic gradients to recruit adaptive immune cells needed to help with the infection. An example of this type of cell is the "Tip-DC" population that plays a key role in early TNF-α and nitric oxide (NO) responses in Listeria infection (Serbina *et al.*, 2003). Tip-DCs most likely represent an example of a specialized macrophage type that is committed to rapid inflammatory responses. Another example is migratory macrophages that enter the lungs and engulf *Mycobacterium tuberculosis* and then transport the bacteria to the draining lymph nodes of the lung to shape the subsequent T-cell response (Skold and Behar, 2008). These macrophages are distinct in function and phenotype from the anti-inflammatory alveolar macrophages discussed above. In the case of cancer inflammation and macrophage infiltration, we do not yet understand what types of macrophages are recruited to the tumor. However, TAMs must arrive from the circulation and seed a growing tumor: TAMs are most likely plastic, and represent immunosuppressive and activated macrophage subtypes.

3.3. Developmental markers and macrophage differentiation

3.3.1. Detection of macrophages within tumors

Cell-surface proteins that have been empirically determined are used to identify tissue macrophages by immunohistochemistry or flow cytometry. Cell-surface markers used to identify macrophages include CD11a, CD11b, F4/80, CD68, Mac2, and Mac3. Unfortunately, not all macrophages express the same cell-surface markers. For instance, white pulp macrophages of the spleen do not express F4/80, while red pulp splenic macrophages are F4/80$^+$ (Austyn and Gordon, 1981; Hume *et al.*, 1983; Morris *et al.*, 1991).

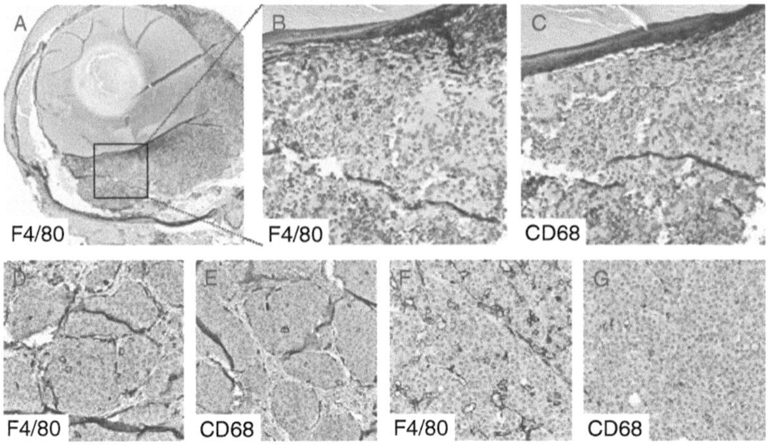

Figure 10.1 Expression of macrophage cell-surface antigens on TAMs. Cross sections from a retinoblastoma (A–C), neuroblastoma (D–E), and EG7 thymoma (F–G) stained by immunohistochemistry for F4/80 (A, B, D, and F) and CD68 (C, E, and G). (See Color Insert.)

This can make it very difficult to define macrophages when analyzing by flow cytometry and/or immunohistochemistry. For this reason, it is imperative to detect macrophages by using combinations of markers for each individual tissue. For example, we routinely stain tissue sections for F4/80, CD68, and Mac3 to ascertain the full spectrum of macrophage infiltration (Fig. 10.1).

3.4. Myeloid-derived suppressor cells

Myeloid derived suppressor cells (MDSCs) encompass a heterogeneous group of mature and immature monocytic and granulocytic lineage cells released from the bone marrow in patients or mouse models undergoing tissue stress. They are linked with solid tumor progression where they are often found in extraordinary numbers in the blood and spleen. MDSCs have been the subject of several recent reviews (Gabrilovich and Nagaraj, 2009; Ostrand-Rosenberg and Sinha, 2009; Peranzoni *et al.*, 2010; Serafini *et al.*, 2006) and will only be briefly mentioned here because of their potential relationship with TAMs. MDSCs are characterized by the coexpression of CD11b and Gr1 (a granulocyte marker) although, like macrophages, no definitive marker combination has yet been found to discriminate MDSCs from their closely related myeloid lineage cousins. The hallmark of MDSCs is their immune suppressive activity: when cocultured with T-cells in *in vitro* experiments, MDSCs potently suppress CD4 and CD8 T-cell proliferation, raising the possibility that MDSCs create a suppressive environment in regions of tissue stress. The evidence for MDSCs suppressing T-cell responses is substantial, since many types of tissue stress cause MDSC elicitation including wounds, burns, vaccination, chronic pathogen infections, inflammatory bowel diseases, and solid tumors (Delano *et al.*, 2007; Haile *et al.*, 2008; Martino *et al.*, 2010). It is important to note that MDSCs are distinct from circulating monocytes and neutrophils, which lack potent suppressive activity toward T-cells. Instead, MDSCs are closer in function to tissue macrophages discussed above. Indeed, a key question concerns the relationship between MDSCs and TAMs. One possibility is that MDSCs are the precursors of TAMs. Since solid tumors must be continuously seeded with macrophages (likely from the blood monocytes), MDSCs may make up a substantial fraction of the tumor infiltrating macrophage pool. Alternatively, MDSCs and TAMs could be separate populations but with overlapping activities. Regardless of any direct lineage link between MDSCs and TAMs, both populations appear to play fundamental roles in shaping the immune environment in solid tumors.

4. Do TAMs Assist Tumors?

Experiments aimed at defining TAM function in tumor progression have generally described them as "protumor," as their depletion results in decreased tumor survival. For instance, mammary tumors in *Csf1*-deficient mice, which have drastically reduced macrophage numbers because of the absence of M-CSF (encoded by *Csf1*), displayed decreased survival and metastasis in comparison to control animals (Lin *et al.*, 2001). This observation was substantiated by additional reports that found decreased tumor progression in models of macrophage dysregulation by depletion, loss of function, and/or reduced recruitment to neoplasms (Aharinejad *et al.*, 2007; Kimura *et al.*, 2007; Luo *et al.*, 2006). Collectively, clinical evidence and mouse models of cancer argue that excessive macrophage infiltration aids tumor growth and survival.

4.1. Recruitment of TAMs to the tumor

Tumors appear to actively recruit TAMs from the circulating blood monocyte population. The tumor-derived chemokines CCL2, CCL3, CCL4, CCL5, CCL8, and CXCL12 have all been implicated in monocyte recruitment to progressing tumors (Coffelt *et al.*, 2009; Sica *et al.*, 2008; Solinas *et al.*, 2009). The growth factors VEGF and PDGF are also produced by tumors and correlate with increased TAM recruitment, and M-CSF production by tumors is critical in aiding the proper development of macrophages (Coffelt *et al.*, 2009; Sica *et al.*, 2008; Solinas *et al.*, 2009). A key question in this area is whether the tumor cells alone make the factors that recruit macrophages to the growing cancer, or if TAMs themselves (or other inflammatory cells) signal further recruitment. In either case, interrupting the recruitment cycle is a potential avenue for therapeutic intervention. Once recruited to the tumor, macrophages undergo differentiation defined by stromal- and tumor-derived factors. While an absolute description of the TAM phenotype is yet to be established, some evidence suggests TAMs develop into a noninflammatory macrophage subtype, commonly referred as an alternatively activated, or an "M2" macrophage (Mantovani *et al.*, 2009, 2002). Alternatively activated macrophages develop *in vitro* by stimulation with either IL-4 or IL-13. They produce little or no proinflammatory cytokines, but express genes that would promote wound healing (Gordon, 2003). The M2 activation state contrasts the "M1," or classically activated, macrophage, which following *in vitro* stimulation with a Toll-like receptor agonist (ex: LPS) and interferons produce high amounts of proinflammatory factors such as NO and TNF-α (Gordon, 2003). Although the M1 and M2 nomenclature can be useful, it greatly oversimplifies the function of TAMs *in vivo* and categorizing this macrophage subtype must be cautiously carried out.

4.2. Hallmarks of cancer and the TAM-mediated contribution

The six hallmarks of cancer, originally proposed by Hanahan and Weinburg, include (1) self-sufficiency in growth signals, (2) evasion of apoptosis, (3) sustained angiogenesis, (4) limitless replicative potential, (5) insensitivity to antigrowth signals, and (6) tissue invasion and metastasis (Hanahan and Weinberg, 2000). Even a decade ago, Hanahan and Weinburg regarded tumors as complex tissues, allowing for the possibility that other "stromal" cells within the tumor microenvironment may assist in tumor progression and survival. More recent work has shown that excessive inflammation can precede, or even lead to cancer, which has resulted in categorizing inflammation as a possible "7th hallmark" of cancer (Colotta *et al.*, 2009; Mantovani, 2009). Macrophages are known for their ability to initiate and control the inflammatory response by secreting cytokines and chemokines (chemotactic cytokines) including IL-1β, IL-6, IL-12, TNF-α, CXCL1, and IL-10, an anti-inflammatory factor, among others. Production of oxygen radicals (superoxide, H_2O_2, etc.) and NO aimed at pathogen control also promotes inflammation. Additionally, macrophages assist with the other hallmarks of cancer, including angiogenesis, evasion of apoptosis, and promotion of metastasis and tissue invasion, each of which are discussed further below (Fig. 10.2).

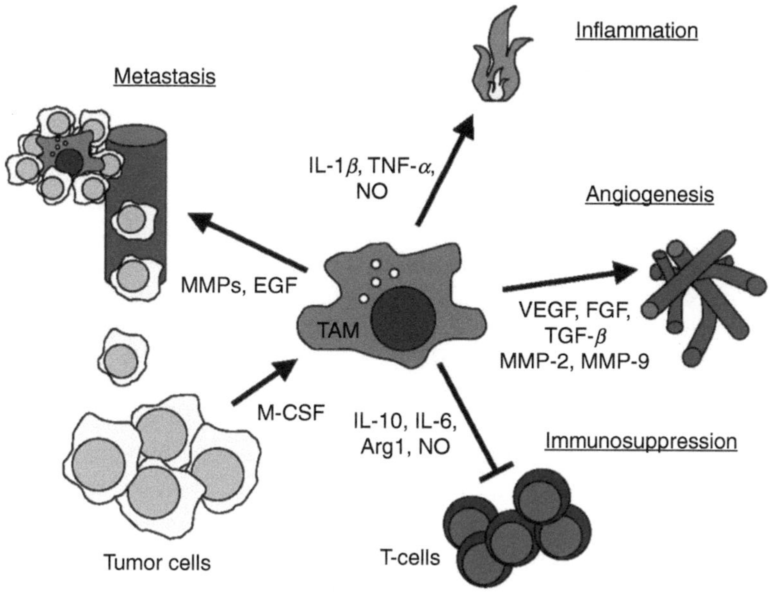

Figure 10.2 TAM-mediated support of tumor progression. TAMs are actively recruited by tumor cell secretion of multiple chemokines, growth factors, and most importantly, M-CSF. Developing neoplasms are benefited by TAM initiation of inflammation, which via NF-kB activation can lead to increased tumor survival. However, a prolonged inflammatory response could prove detrimental to tumors by infiltrating cytotoxic T-cells. Therefore, TAMs also block T-cell activity through production of immunosuppressive factors, including IL-6, IL-10, Arg1, and NO. TAMs induced by hypoxia, and other tumor-derived signals, also promote angiogenesis by producing VEGF, FGF, TGF-β, MMP-2, and MMP-9. This process is closely related to the proposed TAM-mediated assistance for tumor metastasis, involving additional MMPs and growth factors. (See Color Insert.)

4.2.1. Inflammation

Inflammation is an evolutionarily conserved process that is vital to eliminating infection. Yet, continued and dysregulated inflammation is deleterious, as observed in many autoimmune disorders. The correlation between cancer and inflammation is compelling: patients with inflammatory bowel diseases, including ulcerative colitis and Crohn's Disease, are highly susceptible to colon cancers, while continuous inflammatory ulcers initiating from *H. pylori* infection are likely to develop into gastric cancer (Coussens and Werb, 2002). Additionally, reports have established that those infected with particular isolates of hepatitis B or C virus (HBV, HCV) and human papillomavirus (HPV) are more likely to acquire liver and cervical cancers, respectively (Coussens and Werb, 2002). By contrast, long-term users of aspirin and other NSAIDs have a decreased risk in developing cancer (Baron and Sandler, 2000; Garcia Rodriguez and Hernandez-Diaz, 2001).

However, the aforementioned cancers are all associated with infectious agents or the gut flora. The basis for inflammation in cancers that have no known infectious complication remains unknown.

TAM production of IL-1β, TNF-α, reactive oxygen, and nitrogen species promote activation/amplification of the NF-κB pathway that leads to enhancement of the cancer hallmarks, including self-sufficiency in growth signals, decreased apoptosis, angiogenesis, and insensitivity to anti-growth signals (Aggarwal and Gehlot, 2009; Hagemann *et al.*, 2009; Solinas *et al.*, 2009). The role of NF-κB in TAMs is complex. For instance, inhibition of NF-κB activity by blockade of IKKβ signaling resulted in TAMs switching from an immunosuppressive phenotype to a proinflammatory one, which repressed ID8 ovarian cancer progression in mice (Hagemann *et al.*, 2008). Indeed, dysregulation of the NF-κB pathway in TAMs has been linked to multiple cases of aberrant inflammatory responses (Biswas *et al.*, 2006; Saccani *et al.*, 2006; Sica *et al.*, 2008; Torroella-Kouri *et al.*, 2005). While inflammatory factors, such as TNF-α, are important for tumor progression early, ongoing aggressive inflammation may promote antitumor immunity as was found for the effects of Coley's toxins (Balkwill, 2002; Balkwill and Mantovani, 2001; Coley, 1893; McCarthy, 2006). In later stages of tumor development, an inflammatory environment promotes lymphocyte recruitment, seemingly aimed at fighting the tumor. Indeed, reports have indicated, in contrast to TAMs, increased numbers of tumor-infiltrating lymphocytes (TILs) generally correlates with a good prognosis (Clemente *et al.*, 1996; Curiel *et al.*, 2004; Pages *et al.*, 2005). The kinetics of the pro- versus anti-inflammatory environment surrounding developing tumors are clearly complex, and future research should focus on this aspect of tumor progression to provide rationale for targeting inflammation as a potential facet of cancer therapy.

4.2.2. Angiogenesis

Angiogenesis is essential for tumor survival. Many of the rapidly growing solid tumors, even when netted in large vascular networks, have large regions of cell death in hypoxic areas, emphasizing the necessity for continuous angiogenesis. A large number of TAMs localize to areas of necrosis, which are greatly hypoxic (Leek *et al.*, 1999, 1996; Negus *et al.*, 1997). In response to hypoxia, and other tumor-specific signals, TAMs provide support for new vasculature construction by producing growth factors including VEGF, FGF, and TGF-β (Solinas *et al.*, 2009). TAMs are also involved in restructuring the extracellular matrix to an environment more suitable for vasculature formation by producing the matrix metalloproteinases MMP-2, -7, -9, and -12 (Solinas *et al.*, 2009). Additionally, infiltrating MDSCs appear to promote angiogenesis that is dependent on MMP-9 production, which correlates with acquiring an endothelial cell-like phenotype (Yang *et al.*, 2004). As tumor survival relies on delivery of oxygen and other nutrients through the blood vessels, targeted disruption of TAM-mediated (and MDSC-mediated) angiogenesis may prove of therapeutic value.

4.2.3. Evasion of apoptosis/immune evasion

The immune system has evolved to differentiate between the body's "self" proteins from those of foreign pathogens (i.e., bacteria, viruses, worms, etc.) in order to mount an immune response only when recognizing the latter. T-cells play a large role in detecting and eliminating cancerous cells. In general, specific cell-surface receptors on T-cells are used to detect either proteins presented by local macrophages or the cancer cells. However, a paradox becomes apparent when desiring an immune response against malignancies because each arise from "self." Neo-antigens, also known as tumor-antigens, provide a solution to this problem. Neo-antigens arise from oncogenic viruses or mutated self-proteins in neoplasms and are presented by the major histocompatability complexes on the surface of tumor cells. As T-cells recognize the neo-antigens, they quickly proliferate and activate macrophages to become more inflammatory by releasing cytokines, or directly kill cancer cells by cytotoxic activity. Yet, T-cells are often found deactivated in tumor sites overexpressing the CTLA-4 receptor. CTLA-4 blocks T-cell costimulation by interfering with the CD28–B7.1/B7.2 interaction between the T-cell and surrounding antigen presenting cell, respectively. Indeed, it has been shown that blockade of CTLA-4 in multiple cancer models inhibit tumor progression (Curran and Allison, 2009; Curran *et al.*, 2010; Peggs *et al.*, 2009).

While the connection between TAMs and T-cell CTLA-4 remains to be established, TAMs, nonetheless, promote immunosuppression against the T-cell-mediated anticancer response. For instance, TAMs produce arginase, an enzyme that hydrolyzes arginine. T-cells need arginine to proliferate, and are less able to do so when cultured with arginase-expressing macrophages (Pesce *et al.*, 2009). In addition, TAMs have been observed to express another arginine-utilizing enzyme called NO synthase that makes NO by oxidoreductase activity on arginine. NO is normally produced as an antipathogenic agent, yet it is highly toxic to the host tissues at high concentrations. One of the side effects of high NO concentration is that it acts as a mutagenic factor leading toward neoplastic formation. For instance, NO toxicity is associated with increased p53 mutations in ulcerative colitis patients, and positively correlates with an increased incidence of colorectal cancer (Ambs *et al.*, 1999; Hussain *et al.*, 2000). NO also has deleterious effects on T-cells. Peroxynitrite, one of the derivatives of NO, has been shown to modify the T-cell receptor, leading to reduced effector activity against the local tumor (Nagaraj *et al.*, 2007). Additionally, TAMs produce factors that make it more difficult for T-cells to fight cancer, including IL-10, TGF-beta, and prostaglandins which all can suppress T-cell activity

(Balkwill and Mantovani, 2001; Sica *et al.*, 2000; Sinha *et al.*, 2005). In summary, TAMs contain an arsenal of pro- and anti-inflammatory factors that is key in regulating the immune response (Fig. 10.2). Unfortunately, in the case of TAMs, signals from the tumor microenvironment influence the macrophages to operate in a protumor fashion.

4.2.3.1. Involvement of the cytokine-activated transcription factor Stat3 Stat3 signaling has an important role in the tumor microenvironment because of its anti-inflammatory effects on TAMs and procarcinogenic activities on tumor cells. Multiple lines of evidence suggest that Stat3 drives cytokine-mediated autocrine–paracrine loops that enhance cancer growth and progression. Although a dynamic and fast-moving research area, it is worth briefly summarizing several elements of Stat3 function connected to TAM function within the tumor microenvironment.

First, anti-inflammatory signaling from the IL-10 receptor is likely to suppress proinflammatory cytokine and chemokine production from TAMs. Stat3 is necessary and sufficient for the activation of the anti-inflammatory signaling pathway and deletion of Stat3 in myeloid cells causes massive inflammation, especially in the gut (Murray, 2006, 2007). Within a tumor the target cell of IL-10 signaling will be TAMs and other myelophagocytic cells because these cell express the IL-10 receptor (Moore *et al.*, 2001; Murray, 2006; Saraiva and O'Garra, 2010). Since all macrophages and T-cells produce IL-10, including TAMs, it is reasonable to assume that a solid tumor is rich in IL-10, which would drive constitutive suppression of inflammation that may enhance tumor survival. A corollary is the effects of IL-10 in the gut where constitutive expression is required to suppress inflammation driven by the gut flora. When IL-10 or IL-10 signaling is disrupted, excessive inflammation is observed in the intestine (Glocker *et al.*, 2009; Kuhn *et al.*, 1993; Pils *et al.*, 2010). IL-10 within a tumor likely contributes to the blockage of productive inflammation that would be detrimental to tumor cells.

A second example of the procancer effects of Stat3 is associated with suppression of inflammatory genes within tumor cells themselves. This effect is tied into the observation that activated phospho-forms of Stat3 are often observed in cancer cells (Inghirami *et al.*, 2005; Levy and Inghirami, 2006). Knockdown of Stat3 in tumor cells caused a rise in multiple inflammatory mediators and made the tumor cells more immunogenic (Burdelya *et al.*, 2005; Wang *et al.*, 2004). These data substantiate that in some tumors Stat3 suppresses gene activation analogous to the suppressive effects of Stat3 on myelophagocytic cells activated by IL-10 signaling. Remarkably, a Stat3 mutant that bypasses cytokine receptor activation has cell-autonomous oncogenic properties suggesting that blocking Stat3 might be a useful chemotherapeutic strategy (Bromberg *et al.*, 1999). While Stat3 inhibitors have been developed (Burdelya *et al.*, 2005;

Kortylewski *et al.*, 2005), we expect them to have limited utility in cancer chemotherapy because of toxicity that will result from system-wide Stat3 inhibition. Toxicity might be especially acute in the gut where Stat3 signaling from the IL-10 receptor must be maintained.

A third example is that the effects of Stat3 are involved in autocrine–paracrine loops that promote cancer survival and growth. Several stunning examples of Stat3 signaling via IL-6 has recently been described including the links between excessive fat intake and liver carcinogenesis, IL-6 signaling in gastrointestinal cancers, and IL-6 having a central role in oncogene-induced senescence (Bollrath *et al.*, 2009; Grivennikov *et al.*, 2009; Park *et al.*, 2010). Finally, many prosurvival and antiapoptotic genes are directly or indirectly Stat3-dependent including the antiapoptotic Bcl-2 family members Mcl-1 and Bcl-X. Activation of antiapoptotic protein expression by IL-6 may be a major survival mechanism of tumor cells as they expand. Collectively, Stat3 has a panoply of roles in cancer from positive and negative effects on TAMs, cell-autonomous effects on cancer cells, and additional key roles in T-cell development that cannot be ignored when considering the tumor microenvironment (O'Shea and Murray, 2008).

4.2.4. Metastasis and tissue invasion

Malignant cancers have the ability to move from their original location and set up secondary tumors in, at times, distant tissue sites. While the supportive studies on TAM involvement in this process are currently limited, some observations indicate that TAMs likely support tumor metastasis. TAM production of MMPs and growth factors including EGF appear to play an important role in tumor metastasis (Hagemann *et al.*, 2004; Mytar *et al.*, 2003; Pollard, 2008; Solinas *et al.*, 2009). As mentioned above, TAMs increase angiogenesis, which provide multiple exit pathways to initiate metastasis. Moreover, experiments have been performed in which blocking macrophage recruitment and/or function to cancers drastically decrease metastases in multiple tumor models, and secondary tumors appear to more strongly establish a successful niche when in close association with macrophages following tissue invasion (Coffelt *et al.*, 2009; Gorelik *et al.*, 1982; Rolny *et al.*, 2008; Wyckoff *et al.*, 2007). Reports have clearly shown that TAMs possess multiple characteristics that assist in tumor progression (Fig. 10.2). Future studies aimed at defining the mechanisms behind their action will likely provide multiple targets for consideration in combination with proven cancer treatments, including chemotherapy, radiation, and surgery.

4.3. Are TAMs always protumor?

While the majority of studies indicate TAMs are protumor, there are some cases in which the presence of TAMs are correlated with a good prognosis (Funada *et al.*, 2003; Lewis and Pollard, 2006; Ohno *et al.*, 2003; Piras *et al.*,

2005). It is unclear at present the reason for this inconsistency and clearly exemplifies the complexity of the tumor/TAM relationship. From a therapeutic standpoint, this information provides optimism. Conceivably experiments that compare the activity of TAMs from these two tumor subsets (i.e., good vs. bad prognosis) would provide clues on TAM-specific functions that are truly antitumor. The obvious next step would then be to manipulate TAM function in the antitumor fashion as a potential cancer therapy.

5. Manipulating Inflammation as Therapy

How can we therapeutically target the inflammatory activities of TAMs? Since the majority of the evidence about TAM activity argues that TAMs are beneficial for tumor growth and survival, targeting TAMs in addition to the tumor itself is a realistic goal. Nevertheless, selective depletion of TAMs from within a tumor while leaving other tissue macrophages unaffected seems impossible. In fact, systemic attempts at depleting macrophages with agents such as liposomes, clodronate, or monoclonal antibodies that target common macrophage markers would likely create widespread toxicity via effects on resident tissue macrophages required for organ homeostasis. Instead, a better strategy would be to target factors produced by TAMs that encompass protumorigenic activities. For example, humanized mAbs against IL-6, the soluble IL-6 receptor, IL-12/IL-23, TNF-α, and IL-1β have been developed, tested and in the case of anti-TNF-α drugs, deployed in multiple clinical settings involving inflammatory diseases. It is conceivable that the use of a single anticytokine drug at the right time in combination with chemotherapy against the tumor could break a protumorigenic autocrine–paracrine cycle and affect multiple cytokines at once. Such an outcome was found for the first effective anticytokine drug used in clinical medicine, anti-TNF-α for rheumatoid arthritis, where blockade of TNF-α also blocked IL-1β in joints (Feldmann and Maini, 2003). Bringing an anticytokine therapy to bear upon solid tumors will, however, be a formidable challenge because diverse drug combinations such as one or more cytotoxic drugs combined with an anticytokine mAb would take enormous effort to establish baseline toxicities and therapeutic indices.

A second alternative to targeting TAM function is to focus on altering signaling behavior within TAMs. As described above, a drug directed at Stat3 has been developed (Yu and Jove, 2004; Yu *et al.*, 2009). But it seems more promising to attack phosphorylation and dephosphorylation events that are specific for highly active macrophages such as TAMs, and less important for tissue macrophages. Most of the signaling molecules and transcription factors that orchestrate the macrophage inflammatory response have been discovered (Akira *et al.*, 2006; Medzhitov and Horng, 2009; Rakoff-Nahoum and Medzhitov, 2009). What is missing is how these factors work together to produce, for example, high, sustained amounts of IL-6. A second missing element is the posttranscriptional alterations that regulate the signaling cascades in macrophages, and how many of these pathways are specific to TAMs. While we know that many of the molecules involved in inflammatory cascades are phosphorylated and ubiquitinated, the specific players involved and their temporal order of activity remains a vast area for research.

A third area for consideration is to target the signals that recruit macrophages to the tumor. Macrophages and monocytes are likely continuously recruited during tumor expansion. What these signals are and their source remains largely unknown. It is possible that tumors make molecules that summon more macrophages. Alternatively, stressed macrophages may recruit new macrophages to "help" with the remodeling of the cancer. Chemokines seem the likely culprits in macrophage recruitment and we know that TAMs express many distinct classes of chemokines (Biswas *et al.*, 2006). However, to date no single chemokine or combinations of chemokines have been firmly tied to TAM recruitment.

6. Concluding Remarks

The underlying mechanisms that promote macrophage recruitment into tumors, TAM activity within tumors, and the effects of TAMs on cancer outcomes remain a vast area for new investigation. Although much evidence has been put forward to support the notion that TAMs are procancer rather than prohost, we suspect that new insights will show that TAMs have multiple pro- and anticancer effects that change depending on the milieu of the tumor microenvironment. Identifying and harnessing the antitumor effects of TAMs is likely a sound approach to novel cancer therapies.

REFERENCES

Aggarwal, B. B., and Gehlot, P. (2009). Inflammation and cancer: How friendly is the relationship for cancer patients? *Curr. Opin. Pharmacol.* **9,** 351–369.

Aharinejad, S., Sioud, M., Lucas, T., and Abraham, D. (2007). Target validation using RNA interference in solid tumors. *Methods Mol. Biol.* **361,** 227–238.

Akira, S., Uematsu, S., and Takeuchi, O. (2006). Pathogen recognition and innate immunity. *Cell* **124,** 783–801.

Ambs, S., Bennett, W. P., Merriam, W. G., Ogunfusika, M. O., Oser, S. M., Harrington, A. M., Shields, P. G., Felley-Bosco, E., Hussain, S. P., and Harris, C. C. (1999). Relationship between p53 mutations and inducible nitric oxide synthase expression in human colorectal cancer. *J. Natl. Cancer Inst.* **91,** 86–88.

Austyn, J. M., and Gordon, S. (1981). F4/80, a monoclonal antibody directed specifically against the mouse macrophage. *Eur. J. Immunol.* **11,** 805–815.

Balkwill, F. (2002). Tumor necrosis factor or tumor promoting factor? *Cytokine Growth Factor Rev.* **13,** 135–141.

Balkwill, F., and Mantovani, A. (2001). Inflammation and cancer: Back to Virchow? *Lancet* **357,** 539–545.

Baron, J. A., and Sandler, R. S. (2000). Nonsteroidal anti-inflammatory drugs and cancer prevention. *Annu. Rev. Med.* **51,** 511–523.

Biswas, S. K., Gangi, L., Paul, S., Schioppa, T., Saccani, A., Sironi, M., Bottazzi, B., Doni, A., Vincenzo, B., Pasqualini, F., Vago, L., Nebuloni, M., *et al.* (2006). A distinct and unique transcriptional program expressed by tumor-associated macrophages (defective NF-kappaB and enhanced IRF-3/STAT1 activation). *Blood* **107,** 2112–2122.

Bollrath, J., Phesse, T. J., von Burstin, V. A., Putoczki, T., Bennecke, M., Bateman, T., Nebelsiek, T., Lundgren-May, T., Canli, O., Schwitalla, S., Matthews, V., Schmid, R. M., *et al.* (2009). gp130-mediated Stat3 activation in enterocytes regulates cell survival and cell-cycle progression during colitis-associated tumorigenesis. *Cancer Cell* **15,** 91–102.

Bromberg, J. F., Wrzeszczynska, M. H., Devgan, G., Zhao, Y., Pestell, R. G., Albanese, C., and Darnell, J. E., Jr. (1999). Stat3 as an oncogene. *Cell* **98,** 295–303.

Burdelya, L., Kujawski, M., Niu, G., Zhong, B., Wang, T., Zhang, S., Kortylewski, M., Shain, K., Kay, H., Djeu, J., Dalton, W., Pardoll, D., *et al.* (2005). Stat3 activity in melanoma cells affects migration of immune effector cells and nitric oxide-mediated antitumor effects. *J. Immunol.* **174,** 3925–3931.

Clemente, C. G., Mihm, M. C., Jr., Bufalino, R., Zurrida, S., Collini, P., and Cascinelli, N. (1996). Prognostic value of tumor infiltrating lymphocytes in the vertical growth phase of primary cutaneous melanoma. *Cancer* **77,** 1303–1310.

Coffelt, S. B., Hughes, R., and Lewis, C. E. (2009). Tumor-associated macrophages: Effectors of angiogenesis and tumor progression. *Biochim. Biophys. Acta* **1796,** 11–18.

Cole, S. W. (2009). Chronic inflammation and breast cancer recurrence. *J. Clin. Oncol.* **27,** 3418–3419.

Coley, W. B. (1893). The treatment of malignant tumors by repeated inoculations of erysipelas: With a report of ten original cases. *Am. J. Med. Sci.* **105,** 487–551.

Colotta, F., Allavena, P., Sica, A., Garlanda, C., and Mantovani, A. (2009). Cancer-related inflammation, the seventh hallmark of cancer: Links to genetic instability. *Carcinogenesis* **30,** 1073–1081.

Coussens, L. M., and Werb, Z. (2002). Inflammation and cancer. *Nature* **420,** 860–867.

Curiel, T. J., Coukos, G., Zou, L., Alvarez, X., Cheng, P., Mottram, P., Evdemon-Hogan, M., Conejo-Garcia, J. R., Zhang, L., Burow, M., Zhu, Y., Wei, S., *et al.* (2004). Specific recruitment of regulatory T cells in ovarian carcinoma fosters immune privilege and predicts reduced survival. *Nat. Med.* **10,** 942–949.

Curran, M. A., and Allison, J. P. (2009). Tumor vaccines expressing flt3 ligand synergize with ctla-4 blockade to reject preimplanted tumors. *Cancer Res.* **69,** 7747–7755.

Curran, M. A., Montalvo, W., Yagita, H., and Allison, J. P. (2010). PD-1 and CTLA-4 combination blockade expands infiltrating T cells and reduces regulatory T and myeloid cells within B16 melanoma tumors. *Proc. Natl. Acad. Sci. USA* **107**(9), 4275–4280.

Delano, M. J., Scumpia, P. O., Weinstein, J. S., Coco, D., Nagaraj, S., Kelly-Scumpia, K. M., O'Malley, K. A., Wynn, J. L., Antonenko, S., Al-Quran, S. Z., Swan, R., Chung, C. S., *et al.* (2007). MyD88-dependent expansion of an immature GR-1(+)CD11b(+) population induces T cell suppression and Th2 polarization in sepsis. *J. Exp. Med.* **204,** 1463–1474.

Feldmann, M., and Maini, R. N. (2003). Lasker Clinical Medical Research Award. TNF defined as a therapeutic target for rheumatoid arthritis and other autoimmune diseases. *Nat. Med.* **9,** 1245–1250.

Funada, Y., Noguchi, T., Kikuchi, R., Takeno, S., Uchida, Y., and Gabbert, H. E. (2003). Prognostic significance of CD8+ T cell and macrophage peritumoral infiltration in colorectal cancer. *Oncol. Rep.* **10,** 309–313.

Gabrilovich, D. I., and Nagaraj, S. (2009). Myeloid-derived suppressor cells as regulators of the immune system. *Nat. Rev. Immunol.* **9,** 162–174.

Garcia Rodriguez, L. A., and Hernandez-Diaz, S. (2001). Relative risk of upper gastrointestinal complications among users of acetaminophen and nonsteroidal anti-inflammatory drugs. *Epidemiology* **12,** 570–576.

Geissmann, F., Manz, M. G., Jung, S., Sieweke, M. H., Merad, M., and Ley, K. (2010). Development of monocytes, macrophages, and dendritic cells. *Science* **327,** 656–661.

Glocker, E. O., Kotlarz, D., Boztug, K., Gertz, E. M., Schaffer, A. A., Noyan, F., Perro, M., Diestelhorst, J., Allroth, A., Murugan, D., Hatscher, N., Pfeifer, D., *et al.* (2009). Inflammatory bowel disease and mutations affecting the interleukin-10 receptor. *N. Engl. J. Med.* **361,** 2033–2045.

Gordon, S. (2003). Alternative activation of macrophages. *Nat. Rev. Immunol.* **3,** 23–35.

Gorelik, E., Wiltrout, R. H., Brunda, M. J., Holden, H. T., and Herberman, R. B. (1982). Augmentation of metastasis formation by thioglycollate-elicited macrophages. *Int. J. Cancer* **29,** 575–581.

Grivennikov, S., Karin, E., Terzic, J., Mucida, D., Yu, G. Y., Vallabhapurapu, S., Scheller, J., Rose-John, S., Cheroutre, H., Eckmann, L., and Karin, M. (2009). IL-6 and Stat3 are required for survival of intestinal epithelial cells and development of colitis-associated cancer. *Cancer Cell* **15,** 103–113.

Hagemann, T., Robinson, S. C., Schulz, M., Trumper, L., Balkwill, F. R., and Binder, C. (2004). Enhanced invasiveness of breast cancer cell lines upon co-cultivation with macrophages is due to TNF-alpha dependent up-regulation of matrix metalloproteases. *Carcinogenesis* **25,** 1543–1549.

Hagemann, T., Lawrence, T., McNeish, I., Charles, K. A., Kulbe, H., Thompson, R. G., Robinson, S. C., and Balkwill, F. R. (2008). "Re-educating" tumor-associated macrophages by targeting NF-kappaB. *J. Exp. Med.* **205,** 1261–1268.

Hagemann, T., Biswas, S. K., Lawrence, T., Sica, A., and Lewis, C. E. (2009). Regulation of macrophage function in tumors: The multifaceted role of NF-kappaB. *Blood* **113,** 3139–3146.

Haile, L. A., von Wasielewski, R., Gamrekelashvili, J., Kruger, C., Bachmann, O., Westendorf, A. M., Buer, J., Liblau, R., Manns, M. P., Korangy, F., and Greten, T. F. (2008). Myeloid-derived suppressor cells in inflammatory bowel disease: A new immunoregulatory pathway. *Gastroenterology* **135,** 871–881, 881 e1-5.

Hanahan, D., and Weinberg, R. A. (2000). The hallmarks of cancer. *Cell* **100,** 57–70.

Hume, D. A. (2008). Macrophages as APC and the dendritic cell myth. *J. Immunol.* **181,** 5829–5835.

Hume, D. A., Robinson, A. P., MacPherson, G. G., and Gordon, S. (1983). The mononuclear phagocyte system of the mouse defined by immunohistochemical localization of antigen F4/80. Relationship between macrophages, Langerhans cells, reticular cells, and dendritic cells in lymphoid and hematopoietic organs. *J. Exp. Med.* **158,** 1522–1536.

Hussain, S. P., Amstad, P., Raja, K., Ambs, S., Nagashima, M., Bennett, W. P., Shields, P. G., Ham, A. J., Swenberg, J. A., Marrogi, A. J., and Harris, C. C. (2000). Increased p53 mutation load in noncancerous colon tissue from ulcerative colitis: A cancer-prone chronic inflammatory disease. *Cancer Res.* **60,** 3333–3337.

Inghirami, G., Chiarle, R., Simmons, W. J., Piva, R., Schlessinger, K., and Levy, D. E. (2005). New and old functions of STAT3: A pivotal target for individualized treatment of cancer. *Cell Cycle* **4,** 1131–1133.

Kimura, Y. N., Watari, K., Fotovati, A., Hosoi, F., Yasumoto, K., Izumi, H., Kohno, K., Umezawa, K., Iguchi, H., Shirouzu, K., Takamori, S., Kuwano, M., *et al.* (2007).

Inflammatory stimuli from macrophages and cancer cells synergistically promote tumor growth and angiogenesis. *Cancer Sci.* **98,** 2009–2018.

Kortylewski, M., Kujawski, M., Wang, T., Wei, S., Zhang, S., Pilon-Thomas, S., Niu, G., Kay, H., Mule, J., Kerr, W. G., Jove, R., Pardoll, D., *et al.* (2005). Inhibiting Stat3 signaling in the hematopoietic system elicits multicomponent antitumor immunity. *Nat. Med.* **11,** 1314–1321.

Kuhn, R., Lohler, J., Rennick, D., Rajewsky, K., and Muller, W. (1993). Interleukin-10-deficient mice develop chronic enterocolitis. *Cell* **75,** 263–274.

Leek, R. D., Lewis, C. E., Whitehouse, R., Greenall, M., Clarke, J., and Harris, A. L. (1996). Association of macrophage infiltration with angiogenesis and prognosis in invasive breast carcinoma. *Cancer Res.* **56,** 4625–4629.

Leek, R. D., Landers, R. J., Harris, A. L., and Lewis, C. E. (1999). Necrosis correlates with high vascular density and focal macrophage infiltration in invasive carcinoma of the breast. *Br. J. Cancer* **79,** 991–995.

Levy, D. E., and Inghirami, G. (2006). STAT3: A multifaceted oncogene. *Proc. Natl. Acad. Sci. USA* **103,** 10151–10152.

Lewis, C. E., and Pollard, J. W. (2006). Distinct role of macrophages in different tumor microenvironments. *Cancer Res.* **66,** 605–612.

Lin, E. Y., Nguyen, A. V., Russell, R. G., and Pollard, J. W. (2001). Colony-stimulating factor 1 promotes progression of mammary tumors to malignancy. *J. Exp. Med.* **193,** 727–740.

Luo, Y., Zhou, H., Krueger, J., Kaplan, C., Lee, S. H., Dolman, C., Markowitz, D., Wu, W., Liu, C., Reisfeld, R. A., and Xiang, R. (2006). Targeting tumor-associated macrophages as a novel strategy against breast cancer. *J. Clin. Invest.* **116,** 2132–2141.

Mantovani, A. (2009). Cancer: Inflaming metastasis. *Nature* **457,** 36–37.

Mantovani, A., Sozzani, S., Locati, M., Allavena, P., and Sica, A. (2002). Macrophage polarization: Tumor-associated macrophages as a paradigm for polarized M2 mononuclear phagocytes. *Trends Immunol.* **23,** 549–555.

Mantovani, A., Allavena, P., Sica, A., and Balkwill, F. (2008). Cancer-related inflammation. *Nature* **454,** 436–444.

Mantovani, A., Sica, A., Allavena, P., Garlanda, C., and Locati, M. (2009). Tumor-associated macrophages and the related myeloid-derived suppressor cells as a paradigm of the diversity of macrophage activation. *Hum. Immunol.* **70,** 325–330.

Martino, A., Badell, E., Abadie, V., Balloy, V., Chignard, M., Mistou, M. Y., Combadiere, B., Combadiere, C., and Winter, N. (2010). Mycobacterium bovis bacillus Calmette-Guerin vaccination mobilizes innate myeloid-derived suppressor cells restraining in vivo T cell priming via IL-1R-dependent nitric oxide production. *J. Immunol.* **184,** 2038–2047.

McCarthy, E. F. (2006). The toxins of William B. Coley and the treatment of bone and soft-tissue sarcomas. *Iowa Orthop. J.* **26,** 154–158.

Medzhitov, R., and Horng, T. (2009). Transcriptional control of the inflammatory response. *Nat. Rev. Immunol.* **9,** 692–703.

Moore, K. W., de Waal Malefyt, R., Coffman, R. L., and O'Garra, A. (2001). Interleukin-10 and the interleukin-10 receptor. *Annu. Rev. Immunol.* **19,** 683–765.

Morris, L., Graham, C. F., and Gordon, S. (1991). Macrophages in haemopoietic and other tissues of the developing mouse detected by the monoclonal antibody F4/80. *Development* **112,** 517–526.

Murray, P. J. (2006). Understanding and exploiting the endogenous interleukin-10/STAT3-mediated anti-inflammatory response. *Curr. Opin. Pharmacol.* **6,** 379–386.

Murray, P. J. (2007). The JAK-STAT signaling pathway: Input and output integration. *J. Immunol.* **178,** 2623–2629.

Mytar, B., Woloszyn, M., Szatanek, R., Baj-Krzyworzeka, M., Siedlar, M., Ruggiero, I., Wieckiewicz, J., and Zembala, M. (2003). Tumor cell-induced deactivation of human monocytes. *J. Leukoc. Biol.* **74,** 1094–1101.

Nagaraj, S., Gupta, K., Pisarev, V., Kinarsky, L., Sherman, S., Kang, L., Herber, D. L., Schneck, J., and Gabrilovich, D. I. (2007). Altered recognition of antigen is a mechanism of CD8+ T cell tolerance in cancer. *Nat. Med.* **13,** 828–835.

Negus, R. P., Stamp, G. W., Hadley, J., and Balkwill, F. R. (1997). Quantitative assessment of the leukocyte infiltrate in ovarian cancer and its relationship to the expression of C-C chemokines. *Am. J. Pathol.* **150,** 1723–1734.

O'Shea, J. J., and Murray, P. J. (2008). Cytokine signaling modules in inflammatory responses. *Immunity* **28,** 477–487.

Ohno, S., Inagawa, H., Dhar, D. K., Fujii, T., Ueda, S., Tachibana, M., Suzuki, N., Inoue, M., Soma, G., and Nagasue, N. (2003). The degree of macrophage infiltration into the cancer cell nest is a significant predictor of survival in gastric cancer patients. *Anticancer Res.* **23,** 5015–5022.

Ostrand-Rosenberg, S., and Sinha, P. (2009). Myeloid-derived suppressor cells: Linking inflammation and cancer. *J. Immunol.* **182,** 4499–4506.

Pages, F., Berger, A., Camus, M., Sanchez-Cabo, F., Costes, A., Molidor, R., Mlecnik, B., Kirilovsky, A., Nilsson, M., Damotte, D., Meatchi, T., Bruneval, P., *et al.* (2005). Effector memory T cells, early metastasis, and survival in colorectal cancer. *N. Engl. J. Med.* **353,** 2654–2666.

Park, E. J., Lee, J. H., Yu, G. Y., He, G., Ali, S. R., Holzer, R. G., Osterreicher, C. H., Takahashi, H., and Karin, M. (2010). Dietary and genetic obesity promote liver inflammation and tumorigenesis by enhancing IL-6 and TNF expression. *Cell* **140,** 197–208.

Peggs, K. S., Quezada, S. A., Chambers, C. A., Korman, A. J., and Allison, J. P. (2009). Blockade of CTLA-4 on both effector and regulatory T cell compartments contributes to the antitumor activity of anti-CTLA-4 antibodies. *J. Exp. Med.* **206,** 1717–1725.

Peranzoni, E., Zilio, S., Marigo, I., Dolcetti, L., Zanovello, P., Mandruzzato, S., and Bronte, V. (2010). Myeloid-derived suppressor cell heterogeneity and subset definition. *Curr. Opin. Immunol.* **22**(2), 238–244.

Pesce, J. T., Ramalingam, T. R., Mentink-Kane, M. M., Wilson, M. S., El Kasmi, K. C., Smith, A. M., Thompson, R. W., Cheever, A. W., Murray, P. J., and Wynn, T. A. (2009). Arginase-1-expressing macrophages suppress Th2 cytokine-driven inflammation and fibrosis. *PLoS Pathog.* **5,** e1000371.

Pierce, B. L., Ballard-Barbash, R., Bernstein, L., Baumgartner, R. N., Neuhouser, M. L., Wener, M. H., Baumgartner, K. B., Gilliland, F. D., Sorensen, B. E., McTiernan, A., and Ulrich, C. M. (2009). Elevated biomarkers of inflammation are associated with reduced survival among breast cancer patients. *J. Clin. Oncol.* **27,** 3437–3444.

Pils, M. C., Pisano, F., Fasnacht, N., Heinrich, J. M., Groebe, L., Schippers, A., Rozell, B., Jack, R. S., and Muller, W. (2010). Monocytes/macrophages and/or neutrophils are the target of IL-10 in the LPS endotoxemia model. *Eur. J. Immunol.* **40,** 443–448.

Piras, F., Colombari, R., Minerba, L., Murtas, D., Floris, C., Maxia, C., Corbu, A., Perra, M. T., and Sirigu, P. (2005). The predictive value of CD8, CD4, CD68, and human leukocyte antigen-D-related cells in the prognosis of cutaneous malignant melanoma with vertical growth phase. *Cancer* **104,** 1246–1254.

Pollard, J. W. (2008). Macrophages define the invasive microenvironment in breast cancer. *J. Leukoc. Biol.* **84,** 623–630.

Rakoff-Nahoum, S., and Medzhitov, R. (2009). Toll-like receptors and cancer. *Nat. Rev. Cancer* **9,** 57–63.

Rolny, C., Capparuccia, L., Casazza, A., Mazzone, M., Vallario, A., Cignetti, A., Medico, E., Carmeliet, P., Comoglio, P. M., and Tamagnone, L. (2008). The tumor suppressor semaphorin 3B triggers a prometastatic program mediated by interleukin 8 and the tumor microenvironment. *J. Exp. Med.* **205,** 1155–1171.

Saccani, A., Schioppa, T., Porta, C., Biswas, S. K., Nebuloni, M., Vago, L., Bottazzi, B., Colombo, M. P., Mantovani, A., and Sica, A. (2006). p50 nuclear factor-kappaB overexpression in tumor-associated macrophages inhibits M1 inflammatory responses and antitumor resistance. *Cancer Res.* **66,** 11432–11440.

Saraiva, M., and O'Garra, A. (2010). The regulation of IL-10 production by immune cells. *Nat. Rev. Immunol.* **10,** 170–181.

Serafini, P., Borrello, I., and Bronte, V. (2006). Myeloid suppressor cells in cancer: Recruitment, phenotype, properties, and mechanisms of immune suppression. *Semin. Cancer Biol.* **16,** 53–65.

Serbina, N. V., Salazar-Mather, T. P., Biron, C. A., Kuziel, W. A., and Pamer, E. G. (2003). TNF/iNOS-producing dendritic cells mediate innate immune defense against bacterial infection. *Immunity* **19,** 59–70.

Sica, A., Saccani, A., Bottazzi, B., Polentarutti, N., Vecchi, A., van Damme, J., and Mantovani, A. (2000). Autocrine production of IL-10 mediates defective IL-12 production and NF-kappa B activation in tumor-associated macrophages. *J. Immunol.* **164,** 762–767.

Sica, A., Allavena, P., and Mantovani, A. (2008). Cancer related inflammation: The macrophage connection. *Cancer Lett.* **267,** 204–215.

Sinha, P., Clements, V. K., and Ostrand-Rosenberg, S. (2005). Interleukin-13-regulated M2 macrophages in combination with myeloid suppressor cells block immune surveillance against metastasis. *Cancer Res.* **65,** 11743–11751.

Skold, M., and Behar, S. M. (2008). Tuberculosis triggers a tissue-dependent program of differentiation and acquisition of effector functions by circulating monocytes. *J. Immunol.* **181,** 6349–6360.

Solinas, G., Germano, G., Mantovani, A., and Allavena, P. (2009). Tumor-associated macrophages (TAM) as major players of the cancer-related inflammation. *J. Leukoc. Biol.* **86,** 1065–1073.

Swirski, F. K., Nahrendorf, M., Etzrodt, M., Wildgruber, M., Cortez-Retamozo, V., Panizzi, P., Figueiredo, J. L., Kohler, R. H., Chudnovskiy, A., Waterman, P., Aikawa, E., Mempel, T. R., *et al.* (2009). Identification of splenic reservoir monocytes and their deployment to inflammatory sites. *Science* **325,** 612–616.

Torroella-Kouri, M., Ma, X., Perry, G., Ivanova, M., Cejas, P. J., Owen, J. L., Iragavarapu-Charyulu, V., and Lopez, D. M. (2005). Diminished expression of transcription factors nuclear factor kappaB and CCAAT/enhancer binding protein underlies a novel tumor evasion mechanism affecting macrophages of mammary tumor-bearing mice. *Cancer Res.* **65,** 10578–10584.

Wang, T., Niu, G., Kortylewski, M., Burdelya, L., Shain, K., Zhang, S., Bhattacharya, R., Gabrilovich, D., Heller, R., Coppola, D., Dalton, W., Jove, R., *et al.* (2004). Regulation of the innate and adaptive immune responses by Stat-3 signaling in tumor cells. *Nat. Med.* **10,** 48–54.

Wyckoff, J. B., Wang, Y., Lin, E. Y., Li, J. F., Goswami, S., Stanley, E. R., Segall, J. E., Pollard, J. W., and Condeelis, J. (2007). Direct visualization of macrophage-assisted tumor cell intravasation in mammary tumors. *Cancer Res.* **67,** 2649–2656.

Yang, L., DeBusk, L. M., Fukuda, K., Fingleton, B., Green-Jarvis, B., Shyr, Y., Matrisian, L. M., Carbone, D. P., and Lin, P. C. (2004). Expansion of myeloid immune suppressor Gr + CD11b + cells in tumor-bearing host directly promotes tumor angiogenesis. *Cancer Cell* **6,** 409–421.

Yu, H., and Jove, R. (2004). The STATs of cancer–new molecular targets come of age. *Nat. Rev. Cancer* **4,** 97–105.

Yu, H., Pardoll, D., and Jove, R. (2009). STATs in cancer inflammation and immunity: A leading role for STAT3. *Nat. Rev. Cancer* **9,** 798–809.

Index

Contents of Previous Volumes

Volume 49

Volume 50

Volume 51

Volume 52

Volume 53

Volume 54

Volume 55

Volume 56

Volume 57

Volume 58

Volume 59

Volume 60

Volume 61

Volume 62

Volume 63

Volume 64

Volume 65

Volume 66

Volume 67

Volume 68

Volume 69

Volume 70

Volume 71

Volume 72

Volume 73

Volume 74

Volume 75

Volume 76

Volume 77

Volume 80

Volume 81

Volume 82

Volume 85

Volume 86

Volume 87

Volume 88

Volume 89

Volume 90

Volume 91

Volume 92

Volume 93

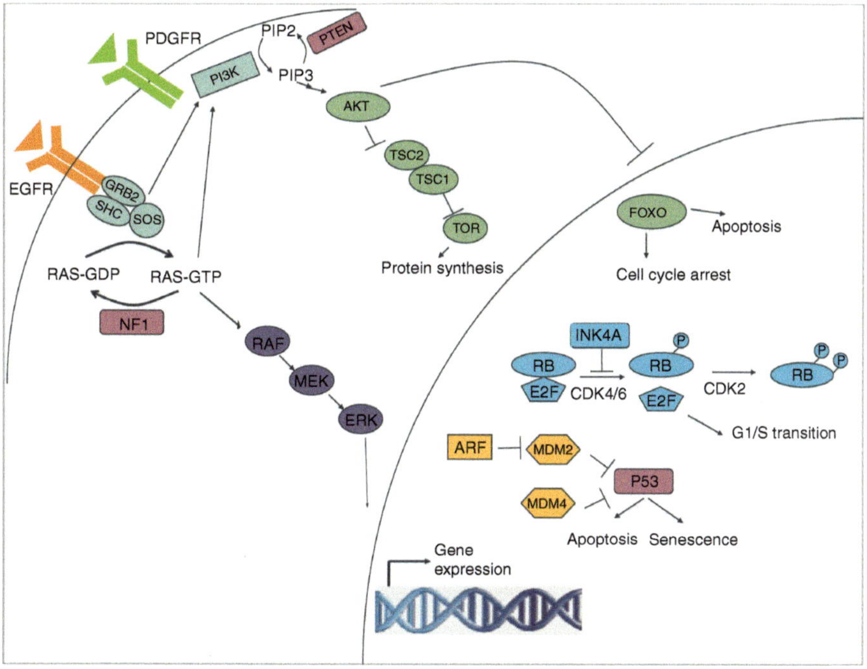

Sheila R. Alcantara Llaguno *et al.*, Figure 2.1 Genetic pathways in malignant gliomas. Simplified schematic of genetic alterations frequently found in malignant gliomas, including growth factor receptor signaling and cell cycle and apoptosis regulation. Overactivation of RTKs such as epidermal growth factor receptor (EGFR) and platelet-derived growth factor receptor (PDGFR) leads to stimulation of its downstream effectors, mainly Ras and phosphatidylinositide-3-kinase (PI3K) pathways, which are frequently activated in malignant gliomas. Ras signals through the Raf–Mek–Erk effector arm, among others, and can also directly activate the PI3K pathway. PI3K-mediated phosphorylation of PIP2 (phosphatidylinositol-4,5-biphosphate) to PIP3 (phosphatidylinositol-3,4,5-triphosphate) activates Akt, leading to activation of TOR (target of rapamycin) and inhibition of FOXO transcription factors, promoting cellular growth and survival. Nf1 and Pten are tumor suppressors that negatively regulate Ras and PI3K signaling, respectively. In the nucleus, G1/S progression is regulated by the retinoblastoma gene Rb, which binds the E2F family of transcription factors and is inhibited by cyclin-dependent kinases Cdk4/6 and Cdk2. On the other hand, the stability of the transcription factor p53, a master regulator of apoptosis and senescence pathways, is controlled by its ubiquitin ligase MDM2 and related gene MDM4. Ink4A and Arf positively regulate the activity of Rb and p53, which are commonly inactivated in astrocytomas.

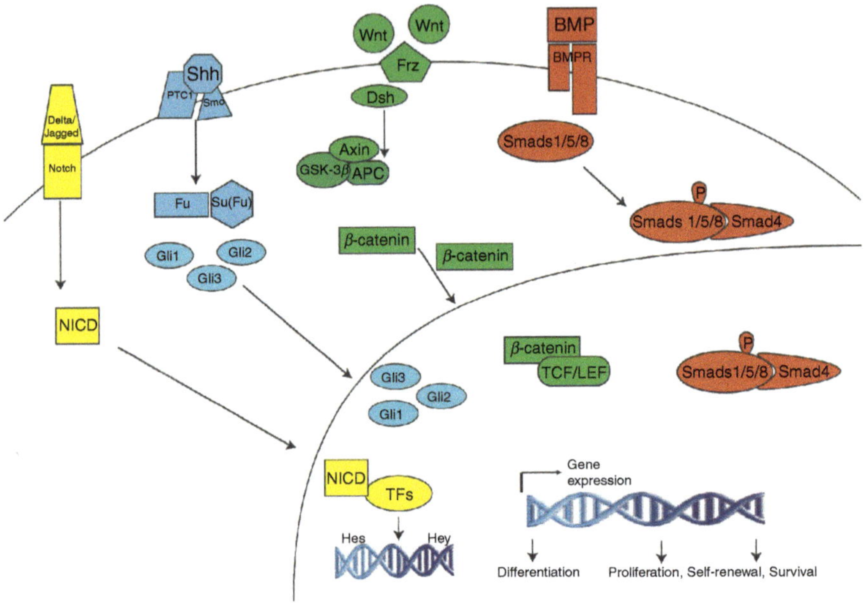

Sheila R. Alcantara Llaguno *et al.*, Figure 2.4 Classical developmental signaling pathways in neurogenesis and gliomagenesis. Simplified schematic diagram of developmental pathways that play a functional role in neural development as well as in adult neurogenesis and glioma formation. Expression of a number of Notch, Sonic Hedgehog (Shh), Wnt, and TGF-β/Smad signaling pathway components has been shown to be dysregulated in glioma, and overexpression or knockdown of some of these components has been shown to alter the properties of glioma stem cells and/or glioma cell lines. These pathways therefore offer potential molecular targets for therapeutic intervention.

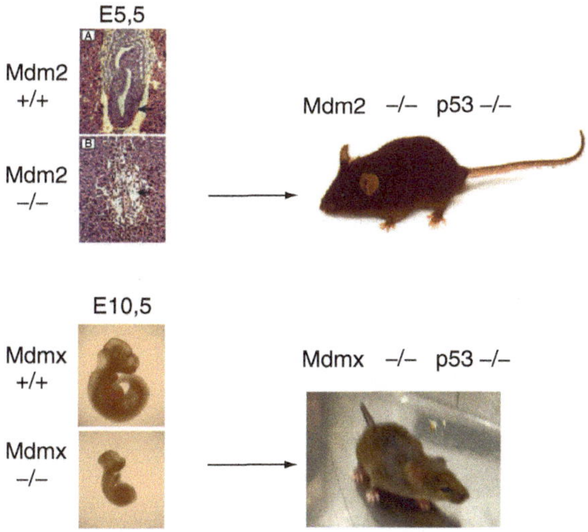

Jean-Christophe Marine, Figure 3.2 A critical role for Mdm2 and Mdmx as negative regulator of p53. Genetic ablation of Mdm2 (top panel) and Mdmx (lower panel) results in embryonic lethal phenotypes. Both of these phenotypes are rescued on a p53-deficient background. (Adapted from Montes de Oca Luna *et al.*, 1995 and Migliorini *et al.*, 2002).

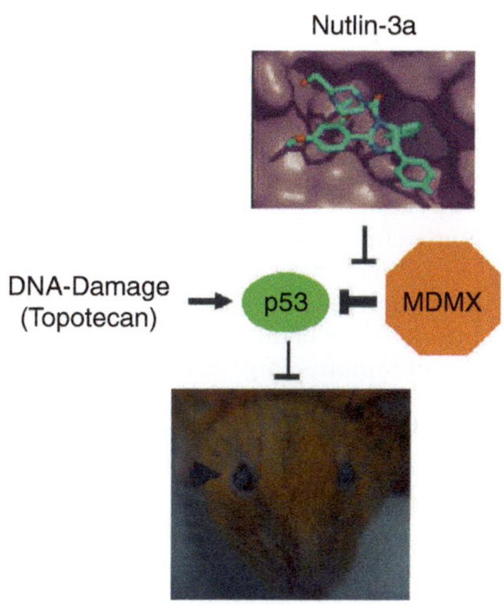

Jean-Christophe Marine, Figure 3.5 Synergistic induction of p53 tumor suppressor function in retinoblastoma by combining Mdm2/Mdmx antagonists and genotoxic drugs.

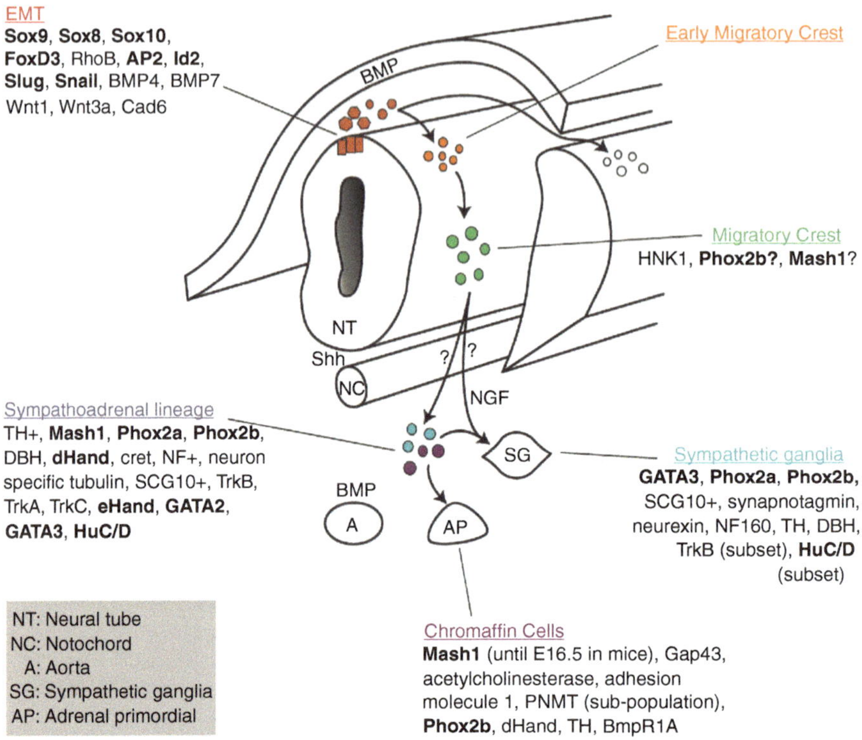

Manrong Jiang *et al.*, Figure 4.2 General Schema of the development of chromaffin cells and sympathetic ganglia. Cells at the dorsal region of the neural tube undergo EMT (red population), delaminate from the neural tube (orange), and migrate ventrally to the aorta (green) where they are commonly referred to as the sympathoadrenal progenitors (blue and purple). From the aortic region, the cells then migrate to the developing adrenal gland (AP) to become chromaffin cells or differentiate to become sympathetic ganglia (SG). As cells begin to differentiate as sympathetic ganglia they upregulate neural markers while chromaffin cells upregulate proteins found in the adrenal gland. Recent studies suggest that the chromaffin cell and sympathetic ganglia may come from divergent lineages rather than a common sympathoadrenal lineage. The question marks between the migrating crest and the sympathoadrenal progenitors address this possibility. A more detailed discussion on the temporal expression of the transcription factors in the sympathoadreanal lineage (including developmental stages) can be found in (Howard *et al.*, 2000) and details about the distinct chromaffin and sympathetic lineages can be found in (Ernsberger *et al.*, 2005). Additional of neurotriphons and their receptors can be found in (Straub *et al.*, 2007). Transcription Factors are shown in bold and factors implicated in neuroblastoma have been underlined. Abbreviations: NT, Neural Tube; NC, notochord; A, Aorta; SG, sympathetic ganglia; AP, adrenal primordial.

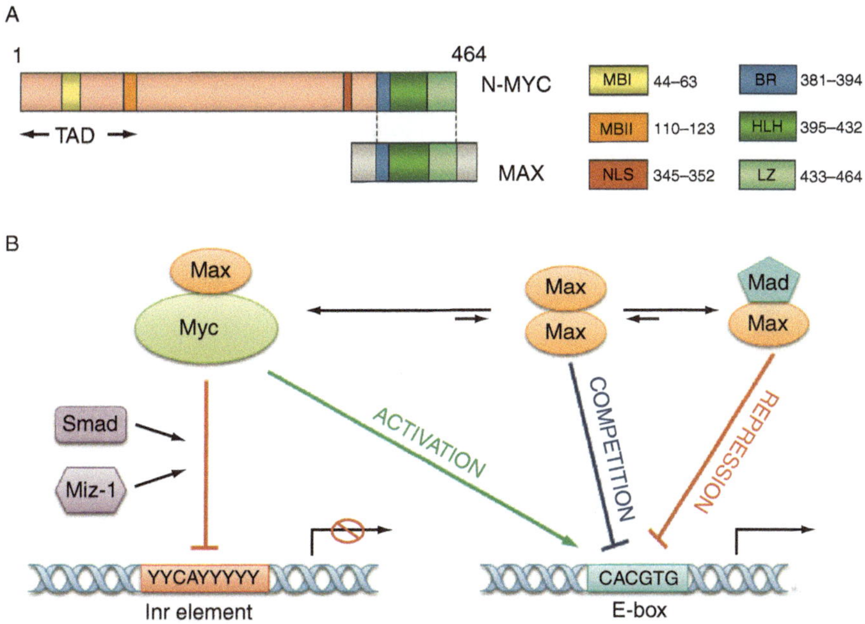

Manrong Jiang *et al.*, Figure 4.3 The structure of N-myc protein and transcription regulation by N-myc. (A) The structure of N-myc protein. The N-terminal transactivation domain (TAD) contains two conserved Myc box I and II (MBI and MBII), which are essential for DNA binding. The C-terminal domain (CTD) harbors basic region (BR), helix-loop-helix (HLH) motif, and leucine zipper (LZ) for dimerization with Max. There is a nuclear localization signal (NLS) before CTD. (B) The model for the transcription regulation by N-myc. Myc–Max heterodimer may bind to E-box element (CACGTG) to activate transcription, however, Myc–Max dimer can associate with other transcription factors such as Miz-1, Smad, and bind to Inr (initiator, weak consensus) element to repress transcription. Max can also form homodimers or heterodimers with Mad to compete or suppress Myc–Max binding to E-box.

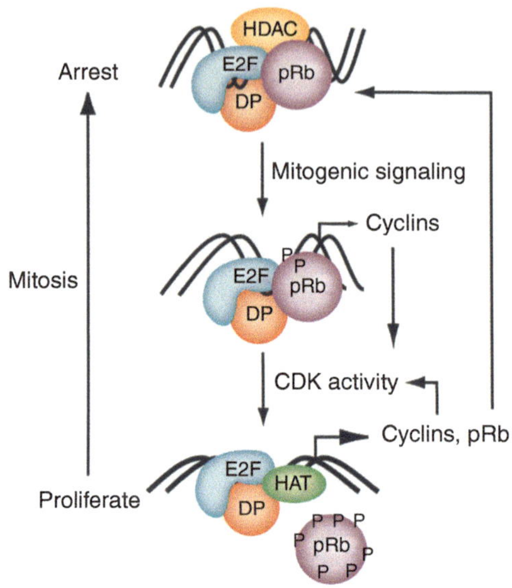

Meenalakshmi Chinnam and David W. Goodrich, Figure 5.2 The pRb/E2F cell proliferation switch. The pRb/E2F switch is bistable, existing primarily on one of two states, either permissive or nonpermissive for cell proliferation. Bistability is established by both positive (red lines) and negative (blue line) feedback loops (see text). In its unphosphorylated state, pRb complexes with E2F transcription factors at gene promoters, blocks the ability of E2F to recruit transcriptional coactivators like histone acetylases (HAT), and recruits corepressors like histone deacetylases (HDAC). These coactivators and corepressors alter chromatin to a more open or condensed states, respectively. Mitogenic kinases phosphorylate pRb, altering its ability to interact with E2F and transcriptional corepressors. The resulting transcriptional derepression of genes like cyclins increases cyclin dependent kinase activity (CDK). Increased CDK activity, in turn, augments pRb phosphorylation thus creating a positive feedback loop that amplifies the mitogenic signal and triggers commitment to a round of cell division. Negative feedback is provided by the fact that the *Rb1* promoter is itself regulated by E2F. As cells transit mitosis, pRb is dephosphorylated, resetting the switch to a nonproliferative state.

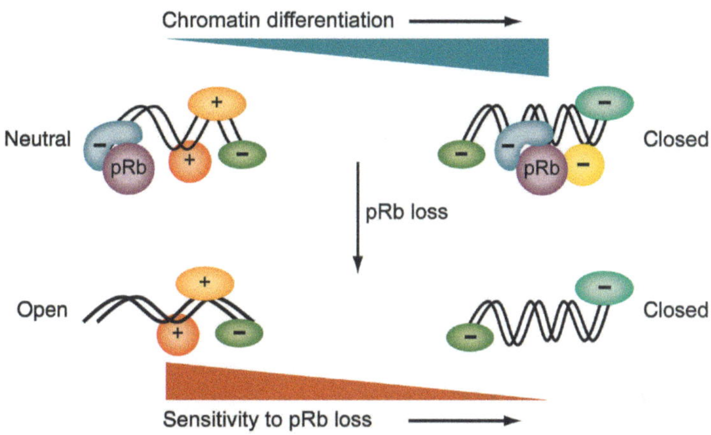

Meenalakshmi Chinnam and David W. Goodrich, Figure 5.3 Stem/progenitor cells and sensitivity to pRb loss. The figure summarizes the hypothesis that there exists an inverse correlation between differentiation and sensitivity to pRb loss. Stem/progenitor cells are relatively undifferentiated and have a neutral chromatin state at key lineage specifying regulatory genes. Chromatin at such genes is characterized by the simultaneous presence of protein complexes and chromatin marks characteristic of both open (shapes marked +) and closed chromatin (shapes marked −). pRb is typically associated with factors that create a closed chromatin state repressive for transcription (although there are notable exceptions). Loss of pRb in such cells is expected to have a large effect on gene expression as it will tip the balance in favor gene expression. In more differentiated cells, chromatin is more rigidly committed to an open or closed state by multiple, redundant mechanisms. Loss of pRb in such cells is expected to have less effect on the expression of stably repressed genes as there are redundant mechanisms in place to maintain the closed chromatin state.

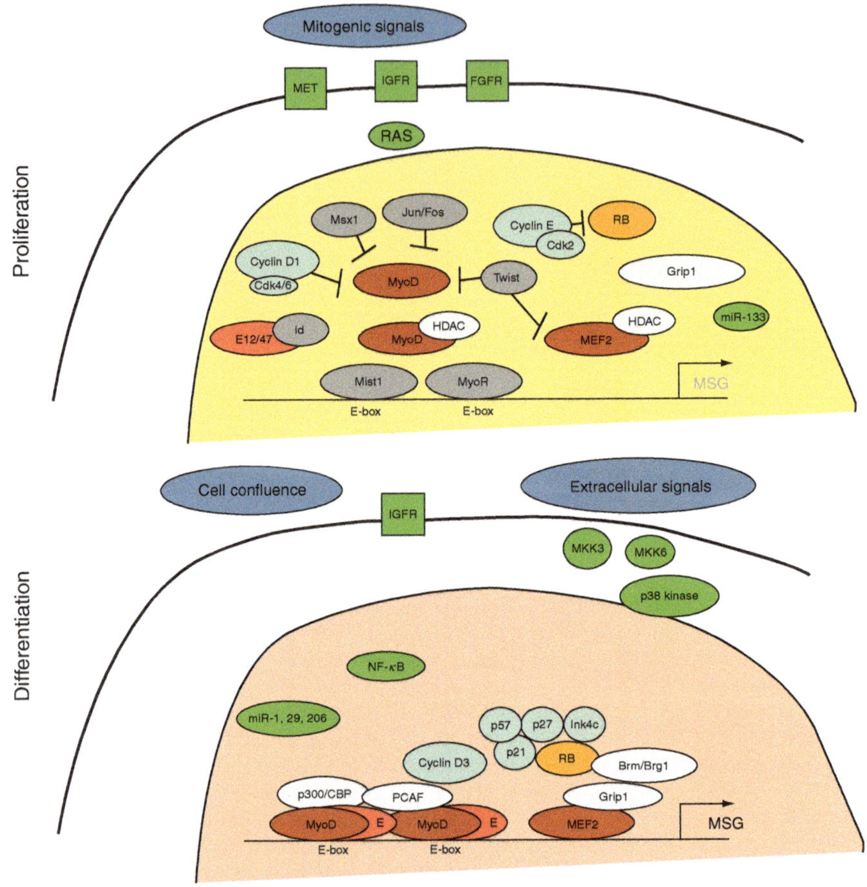

Raya Saab *et al.*, Figure 7.6 Schematic diagram depicting proteins and certain functional or physical interactions among them to either block the expression of muscle-specific genes (MSG) in proliferating myoblasts (top) or to increase MSG expression in differentiating myoblasts. MyoD is depicted as representative of other myogenic bHLH proteins. (See text for additional details.)

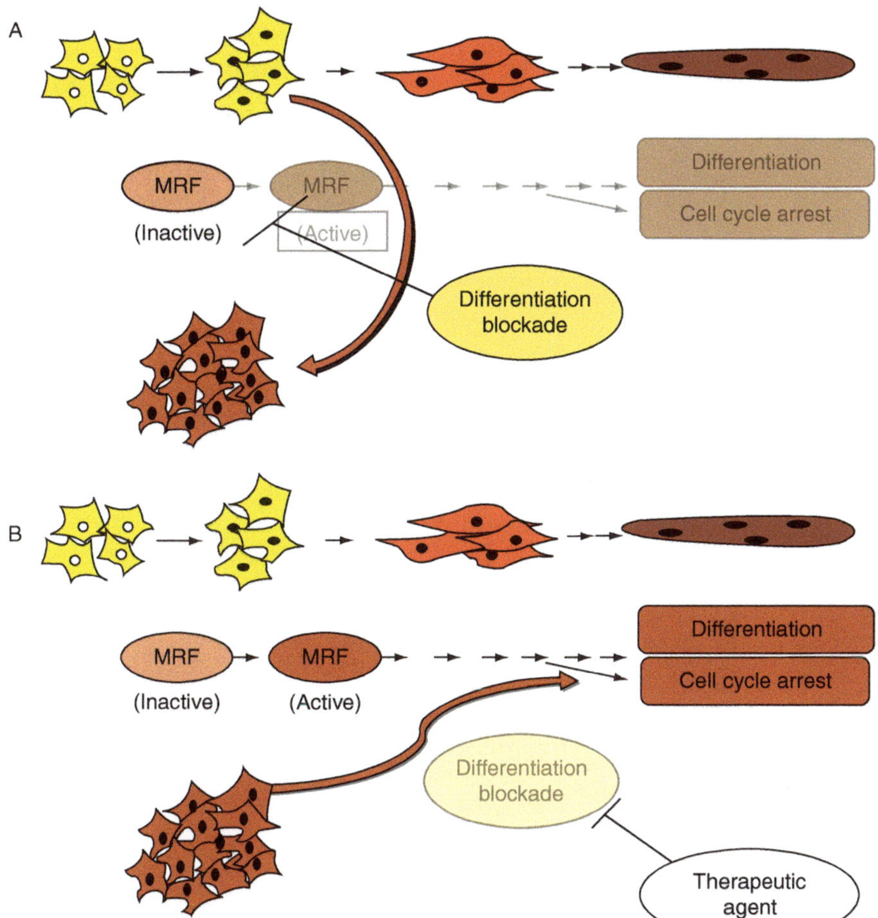

Raya Saab *et al.*, Figure 7.7 Schematic diagram depicting the general concept that a component of the phenotype of rhabdomyosarcoma is cellular and molecular evidence of arrested (or dedifferentiation) to a myoblast-like state (A). This likely correlates with a cell-intrinsic molecular change that blocks the continued, feed-forward differentiation program rendering muscle regulatory factors (MRF) incapable of promoting terminal differentiation. In principle, therapeutic interventions primarily aimed at reactivating differentiation could culminate in terminal differentiation and irreversible cell proliferation arrest. (See text for additional details.)

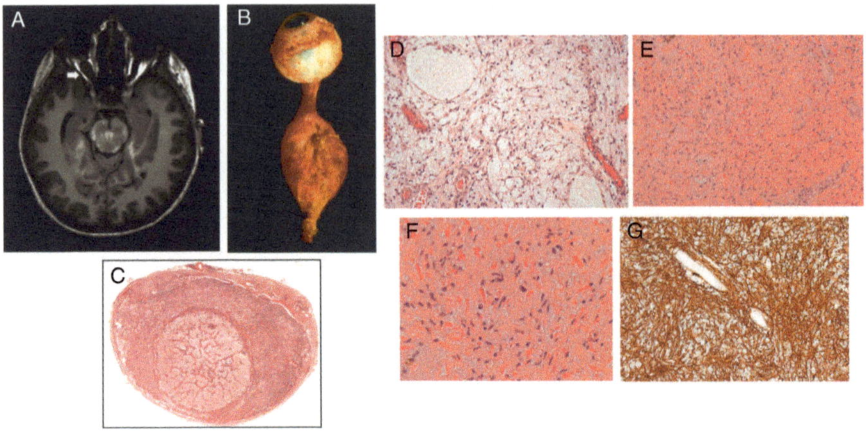

Nikkilina R. Crouse *et al.*, Figure 9.1 NF1-associated gliomas. (A) Magnetic resonance imaging from a young patient with a right optic nerve glioma. The arrow points to the glioma expanding the right optic nerve. (B) Enucleation specimen demonstrates a fusiform enlargement of optic nerve by a pilocytic astrocytoma. (C) Cross-section of this tumor highlights the expansile and solid nature of a pilocytic astrocytoma at this location. These tumors contain (D) loose microcystic and (E) compact areas rich in Rosenthal fibers (F). (G) The piloid neoplastic glial cells strongly express glial fibrillary acidic protein (GFAP). Images are courtesy of Dr. Robert E. Schmidt, Neuropathology Division, Washington University.

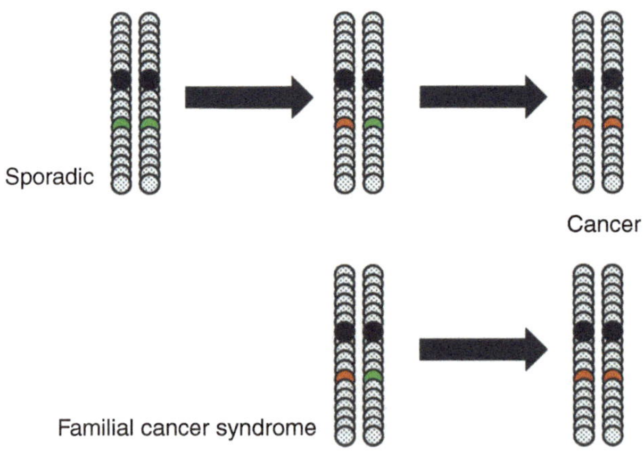

Nikkilina R. Crouse *et al.*, Figure 9.2 The two-hit hypothesis. Cancer in the general population (sporadic) requires the sequential inactivation of a given tumor suppressor gene in the same cell. In contrast, individuals with inherited cancer syndromes begin life with one nonfunctional copy of a given tumor suppressor gene, and develop cancer after the one remaining functional allele is inactivated.

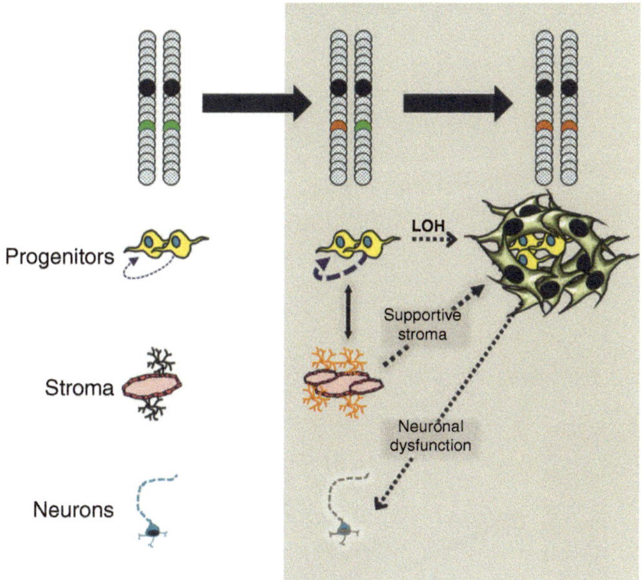

Nikkilina R. Crouse *et al.*, Figure 9.7 Expanded two-hit hypothesis model. In the original two-hit hypothesis, tumor suppressor gene heterozygosity resulted in a statistically increased frequency of biallelic tumor suppressor gene inactivation (second hit). In addition, tumor suppressor gene heterozygosity changes the proliferative and self-renewal potential of progenitor cells, thus facilitating this loss of heterozygosity. We now propose that tumor suppressor gene heterozygosity in non-neoplastic cells in the tumor microenvironment establishes a supportive niche for tumor formation and maintenance, which in some cases, like NF1, is required for tumorigenesis. Finally, tumor suppressor gene heterozygosity may also lead to increased neuronal dysfunction as a consequence of glioma formation and growth.

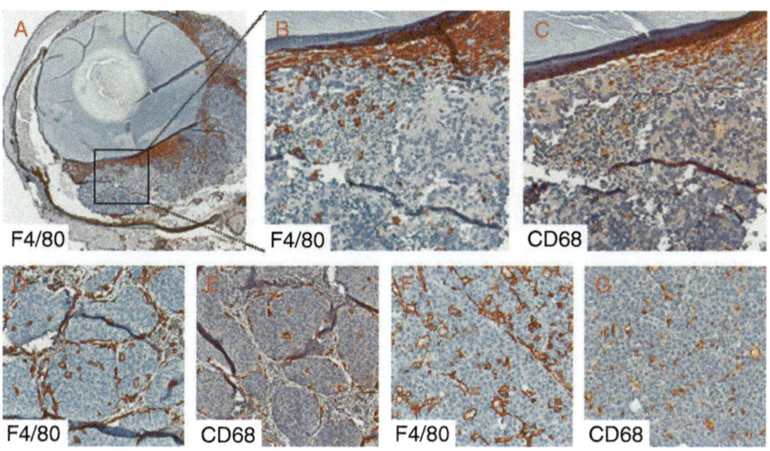

Joseph E. Qualls and Peter J. Murray, Figure 10.1 Expression of macrophage cell-surface antigens on TAMs. Cross sections from a retinoblastoma (A–C), neuroblastoma (D–E), and EG7 thymoma (F–G) stained by immunohistochemistry for F4/80 (A, B, D, and F) and CD68 (C, E, and G).

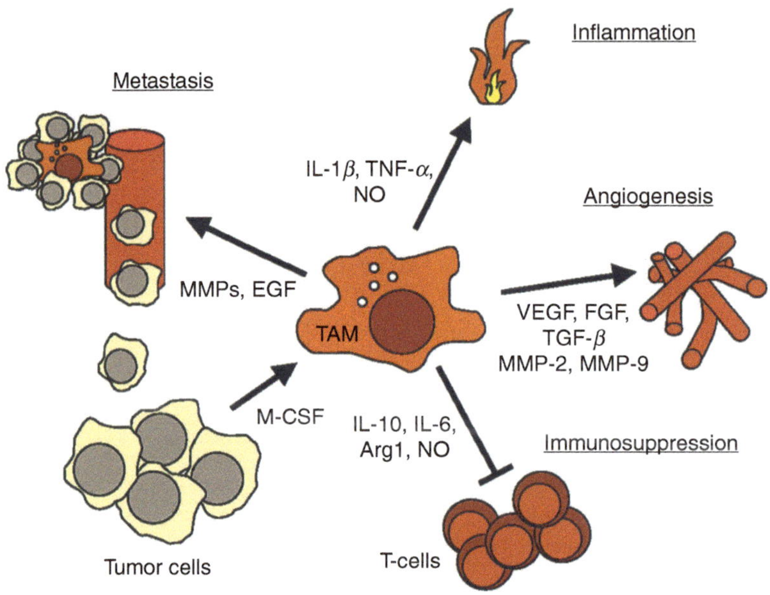

Joseph E. Qualls and Peter J. Murray, Figure 10.2 TAM-mediated support of tumor progression. TAMs are actively recruited by tumor cell secretion of multiple chemokines, growth factors, and most importantly, M-CSF. Developing neoplasms are benefited by TAM initiation of inflammation, which via NF–kB activation can lead to increased tumor survival. However, a prolonged inflammatory response could prove detrimental to tumors by infiltrating cytotoxic T-cells. Therefore, TAMs also block T-cell activity through production of immunosuppressive factors, including IL-6, IL-10, Arg1, and NO. TAMs induced by hypoxia, and other tumor-derived signals, also promote angiogenesis by producing VEGF, FGF, TGF-β, MMP-2, and MMP-9. This process is closely related to the proposed TAM-mediated assistance for tumor metastasis, involving additional MMPs and growth factors.

Kolibri 360 GmbH
Ferdinand-Jühlke-Straße 7
99095 Erfurt, Deutschland
produktsicherheit@kolibri360.de